Y0-DDA-061

Modern Industrial
Plastics

Modern Industrial

Plastics

Terry A. Richardson

Northern State College,
Aberdeen, South Dakota

HOWARD W. SAMS & CO., INC.

INDIANAPOLIS · KANSAS CITY · NEW YORK

FIRST EDITION

FIRST PRINTING—1974

International Standard Book Number: 0-672-20948-2
Library of Congress Catalog Card Number 72-92621

Cover: *Courtesy of Andretti Firestone High Performance Center and P.R. Mallory & Co., Inc., Indianapolis, Ind.*

Preface

Of all industrial materials none has the virtually unlimited prospects as do plastics. Indeed, according to the figures of the World Chemical Engineering Congress, by the year 2000 the production of plastics will be nearly three times that of all other hard goods and dry goods put together. Plastics production may well become the basis of American economy. Even now there are not enough personnel trained or knowledgeable in industrial plastics to meet the needs of industry, and such needs are bound to multiply.

The purpose of this book is to provide useful information about the fundamental processes, concepts, and aspects of industrial plastics. To this end the historical development of plastics, their basic chemistry, and the various processing methods used in industry have been emphasized. Each of these areas may be studied independently and not in the order presented in this book.

The early chapters develop the basic chemistry of materials rather rapidly. The reader or student should have had, or be taking concurrently, a standard course in high-school chemistry. Significant new terms are introduced in each chapter, and these are restated in the Glossary of New Terms at the end of the chapter. A set of review questions is also included at the end of each chapter, for self-testing or classroom assignment. A bibliography of books for further reference and a list of tradenames and manufacturers of plastics are given in Appendixes A and B, at the end of the book.

The multitude of illustrations, applications, and properties of plastics covered in this book should provide the reader with a good, basic understanding of some of the most versatile and useful materials on earth.

Terry A. Richardson

Contents

1

Introduction to Plastics

Early man probably obtained his knowledge of fire from the world of nature around him. His earliest tools were made of stone. His earliest fuel was wood; his transportation—his legs.

The materials needed by man were found occurring naturally on earth. Copper, gold, and iron ores were easily obtained and used as man developed more sophisticated methods of processing the raw materials.

As man began to industrialize, more goods were needed that were strong, durable, malleable, and practical. Neither the surface nor the bowels of the earth, however, yielded a new natural material for the production of needed goods.

Men had to develop a new type of material called *synthetics*. There were many technical achievements to accomplish before these synthetic or man-made materials could become a reality.

Chemistry

The medieval alchemists tried in vain to produce gold from such inexpensive metals as tin, iron, and lead. It is from the medieval Latin word *alchimista* that the modern words "chemistry" and "chemist" are derived.

Chemistry is a science dealing with the composition of matter and the changes in the composition of matter. There are two broad classes of chemical study: *inorganic* and *organic*.

Inorganic chemistry deals with the study of matter which is of mineral origin.

Organic chemistry is the chemistry of matter which contains the element carbon. The term "organic" was used originally to designate compounds of plant or animal origin; contemporary usage includes many synthetic materials as well.

There are many specialized branches of chemistry—biochemistry, medical chemistry, atomic chemistry, agricultural chemistry, and polymer chemistry, to name only a few.

It is this latter field of chemistry which has so profoundly affected our daily living by producing synthetic materials called *plastics*.

The Age of Plastics

What material may be as hard as stone, transparent as glass, elastic as rubber, strong, light in weight, moisture and chemical resistant, and come in a variety of colors? Answer: Plastics. According to the Society of Plastics Industry, Incorporated (SPI), this material will surpass the production of steel, wood, and all other materials combined by the year A.D. 2000. (See Figs. 1-1 and 1-2.) Yes, today may truly be called the "Plastics Age."

According to *Webster's New World Dictionary*, the word "plastics" is derived from the Greek word *plastikos*, meaning "to form or fit for molding." The Society of Plastics Industry has defined "plastics" as follows:

> Any one of a large and varied group of materials consisting wholly or in part of combinations of carbon with oxygen, nitrogen, hydrogen, and other

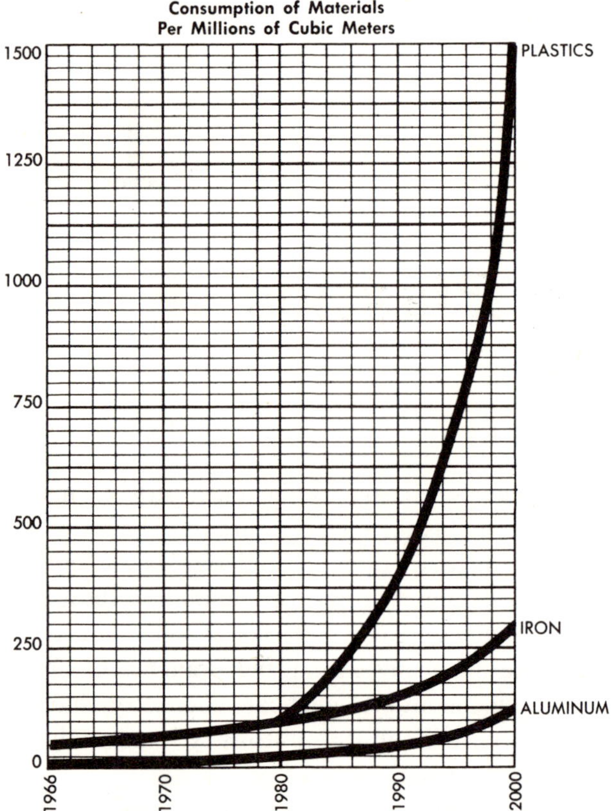

Fig. 1-1. Projected annual world consumption of materials. *Figures based on presentations at the World Chemical Engineering Congress.* (SPI)

organic or inorganic elements which, while solid in the finished state, at some stage in its manufacture is made liquid, and thus capable of being formed into various shapes, most usually through the application, either singly or together, of heat and pressure.

Plastics are closely related to resins, with which they are often confused. Resins are gumlike solid or semisolid substances used in the manufacture of paints, varnishes, plastics, and similar substances. Natural resins are products from the secretion of the sap of certain trees and plants. Synthetic resins are resinous compounds made from synthetic materials.

Plastics may be man-made (synthetic) resins, or they may be compositions formed from such natural resins as shellac, bitumen, rosin, asphalt, pitch, copal, and rubber. According to the preceding definition by the Society of Plastics Industry, resin is not plastic until the resin has become "solid in the finished state." Plastics products, for ex-

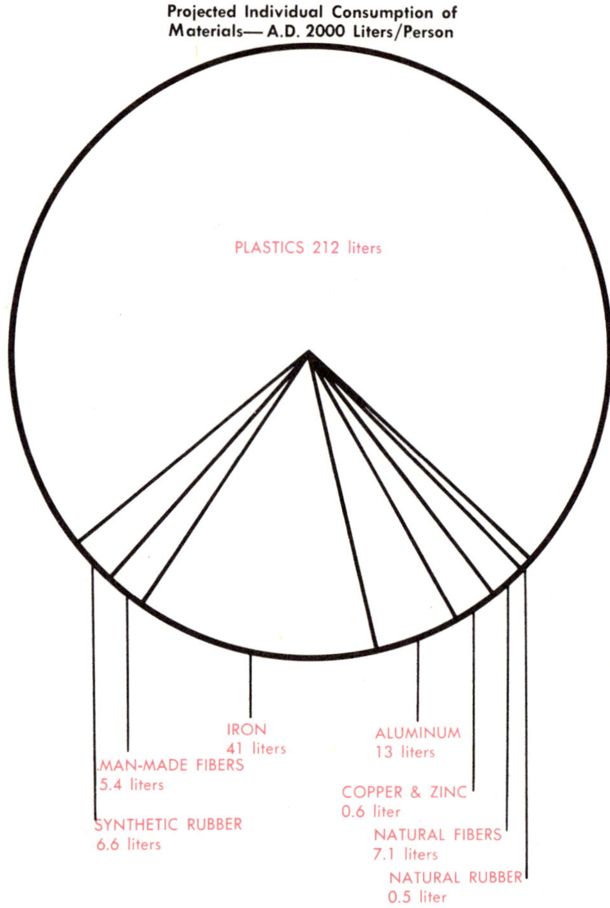

Fig. 1-2. Projected annual consumption of materials by individual by A.D. 2000. *Figures based on presentations at the World Chemical Engineering Congress.* (SPI)

ample, are produced from resins which are processed and caused to become solid. Cellulose and protein derived plastics are sometimes included with the natural resins; however, they should be classed with the synthetic resins. Both types of material, natural and synthetic, are composed of a series of molecules bonded together. Some of the resins are composed of several thousand of these bonded molecules.

History of Plastics

Like many of the materials used by early man, those materials found in nature were his first sources. Natural copper and gold could be formed into implements of labor or adornment. Shellac, bitumen, rosin, asphalt, pitch, and rubber resins could also be used for building and ornamental purposes. Prime raw materials today are coal, petroleum, limestone, salt, sulfur, air, water, cellulose, silicone, chlorine, fluorine, and other elements.

Early observations in 1833 by the noted chemist J. J. Berzelius led to the production of one of the first synthetic compounds containing several thousand molecules.

The term "polymer" was used by chemist H. V. Regnault in 1835 when he synthesized the plastic vinyl chloride. *Polymer* is the name for a substance composed of large molecules made by joining many small molecules of one or several substances. Commercial production of this vinyl, however, did not begin until 1925, in Germany.

The first commercial use of a polymeric material was in 1843 by Dr. George IV William Montgomerie. Dr. Montgomerie, a Malayan surgeon, discovered that the natives were using a natural polymer material ("gutta percha") to make knife handles and whips. Resins from the gutta percha trees were collected and sent to England to interested scientists and manufacturers.

Michael Faraday, a pioneer in electricity, discovered that gutta percha had good electrical insulating properties when immersed in water. Consequently, this material was used as insulation in laying the first transatlantic ocean cable.

When she sailed from England to New York in June, 1860, the *Great Eastern* was the largest vessel ever built. Although she was also the largest white elephant in maritime history, she was the only vessel capable of holding enough cable to reach across the Atlantic. The *Great Eastern* displaced 18,900 gross tons and provided space for 4000 passengers and a crew of 400. After a series of near catastrophes, she was withdrawn from passenger service and used to lay the transatlantic cable (Fig. 1-3).

Modern undersea cables still incorporate plastics in forms, bases, housings, shields, and cores. Fig. 1-4 shows an undersea telephone cable amplifier making such use of plastics. It is designed for continuous performance for over 20 years at depths of up to 2½ miles. This amplifier is 13 inches in diameter, 36 inches long, and contains approximately 5000 precision parts.

When the nitrogen content of nitrocellulose, a combination of nitric acid and cellulose, is more than 13 percent, this material is referred to as

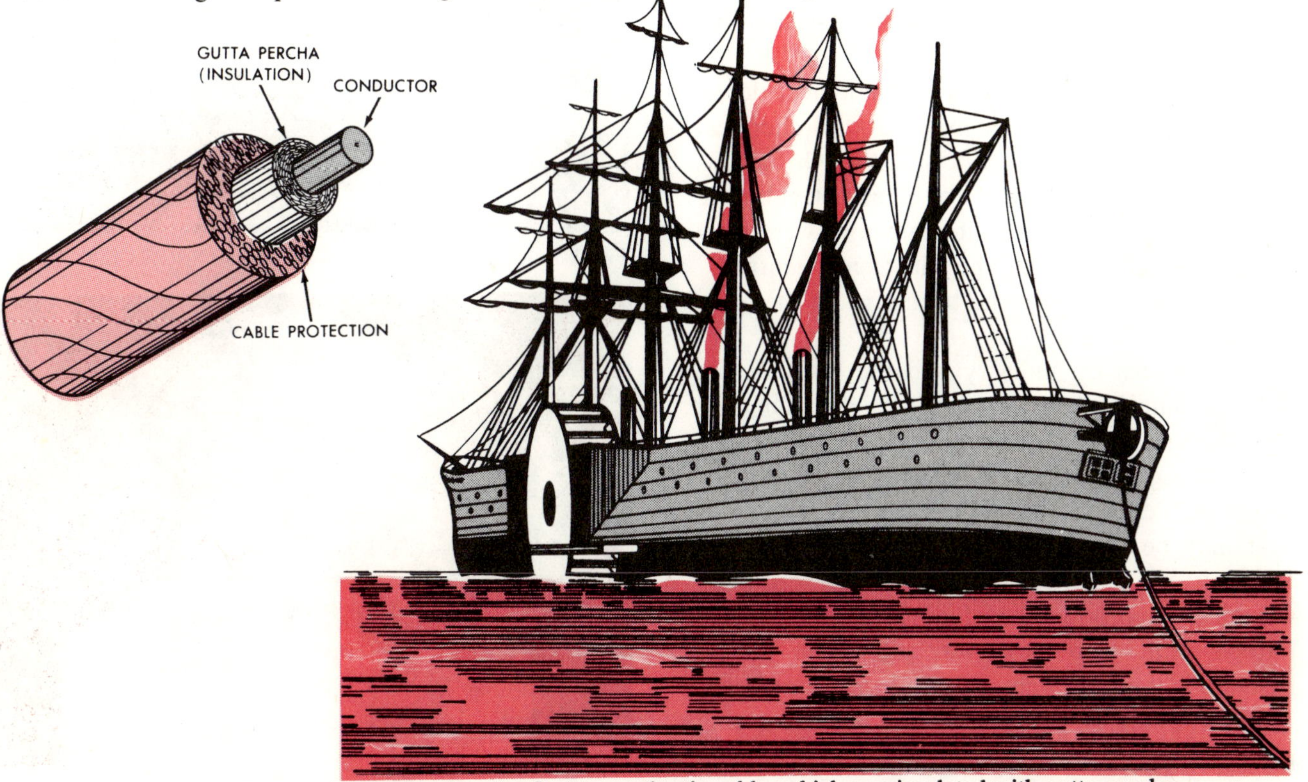

Fig. 1-3. The *Great Eastern* depicted laying the first transatlantic cable, which was insulated with gutta percha.

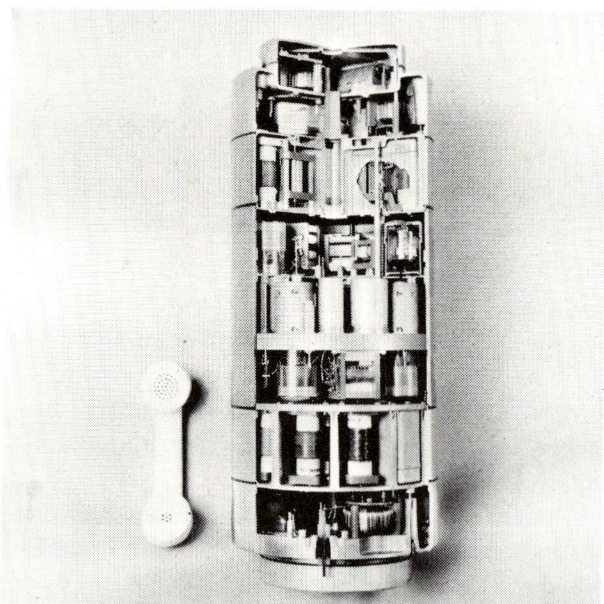

Fig. 1-4. Modern application of plastics for continuous performance on the ocean floor. (Western Electric Co.)

"guncotton," an explosive used in the Civil War and World War I.

In 1862, Alexander Parks of Birmingham, England, at the International Exhibition in London, displayed a new plastics, "Parkesine," which was made from nitrocellulose containing less than 12 percent nitrogen.

There are many significant technological developments which paved the way for the chemical production of plastics products. Although there is documented proof of the use of oil and steel before the birth of Christ, only limited amounts were used until the second half of the nineteenth century.

One authority places the beginning of the *Steel Age* at about 1850.

The modern oil industry and the birth of the *Oil Age* date back to August 29, 1859. On that date the first deliberate drilling for oil was supervised by Col. Edwin L. Drake in the region of Titusville, Pennsylvania.

It may be difficult to imagine that during the early history of petroleum processing, seemingly worthless byproducts including gasoline were burned or dumped into rivers and lakes. Today, gasoline is the most valuable product of crude oil; however, many other synthetic products, including plastics, may be synthesized from crude petroleum chemicals.

It was John W. Hyatt, a New York printer, whose efforts actually marked the beginning of synthetic plastics production in the United States in 1868. Mr. Hyatt responded to an advertisement to find a substitute for gutta percha and ivory billiard balls. There is no recorded evidence that Hyatt ever received the reward. He did succeed in producing a new plastics. Fig. 1-5 is a photograph of Mr. Hyatt, and Fig. 1-6 shows the "Celluloid" billiard ball produced by John Hyatt and his brother, Isaiah.

Fig. 1-5. John Wesley Hyatt. (Celanese Plastics Co.)

Fig. 1-6. The Hyatt billiard ball. (Celanese Plastics Co.)

According to legend, Hyatt experimentally mixed camphor with some pyroxylin (a nitrocellulose with low nitrogen content), consequently starting the development and production of a new plastics called "Celluloid." This new plastics was easily molded and was less explosive than previous nitrocellulose plastics (Fig. 1-7).

Fig. 1-7. Nitrocellulose plastics were sometimes dangerous.

With the end of the Civil War, huge quantities of nitrocellulose in the form of guncotton were left as surplus. John Hyatt and his brother Isaiah Smith Hyatt purchased large quantities of this surplus nitrocellulose and were granted over seventy-five United States patents for the production of plastics.

Fig. 1-8 shows the oldest plastics plant in the United States. This plant is still in operation at Henry Street in Newark, New Jersey. It began production in 1873 with 150 employees. It was called the Celluloid Manufacturing Company. In 1941 it merged with the Celanese Company.

An elaborate molded comb and brush produced in the early years of the plastics industry is shown in Fig. 1-9.

W. Krische in 1897 found that by using the protein from milk a new plastics called *casein* could be produced. Adolf Spittler, of Bavaria, found that

Fig. 1-8. The first American plastics plant and some personnel. (Celanese Plastics Co.)

by treating the compressed protein sheets with formaldehyde a more water resistant plastics was developed. Casein plastics found little favor in the United States. Consequently they are of relatively little industrial significance except as adhesives. Less than 2000 tons of casein plastics were used in the United States in 1973.

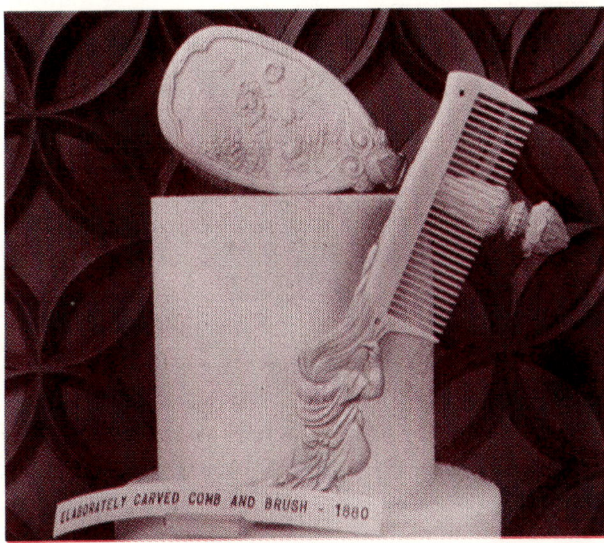

Fig. 1-9. The comb and brush were produced from celluloid in 1880. (Celanese Plastics Co.)

From papers prepared by Adolf von Baeyer some 35 years earlier, Leo Hendrik Baekeland of the United States produced a new resin that advanced the commercial use and acceptance of plastics. Baekeland, in 1909, secured the necessary patents and began producing plastics products under the trade name of "Bakelite."

The chronology of plastics shown in Table 1-1 will help to illustrate other commercial developments in plastics.

Table 1-1. Chronology of Plastics.*

DATE	MATERIAL	EXAMPLE
1868	Cellulose Nitrate	Eyeglass Frames
1909	Phenol-Formaldehyde	Telephone Handset
1909	Cold Molded	Knobs and Handles
1919	Casein	Knitting Needles
1926	Alkyd	Electrical Bases
1926	Analine-Formaldehyde	Terminal Boards
1927	Cellulose Acetate	Toothbrushes, Packaging
1927	Polyvinyl Chloride	Raincoats
1929	Urea-Formaldehyde	Lighting Fixtures
1935	Ethyl Cellulose	Flashlight Cases
1936	Acrylic	Brush Backs, Displays
1936	Polyvinyl Acetate	Flash Bulb Lining
1938	Cellulose Acetate Butyrate	Irrigation Pipe
1938	Polystyrene or Styrene	Kitchen Housewares
1938	Nylon (Polyamide)	Gears
1938	Polyvinyl Acetal	Safety Glass Interlayer
1939	Polyvinylidene Chloride	Auto Seat Covers
1939	Melamine-Formaldehyde	Tableware
1942	Polyester	Boat Hulls
1942	Polyethylene	Squeezable Bottles
1943	Fluorocarbon	Industrial Gaskets
1943	Silicone	Motor Insulation
1945	Cellulose Propionate	Automatic Pens and Pencils
1947	Epoxy	Tools and Jigs
1948	Acrylonitrile-Butadiene-Styrene	Luggage
1949	Allylic	Electrical Connectors
1954	Polyurethane or Urethane	Foam Cushions
1956	Acetal	Automotive Parts
1957	Polypropylene	Safety Helmets
1957	Polycarbonate	Appliance Parts
1959	Chlorinated Polyether	Valves and Fittings
1962	Phenoxy	Bottles
1962	Polyallomer	Typewriter Cases
1964	Ionomer	Skin Packages
1964	Polyphenylene Oxide	Battery Cases
1964	Polyimide	Bearings
1964	Ethylene-Vinyl Acetate	Heavy Gauge Flexible Sheeting
1965	Parylene	Insulating Coatings
1965	Polysulfone	Electrical/Electronic Parts

*From the booklet **The Story of the Plastics Industry** (SPI).

Modern Industrial Plastics

The first commercial plastics, celluloid, was developed just a little over 100 years ago. But the explosive growth and diversification in the industry has occurred since World War II. Since World War II the plastics industry has been growing at a rate double that of all manufacturing, or about 15 percent per year over the Gross National Product. Fig. 1-10 shows a modern plastics plant, which may be compared with the first plastics plant, shown in Fig. 1-8.

Fig. 1-10. A modern plastics producing plant. (Chemplex Company.)

Plastics already are put to a great variety of uses in everyday living, in industry, agriculture, science, recreation, and at the frontiers of the technological revolution (Fig. 1-11). From housewares to automobiles, buildings, boats, packaging, exotic aerospace applications, plastics are performing the tasks of the more traditional materials and creating new uses of their own.

The Society of Plastics Industry indicates that this material will surpass the production of steel, wood, and all other materials combined by A.D. 2000. By 1983 the volume of plastics consumed will exceed the volume of metals consumed. Fig. 1-12 shows the estimated consumption of raw materials to A.D. 2000.

The layman's image of plastics may be largely associated with such obvious products as housewares or toys, but the biggest growth areas are elsewhere (Fig. 1-13). Construction is the number one market for plastics, and only limited use has begun to be accepted by labor unions as well as

(A) *Autos with plastics bodies and trim. (Marbon Div., Borg-Warner)*

(B) *Solid vinyl siding doesn't rot, peel, or blister like wood. (Bird & Son, Inc.)*

(C) *U.S.A.F. C-5A cargo airplane has epoxy film adhesive binding in leading edge slats. (Lockheed-Georgia Co.)*

Fig. 1-11. Diverse uses of plastics.

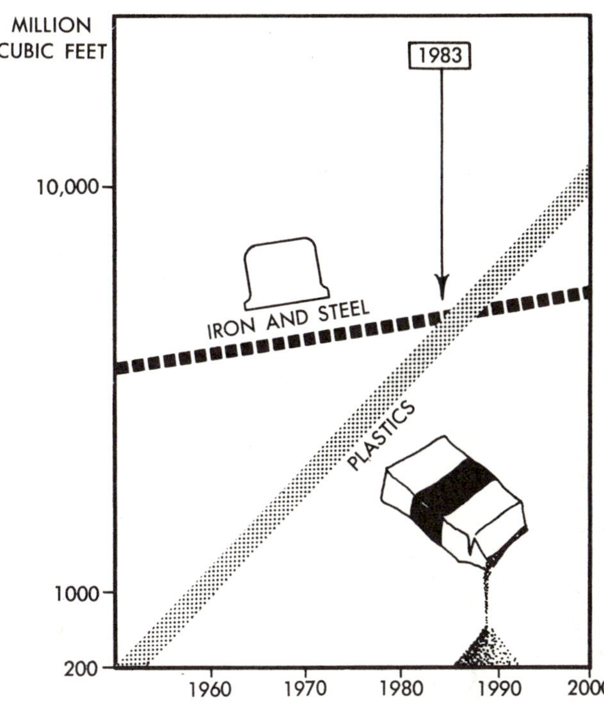

Fig. 1-12. World consumption of raw material by volume. (SPE/SPI Education Task Force pamphlet *The Need For Plastics Education)*

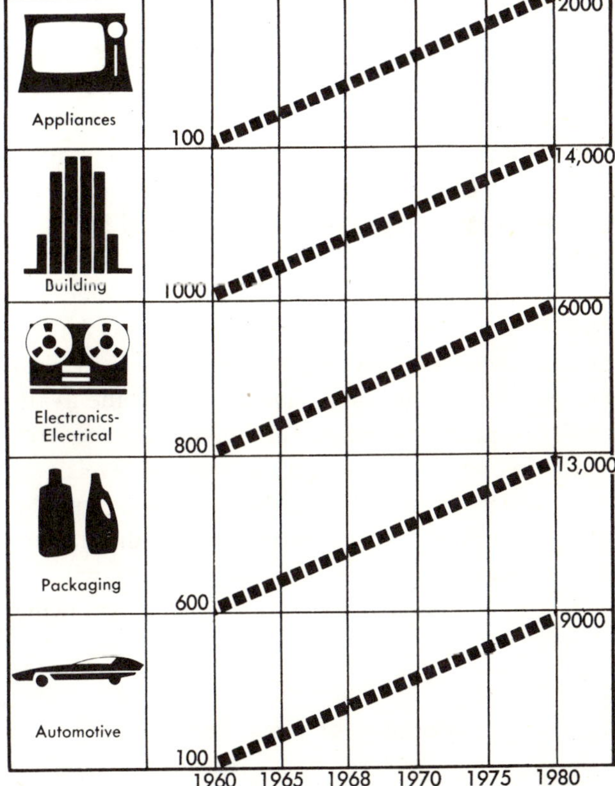

Fig. 1-13. Estimated U.S. consumption of plastics in millions of pounds by selected industry. (SPE/SPI Education Task Force pamphlet *The Need for Plastics Education)*

the public. Electronics now uses 1.4 billion pounds of plastics, worth one billion dollars, in the manufacture of electronic components. The growing use of plastics in transportation is illustrated by the consumption of plastics by the automobile industry.

Because of their inherent processability, plastics materials have become the perfect substitutes. Even better results may be gained by using plastics as original design materials, rather than in substitute roles.

What is the extent of the plastics industry? With each passing day, it becomes increasingly difficult to define plastics as a discrete industry. Plastics are products of practically every industry—steel, paper, chemicals, petroleum, and electronics, to name only a few.

The plastics industry may be divided into three large categories which sometimes overlap: (1) the plastics material manufacturer who produces the basic plastics from chemicals (resins); (2) the processor who converts the basic plastics into solid shapes; and (3) the fabricator and the finisher who further fashions and decorates the plastics.

Plastics are known by many names. Some of these names are difficult to pronounce. Thus manufacturers have used tradenames to indicate their brand of resin and at the same time make it easier for the consumer to pronounce and remember. Acrylite, Lucite, and Plexiglas are tradenames for the family of acrylic resins manufactured by three different companies.

Because plastics could potentially become the basis of the American economy in the near future, and because there is a dire shortage of trained plastics personnel, educators should increase the availability of plastics educational facilities and courses at the high-school and post-high-school levels.

In order to understand the processing of the plastic materials, an understanding of the atomic and molecular composition of the resin will be necessary. This will be given in the following chapter.

Glossary of New Terms

Celluloid—An elastic and very strong plastic made from nitrocellulose, camphor, and alcohol. It is used as a tradename for some plastics.

Gutta percha—A rubberlike product obtained from certain Malayan trees.

Inorganic—Applies to the chemistry of all elements and compounds not classified as organic; matter other than animal or vegetable, such as earthy or mineral matter.

Nitrocellulose (cellulose nitrate)—Material formed by the action of a mixture of sulphuric acid and nitric acid on cellulose. The cellulose nitrate used for celluloid manufacture usually contains 10.8 to 11.1 percent of nitrogen.

Organic—Designating or composed of matter originating in plant or animal life or composed of compounds of hydrogen and carbon (hydrocarbons), either natural or synthetic.

Plastic—An adjective, meaning pliable and capable of being shaped by pressure. "Plastic" is incorrectly used as the generic word for the plastics industry and its products.

Plastics—An organic substance, usually synthetic or semisynthetic, which can be formed into various shapes by heat and pressure and retain these shapes after heat and pressure have been removed. In its finished state it is a flexible or rigid but not elastic solid containing a polymer of high molecular weight.

Polymer—A high molecular weight compound, natural or synthetic, whose structure can be represented by a repeated small unit (the *mer*). Some polymers are elastic and some are plastic.

Resin—Gumlike solid or semisolid substances which may be obtained from certain plants and trees or made from synthetic materials.

Tradename—A name given to a product to make it easier to spell, pronounce, and recognize. In the plastics industry it is used by the manufacturer to identify his particular resin or product.

Review Questions

1-1. Name the first commercially important plastics and its inventor.

1-2. Give the reason for the great rise in the development and use of plastics.

1-3. Give two definitions of the word "plastics."

1-4. What is the difference between synthetic and natural resins? Name some natural resins.

1-5. What significance did the polymer gutta percha play in about 1843?
1-6. Name several technological developments which paved the way for the invention and chemical production of plastics products.
1-7. Whose efforts marked the beginning of the synthetic plastics production in the United States in 1868?
1-8. What was the name of the plastics material Parks produced?
1-9. What chemical was added to nitrocellulose to produce celluloid?
1-10. What plastics material is produced from the protein of milk?
1-11. What is the estimated future for the applications plastics by the year A.D. 2000?
1-12. Name six broad areas of industry which consume large amounts of plastics. Which industry may have the largest consumption by the year 2000, and which areas of industry will expand in plastics consumption in the near future?
1-13. List some of the typical applications or uses of plastics production.
1-14. Name the three overlapping categories in which the plastics industry is divided.
1-15. Why are tradenames used for plastics?
1-16. Why should people know about the plastics industry and plastics materials?
1-17. What are some of the semiskilled jobs in the plastics industry?
1-18. What type of education is necessary to become an engineer in the plastics industry?
1-19. What are the future prospects for employment in the plastics industry?
1-20. How does the study of organic chemistry differ in content from the study of inorganic chemistry?

2

The Atom

Nearly 2000 years after the Greek philosopher-scientist Democritus stated that everything can be broken down into small particles called *atoms* did other scientists give atomic theory serious thought again. It was Democritus who gave us the word "atom." He used the Greek word *atomos,* which means something that cannot be cut or something that is indivisible, to describe these small particles of matter.

The noted Greek philosopher Aristotle was born in 384 B.C., when Democritus was still teaching. Aristotle, however, instructed his students that atoms did not exist and that there were only four basic qualities: hot, cold, moist, and dry. Although this basic philosophy of Aristotle's was disarmingly simple, his ideas governed the mind of Western man for almost 2000 years.

Famous scientists like Galileo Galilei, Pierre Gassendi, and Sir Isaac Newton disputed the accepted teachings of Aristotle but the experimental proof of the atom's existence still lay a long time in the future.

In approximately 1650, Irish researcher Robert Boyle observed that alchemists and others had searched for many years for ways of making gold by combining other less expensive materials. Boyle concluded that gold must be made of a limited number of simple substances since it cannot be broken into simpler materials. These simple substances he called *elements* from an ancient Greek name. There were very few elements known in Boyle's time, as shown in Table 2-1.

Table 2-1. Elements Known in Boyle's Time.

Elements Known A.D. 100	Elements Added by A.D. 1600	Elements Added by A.D. 1700
Gold	Zinc	Phosphorous
Silver	Antimony	Oxygen
Tin	Bismuth	Hydrogen
Lead	Arsenic	
Copper	Reference made in 1557 to	
Mercury	"unmeltable metal—Platinum	
Iron	(not then named)	
Carbon		
Sulphur		

Two British scientists discovered the elements hydrogen and oxygen. A Frenchman, Lavoisier, found nitrogen; a Swede, Scheele, found chlorine.

In 1808, with the aid of the microscope, the British schoolmaster, physicist, and chemist John Dalton noted that crystals of the known elements appeared different from one another. Crystals of copper always looked alike but the crystals of gold did not resemble those of copper. Dalton then concluded that atoms of these substances must also have corresponding characteristics. Dalton even drew pictures of his atoms and claimed that an unknown force held atoms together.

Although Dalton is associated with modern atomic theory, he did make a basic mistake in assuming that one ounce of hydrogen and eight ounces of oxygen contain the same number of atoms, producing nine ounces of water. In chemical symbols, he said

$$\underset{(1\text{ ounce})}{H} + \underset{(8\text{ ounces})}{O} = \underset{(9\text{ ounces})}{HO}$$

Three years after John Dalton announced his atomic theory, the Italian physicist Amadeo Avogadro discovered the law that has been linked to his name:

Equal volumes of gases, measured under the same conditions of temperature and pressure, contain the same number of gas particles.

Using a law discovered by Robert Boyle, Avogadro solved the problem that faced Dalton in his water formula. "Boyle's law" states that any amount of gas enclosed in a container will double in pressure if squeezed into half the space.

Avogadro only had to weigh a gallon of hydrogen and a gallon of oxygen to prove that a gallon of oxygen weighs sixteen times as much as a gallon of hydrogen. There were the same number of atoms in each of the gallon containers. Thus there is the same number of atoms in one-half ounce of of hydrogen as in eight ounces of oxygen. When these atoms compounded or combined, Avogadro called them *molecules*, which means "little masses."

According to Avogadro, (1) chemical elements are composed of single atoms or molecules with like atoms; (2) chemical compounds are molecules composed of two or more different atoms.

The Elements

The current theory explaining the structure of matter states that all matter is composed of atoms.

Matter is anything that has weight or mass and occupies space. It may exist in the form of a solid, liquid, or gas.

There are 92 natural elements which make up all matter and 13 man-made or artificial elements. (See Table 2-2.) Element 104 is claimed by both Russia and America, but neither claim has been verified. Element 105 is also unofficial.

When two or more elements combine, they produce a new material called a *compound* (Fig. 2-1). Avogadro's water formula (H_2O) is a compound containing one oxygen and two hydrogen atoms. He also pointed out that a *molecule* is the smallest particle that a compound can be reduced to before it breaks down into its elements.

The atom is the smallest particle that an element can be reduced to and still keep the properties of the element. The atom may be broken down into different small subatomic particles.

Table 2-2. The Elements.

Natural Elements

Atomic Number	Name	Symbol	Atomic Number	Name	Symbol
1	Hydrogen	H	47	Silver	Ag
2	Helium	He	48	Cadmium	Cd
3	Lithium	Li	49	Indium	In
4	Beryllium	Be	50	Tin	Sn
5	Boron	B	51	Antimony	Sb
6	Carbon	C	52	Tellurium	Te
7	Nitrogen	N	53	Iodine	I
8	Oxygen	O	54	Xenon	Xe
9	Fluorine	F	55	Cesium	Cs
10	Neon	Ne	56	Barium	Ba
11	Sodium	Na	57	Lanthanum	La
12	Magnesium	Mg	58	Cerium	Ce
13	Aluminum	Al	59	Praseodymium	Pr
14	Silicon	Si	60	Neodymium	Nd
15	Phosphorus	P	61	Promethium	Pm
16	Sulfur	S	62	Samarium	Sm
17	Chlorine	Cl	63	Europium	Eu
18	Argon	A	64	Gadolinium	Gd
19	Potassium	K	65	Terbium	Tb
20	Calcium	Ca	66	Dysprosium	Dy
21	Scandium	Sc	67	Holmium	Ho
22	Titanium	Ti	68	Erbium	Er
23	Vanadium	V	69	Thulium	Tm
24	Chromium	Cr	70	Ytterbium	Yb
25	Manganese	Mn	71	Lutetium	Lu
26	Iron	Fe	72	Hafnium	Hf
27	Cobalt	Co	73	Tantalum	Ta
28	Nickel	Ni	74	Tungsten	W
29	Copper	Cu	75	Rhenium	Re
30	Zinc	Zn	76	Osmium	Os
31	Gallium	Ga	77	Iridium	Ir
32	Germanium	Ge	78	Platinum	Pt
33	Arsenic	As	79	Gold	Au
34	Selenium	Se	80	Mercury	Hg
35	Bromine	Br	81	Thallium	Tl
36	Krypton	Kr	82	Lead	Pb
37	Rubidium	Rb	83	Bismuth	Bi
38	Strontium	Sr	84	Polonium	Po
39	Yttrium	Y	85	Astatine	At
40	Zirconium	Zr	86	Radon	Rn
41	Niobium (Columbium)	Nb	87	Francium	Fr
			88	Radium	Ra
42	Molybdenum	Mo	89	Actinium	Ac
43	Technetium	Tc	90	Thorium	Th
44	Ruthenium	Ru	91	Protactinium	Pa
45	Rhodium	Rh	92	Uranium	U
46	Palladium	Pd			

Artificial Elements

Atomic Number	Name	Symbol	Atomic Number	Name	Symbol
93	Neptunium	Np	100	Fermium	Fm
94	Plutonium	Pu	101	Mendelevium	Mv
95	Americium	Am	102	Nobelium	No
96	Curium	Cm	103	Lawrencium	Lw
97	Berkelium	Bk	104	Not Named	
98	Californium	Cf	105	Not Named	
99	Einsteinium	E			

Atomic Particles

The atoms of each basic element differ in their number of the three basic subatomic particles: neutrons, protons, and electrons. No two elements have identical atoms.

The *neutron*, a particle with no electrical charge, and the *proton*, a positively (+) charged electrical

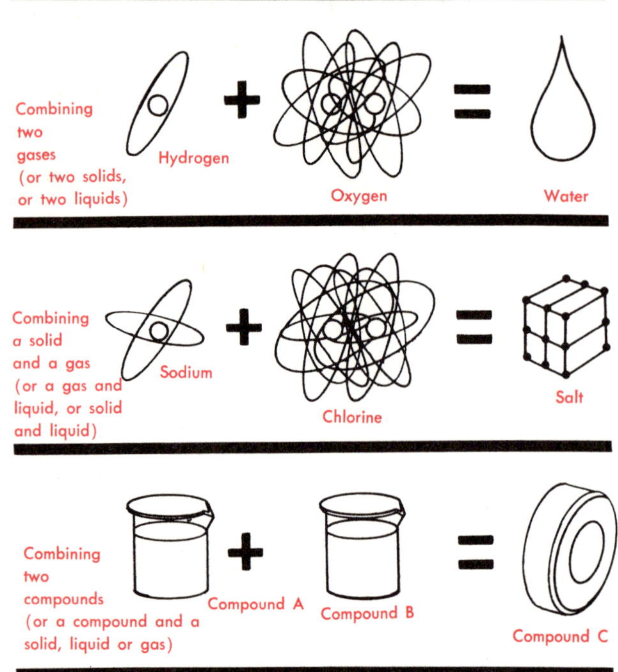

Fig. 2-1. Ways compounds may be produced.

particle, occupy a very dense compact area in the central part of the atom called the *nucleus.* Protons are very small dense particles estimated to be 0.07 trillionth of an inch in diameter.

Electrons are negatively (—) charged electrical particles and are located in orbits or shells around the nucleus. Although they are nearly three times larger in diameter than protons, they are about 1840 times lighter. This is one factor which makes the protons difficult to move and the electrons easy to move.

The simplest atom is the hydrogen atom. It has one proton (as nucleus), one electron, and no neutrons (Fig. 2-2).

The atom itself is electrically neutral because the negative charge of the electrons moving around the nucleus is equal to the positive charge of the protons in the nucleus and because the neutron is electrically neutral. Should an atom gain or lose electrons, it would become *electronegative* (a negative ion) or *electropositive* (a positive ion), respectively. A simple *ion* is an atom which has either gained or lost one or more electrons.

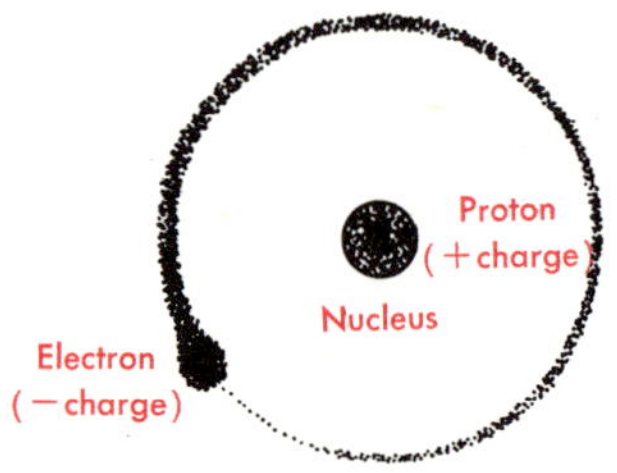

Fig. 2-2. Diagram of a hydrogen atom.

The Periodic Table

Elements are classified in the periodic table (Table 2-3) of the elements according to their *atomic number,* which is the number of protons in

Table 2-3. Periodic table of the elements. Note position of elements used extensively to produce plastics (C, N, O, F, Si, and Cl).

Periods Principal Quantum Number n ↓	IVA	VA	VIA	VIIA	O Rare Gases	IA	IIA	IIIA	IVB	VB	VIB	VIIB	VIII	VIII	VIII	IB	IIB	IIIB
								← Transition Elements									→	
n = 1 to He						1.00797 H 1												
n = 2 to Ne					4.0026 He 2	6.939 Li 3	9.0122 Be 4											10.811 B 5
n = 3 to Ar	12.01115 C 6	14.0067 N 7	15.9994 O 8	18.9984 F 9	20.183 Ne 10	22.9898 Na 11	24.312 Mg 12											26.9815 Al 13
n = 4 to Kr	28.086 Si 14	30.9738 P 15	32.064 S 16	35.453 Cl 17	39.948 Ar 18	39.102 K 19	40.08 Ca 20	44.956 Sc 21	47.90 Ti 22	50.942 V 23	51.996 Cr 24	54.9380 Mn 25	55.847 Fe 26	58.9332 Co 27	58.71 Ni 28	63.54 Cu 29	65.37 Zn 30	69.72 Ga 31
n = 5 to Xe	72.59 Ge 32	74.9216 As 33	78.96 Se 34	79.909 Br 35	83.80 Kr 36	85.47 Rb 37	87.62 Sr 38	88.905 Y 39	91.22 Zr 40	92.906 Nb 41	95.94 Mo 42	(97) Tc 43	101.07 Ru 44	102.905 Rh 45	106.4 Pd 46	107.870 Ag 47	112.40 Cd 48	114.82 In 49
n = 6 to Rn	118.69 Sn 50	121.75 Sb 51	127.60 Te 52	126.9044 I 53	131.30 Xe 54	132.905 Cs 55	137.34 Ba 56	138.91 La 57	178.49 Hf 72	180.948 Ta 73	183.85 W 74	186.2 Re 75	190.2 Os 76	192.2 Ir 77	195.09 Pt 78	196.967 Au 79	200.59 Hg 80	204.37 Tl 81
n = 7 to Lw	207.19 Pb 82	208.980 Bi 83	(209) Po 84	(210) At 85	(222) Rn 86	(223) Fr 87	(226.05) Ra 88	(227) Ac 89										

Lanthanide Series

140.12 Ce 58	140.907 Pr 59	144.24 Nd 60	(145) Pm 61	150.35 Sm 62	151.96 Eu 63	157.25 Gd 64	158.924 Tb 65	162.50 Dy 66	164.930 Ho 67
167.26 Er 68	168.934 Tm 69	173.04 Yb 70	174.97 Lu 71						

Actinide Series

232.038 Th 90	(231) Pa 91	238.03 U 92	(237) Np 93	(244) Pu 94	(243) Am 95	(247) Cm 96	(247) Bk 97	(251) Cf 98	(254) Es 99
(257) Fm 100	(256) Md 101	(255) No 102	(257) Lw 103	(260) 104					

the nucleus of the atom. The atomic number of carbon, for example, is 6, the number of protons in its nucleus, or the number of electrons around the nucleus (Fig. 2-3). Of course the atom is really three-dimensional, not two-dimensional as shown in Fig. 2-3.

Hydrogen, symbolized *H*, with an atomic number of 1, is located in the upper middle corner of the periodic table. It is the first element in Group IA and is followed by Li, Na, K, Rb, Cs, and Fr. All of the elements in Group IA have one electron in their outer shell and have similar chemical properties. The elements in Group I are highly active chemically while the elements in Group VIII are inert elements and will not readily combine chemically. Carbon, chlorine, fluorine, oxygen, hydrogen, nitrogen, and silicon are the primary elements that are of concern in the study of plastic materials. Note the positions of these elements in the periodic table.

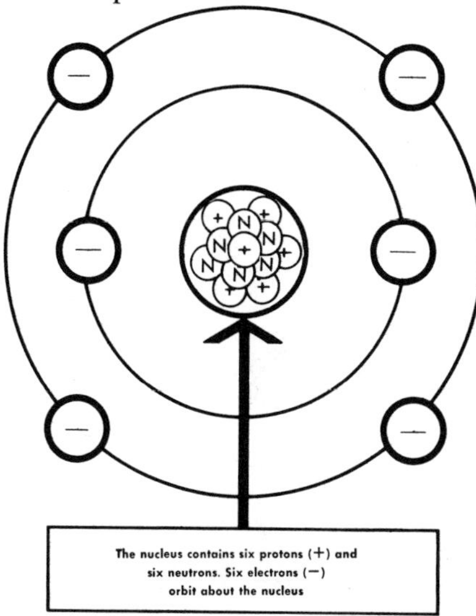

The nucleus contains six protons (+) and six neutrons. Six electrons (−) orbit about the nucleus

Fig. 2-3. Two-dimensional representation of carbon atom.

The atom is chiefly empty space, very much like our solar system, but it is actually very heavy for its size. One author states that if 5000 battleships could be placed in the volume of a baseball and that baseball weighed as much as the 5000 battleships, it would compare with the density and relative weight of one atom.

The *atomic weight* of an element is the weight of an atom of the element compared to the weight of an atom of carbon. As we will see, the weight of a carbon atom is very inconvenient to use. So a system of relative atomic weights of the elements is used, setting carbon at 12 g. This relative atomic weight of an element in grams is called a *gram-atom*; so one gram-atom of carbon is 12 g. Experimentally is is known that one gram-atom of any element contains 6.02×10^{23} atoms. Therefore the atomic weight of an element in grams (one gram-atom) divided by the number of atoms per gram-atom gives the weight of an atom in grams:

$$\text{weight of atom in grams} = \frac{\text{atomic weight in grams}}{6.02 \times 10^{23}}$$

For carbon, then,

$$\text{weight of carbon atom in grams} = \frac{12\text{ g}}{6.02 \times 10^{23}}$$
$$\cong 2 \times 10^{-23}\text{ g}$$

The number 6.02×10^{23} is referred to as *Avogadro's number*.

In 1962, by international agreement, carbon was assigned the number 12 to serve as the standard for all atomic weights. The average weight of hydrogen is about one-twelfth that of carbon, making the atomic weight of hydrogen approximately 1. Remember, these weights are only approximate. To find the actual weight of the atom, the weight of neutrons, protons, and electrons should be included.

Electrons and Electron Shells

The arrangement of the electrons about the nucleus of an atom greatly influences the chemical and physical properties of an element. Electrons may arrange themselves in a number of possible shells or energy levels around the nucleus.

It should be noted that an electron spins like a top and vibrates as it orbits the nucleus in various shell or energy levels. The exact position of any electron, at any given instant, cannot be known. The probability of its location in a specific subshell, however, can be determined. In Fig. 2-4 the density of the electron cloud is proportional to the probability of an electron being at a given spot. For example, an electron is twice as likely to be in the Y ring as in the X ring.

The atom should also be thought of as being a three-dimensional object like a ball and not flat

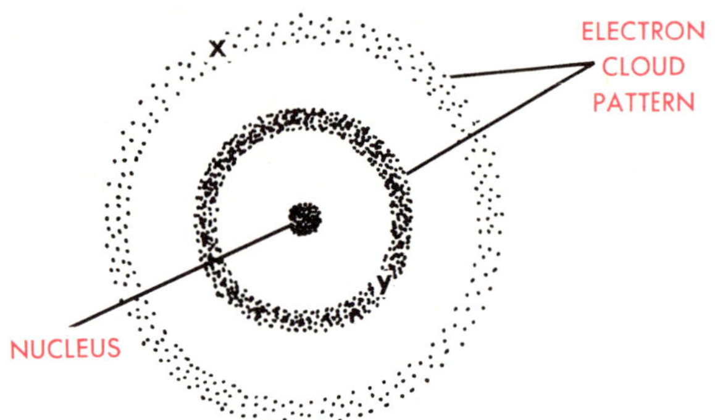
ELECTRON
CLOUD
PATTERN
NUCLEUS
X
Y

Fig. 2-4. The density of the electron cloud at any point is proportional to the probability of an electron being at that point.

or in one plane as pictured on many schematic diagrams. Fig. 2-5 shows a three-dimensional representation of a lithium atom, which has three spinning electrons in two shells. Note the spirals indicating electron spin. The lithium nucleus contains three protons and four neutrons.

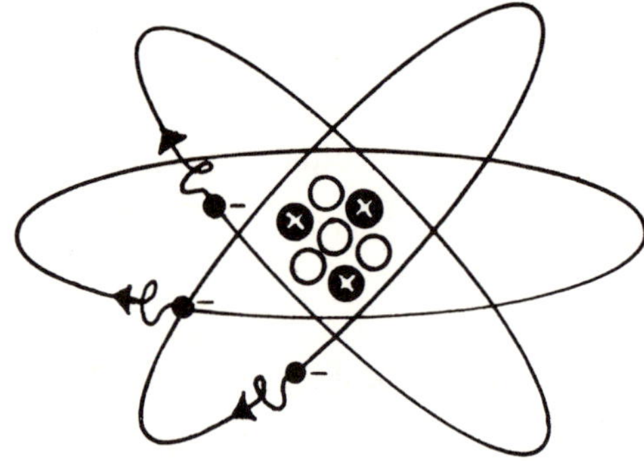

Fig. 2-5. A lithium atom with spinning electrons in shell orbits around the nucleus.

Atoms may have up to seven shells or orbital paths for the electrons. The first or lowest energy level of an atom and the shell nearest the nucleus is labeled the *K* shell. The first or lowest energy level shell (*K*) contains a maximum number of two electrons. Electrons seek the lowest energy level possible until the lower energy levels are filled.

The remaining, unattracted electrons will assume positions in progressively higher energy levels. This atomic model has one important weakness. It does not account for the geometry of molecules which are very important for polymers.

There are a few exceptions where higher energy levels or shells fill with electrons before a lower one. The scandium to nickel series and the yttrium to palladium series are examples.

The common shells in the order of increasing electron energy are the *K, L, M, N,* and *O* shells with energy levels identified as the 1, 2, 3, 4, and 5 energy levels in place of letter designations. More energy would be required to move electrons from the *L* energy level than the *K* energy level.

The formula $2n^2$, where n is the energy level number, can be used to determine the maximum number of electrons possible in any energy level. The greatest possible number of electrons in each energy level is therefore as follows:

First Shell (K) 2
Second Shell (L) .... 2 + 6 = 8
Third Shell (M) .... 2 + 6 + 10 = 18
Fourth Shell (N) .. 2 + 6 + 10 + 14 = 32

See Fig. 2-6.

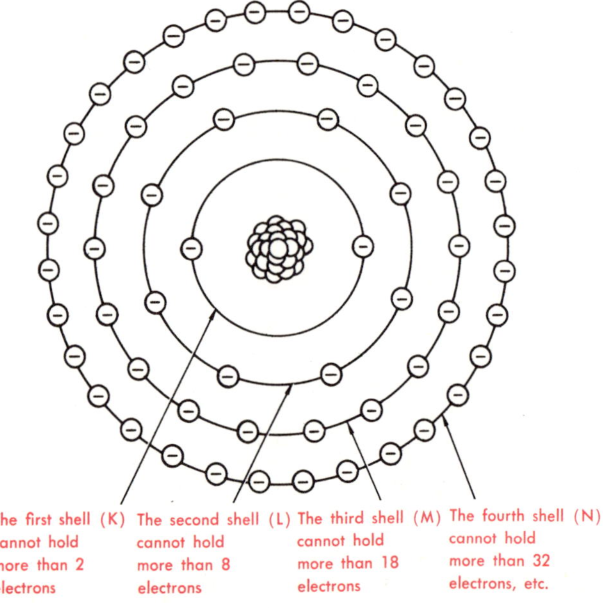
The first shell (K) cannot hold more than 2 electrons
The second shell (L) cannot hold more than 8 electrons
The third shell (M) cannot hold more than 18 electrons
The fourth shell (N) cannot hold more than 32 electrons, etc.

Fig. 2-6. Maximum possible number of electrons in energy shells around the nucleus.

Numbers 2, 8, 18, and 32 are identical with the numbers which comprise the length of the various periods in the periodic table system (Table 2-3). Period I (n = 1) begins with hydrogen (H) and ends with helium (He). Period II (n = 2) begins with lithium (Li) and ends with neon (Ne), or exactly eight elements. Period III has eight, period IV eighteen, period V eighteen, period VI thirty-two, and period VII thirty-two elements.

The electrons in the highest energy level of an atom are known as *valence electrons.* Valence electrons are the primary determiner of an atom's chemical properties. Nonvalence electrons are closely associated with the nucleus of the atom and normally do not enter into reactions with other atoms.

Shells or energy levels are divided into subshells labeled *s, p, d,* and *f*. These subshells, which are listed in order of increasing electron energy, contain a maximum of 2, 6, 10, and 14 electrons apiece.

Electronic notation may be used to describe the energy level placement of electrons around the nucleus. The notation $1s^2 2s^2 2p^6 3s^2 3p^5$ or $Ks^2 Ls^2 Lp^6 Ms^2 Mp^5$ is an example of electron notation and indicates that two electrons are in the *K* shell (only one level), eight electrons in the *L* shell (with two in its lower (*s*) subshell and six in the next higher (*p*) subshell), and seven electrons in the *M* shell (with two in its lower (*s*) subshell and five in its next higher (*p*) subshell). This electronic notation indicates that there are 17 electrons in orbit. If the atom is in balance (not an ion) there should be 17 protons in the nucleus. The number indicated in the lower left of the element squares on the periodic chart is the number of protons, or the atomic number of the element. Atomic number 17 indicates an atom of chlorine (Fig. 2-7).

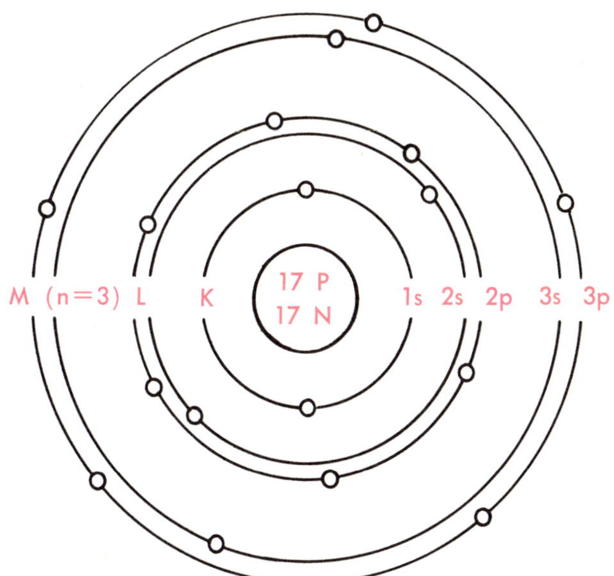

Fig. 2-7. Chlorine atom has notation $1s^2 2s^2 2p^6 3s^2 3p^5$ or $Ks^2 Ls^2 Lp^6 Ms^2 Mp^5$.

The extension of the electron quantum numbers theme is shown in Table 2-4. The electronic notation for a single iron atom would then be Fe = $1s^2 2s^2 2p^6 3s^2 3p^6 3d^6 4s^2$.

The *outer* shell of any atom will have no more than eight electrons. Atoms are stable when they have their outer valence shell filled with eight electrons. That is, there are never more than two electrons in the first (*K*) level and never more than eight electrons in the outermost shell when filled.

Because energy is shared or absorbed among electrons, less energy is required to attract or repel electrons from atoms with only one electron in its outer shell. These atoms are conductors of electricity. Fig. 2-8 shows atoms of three elements which are good conductors, namely the elements copper, gold, and silver.

If the outer shell of an atom is more than half filled but has less than eight electrons, then that

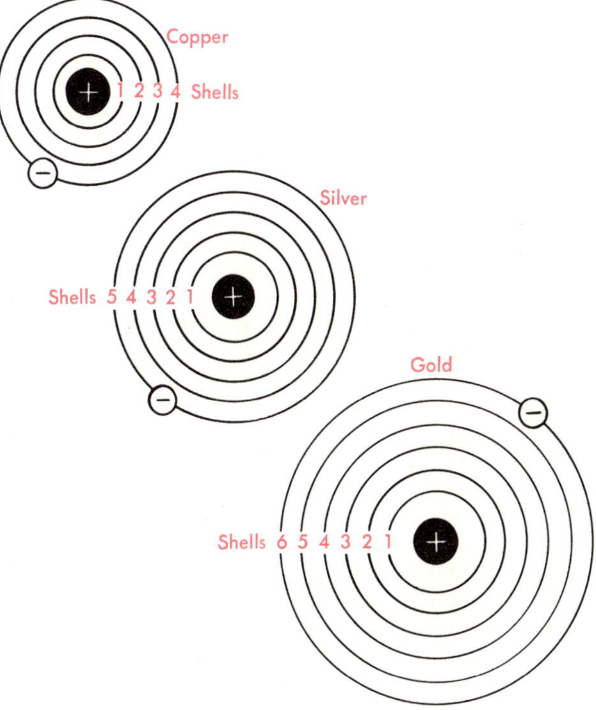

Fig. 2-8. Atoms with only one valence electron are good conductors. (Only valence electrons are shown.)

Table 2-4. Subshell Electrons of Atoms.

Element		K (n=1)	L (n=2)		M (n=3)			N (n=4)				O (n=5)				P (n=6)			Q (n=7)
Symbol	Number	1s	2s	2p	3s	3p	3d	4s	4p	4d	4f	5s	5p	5d	5f	6s	6p	6d	7s
H	1	1																	
He	2	2																	
Li	3	2	1																
Be	4	2	2																
B	5	2	2	1															
C	6	2	2	2															
N	7	2	2	3															
O	8	2	2	4															
F	9	2	2	5															
Ne	10	2	2	6															
Na	11	2	2	6	1														
Mg	12	2	2	6	2														
Al	13	2	2	6	2	1													
Si	14	2	2	6	2	2													
P	15	2	2	6	2	3													
S	16	2	2	6	2	4													
Cl	17	2	2	6	2	5													
Ar	18	2	2	6	2	6													
K	19	2	2	6	2	6		1											
Ca	20	2	2	6	2	6		2											
Sc	21	2	2	6	2	6	1	2											
Ti	22	2	2	6	2	6	2	2											
V	23	2	2	6	2	6	3	2											
Cr	24	2	2	6	2	6	5	1											
Mn	25	2	2	6	2	6	5	2											
Fe	26	2	2	6	2	6	6	2											
Co	27	2	2	6	2	6	7	2											
Ni	28	2	2	6	2	6	8	2											
Cu	29	2	2	6	2	6	10	1											
Zn	30	2	2	6	2	6	10	2											
Ga	31	2	2	6	2	6	10	2	1										
Ge	32	2	2	6	2	6	10	2	2										
As	33	2	2	6	2	6	10	2	3										
Se	34	2	2	6	2	6	10	2	4										
Br	35	2	2	6	2	6	10	2	5										
Kr	36	2	2	6	2	6	10	2	6										
Rb	37	2	2	6	2	6	10	2	6			1							
Sr	38	2	2	6	2	6	10	2	6			2							
Y	39	2	2	6	2	6	10	2	6	1		2							
Zr	40	2	2	6	2	6	10	2	6	2		2							
Nb	41	2	2	6	2	6	10	2	6	4		1							
Mo	42	2	2	6	2	6	10	2	6	5		1							
Te	43	2	2	6	2	6	10	2	6	6		1							
Ru	44	2	2	6	2	6	10	2	6	7		1							
Rh	45	2	2	6	2	6	10	2	6	8		1							
Pd	46	2	2	6	2	6	10	2	6	10									
Ag	47	2	2	6	2	6	10	2	6	10		1							
Cd	48	2	2	6	2	6	10	2	6	10		2							
In	49	2	2	6	2	6	10	2	6	10		2	1						
Sn	50	2	2	6	2	6	10	2	6	10		2	2						
Sb	51	2	2	6	2	6	10	2	6	10		2	3						
Te	52	2	2	6	2	6	10	2	6	10		2	4						
I	53	2	2	6	2	6	10	2	6	10		2	5						
Xe	54	2	2	6	2	6	10	2	6	10		2	6						
Cs	55	2	2	6	2	6	10	2	6	10		2	6			1			
Ba	56	2	2	6	2	6	10	2	6	10		2	6			2			
La	57	2	2	6	2	6	10	2	6	10		2	6	1		2			

Table 2-4. Subshell Electrons of Atoms. (Continued)

Element Symbol	Number	K (n=1) 1s	L (n=2) 2s	2p	M (n=3) 3s	3p	3d	N (n=4) 4s	4p	4d	4f	O (n=5) 5s	5p	5d	5f	P (n=6) 6s	6p	6d	Q (n=7) 7s
Ce	58	2	2	6	2	6	10	2	6	10	2	2	6			2			
Pr	59	2	2	6	2	6	10	2	6	10	3	2	6			2			
Nd	60	2	2	6	2	6	10	2	6	10	4	2	6			2			
Pm	61	2	2	6	2	6	10	2	6	10	5	2	6			2			
Sm	62	2	2	6	2	6	10	2	6	10	6	2	6			2			
Eu	63	2	2	6	2	6	10	2	6	10	7	2	6			2			
Gd	64	2	2	6	2	6	10	2	6	10	7	2	6	1		2			
Tb	65	2	2	6	2	6	10	2	6	10	8	2	6	1		2			
Dy	66	2	2	6	2	6	10	2	6	10	10	2	6			2			
Ho	67	2	2	6	2	6	10	2	6	10	11	2	6			2			
Er	68	2	2	6	2	6	10	2	6	10	12	2	6			2			
Tm	69	2	2	6	2	6	10	2	6	10	13	2	6			2			
Yb	70	2	2	6	2	6	10	2	6	10	14	2	6			2			
Lu	71	2	2	6	2	6	10	2	6	10	14	2	6	1		2			
Hf	72	2	2	6	2	6	10	2	6	10	14	2	6	2		2			
Ta	73	2	2	6	2	6	10	2	6	10	14	2	6	3		2			
W	74	2	2	6	2	6	10	2	6	10	14	2	6	4		2			
Re	75	2	2	6	2	6	10	2	6	10	14	2	6	5		2			
Os	76	2	2	6	2	6	10	2	6	10	14	2	6	6		2			
Ir	77	2	2	6	2	6	10	2	6	10	14	2	6	7		2			
Pt	78	2	2	6	2	6	10	2	6	10	14	2	6	8		2			
Au	79	2	2	6	2	6	10	2	6	10	14	2	6	10		1			
Hg	80	2	2	6	2	6	10	2	6	10	14	2	6	10		2			
Tl	81	2	2	6	2	6	10	2	6	10	14	2	6	10		2	1		
Pb	82	2	2	6	2	6	10	2	6	10	14	2	6	10		2	2		
Bi	83	2	2	6	2	6	10	2	6	10	14	2	6	10		2	3		
Po	84	2	2	6	2	6	10	2	6	10	14	2	6	10		2	4		
At	85	2	2	6	2	6	10	2	6	10	14	2	6	10		2	5		
Rn	86	2	2	6	2	6	10	2	6	10	14	2	6	10		2	6		
Fr	87	2	2	6	2	6	10	2	6	10	14	2	6	10		2	6		1
Ra	88	2	2	6	2	6	10	2	6	10	14	2	6	10		2	6		2
Ac	89	2	2	6	2	6	10	2	6	10	14	2	6	10		2	6	1	2
Th	90	2	2	6	2	6	10	2	6	10	14	2	6	10		2	6	2	2
Pa	91	2	2	6	2	6	10	2	6	10	14	2	6	10	2	2	6	1	2
U	92	2	2	6	2	6	10	2	6	10	14	2	6	10	3	2	6	1	2
Np	93	2	2	6	2	6	10	2	6	10	14	2	6	10	5	2	6		2
Pu	94	2	2	6	2	6	10	2	6	10	14	2	6	10	6	2	6		2
Am	95	2	2	6	2	6	10	2	6	10	14	2	6	10	7	2	6		2
Cm	96	2	2	6	2	6	10	2	6	10	14	2	6	10	7	2	6	1	2

element will be a good insulator. The atom tries to become stable by filling its valence shell and it is very difficult to free an electron from the valence shell.

When the outer shell is completely filled with eight electrons, it is said to be *stable*. Atoms, if necessary, will attract electrons to their valence shell or repel electrons from their valence shell in an effort to have eight electrons in their outer shell.

It is this property of atoms which determines how atoms combine with one another to form molecules.

Glossary of New Terms

Atomic number (atomic mass number)—The number equal to the number of protons in a nucleus of the atom of an element.

Atomic weight—Relative weight of an atom of any element as compared with that of one atom of carbon taken at 12 g.

Atoms—Smallest particles of an element that combine with particles of other elements to produce compounds. Atoms combine to form molecules and consist of a complex arrangement of electrons revolving about a positively charged nucleus containing particles called protons and neutrons.

Avogadro's number—The number of atoms in a gram-atom weight of an element: 6.02486×10^{23}.

Compound—A substance composed of two or more elements joined together in definite proportions.

Electron—A negatively charged particle which is present in every atom.

Inert (rare) gases—Helium, argon, neon, krypton, xenon, and radon (Group O of the periodic table). These gases have practically no tendency to combine with other elements.

Ion—An atom with a net electrical charge.

Molecule—The smallest particle of a substance that can exist independently, retaining the chemical properties of the substance.

Periodic table—An arrangement of the elements in order of increasing atomic weight, forming groups the members of which show similar physical and chemical properties. Table 2-3, on p. 12, is the periodic table.

Proton—A positively charged particle in the nucleus of an atom. Its charge is equal but opposite to that of the electron.

Shell, electron—A region about the nucleus of an atom in which electrons move; each electron shell corresponds to a definite energy level.

Valence electrons—The electrons in the outermost shell of an atom.

Review Questions

2-1. What significance does the discovery and study of elements, atoms, and molecules have to the study of plastics?

2-2. What elements are primarily used in the production of plastics?

2-3. What significance does electronic notation have to the study of plastics?

2-4. What is the element with the electronic notation $1s^2 2s^2 2p^6 3s^2 3p^5$?

2-5. What is Avogadro's law?

2-6. Why does the element carbon have the atomic weight of 12?

3

Bonding

Only the gas elements not easily entering into chemical reactions ("rare" or "inert" gases: He, Ne, Ar, Xe, Rn) are found occurring alone in nature. The majority of the elements are combinations of two or more atoms bonded together forming molecules or crystals in specified geometric patterns. Since the attractive forces of the atoms form invisible bonds and influence the properties of all materials, a better understanding of the fundamental principles of bonding must be gained.

Bonding in Plastics

Ionic, covalent, and van der Waals' forces are all-important considerations in the study of plastics because these forces or bonds determine many of the physical characteristics of the finished plastics product.

All bonds between atoms and molecules are electrical in nature. Probably the most important force holding the atom together is the electrical forces of the atom's individual particles. It is these same forces of electrical charges, either negative or positive, which bond atoms into the composite of the molecule. It is the valence formed by the electrons that provides the connecting link or bond that creates a molecule.

Since most elements except the inert gases are constantly trying to reach a stable state with eight electrons in their outer valence shell, they must become stable by one of the following procedures: (1) receiving extra electrons, (2) releasing electrons, or (3) sharing electrons.

Primary Bonds

Ionic and covalent bonding forces are known as *primary* bonds. The simplest bond to understand is the *ionic* bond, where there is an actual lending of electron(s) from one atom to another. For example, when sodium and chlorine bond, the sodium has only one electron in its outermost shell. (See Fig. 3-1.) The chlorine has only seven electrons in its outer shell but tends to become stable with eight. Since there is less energy required to lend one electron than seven, the sodium electron jumps over to the outer shell of the chlorine atom, thereby converting the previously neutral atoms to oppositely charged ions. A strong electrical attraction between the ions provides the bond which holds together the sodium chloride compound (common table salt) as an NaCl molecule.

The major requirement in an ionically bonded material is that the number of positive charges equal the number of negative charges.

Another primary force of attraction is the *covalent* bond. Covalent bonding exists where electrons are shared and the valence orbitals overlap between adjacent atoms (Fig. 3-2). Covalent bonding makes it possible for elements adjacent to each other on the periodic table to bond together, whereas ionic bonding can only occur

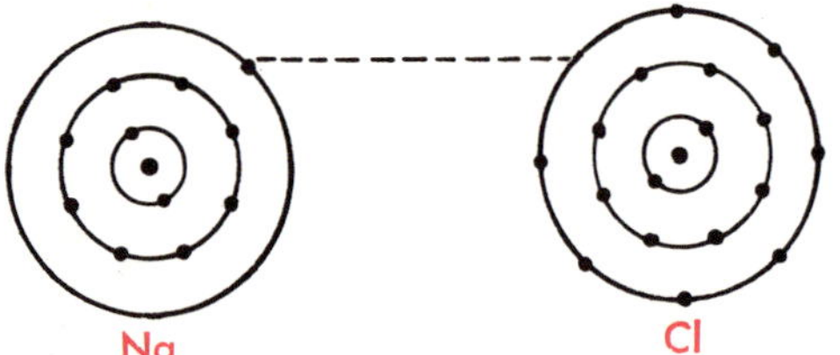

(A) Sodium valence electron is easily lost to chlorine atom.

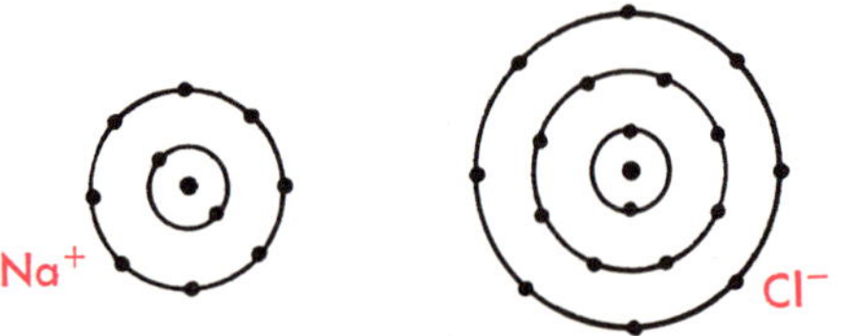

(B) Sodium atom is now charged positively, and chlorine atom is charged negatively.

(C) Electrical or ionic attraction occurs to form NaCl molecule.

Fig. 3-1. An ionic bond joins atoms of sodium and chlorine in sodium chloride (NaCl).

between elements on opposite sides of the periodic table, e.g., from group IA to group VIIA as for NaCl molecules.

The simplest example of sharing electrons with an adjacent atom is found in the H_2 hydrogen molecule. The element hydrogen has only one electron in its *K* shell as shown in Table 2-4 of Chapter 2.

Hydrogen rarely occurs this way in nature because the *K* shell level is seeking to be completed. Remember, only two electrons are required in the *K* level.

When two hydrogen atoms combine and each shares an electron, a single covalent bond is formed as shown in Fig. 3-2. Note that, unlike sodium and chlorine, which are oppositely charged when they are combined, hydrogen atoms remain electrically neutral even after they have joined forces to make a molecule. The atoms merely draw close enough together so that their electrons begin orbiting around both nuclei. The strong bond resulting from this sharing of electrons is the type that holds most giant molecules together.

When two electrons from each atom are shared, a double bond is formed. Some of the more com-

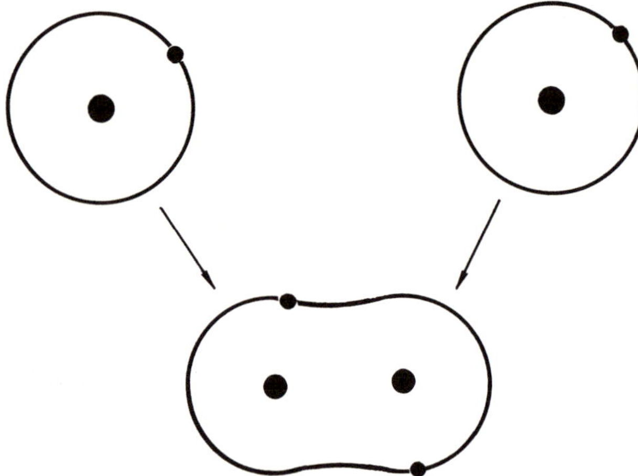

Fig. 3-2. A covalent bond is formed between hydrogen atoms when they share a pair of electrons.

mon elements encountered in the study of plastics are molecules having the following number of bonds: H, F, Cl (one each); O, S (two each); N (three); and C, Si (four each). With the exception of hydrogen and helium, the number of covalent bonds may be expressed by the general formula $N = 8 - G$, where N is the number of bonds, 8 the maximum number of electrons in the outer valence shell, and G is the group number of the atom within the periodic table.

Using the molecule of methane as an example, a carbon atom, which is in Group IV in the periodic table (Table 2-3), is surrounded by four hydrogen atoms sharing electrons. Using the formula $N = 8 - G$ gives $8 - 4 = 4$ the number of covalent bonds (Fig. 3-3).

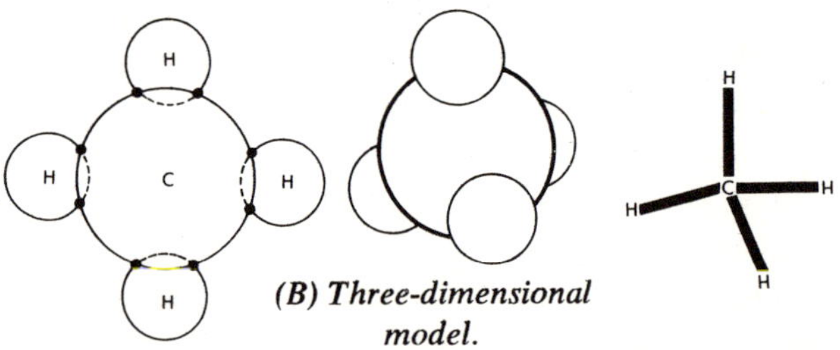

(A) Two-dimensional representation.

(B) Three-dimensional model.

(C) Covalent bonds of atoms.

Fig. 3-3. Models of methane (CH_4).

The schematic arrangement of electrons in diatomic (two-atom) molecules is shown in Fig. 3-4. It is worthy to note that a closer bond is produced when more electrons are shared. The bonding lengths, angles, and energy levels of bonding are

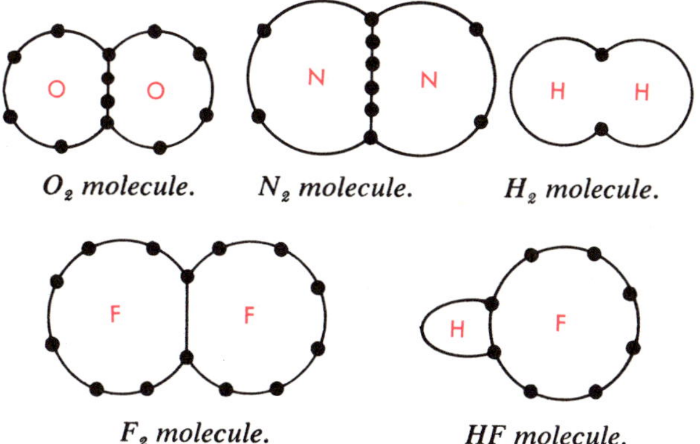

Fig. 3-4. Diatomic molecules.

beyond the scope of this text, and the third type of primary bonding, *metallic bonding,* will not be discussed.

Secondary Bonds

Secondary bonding forces, or van der Waals' forces, are much weaker than the primary bonds. The forces which bind the atoms of inert gases together to form a liquid or solid are called *van der Waals' forces* from a well-known gas equation of van der Waals.

These invisible bonds that hold molecules together are essentially electrical forces and are closely related to the electrical forces that hold atoms together in a molecule.

Ionic and covalent bonds are then between atoms which make molecules while van der Waals' forces are the bonds between the molecules or between the atoms in different molecules. Most van der Waals' forces of attraction arise from three sources: (1) dipole bonds, (2) dispersion bonds, and (3) hydrogen bonds.

The *dipole* bonding effect may be seen in Fig. 3-5. When the hydrogen and fluorine atoms combine they are left with an electrical imbalance even though the *K* shell of the hydrogen and the *L* shell of the fluorine are filled. This imbalance occurs because the fluorine nucleus is completely surrounded by electrons while the covalently shared hydrogen electrons are off center. This leaves the hydrogen nucleus exposed at the end of the bond; consequently the heavy positive nucleus is exposed. This imbalance produces a molecule of which one end is in effect slightly positive and the other end is slightly negative. Such an imbalance of charge is called an *electric dipole.* It is these electrical imbalances or *secondary* forces of attraction which cause many molecules to join (see Fig. 3-5).

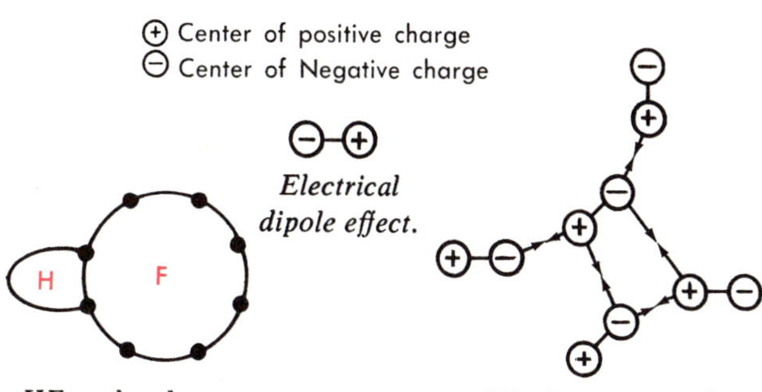

Fig. 3-5. Dipole bonding effect.

An important bonding force between molecules of many organic solids is a *dispersion bond.* As the random movement of electrons rotate around the nucleus of each atom, a momentary dipole effect (polarization) is established in the atom. With a molecule or atom of hydrogen, it is easy to see how one side of the atom can become more positive for a moment when only one electron is in the *K* shell level. Fig. 3-6 shows the resulting attraction between the electron of a neighboring hydrogen atom and the nucleus of another hydrogen atom.

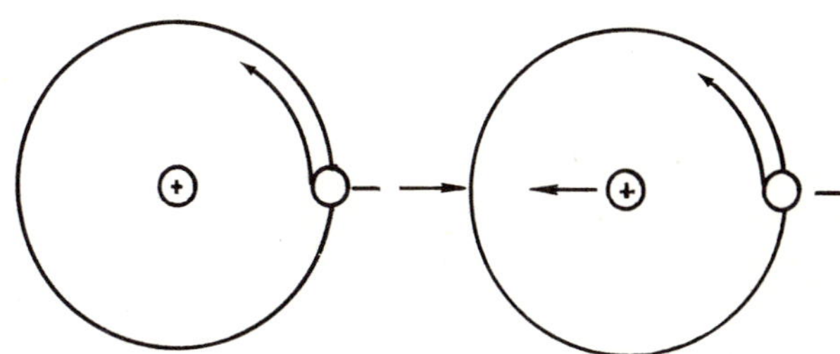
Fig. 3-6. Dispersion bonding of hydrogen atoms.

A weak secondary van der Waals' force similar to the dipole bond is the *hydrogen bond.* This attractive force is produced by the exposed nucleus (proton) of the hydrogen atom to the unshared electrons of nearby molecules. See Fig. 3-7.

Because all van der Waals' forces are relatively weak bonding forces, plastics and other materials utilizing these forces generally have low melting points and are comparatively soft.

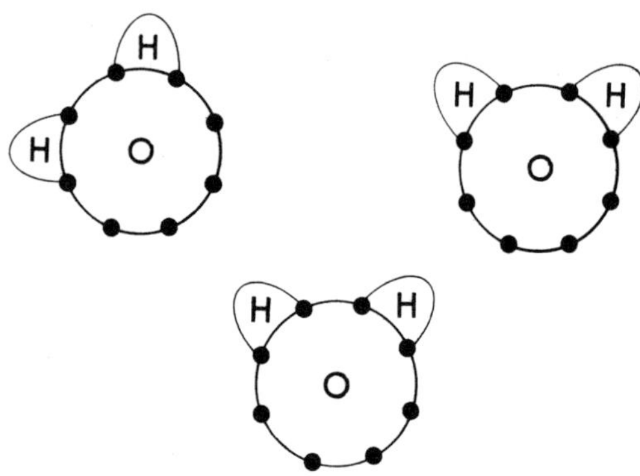

Fig. 3-7. Hydrogen bonding between molecules.

Glossary of New Terms

Covalent bonding—Atomic bonding by sharing electrons.

Electrical dipole—A nonuniform charge distribution one end of which is positive and the other negative.

Ionic bonding—Atomic bonding by electrical attraction of unlike ions.

Polarization—The separation of the positive and negative charges in a molecule.

Primary bonds—A strong association (interatomic attraction) between atoms.

Secondary bonds—Forces of attraction, other than primary bonds, which cause many molecules to join.

Van der Waals' forces—Weak secondary interatomic attraction arising from internal dipole effects.

Review Questions

3-1. Name three kinds of atomic bonding methods used in plastics.

3-2. What does the valence of the individual atom have to do with ionic and covalent bonding forces?

3-3. How many bonding sites do the following elements have: H, F, Cl, O, S, N, C, and Si?

3-4. Why is molecular bonding an important consideration in the manufacture of plastics?

3-5. What significance is the group number *G* of an element in the periodic table?

3-6. Give three bonds that arise from van der Waals' forces.

3-7. Give the three ways that atoms may become stable with eight electrons in their outer electron shells.

4

Polymeric Molecules

Molecules of organic materials are composed of carbon as a base element and are most commonly joined to hydrogen by covalent bonds. These molecules are referred to as *hydrocarbons*. As the name suggests, hydrocarbons are composed of hydrogen and carbon. Because of the ability of carbon atoms to link up with other atoms, forming chains, rings, and other complex molecules, thousands of hydrocarbons are known to the organic chemist.

Historically, as mentioned previously, all organic compounds at one time came from plants or animals. Coal, oil, and natural gas are good sources of the carbon for the chemicals used in making plastics (Fig. 4-1). With new and better technology in the petroleum and natural gas industry, more hydrocarbon compositions are produced.

Many organic hydrocarbons may be obtained directly from plant or animal sources. The oils from cotton seed, linseed, soybean, lard, and other similar products all yield hydrocarbons for plastic production. Corncobs and oathulls are even utilized. New advances in the collection of simple methane and other gases from sewage sludge will yield additional sources for the chemist.

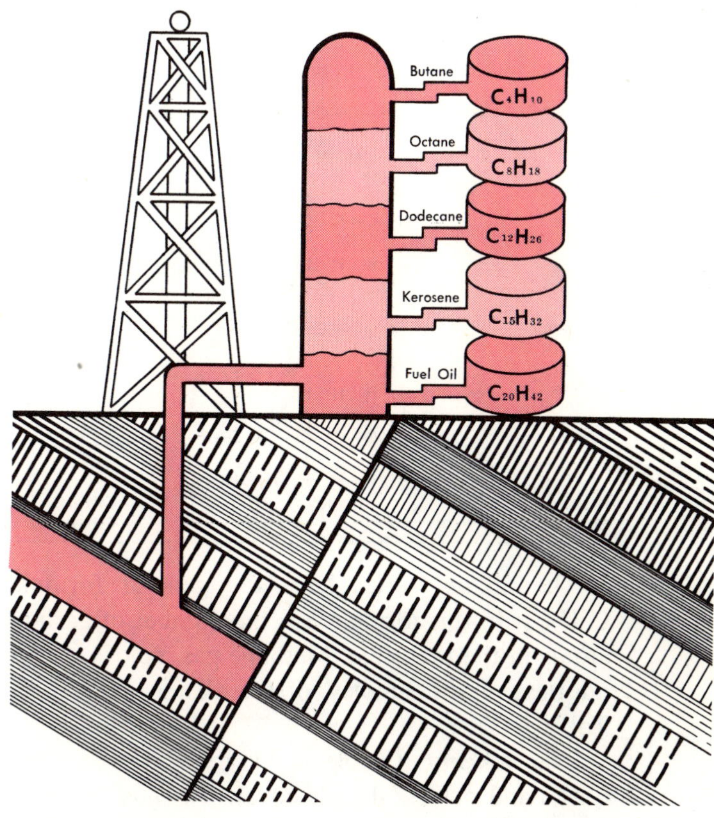

Fig. 4-1. Petroleum is found in trapped underground deposits. It is taken to a fractionating tower where different compounds are produced from the hot vaporized petroleum.

Organic Nomenclature

In the study of hydrocarbons, there are two major organic groups to consider: (1) *aliphatic* or straight-chain molecules, and (2) *cyclic* or ring-shaped molecules which do not have a general formula.

The aliphatic hydrocarbons may be further reduced into three hydrocarbon series: (1) *alkanes* or paraffins, (2) *alkenes* or olefins, and (3) *alkynes* or acetylene.

The aliphatic alkanes are linked together by a single covalent bond. Common shared electrons are usually depicted by a single line as shown in the alkanes in Table 4-1. Molecules of this type possess strong *intra*molecular covalent bonds and, in addition, the weaker *inter*molecular van der Waals' bonds.

Table 4-1. Alkanes.

Name	Structure	Composition
Methane	H | H—C—H | H	or CH_4
Ethane	H H | | H—C—C—H | | H H	or C_2H_6
Propane	H H H | | | H—C—C—C—H | | | H H H	or C_3H_8
Butane	H H H H | | | | H—C—C—C—C—H | | | | H H H H	or C_4H_{10}
Pentane	H H H H H | | | | | H—C—C—C—C—C—H | | | | | H H H H H	or C_5H_{12}
Hexane	H H H H H H | | | | | | H—C—C—C—C—C—C—H | | | | | | H H H H H H	or C_6H_{14}
Heptane	etc.	C_7H_{16}
Octane	↓	C_8H_{18}

The alk*ane* series have the general formula C_nH_{2n+2}. In this formula n is the number of carbon atoms in the molecule. Methane gas, colorless and odorless, is the simplest of the alkane series (see Table 4-2).

The alk*enes* (Table 4-3) have the formula C_nH_{2n} and possess double covalent bonds. When a molecule has a multiple carbon-to-carbon bond, it is considered to be a *unsaturated hydrocarbon*. There occurs a hydrogen shortage because an extra bond that might otherwise be used for the hydrogen atoms are used to hold the carbon atoms together. With the al*kanes* there is only one covalent bond and the molecule has all the hydrogen atoms it can hold.

Members of the al*kynes* (Table 4-4) are known by a triple bond between two carbon atoms;

Table 4-2. Alkane Series with Melting and Boiling Points.

Formula	Name	Melting Point in °C	Boiling Point in °C
CH_4	Methane	−182.5	−161.5
C_2H_6	Ethane	−183.3	−88.6
C_3H_8	Propane	−187.7	−42.1
C_4H_{10}	Butane	−138.4	−0.5
C_5H_{12}	Pentane	−129.7	+36.1
C_6H_{14}	Hexane	−95.3	68.7
C_7H_{16}	Heptane	−90.6	98.4
C_8H_{18}	Octane	−56.8	125.7
C_9H_{20}	Nonane	−53.5	150.8
$C_{10}H_{22}$	Decane	−30	174
$C_{11}H_{24}$	Undecane	−26	196
$C_{12}H_{26}$	Dodecane	−10	216
$C_{15}H_{32}$	Pentadecane	+10	270
$C_{20}H_{42}$	Eicosane	36	345
$C_{30}H_{62}$	Triacontane	66	distilled at reduced pressure to avoid decomposition
$C_{40}H_{82}$	Tetracontane	81	
$C_{50}H_{102}$	Pentacontane	92	
$C_{60}H_{122}$	Hexacontane	99	
$C_{70}H_{142}$	Heptacontane	105	

Table 4-3. Alkene Series with Melting and Boiling Points.

Formula	Name	Melting Point in °C	Boiling Point in °C
$CH_2{=}CH_2$	Ethene (Ethylene)	−169	−103.7
$CH_3{-}CH{=}CH_2$	Propene (Propylene)	−185	−47.7
$C_2H_5{-}CH{=}CH_2$	1-Butene	−185	−6.3
$C_3H_7{-}CH{=}CH_2$	1-Pentene	−165	+30.0
$C_4H_9{-}CH{=}CH_2$	1-Hexene	−140	64.6
$C_5H_{11}{-}CH{=}CH_2$	1-Heptene	−119	93.6
$C_6H_{13}{-}CH{=}CH_2$	1-Octene	−102	121.3
$C_7H_{15}{-}CH{=}CH_2$	1-Nonene	−81	146.9
$C_{10}H_{20}$	1-Decene	−66	170.5
$C_{15}H_{30}$	1-Pentadecene	−3.3	268.6
$C_{20}H_{40}$	1-Eicosene	+30.1	344

Table 4-4. Alkyne Series with Melting and Boiling Points.

Formula	Name	Melting Point in °C	Boiling Point in °C
$CH{\equiv}CH$	Ethyne (Acetylene)	−81 *	−84 †
$CH_3{-}C{\equiv}CH$	Propyne (Methylacetylene)	−102.7	−23.2
$C_2H_5{-}C{\equiv}CH$	1-Butyne	−125.7	+8.1
$C_3H_7{-}C{\equiv}CH$	1-Pentyne	−106	40.2
$C_4H_9{-}C{\equiv}CH$	1-Hexyne	−132	71.4
$C_5H_{11}{-}C{\equiv}CH$	1-Heptyne	−81	99.7
$C_6H_{13}{-}C{\equiv}CH$	1-Octyne	−79.4	126.2
$C_7H_{15}{-}C{\equiv}CH$	1-Nonyne	−58	151.0
$C_{10}H_{18}$	1-Decyne	−40	174.2
$C_{15}H_{28}$	1-Pentadecyne	+9	270.7
$C_{20}H_{38}$	1-Eicosyne	39	345

*Pressure 891 mm, rather than one standard atmosphere.

†Sublimation temperature, one atmosphere pressure.

consequently they are unsaturated hydrocarbons. Alkynes have the general formula C_nH_{2n-2}.

The alkanes, alkenes, and alkynes are very much alike. They are colorless compounds with low molecular weights. In fact, with only one or five carbon atoms of length, these molecules are gases. With five to ten carbon atoms they are liquids. Only when the carbon content is above ten, do these molecules become solids:

C_1 to C_5 = gases
C_5 to C_{10} = liquids
C_{10} and up = solids

Their melting point and boiling points vary according to: (1) number of carbon atoms, (2) number of bonds, either double or triple, and (3) number and kind of chain branches among the long molecules.

Several compounds that have the same molecular formula but different structured properties are called *isomers*. For example, here are two different compounds with the same molecular formula, C_4H_{10}:

```
   H  H  H  H               H  H  H
   |  |  |  |               |  |  |
H—C—C—C—C—H   and   H—C—C—C—H
   |  |  |  |               |  |  |
   H  H  H  H               H  |  H
                               |
                            H—C—H
                               |
                               H
```

Butane *Methyl Propane*

The hydrogen atoms may be replaced by atoms from the very active, nonmetallic chemical elements, fluorine, chlorine, bromine, astatine, and iodine.

Ethylene C_2H_4 may be formed into vinyl chloride with the replacement of one atom of hydrogen by one of chlorine.

```
H  H      H  H      H  Cl     F  F
|  |      |  |      |  |      |  |
C══C      C══C      C══C      C══C
|  |      |  |      |  |      |  |
H  H      H  Cl     H  Cl     F  F
```

Ethylene *Vinyl Chloride* *Vinylidene Chloride* *Tetrafluoro-ethylene*

As a result of the double bonding, long chains of molecules may link together. If many of these unsaturated hydrocarbons link together, a *polymer* is formed. This linking together is called *polymerization*, and the vinyl chloride polymer becomes polyvinyl chloride.

Cyclic Hydrocarbons

There is nothing to prevent the end of molecules from joining if the bonding site is acceptable. In fact, propane from the alkane series may join carbon atoms and produce a molecule called *cyclopropane*. The carbon atoms of the propane molecule have joined ends forming a circle.

In general, compounds containing rings of atoms are called *cyclic* or *aromatic* hydrocarbons. The

```
        H
        |
        C
      //  \\
   HC       C—CH3
   ||       |
   HC       CH
      \   //
        C
        |
        H
```

Toulene: an aromatic hydrocarbon.

most important "cycle" carbon ring is the molecule of *benzene*. Do not confuse benzene with the common solvent, benzine, which is a base for aspirin and insecticides. Benzene has the molecular formula C_6H_6 with alternating double and single bonds.

It would appear that the benzene ring should be highly unsaturated and easily undergo addi-

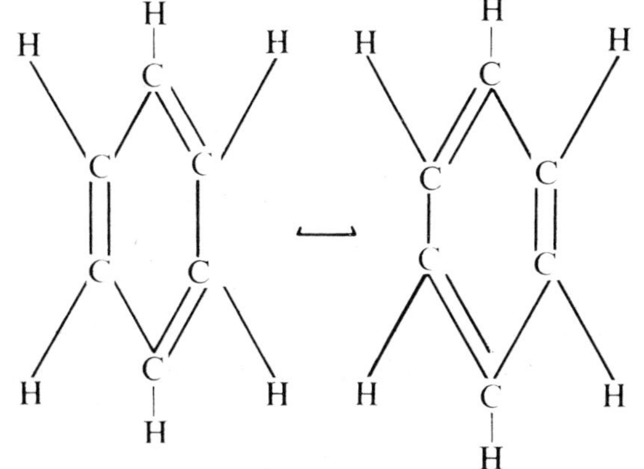

where the symbol ⟷ means oscillation between two structures.

tional reactions. This hexagonal ring has unusual bonds in that they are neither single nor double but a cross between single and double bonding. One authority indicates that the electrons are probably shared by all the carbon atoms in the ring, rather than a simple single or double bond. Another postulate suggests that the double bonds oscillate rapidly between the positions.

Because of the number of possible bonding sites and elements, there are a number of cyclic compounds possible. The compound, chlorobenzene, with the formula C_6H_5Cl, may be illustrated two ways. A simple hexagon figure is often used for simplicity.

Benzene

Chlorobenzene (there is no difference between symbols)

A dichlorobenzene atom with the formula $C_6H_4Cl_2$ may be shown several ways, depending on the position of the bonding site:

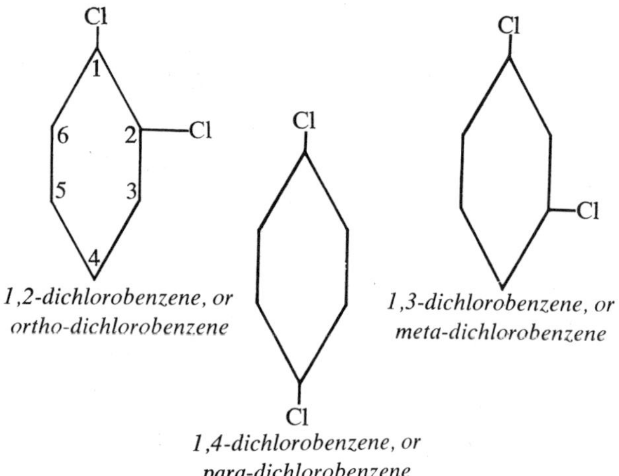

1,2-dichlorobenzene, or ortho-dichlorobenzene

1,4-dichlorobenzene, or para-dichlorobenzene

1,3-dichlorobenzene, or meta-dichlorobenzene

These ring structures are found in the compounds which help form such synthetic plastics as epoxy, polystyrene, polyurethane, and all of the polyesters.

Table 4-5 gives the classification of aromatic and aliphatic hydrocarbons for easy reference.

Table 4-5. Classification of Hydrocarbons.

Name	Description	Examples
Aromatic hydrocarbons	Have structures and properties related to the compound benzene	Benzene Naphthalene
Aliphatic hydrocarbons	Structures not related to benzene	
Saturated Alkanes (paraffins)	Have only single bonds A general name for saturated hydrocarbons	Ethane H—C—C—H
Unsaturated Alkenes (olefins)	Have multiple bonds Contain at least one —C=C—	Ethene (ethylene) H—C=C—H
	Contain at least one —C≡C—	
Alkynes (acetylenes)		Ethyne (acetylene) H—C≡C—H

Polymers

When a large number of molecules bond together in a regular pattern a *polymer* is formed. "Poly" means many and "mer" means unit; thus a polymer is composed of many units which are repeated in a pattern.

In the example below, many "mers" are repeated to form a long molecule of a single bonded polymer. These mers cannot stand alone because of the requirement of the carbon's four binding sites. This mer of the linear polymer is not a molecule but only a segment of the molecular chain.

Monomer

Mer

A *monomer* is the initial molecule in forming a polymer. It may occur alone because it is a molecule with a single mer. Most monomers are also liquids.

MONOFUNCTIONAL MONOMERS AND THEIR POLYMERS

MONOMER	POLYMER	PLASTIC PRODUCTS
$CH_2{=}CH_2$ Ethylene	$-CH_2-CH_2-CH_2-CH_2-CH_2-CH_2-CH_2-CH_2-\ldots$ Polyethylene	Toys, containers, film, etc.
$CH_2{=}CH(O-COCH_3)$ Vinyl acetate	$-CH_2-CH(O-COCH_3)-CH_2-CH(O-COCH_3)-CH_2-CH(O-COCH_3)-CH_2-CH(O-COCH_3)-\ldots$ Polyvinyl acetate	Film, packaging, etc.
$CH_2{=}CH(Cl)$ Vinyl chloride	$-CH_2-CH(Cl)-CH_2-CH(Cl)-CH_2-CH(Cl)-CH_2-CH(Cl)-\ldots$ Polyvinyl chloride	Pipe, ducts, film, etc.
$CH_2{=}CH(C_6H_5)$ Styrene (Vinyl benzene)	$-CH_2-CH(C_6H_5)-CH_2-CH(C_6H_5)-CH_2-CH(C_6H_5)-CH_2-CH(C_6H_5)-\ldots$ Polystyrene	Toys, containers, packages, etc.
$CH_2{=}C(Cl)_2$ Vinylidene chloride	$-CH_2-C(Cl)_2-CH_2-C(Cl)_2-CH_2-C(Cl)_2-CH_2-C(Cl)_2-\ldots$ Polyvinylidene chloride	Film, packaging, etc.
$CH_2{=}CH(COOH)$ Acrylic acid	$-CH_2-CH(COOH)-CH_2-CH(COOH)-CH_2-CH(COOH)-CH_2-CH(COOH)-\ldots$ Polyacrylic acid	Coating, castings, etc. Textiles
$CH_2{=}C(COOH)(CH_3)$ Methacrylic acid	$-CH_2-C(COOH)(CH_3)-CH_2-C(COOH)(CH_3)-CH_2-C(COOH)(CH_3)-CH_2-C(COOH)(CH_3)-\ldots$ Polymethacrylic acid	Glazing, coatings, etc. Textiles

The mechanism or chemical reaction which joins these monomers together is called *polymerization* and falls into two general categories: *addition* and *condensation*. By these two general types of polymerization and some modifications of each, small molecules are linked together in a variety of patterns forming many kinds of giant molecules.

Molecular Structure

When monomers are initially being linked together to form the basic resins which make up the polymer, there are three basic arrangements that the monomers take to form the molecular structure: (1) *Linear structures* are long chains of molecules without appendages, something like a rope (Fig. 4-2A). (2) *Crosslinked structures* refer to the chemical links between the molecular chains (Fig. 4-2B). For crosslinking to be very extensive, there must be a number of unsaturated carbon atoms. This crosslinking restricts movement between adjacent chains. This type of polymer is commonly referred to as a *thermosetting material*. Thermosetting and thermoplastic materials will be discussed later. (3) *Branched structures* are three-dimensional "branchings" of the linear chain (Fig. 4-2C). These side chains permit greater interlocking of the structure and a wide variety of *molecular weights*. Because the molecular structure may vary in shape and length, a molecular weight average is used when referring to polymers.

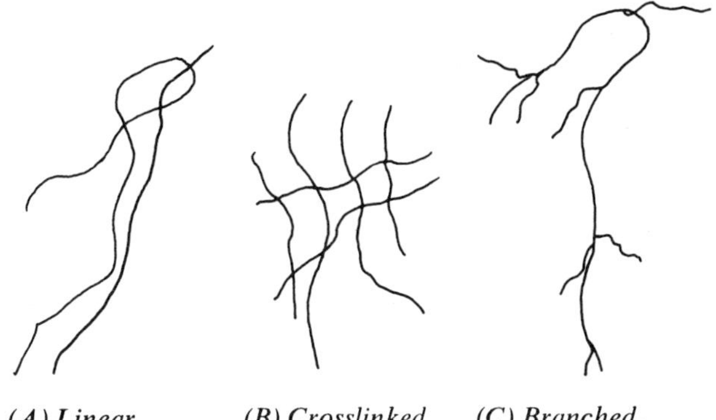

(A) Linear. *(B) Crosslinked.* *(C) Branched.*

Fig. 4-2. Molecular structure arrangements of monomers.

The molecular weight of most polymers falls in the 10,000 to 1,000,000 range. Molecular weights of polymers have the greatest effect on properties such as viscosity.

The weight of a molecule is determined by adding the atomic weights of the various atoms making up the molecule. The atomic weight of hydrogen is about 1 and the atomic weight of oxygen is about 16. Thus the molecular weight of one molecule of water, H_2O, is 1 plus 1 plus 16, or 18. Molecular weights will be discussed in greater detail later.

Only a few polymers occur in nature (all proteins are polymers). Consequently man must cause these monomers to join in various combinations and arrangements to meet his needs. A *catalyst* is used to bring about the union of the monomers from the complex plastics polymer. A catalyst is a substance which can change the speed of reaction between two or more materials, without being permanently changed or used up. Actually this changing or joining of the monomers will occur without a catalyst but these actions are hardly perceptible because they are so slow. A catalyst speeds up the reaction enormously.

Because the mechanisms of catalysts are complicated and somewhat speculative, two theories are commonly expressed. One theory speculates that a catalyst "grabs" molecules in such a way as to distort their structure so that the strained molecules will react when hit by other reactants. The second theory states that the catalyst actually forms an intermediate compound with one of the reactants which is thus vulnerable when they collide.

These catalysts may react with other chemical agents sometimes called *promoters.* Promoters are weak solutions or reactive agents which help in the polymerization process. Actually, any energy-causing agent could be classed as a catalyst in joining monomers. Heat energy, for example, is used to cause polymerization in both natural and synthetic polymers. Ionic or electrical energy is yet another method.

In *addition* polymerization, unsaturated hydrocarbons are caused to form large molecules. The mechanism of the reaction involves three basic phases: (1) *Initiation* is a phase where the monomers are caused to join by the aid of some high-energy source such as a catalyst or heat. A typical example is that of the gas, ethylene, C_2H_4. Since a double bond is present, two bonding sites (R) can be made available:

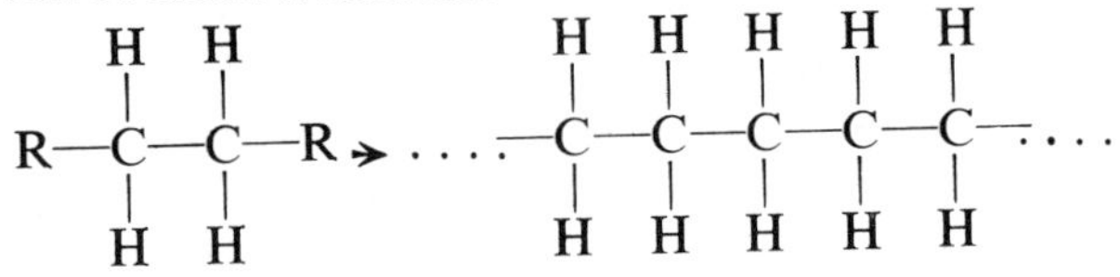

```
   H  H              H   H   H   H   H
   |  |              |   |   |   |   |
R—C—C—R → . . . —C—C—C—C—C— . . .
   |  |              |   |   |   |   |
   H  H              H   H   H   H   H
```

It must be remembered that energy must be supplied to break the double bonds. (2) The *propagation phase* is the phase in which the addition process continues and the final product is the polymer. In the example above, it would be polyethylene. (3) The *termination phase* is exactly what the name implies. The polymerization process is completed or caused to be stopped. Termination of growth can occur in several ways. In one method, the molecular weight depends on the catalyst reaction by causing many or few mers to form the molecular chain growth.

A special ionic catalyst may be used to bond with the free ends of the polymer chain and thus completely stabilize the growth. This process terminates the bonding of individual mers into a repeating molecular pattern.

Theoretically there is no limit to the length of the linear addition polymer; however, it seldom exceeds several thousand bonded atoms. If each chain could be so positioned that the available bonding site was available for the next monomer chain end, an endless molecular chain would form. Unfortunately, it is not possible for this reaction to occur because the monomers are constantly vibrating and moving about, becoming entangled like a plate of spaghetti. This long chain growth is often referred to as a *macromolecule.* "Macro" means long and large.

In commercial plastics production a polymer normally falls in the range of 75 to 750 mers per molecule. Polyvinyl chloride, for example, may contain 1000 carbon, 1500 hydrogen, and 500 chlorine atoms per 500 mers and have a molecular weight of more than 31,000. Because not all of the molecules in a polymer are identical in length, only average weights are usually expressed when figuring molecular weight. The molecular weight of most polymers falls in the 10,000 to 1,000,000 range.

To actually figure the average molecular weight of a polymer the *degree of polymerization* must be known. This degree of polymerization, or DP, is the number of the mers in a polymer molecule. In chemical laboratories this (DP) number is measured by several methods: melt index, viscosity, osmotic pressure, and light scattering.

An example of the molecular weight of C_2H_3Cl, polyvinyl chloride, is:

$C_2H_3Cl = 24 + 3 + 35 = 62$ grams per mer weight, more accurately

$$\frac{24.02 + 3.01 + 35.45}{6.02 \times 10^{23}}$$

Formula: molecular weight of polymer = DP × mer weight, or

$$DP = \frac{\text{Molecular Weight}}{\text{Mer Weight}}$$

Thus,

$$\text{Molecular Weight} = 500 \times 62 = 31{,}000$$

Remember, the normal degree of polymerization falls in the 75 to 750 mers per molecule range. The number 500 was measured by the polymer chemist by one of the methods mentioned above. The 500 number was figured as an average DP number for an "average" polyvinyl chloride polymer.

The important thing to remember about average molecular weights is that they have the greatest effect on properties such as *viscosity*.

Probably one of the most important aspects of polymer growth is the ability to modify the existing polymers and form new properties. When two or more different monomers are joined together in a regular sequence, it is called *copolymerization* or a *copolymer plastics*. ABS is a good example of copolymerization. A copolymer of styrene, acrylonitrile, and butadiene rubber forms this plastics, acrylonitrile-butadiene-styrene, or ABS.

Through combinations of this type, polymers can be custom made to fit specific requirements of our spiraling technology.

Block and Graft Polymers

Block and graft polymers are important considerations to copolymerization study. The term *block polymer* refers to the long sequences or blocks of repeating monomers. For example, when forming two monomers, with *A* and *B* representing the different monomers, a block pattern is formed: *AAAAAAAAAAAAA . . BBBBBBBBB . . AA AAAAAAA . . BBBBBBB . . .,* etc.

Graft polymers are related to the crosslinked and branch structured copolymers. In the formation of graft polymers a backbone of one monomer has branches of another:

AAAAAAAAAAAAAAAAAAAAAAAAAAAAAAA

```
  |                        |
  B                        B
  B                        B
  B                        B
  B                        B
```

The main difference between block and graft polymers is the point of attachment of the two monomers. With block polymers the attachment point is at the end of the *A* structure. In graft polymers it occurs periodicaly along *A's* length.

There are several other important types of copolymer systems. *Random* copolymers are situated along the chain: *ABBAABAAABABBB*. An *alternating* copolymer has a definite ordered alternation of the *A* and *B* monomer units: *ABABABABAB-ABAB*.

The second general process of polymerization is called *condensation* polymerization. The greatest bulk of the polymers are presently produced utilizing the addition method of production. However, a different polymer, with a variety of properties, may be produced with the condensation process. The condensation process usually involves a chemical reaction in which a small molecule, often water, is eliminated from the polymer. For this reason it is known as a "condensation" action.

Probably one of the best examples of condensation polymerization is Bakelite. When polymerization is in progress between formaldehyde (CH_2O) and phenol (C_6H_5OH) a bridge occurs between the benzene rings and a water molecule is left as condensation:

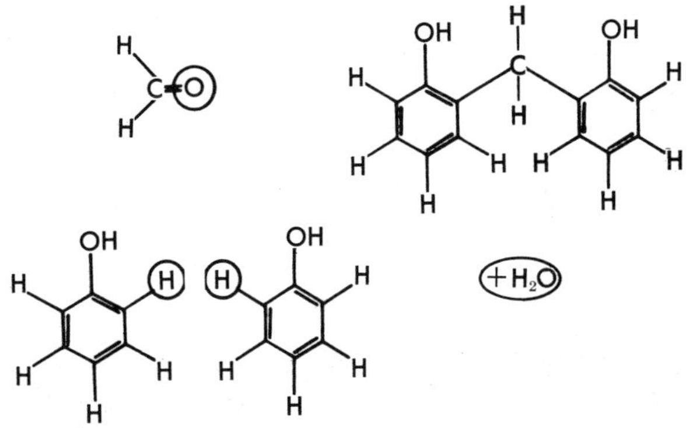

In addition polymers, unsaturated monomers are completely utilized in the final product. In condensation polymers, however, a by-product is always produced during polymerization. Dacron, Melmac, Mylar, and Nylon are only a few of the many tradenames of plastics which are polymerized by condensation. There are other modifications to the polymerization process. It is not difficult to see, however, that simply placing monomers closely together will not ensure automatic polymerization. The application of energy in the form of heat, light, pressure, or a chemical catalyst may be needed.

Polymer Structures

The properties of plastics are affected by the arrangement of the atoms and molecules of the material. The atomic arrangement of polymers can be classified as crystalline, amorphous, and molecular.

Crystallinity is the three-dimensional arrangement of atoms, ions, or molecules in a regular pattern. Diamonds and table salt are two common materials with a crystalline structure. Because of the great length of the polymer there is less crystallization in its structure.

It must be remembered that only weak van der Waals' forces are available to align the molecules and a great number of atoms must be maneuvered to produce any degree of crystallization in the polymer.

Because the molecular chains are partially ordered and others disordered, most of these plastics are seldom transparent in the solid state. Partially crystalline polymers, like the linear polyolefins, polyacetals, and polyamides, are usually translucent to opaque plastics.

The crystallization of polymers is usually characterized by the formation of large crystalline aggregates referred to as *spherulites*. These spherulites are crystalline structures in polymers which scatter light and give polyethylene and polypropylene its milky appearance (Fig. 4-3). The electron microscope, X-ray defraction, or the naked eye may be able to recognize these spherulite crystals.

Fig. 4-3. Fine fibrous links bridge the gap between radial arms of a single spherulite of polyethylene. *Magnification 10,000 diameters.* (Bell Telephone Laboratories)

While in the molten state, polymers are amorphous and do not possess any chain order. A rapid quenching may preserve most of this amorphous state. With normal processing, however, a crystalline polymer may begin to form many spherulite areas as it continues to cool. The actual processing may help to align or orient the molecular chains for greater crystallinity and tensile strength (see Fig. 4-4).

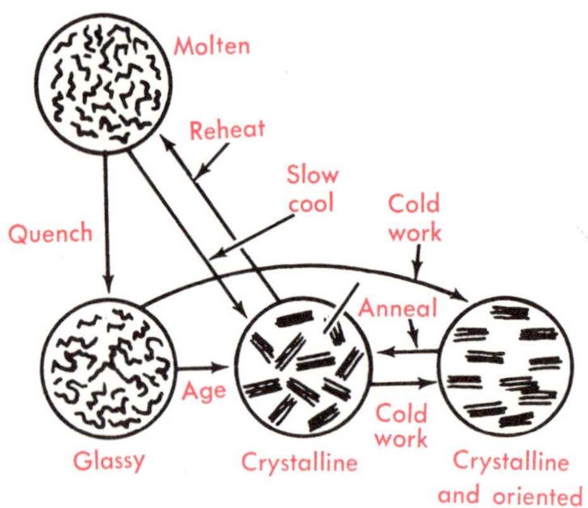

Fig. 4-4. Deformation of linear polymers.

There are few hard and fast rules when it comes to crystallinity in polymers. Yet it should be evident that crystallization and controlling the crystallization process is an important consideration in determining the physical properties of the polymer.

Materials which are amorphous, meaning "without form," have atoms or molecules which are noncrystalline and without long-range order (Fig. 4-5A). One analogy of the shape of an amorphous polymer indicates these linked molecules look like a mass of cooked spaghetti. Amorphous polymers are usually less rigid than crystalline ones but are transparent if most crystallinity can be prevented. Low-density polyethylene is amorphous and transparent if crystallinity is reduced, while high-density polyethylene is crystalline and has a higher melting point.

(A) Amorphous macromolecular structure.

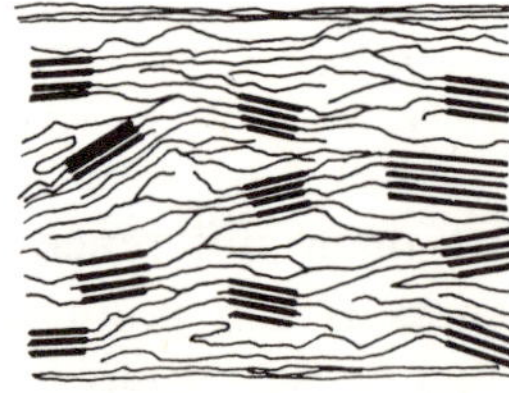

(B) Oriented crystalline macromolecular structure.

(C) Mixed amorphous crystalline macromolecular polymer structure.

Fig. 4-5. Ordered and disordered polymers.

Molecular Chains

The effect of the molecular chain structure may be exemplified by viewing the examples of polymer shapes in Fig. 4-6. The polyethylene chain shown in Fig. 4-6A depicts only the carbon atoms with simple uniform structure. The second polymer chain is polyvinyl chloride (Fig. 4-6B). Movement of the molecular chains will be more difficult because of the addition of the chlorine atoms. The third structure is of polystyrene (Fig. 4-6C) and has a benzene structure attached to the molecular chain. It should become evident that these last two polymers would not be as flexible as polyethylene because the van der Waals' forces are greater and the additional chlorine or benzene structure makes molecular movement in these polymers more difficult to take place.

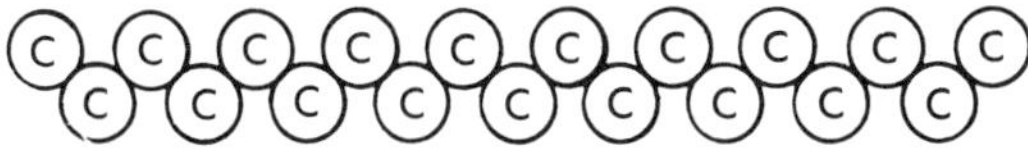

(A) Polyethylene.

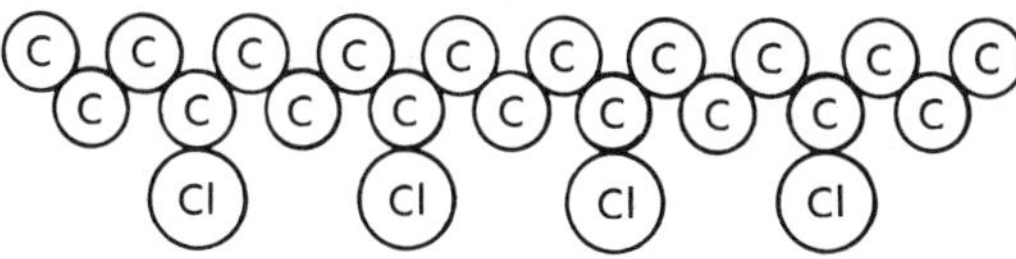

(B) Polyvinyl chloride.

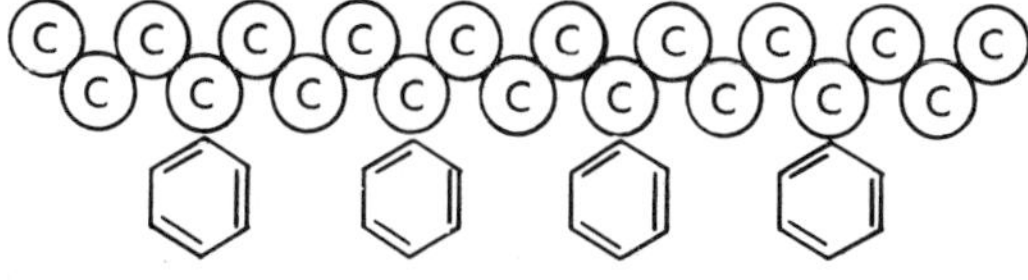

(C) Polystyrene.

Fig. 4-6. Molecular chains.

Stereoisomerism

There are several general molecular chain arrangements which permit close packing of the molecules and which may lead to high crystallinity.

An *isotactic* polypropylene chain is shown in Fig. 4-7 with an ordered placement of the CH_3 groups. This type arrangement may lead to close molecular chain packing and to crystallinity. *Syndiotactic* arrangement of the same CH_3 group will also lead toward a tougher, more closely packed crystalline structure. By proper polymerization control the polypropylene polymer formed has a random placement of the CH_3 groups called *atactic*. This random placement prevents chain packing, and a noncrystalline polymer is formed.

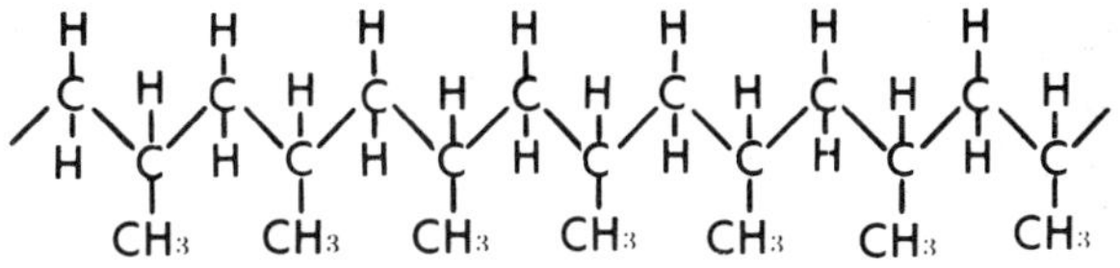

(A) Isotatic.

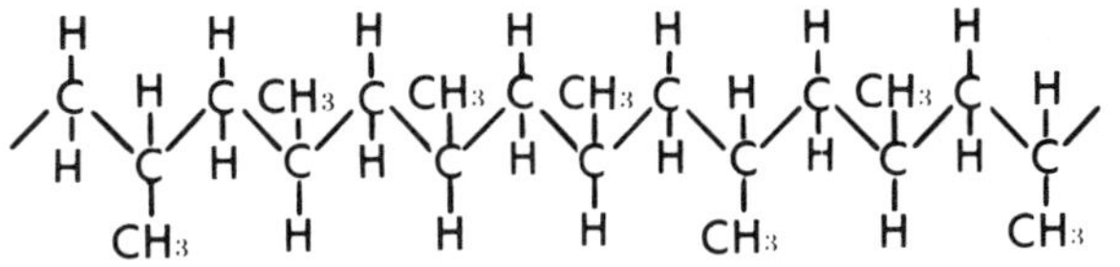

(B) Syndiotactic.

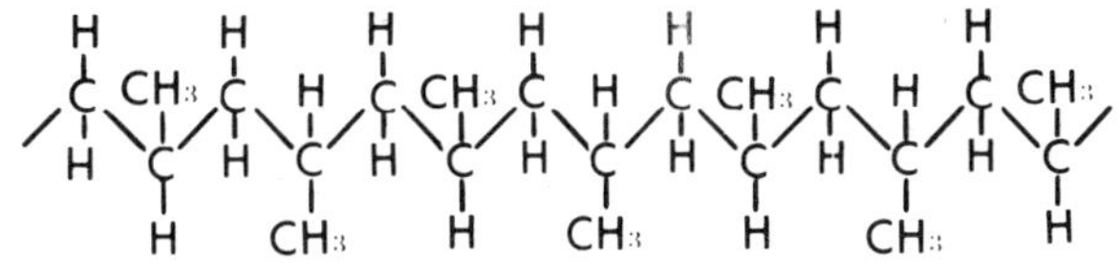

(C) Atactic.

Fig. 4-7. Stereotactic arrangements (polypropylene).

Resins

There are a number of ingredients of plastics. Recall that a plastics material is solid in the finished state. Resins are the basic organic materials from which plastics are formed. A resin is the polymeric material which helps impart many of the physical characteristics of the solid plastic. Fillers, solvents, plasticizers, stabilizers, and colorants also influence many of the principal characteristics of the plastics. It is the resin, however, that determines whether a plastics is *thermosetting* or *thermoplastic.*

Thermoplastic materials increase in plasticity with temperature. Thermoplastic plastics become soft when heated and become solid when cooled to room temperature. One analogy is a dish of butter: when heated, the butter becomes soft, but when cooled to room temperature it solidifies. These plastics are easily deformed into useful products because only weak van der Waals' forces are holding the molecules together; consequently, slip may occur between polymer molecules (Fig. 4-8). The molecules themselves are held by the stronger covalent bonds.

There is a practical limit to the number of times a thermoplastic material may be formed. Repeated heating and forming may cause some of the additives in the basic resin to be lost. Selected important members of the thermoplastic family are acrylics, cellulosics, polyamide, polystyrene, polyethylene, fluorocarbons, polyvinyls, polycarbonate, and polysulfone.

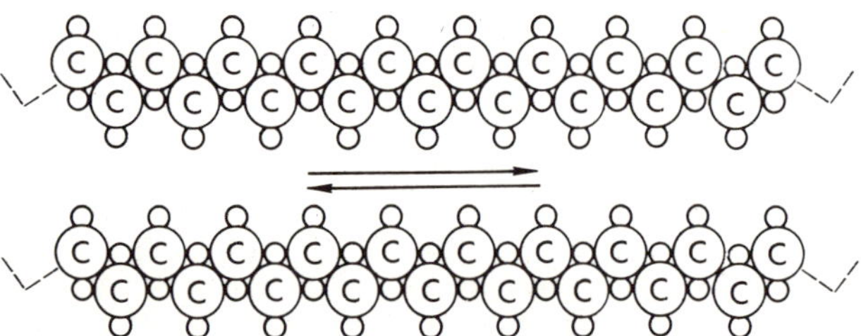

Fig. 4-8. Slip between polymer molecules (polyethylene).

Thermosetting plastics are polymeric materials with structural frameworks which do not allow deformation or slip to occur between molecules. They are composed of strong, primary covalent bonds and may be thought of as one large molecule (Fig. 4-9). Once plastics, these materials may not be reshaped or formed by applying heat. These plastics have a permanent "set" once they have been polymerized.

Selected important members of the thermosetting plastics are aminos, casein, epoxies, phenolics, polyesters, silicones, and polyurethanes.

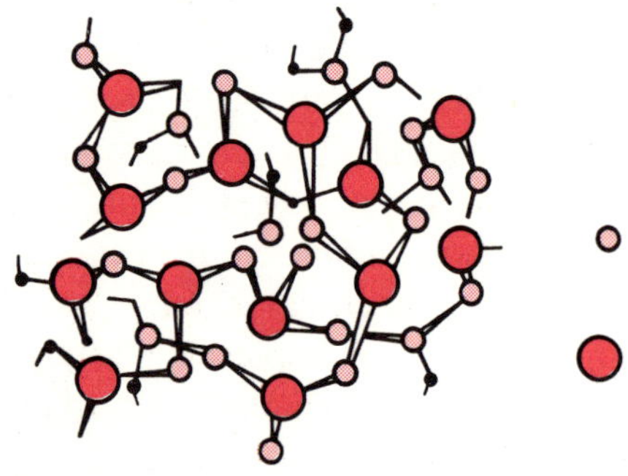

Fig. 4-9. Framework structure of thermosetting plastic, where slip does not occur as readily as in thermoplastics.

In general, thermosetting plastics are stronger than thermoplastic plastics and have a higher product service temperature. Thermoplastics may hold many process advantages over thermosets and may be reground and recycled into useful products.

Inorganic Polymers

Because of the growing research in the field of inorganic polymers, chemists have turned to many inorganic materials based on silicon, nitrogen, phosphorus, boron, and a number of other elements, including tin and germanium. One of the most common polymer materials with covalent bonding is the silicone plastics family. The bonding capacity of silicon is comparable to that of carbon. Silicon itself constitutes 28 percent of the earth's crust.

The silicone plastics may be based on chains, rings and branching networks of alternate silicon and oxygen atoms. Two common silicon polymers known as polydi*methyl*siloxane and polydi*phenyl*-siloxane are illustrated below:

$$\begin{array}{cccccccccccc} & CH_3 & & CH_3 & & CH_3 & & CH_3 & & CH_3 & & CH_3 \\ & | & & | & & | & & | & & | & & | \\ - & Si & -O- & Si & -O- & Si & -\text{and}- & Si & -O- & Si & -O- & Si-O- \\ & | & & | & & | & & | & & | & & | \\ & CH_3 & & CH_3 & & CH_3 & & C_6H_5 & & C_6H_5 & & C_6H_5 \end{array}$$

Polydimethylsiloxane and Polydiphenylsiloxane

Many authorities consider the inorganic polymers as a kind of organic hybrid; however, increasing importance and interest along with newer technologies have given rise to this important area of study.

Glossary of New Terms

Addition polymerization—Polymers formed by the combination of monomer molecules without the splitting-off of low molecular weight by-products such as water.

Aliphatic molecules—Organic compounds whose molecules do not have their carbon atoms arranged in a ring structure.

Alkanes—Important hydrocarbons with the general formula C_nH_{2n+2}.

Alkenes—Important hydrocarbons with the general formula C_nH_{2n} and possess double covalent bonds.

Alkynes—Important hydrocarbons with the general formula C_nH_{2n-2} with triple bonds between two carbon atoms.

Amorphous—Noncrystalline; without descriptive physical form or selective structure.

Aromatic hydrocarbons—Derived from or characterized by presence of unsaturated ring structures.

Benzene—A clear, inflammable liquid, C_6H_6. It is the most important aromatic chemical.

Block and graft polymers—Block polymer: A polymer whose molecule is made up of comparatively long sections that are of one chemical composition, those sections being separated from one another by segments of different chemical character.

Graft polymers: A chain of one type of polymer to which side chains of a different type are attached or grafted.

Branching—Refers to side chains attached to the main chain of the polymer. Side chains may be long or short.

Catalyst—A chemical substance added in minor quantity as compared to the amounts of primary reactants which markedly speeds up the cure (polymerization) of a compound.

Condensation polymerization—Polymerization by chemical reaction which also produces a by-product.

Copolymerization—Addition polymerization involving more than one type of mer.

Crosslinked—The tying together of adjacent polymer chains.

Crystallization—The process or state of molecular structure in some plastics which denotes uniformity and compactness of the molecular chains forming the polymer. Normally can be attributed to the formation of solid crystals having a definite geometric form.

Cyclic hydrocarbons—Referring to cyclic or ringed compounds with benzene (C_6H_6) being one of the most important aromatic hydrocarbons with this schematic shape.

Degree of polymerization (DP)—Number of structural units per average molecular weight. In most plastics, the DP must reach several thousand if worthwhile physical properties are to be had. The average number of monomer units per polymer molecule, a measure of molecular weight.

Hydrocarbon plastics—Plastics based on resins made by the polymerization of monomers composed of carbon and hydrogen only.

Hydrocarbons—Compounds composed of hydrogen and carbon.

Initiation phase—The first of three steps of addition polymerization. Referring to the reactive state of molecules, usually by some high-energy sources, such as peroxides, thermal energy, catalysts, or radiation.

Isomers—Molecules with the same composition but different structures.

Linear—Referring to a long chain molecule as contrasted to one having many side chains or branches.

Macromolecules—The large ("giant") molecules which make up the high polymers.

Mer—The smallest repetitive unit in a polymer.

Molecular weight—The sum of the atomic weights of all atoms in a molecule. In high polymers, the molecular weights of individual molecules vary widely so that they must be expressed as averages. Average molecular weights of polymers may be expressed as number-average molecular weight (Mn) or weight average molecular weight (Mw). Molecular weight measurement methods include: osmotic pressure, light scattering, solution viscosity, sedimentation equilibrium, etc.

Monomer—A molecule with a single mer.

Polymerization—Process of growing large molecules from small ones.

Promoters—A chemical, itself a feeble catalyst, that greatly increases the activity of a given catalyst.

Propagation phase—The second step in addition polymerization. Referring to the rapid growth or addition of monomer units to the molecular chain.

Spherulites—A rounded aggregate of radiating crystals with fibrous appearance. Spherulites are present in most crystalline plastics. Spherulites may range in diameter from a few tenths of a micron to several millimeters.

Stereoisomerism—The arrangement of molecular chains in a polymer. *Attactic* pertains to an arrangement which is more or less random. *Isotactic* pertains to a type of polymeric molecular structure containing a sequence of regularly spaced asymmetric atoms arranged in like configuration in a polymer chain. *Syndiotactic* pertains to a polymer molecule in which groups of atoms that are not part of the primary backbone structure alternate regularly on opposite sides of the chain.

Termination phase—The last of three steps in addition polymerization. Referring to the end of molecular growth in polymers by adding chemicals to stop growth.

Thermoplastic—(adj.) Capable of being repeatedly softened by heat and hardened by cooling. (*n.*) A material that will repeatedly soften when heated and harden when cooled.

Thermosetting—(adj.) Material that will undergo or has undergone a chemical reaction by the action of heat, catalysts, ultraviolet light, etc., leading to a relatively unmeltable state.

Review Questions

4-1. Why may a monomer or polymer occur or exist alone but a mer may not?

4-2. Name two general categories or chemical reactions into which polymerization may be divided.

4-3. Why is the molecular weight average used when referring to polymers?

4-4. Name three basic arrangements that monomers form in the molecular structure of plastics.

4-5. How is the weight of a molecule determined, and what property does it affect?

4-6. What function does the catalyst have on monomers?

4-7. Name several forms of energy used to cause polymerization.

4-8. What is the purpose of block and graft polymer systems?

4-9. Why are many plastics translucent to opaque?

4-10. What is the difference between addition and condensation polymerization?

4-11. What property do crystallization and spherulites have on plastics?

4-12. What is the difference between crystalline and amorphous plastics?

4-13. What is the difference between isotactic and atactic arrangements of molecular groups?

4-14. What properties may be affected by molecular chain arrangements?

4-15. What is the difference between resins and plastics?

4-16. What is the difference between thermoplastic and thermosetting plastics?

4-17. Use two sketches to compare molecular structures of thermosets and thermoplastics.

4-18. What is the backbone element used in the manufacture of the inorganic plastics silicone?

4-19. What is addition polymerization? Condensation polymerization?

4-20. What do the terms *atactic* and *isotactic* mean?

4-21. What do the terms *branching* and *graft* polymers mean?

4-22. What do the terms *homopolymer* and *copolymer* mean?

4-23. What does *crosslinking* mean?

4-24. What is the importance of the degree of polymerization?

4-25. Can you define *plastics?*

4-26. What is the difference between macromolecules and micromolecules?

4-27. What is the importance of molecular weight? How is it measured?

4-28. Use two sketches to compare the molecular structure of thermoplastics and thermoset plastics.

5

Fundamentals for Selecting Polymers

Polymers may be classified according to three broad categories: (1) relative to source, (2) relative to light penetration, and (3) relative to heat reaction.

Plastics Relative to Source

When classifying plastics relative to source, only three principal categories of origin are included: natural, modified natural, and synthetic.

The natural resins are derived from a number of natural sources including animal, vegetable, and mineral. Some of the more common examples are as follows: (1) *Rosin* is a by-product of the distillation of turpentine. This resinous material may be seen oozing out of pine tree stumps and lumber. At one time it was widely used in making linoleum and numerous electrical insulating compounds. (2) *Asphalt,* sometimes called "pitch," occurs naturally and was formed from the decaying remains of plants and animals. Today, most asphalt is a by-product of the petroleum industry. Asphalt was once used for molding battery cases and electrical insulators. (3) *Tar* is obtained by the distillation of organic materials such as wood, waste fats, petroleum, coal, and peat. Tar is still used on road surfaces, for sealing roofs, and in making color dyes. (4) *Amber* is a fossilized resin produced from the ancient oily sap of cone-bearing trees. It was once used for molding knife handles and other ornamental objects. (5) *Gelatin* is a protein mixture made from bones, hoofs, and skins of animals. It is used as an odorless, tasteless, and transparent filler in candies, meats, ice cream, and pharmaceutical products. (6) A very important natural polymer is *rubber.* It is made from a milky juice called *latex* (from the Latin word for milk, which it resembles).

The term "natural rubber" usually refers to elastomers produced from the latex of the hevea tree. A polymer related chemically to rubber is gutta percha. The gutta percha tree of Malaya, Borneo, and Sumatra yields the milky latex for this polymer.

A mature rubber tree will yield above five grams of latex per day or about four pounds of rubber per year. Though there is still a large market for natural rubber products, synthetic rubbers have largely replaced natural resins.

The study of synthetic rubber or elastomers is often considered a separate branch of study not included with the plastics. (7) Probably the oldest of the natural resins used by man are the *copal* resins. Copal resins were once used in the production of paints, linoleum, and varnishes. This resin is obtained from several areas of the world and varies from white to brown in color. (8) *Lignin* is the resinous binder which surrounds each wood cell. The lignin content in various woods varies from 35 to 90 percent. There are only two general uses of lignin today; most of it is used as filler for other plastics and as an adhesive in binding wood chips together under pressure. (9) Probably one of the first natural resins to be molded was *shellac*. (See Figs. 5-1, 5-2, and 5-3.) At one time large quantities of shellac were molded into phonograph records. Today shellac is used as a filler, finish, or electrical insulation.

Fig. 5-1. These natives are scraping shellac from branches and twigs. (Wm. Zinser & Co., Inc.)

Fig. 5-2. In these jars a stomper treads and stirs the sifted lac to wash out the dye. As more water is then poured into the jars, some impurities float to the top and are skimmed off. This is done several times before the lac is pronounced "clean." (Wm. Zinsser & Co., Inc.)

The modified natural resins include the cellulose and protein resins. Cellulose is a major constituent of all plants. The supply of cellulose is virtually inexhaustible as a source for making cellulose derived plastics.

One of the purest forms of cellulose is obtained from cotton linters. Linters are the short fibers which remain stuck to the cotton seeds after "ginning." Polymers made from this material have freedom from yellowness and possess maximum transparency. The second largest source of cellulose and one which possesses great possibilities for plastics production is wood pulp.

Fig. 5-3. Note the stretching quality of this natural plastics, shellac.

There are numerous other cellulose sources. A large potential source may be wastes and residues from such agriculture products as straw, cornstalks, corncobs, grass, weeds, and many others. Although all of these materials have proved useful, the economical collection and processing are the prime inhibiting factors.

Although cellulose is an extremely complex material, numerous types of plastics are produced from it. There are more than ten basic resins in the cellulose family. Cellulose acetate, cellulose nitrate, cellulose acetate butyrate, and cellulose propionate are the most familiar. If you will recall, it was from cellulose nitrate that John Wesley Hyatt began producing many of the first commercial plastics products in the United States.

The second modified natural resin source has not been commercially exploited to a large degree of success. This source is from protein and may be obtained from milk, soy beans, peanuts, coffee beans and corn. Only one protein derived plastics has had limited commercial success—casein.

Casein is obtained from skim milk. Rennet, an enzyme, is added, causing the protein in the milk to coagulate. This coagulated "curd" is the material from which casein plastics are produced. There are numerous other protein sources, including human and animal hair, feathers, bones, and other industrial wastes; however, there has been little interest in these sources because of collection and limited use of protein derived plastics.

The third and most important source of material for plastics production comes from synthetic sources. There are three main chemical sources from which synthetic plastics materials may be made: agriculture, petroleum, and coal.

By far the most important of these sources is petroleum. As this "oil" is extracted from the earth, it is mainly a mixture of solid, liquid, and gaseous hydrocarbons. Natural gas is composed of propane, butane, and other hydrocarbons.

Distillation and refining of petroleum yields various components including the heavy based asphalt, tar, and oil. The lighter fuels, kerosene, oil, diesel fuel, gasoline, and benzene are extracted near the top of a fractionating tower. The lightweight gases methane, ethylene, propane, and propylene are by-products of further refining of the lightweight "oils" and solvents. It is from these basic materials that chemists produce the man-made or synthetic plastics. Combining in varying proportions of hydrocarbons, salt, chlorine, formaldehyde, nitrogen air, and numerous other chemicals may create a great variety of polymers.

Other sources of hydrocarbons, the building blocks of polymers, may be obtained from agriculture products such as cottonseed, linseed, soybean, lard, and safflower. Methane and other gases are obtained from sewage sludge and pulverized coal.

As man depletes his supply of oil and coal, he will have to turn to other sources of hydrocarbon for the production of polymers. Agriculture and human waste products will probably be utilized to greater extent than they are today. As man unlocks the secrets of photosynthesis, the process by which plants convert sunshine and carbon dioxide into sugar, he may find new ways to produce simple organic chemicals. The fermentation of sugars by bacteria or small plants may produce alcohol, a raw material for many plastics.

Plastics Relative to Light Penetration

Because many plastics possess unique optical properties in addition to ease of fabrication, they are often classified relative to light penetration. The following optical properties will help to summarize this classification.

1. *Opaque*—Light will not pass through. You cannot see through.
2. *Transparent*—Light will pass through. You can see through.
3. *Translucent*—Light will pass through. You cannot see through.
4. *Luminescence*—(a) *fluorescence*: Emits light only when electrons are being excited. Usually transparent.

   (b) *phosphorescent*—Gives off light energy slower than it takes on light. Translucent.

Plastics Relative to Heat Reaction

There are only two broad categories into which all plastics fall relative to heat reaction. As previously stated, thermoplastic materials become soft when heated and become solid when cooled to room temperature. Only weak van der Waals' forces hold the molecules together. This softening and setting may be repeated many times, something like melting, and then cooling and hardening wax. When cooled the plastics becomes firm or hard.

Thermosetting materials may not be reheated and resoftened. Once the structural framework of bonds is completed in the polymerization process, these plastics cannot be reformed. Baking a cake or boiling an egg may be a simplified analogy of polymerization in thermosetting plastics.

A selected list of plastics, both thermosetting and thermoplastic, is discussed in greater detail in Chaps. 13 and 14.

Glossary of New Terms

Elastomer—A rubberlike substance that can be stretched to several times its length and that, on release of the stress, returns rapidly to almost its original length.

Lac—A dark-red resinous substance deposited by scale insects on the twigs of trees; used in making shellac.

Latex—An emulsion of natural or synthetic resin particles dispersed in a watery medium.

Shellac—A natural polymer; refined lac, a resin usually produced in thin, flaky layers or shells and used in varnish and insulating materials.

Synthetic—Materials produced by chemical synthesis, rather than natural origin.

Review Questions

5-1. What are the three broad categories in which polymers may be classified?

5-2. Name the most important natural polymer used today.

5-3. What is the most important source for the production of synthetic plastics? What are some other sources of hydrocarbons for the production of plastics?

5-4 Give the four classifications of plastics relative to light penetration.

5-5. What are the two classifications of plastics relative to heat reaction?

5-6. What are the two largest sources of cellulose for production of plastics?

5-7. What is the only protein derived plastics material used commercially?

5-8. How may the study of photosynthesis be of possible benefit to the plastics industry?

6

Ingredients of Plastics

Various metals are often added to iron ore to produce special alloys of steel. Similarly, there are numerous ingredients which are added to plastics in order to provide each plastics with special or outstanding physical characteristics.

Resins, Fillers, and Reinforcements

(1) The *resin* itself determines whether a plastics is thermosetting or thermoplastic. The nouns "plastics" and "resin" are often used interchangeably; however, resins do not become plastics until polymerization has occurred and the product is in the "finished state." Plastics do not include the wide range of elastomers, such as synthetic rubbers. Polymers, on the other hand, include plastics and elastomers.

(2) *Fillers* may be either organic or inorganic additives in resins. Fillers are added to resins to increase bulk, replace more expensive ingredients, and improve the physical properties of the molded article.

The principal types of fillers and their functions are shown in Table 6-1.

Although many fillers are added to the resin to make it less costly, they may improve processability, product appearance, and other properties. One of the most widely used fillers is wood flour. Wood flour is obtained by grinding waste wood stock into a fine granular state. This powdered filler is commonly added to phenolic resins in order to reduce brittleness, reduce resin cost, and improve product finish.

Table 6-1. Principal Types of Fillers.

FILLER ↓ / FUNCTION →	Bulk	Processibility	Thermal Resistance	Electrical Resistance	Stiffness	Chemical Resistance	Hardness	Reinforcement	Electrical Conductivity	Thermal Conductivity	Lubricity	Moisture Resistance	Impact Strength	Tensile Strength	Dimensional Stability
(Organic)															
Wood Flour	X	X												X	X
Shell Flour	X	X										X		X	X
Alpha Cellulose (wood pulp)	X			X	X									X	
Sisal Fibers	X			X	X	X	X	X				X	X	X	X
Macerated Paper	X			X									X		
Macerated Fabric	X				X								X		
Lignin	X	X													
Keratin (feathers, hair)	X				X								X		
Chopped Rayon		X	X	X		X	X	X				X	X	X	X
Chopped Nylon		X	X	X	X	X	X	X			X		X	X	X
Chopped Orlon		X	X	X	X	X	X	X				X	X	X	X
Powdered Coal	X		X		X	X						X			
(Inorganic)															
Asbestos	X		X	X	X	X	X	X					X		X
Mica	X		X	X	X	X	X				X	X			X
Quartz			X	X	X		X					X	X		
Glass Flakes		X	X	X	X	X	X	X			X	X	X		
Chopped Glass Fibers			X	X	X	X	X	X			X	X	X	X	
Milled Glass Fibers	X	X	X	X	X	X	X	X			X	X	X	X	X
Diatomaceous Earth	X	X	X	X	X		X						X		X
Clay	X	X	X	X	X		X						X		X
Calcium Silicate		X	X		X		X					X	X		X
Calcium Carbonate		X	X		X		X								
Alumia Trihydrate		X		X	X		X					X			
Aluminum Powder					X		X	X	X	X			X		
Bronze Powder					X		X	X	X	X			X		
Talc	X	X	X	X	X	X	X			X		X			X

This Table does not indicate the degree of improvement of function. The prime function will also vary between thermosetting and thermoplastic resins. This Table is to be used only as a guide to the selection of fillers.

In most molding operations, the volume of fillers does not exceed 40 percent; but as little as ten percent resin is commonly used for molding large desk tops, trays, and particle boards.

In the foundry industry, as little as three percent resin is used to adhere the sand together in the shell molding process.

One product, called "cultured marble," utilizes inorganic marble dust and polyester resin to produce articles which have the appearance of genuine marble. This product has the advantage of being stain resistant, easily produced in a variety of colors, and can be molded into a variety of shapes and sizes.

Many of the organic fillers cannot withstand high processing temperatures, and have low heat resistance. For added heat resistance a silica filler is used. Sand, quartz, tripoli, diatomaceous earth, and asbestos are common examples.

Diatomaceous earth is obtained from the fossilized remains of microscopic organisms called "diatoms." The addition of this type filler provides improved compressive strength in rigid polyurethane foam.

Asbestos products are added to resins to improve heat resistance and reduce burning rate in flammable plastics. Asbestos is commonly used as a filler in plastics floor tile.

A filler as fine as cigarette smoke, or 0.007 to 0.050 micron in diameter, is fumed silica (Fig. 6-1). This submicroscopic silica is added to resins to provide *thixotropy*. Thixotropy is a state of a material that is gel-like at rest but fluid when agitated. Thixotropic fillers increase viscosity (internal resistance to flow) and are desirable in paints, adhesives, and in applying compounds to vertical surfaces. These fillers may be added to polyester resin in fabrication of inclined or vertical surfaces. Thixotropic fillers may be added to either thermosetting or thermoplastic resins to thicken the resin, provide additional strength, suspend other additives, improve the flow properties of powders, and lower costs.

Fig. 6-1. Fume silica is added to vinyl compound as it comes from the blender, greatly increasing the body of the mix and giving it a drier, more solid consistency. (Cabot Corporation)

Thixotropic fillers may be made from a number of materials including: extremely fine powders of polyvinyl chloride, china clay, silicon dioxide, alumina, calcium carbonate and other silicates. Cab-O-Sil and Sylodex are tradenames for two commercial thixotropic fillers.

Thixotropic fillers sometimes act as emulsifiers by preventing separation of two or more liquids. Oil and water based additives may thus be added and held in an emulsified state.

Metal fillers such as steel, brass, and aluminum are added to resins to produce conductive moldings which may be electroplated, or to give added strength to the plastics. Powdered lead is used as a neutron and gamma-ray shield when added to resins.

Wax, graphite, brass, and glass are sometimes added to provide self-lubricating qualities to such plastics products as gears, bearings, and slides.

Glass, as a filler, is finding a growing market in plastics because it is easily added, relatively inexpensive, improves physical properties and may be colored. Colored glass has optical advantages over other chemical colorants and has shown exceptional color stability in our environment.

(3) *Reinforcements* are sometimes confused with fillers. Fillers are small particles which contribute only slightly to strength. Reinforcements are long fibrous ingredients which increase strength, impact resistance, and stiffness. One reason for confusion is that fillers like asbestos and glass may act as either a filler or a reinforcement or both.

One of the most important reinforcing materials is glass fibers. These fibers are composed of lime-aluminum-borosilicate glass which has been produced by several different methods. One method of fiber glass production is to force molten glass through small openings or orifices in a die and mechanically pulling the hot fibers until they are drawn to the desired diameter. These filament fibers may be as small as 0.00075 of an inch in diameter; they may possess tensile strengths of about 500,000 pounds per square inch (psi), however. Handling and fabrication of the finished product normally reduces these values to approximately 250,000 psi. (See Tables 6-2 and 6-3.)

Table 6-2. Thermoplastics Property Comparisons: Unreinforced and Reinforced Materials.

(Columns marked "U" = unreinforced, "R" = reinforced)

Property	Polyamide		Polystyrene*		Polycarbonate		Styrene Acrylonitrile†		Polypropylene		Acetal		Linear Polyethylene	
	U	R	U	R	U	R	U	R	U	R	U	R	U	R
Tensile Strength (1000 psi)	11.8	30.0	8.5	14.0	9.0	20.0	11.0	18.0	5.0	6.6	10.0	12.5	3.3	11.0
Impact Strength, Notched (ft-lb/in) At 73°F	0.9	3.8	0.3	2.5	2.0§	4.0§	0.45	3.0	1.3-2.1	2.4	60.0	3.0	—	4.5
At −40°F	0.6	4.2	0.2	3.2	1.5§	4.0§	—	4.0	—	2.2	—	3.0	—	5.0
Tensile Modulus (10^5 psi)	4.0	—	4.0	12.1	3.2	17.0	5.2	15.0	2.0	4.5	4.0	8.1	1.2	9.0
Shear Strength (1000 psi)	9.6	14.0	—	9.0	9.2	12.0	—	12.5	4.6	4.7	9.5	9.1	—	5.5
Flexural Strength (1000 psi)	11.5	37.0	11.0	20.0	12.0	26.0	17.0	26.0	6 to 8	7.0	14.0	16.0	—	12.0
Compressive Strength (1000 psi)	4.9††	24.0	14.0	17.0	11.0	19	17.0	22.0	8.5	6.0	5.2	13.0	2.7 to 3.6	6.0
Deformation, 4000 psi load (%)	2.5	0.4	1.6	0.6	0.3	0.1	—	0.3	—	6.0	—	1.0	—	0.4‡
Elongation (%)	60.0	2.2	2.0	1.1	60-100	1.7	3.2	1.4	>200	3.6	9-15	1.5	60.0	3.5
Water Absorption in 24 hr (%)	1.5	0.6	0.03	0.07	0.3	0.09	0.2	0.15	0.01	0.05	0.20	1.1	0.01	0.04
Hardness, Rockwell	M79	E75 to 80	M70	E53	M70	E57	M83	E65	R101	M50	M94	M90	R64	R60
Specific Gravity	1.14	1.52	1.05	1.28	1.2	1.52	1.07	1.36	0.90	1.05	1.43	1.7	0.96	1.30
Heat Distortion Temp at 264 psi (°F)	150	502	190	220	280	300	200	225	155	280	212	335	126	260
Coef. of Thermal Expansion (per °F × 10^{-5})	5.5	0.9	4.0	2.2	3.9	0.9	4.0	1.9	4.7	2.7	4.5	1.9	9.0	1.7
Dielectric Strength, short time (v/mil)	385	480	500	396	400	482	450	515	750	—	500	—	—	600
Volume Resistivity (ohm-cm) × 10^{15}	450	2.6	10.0	36.0	20.0	1.4	10^{16}	43.5	17.0	15.0	0.6	38.0	10^{15}	29.0
Dielectric Constant at 60 Hz	4.1	4.5	2.6	3.1	3.1	3.8	3.0	3.6	2.3	—	—	—	2.3	2.9
Power Factor at 60 Hz	0.0140	0.009	0.0030	0.0048	0.0009	0.0030	0.0085	0.005	—	—	—	—	—	0.001
Approximate Cost (¢/in³)	3.0	8.0	0.5	2.5	3.6	6.5	0.9	3.5	0.6	2.1	3.3	7.8	0.7	3.1

*Medium-flow, general-purpose grade. †Heat-resistant grade. §Impact values for polycarbonates are a function of thickness. ‡1000-psi load. ††At 1% deformation.

Machine Design: Plastics Reference Issue

Table 6-3. Thermosetting Plastics Properties (Glass Fiber Reinforced Resins).

Property	Base Resin				
	Polyester	Phenolic	Epoxy	Melamine	Polyurethane
Molding Quality	Excellent	Good	Excellent	Good	Good
Compression Molding Temperature (°F)	170 to 320	280 to 350	300 to 330	280 to 340	300 to 400
Pressure (psi)	250 to 2000	2000 to 4000	300 to 5000	2000 to 8000	100 to 5000
Mold Shrinkage (in./in.)	0.0 to 0.002	0.0001 to 0.001	0.001 to 0.002	0.001 to 0.004	0.009 to 0.03
Specific Gravity	1.35 to 2.3	1.75 to 1.95	1.8 to 2.0	1.8 to 2.0	1.11 to 1.25
Tensile Strength (1000 psi)	25 to 30	5 to 10	14 to 30	5 to 10	4.5 to 8
Elongation (%)	0.5 to 5.0	0.02	4	....	10 to 650
Modulus of Elasticity (10^{-5} psi)	8 to 20	33	30.4	24	....
Compression Strength (1000 psi)	15 to 30	17 to 26	30 to 38	20 to 35	20
Flexural Strength (1000 psi)	10 to 40	10 to 60	20 to 26	15 to 23	7 to 9
Impact, Izod (ft-lb/in of notch)	2 to 10	10 to 50	8 to 15	4 to 6	No break
Hardness, Rockwell	M70 to M120	M95 to M100	M100 to M108	....	M28 to R60
Thermal Expansion (per °C)	2 to 5×10^{-5}	1.6×10^{-5}	1.1 to 3.0×10^{-5}	1.5×10^{-5}	10 to 20×10^{-5}
Volume Resistivity at 50% RH, 23 C (ohm-cm)	1×10^{14}	7×10^{12}	3.8×10^{15}	2×10^{11}	2×10^{11} to 10^{14}
Dielectric Strength, ⅛ in thickness (v/mil)	350 to 500	140 to 370	360	170 to 300	330 to 900
Dielectric Constant at 60 Hz	3.8 to 6.0	7.1	5.5	9.7 to 11.1	5.4 to 7.6
at 1 kHz	4.0 to 6.0	6.9	....	....	5.6 to 7.6
Dissipation Factor at 60 Hz	0.01 to 0.04	0.05	0.087	0.14 to 0.23	0.015 to 0.048
at 1 kHz	0.01 to 0.05	0.02	....	....	0.043 to 0.060
Water Absorption (%)	0.01 to 1.0	0.1 to 1.2	0.05 to 0.095	0.09 to 21	0.7 to 0.9
Sunlight (change)	Slight	Darkens	Slight	Slight	None to slight
Chemical Resistance	Fair*	Fair*	Excellent	Very good†	Fair
Machining Qualities	Good	....	Good	Good	Good

*Attacked by strong acids or alkalies. †Attacked by strong acids. §Attacked by strong alkalies. (Machine Design: Plastics Reference Issue)

Fig. 6-2B shows a reinforced plastics gear housing which is lighter and stronger than the metal housing it has replaced (Fig. 6-2A). Fig. 6-3 shows the glass-filled diallyl phthalate drum pod and cap of a coffee vending machine. Brewing can be performed in a matter of seconds due to the extreme heat the resin-molded drum can withstand. The drum also resists staining than can affect the coffee flavor.

Fibrous glass comes in a variety of forms. *Rovings* are long strands of fibrous glass which may be easily cut and applied to resins. Rovings (Fig. 6-4) and *chopped strands* (Fig. 6-5) are among the least expensive to use as glass reinforcements. Chopped strands range in length from ⅛ to 2 inches. Fig. 6-6 shows the production of chopped strands from rovings.

Most milled fibers are less than 1/16 of an inch long and are produced by hammer milling the glass strands (Fig. 6-7). Milled fibers are added to the resin as a premix to increase viscosity and increase product strength (Fig. 6-8).

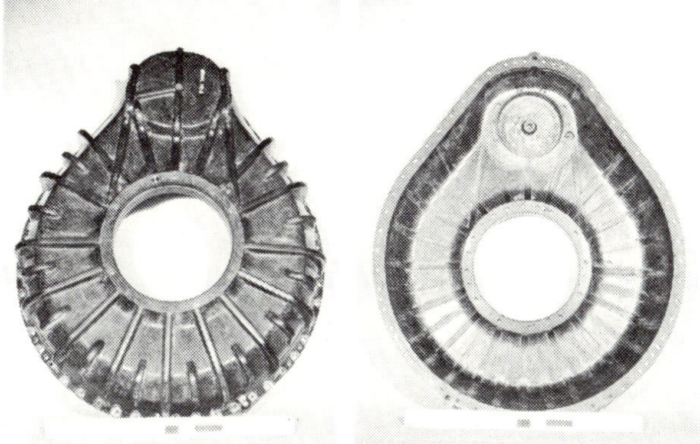

(A) Original metal gear housing.

(B) Boron-epoxy replacement gear housing.

Fig. 6-2. Use of reinforced plastics (Detroit Diesel Allison Division)

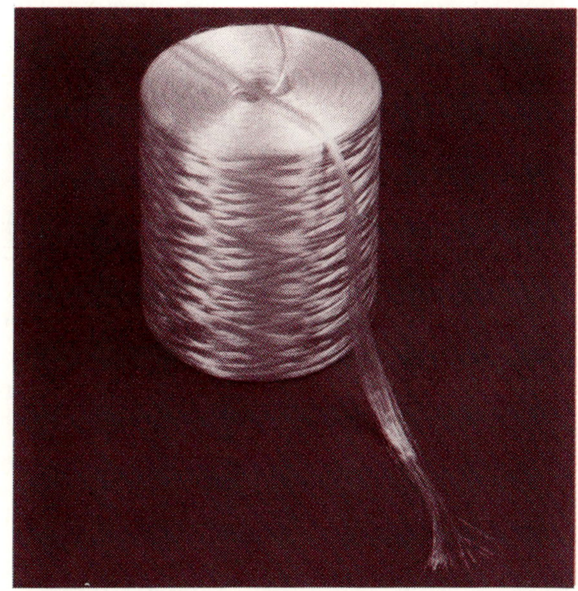

Fig. 6-3. Drum pod and cap of coffee vending machine. (FMC Corporation)

Fig. 6-4. Fibrous glass roving. (FERRO)

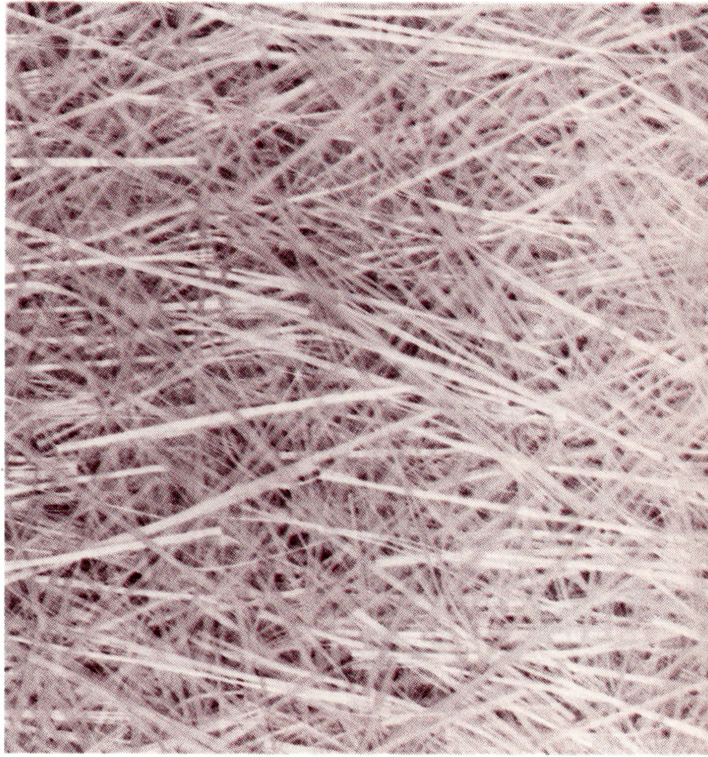

Fig. 6-5. Fibrous glass chopped strands. (FERRO)

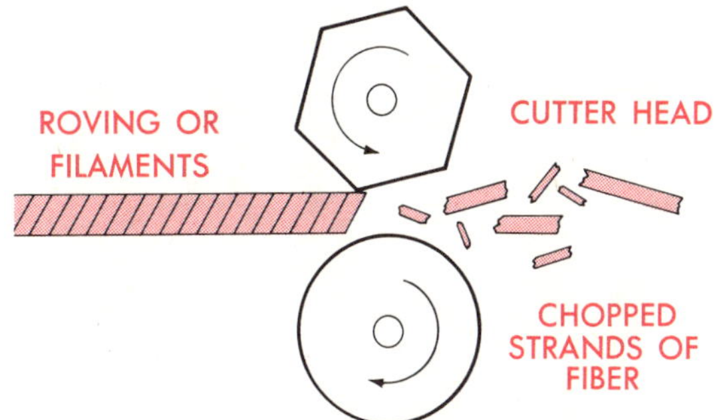

Fig. 6-6. Production of chopped glass strands.

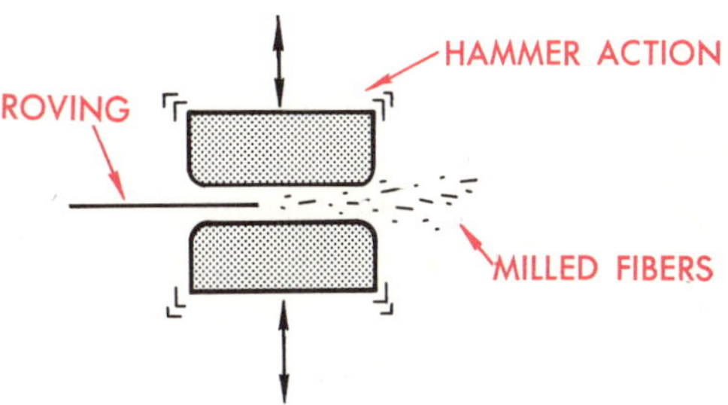

Fig. 6-7. Production of milled glass fibers.

Fig. 6-8. Milled glass fibers may be used as reinforcements or fillers. (FERRO)

Yarns are something like rovings except they are twisted like a rope (Fig. 6-9). Yarns have found use in special applications of reinforcement such as fabrication and large liquid tank containers.

(A) Single monofilament.

(B) Multifilament yarn.

(C) Woven yarn fabric.

Fig. 6-9. Yarns.

Reinforcing mats are nondirectional pieces of chopped strands held together by a resinous binder or by mechanical stitching called *needling* (see Fig. 6-10).

Woven cloth can provide the highest all-round physical strength of all the fibrous forms. This woven form is also about 50 percent more costly. Standard rovings may be woven into a fabric form normally used for thick reinforcements.

There are several types of woven glass fabrics. Fiber glass yarns are used and woven into several basic patterns including the plain or square weave. (See Fig. 6-11.)

Fig. 6-12 shows five different forms of fibrous glass reinforcements.

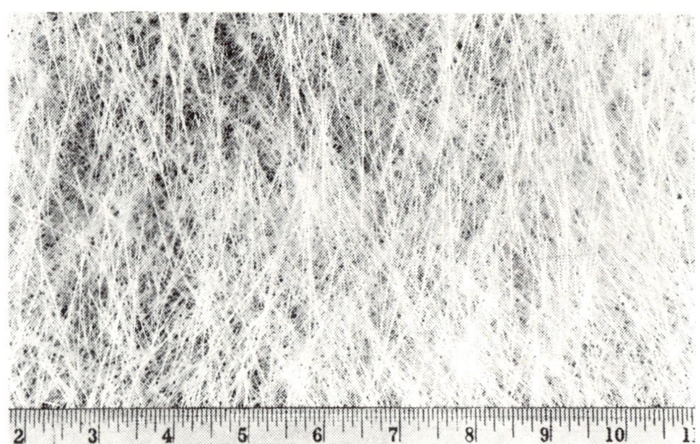

(A) Resin bonded.

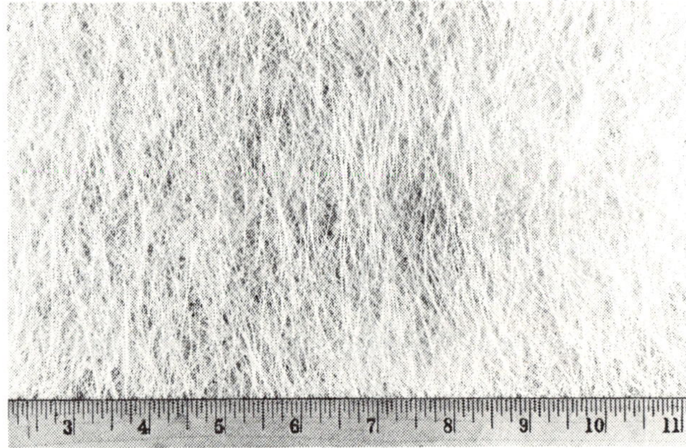

(B) Stitched (needled).

Fig. 6-10. Fibrous glass mats. (Owens/Corning)

Fig. 6-13 illustrates the relationship between the amount of glass used and the arrangement of the reinforcement.

For maximum strength in one direction, fibers are placed in a unidirectional or single direction. A random arrangement of the fibers will develop strength in all directions. This random configuration has less strength than woven cloth.

A hollow fiber nearly 40 percent lighter is also produced for special applications. Thermosetting and thermoplastic resins utilize fibrous glass reinforcements.

There is an increasing interest in and use of other reinforcement fibers including exotic fibers like boron, polyvinyl alcohol, crystal whiskers, graphite, ceramic, and metal fibers.

Long fibers of sisal, cotton, nylon, asbestos, and rayon are also effective reinforcing materials.

Boron filaments may be used to supplement existing structures, including aircraft floor beams, flooring, and struts. Polyvinyl alcohol (PVA) fibers

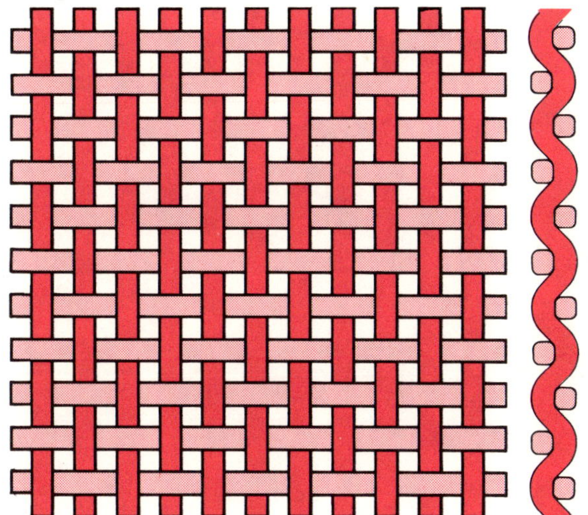

(A) Plain (or square) weave.

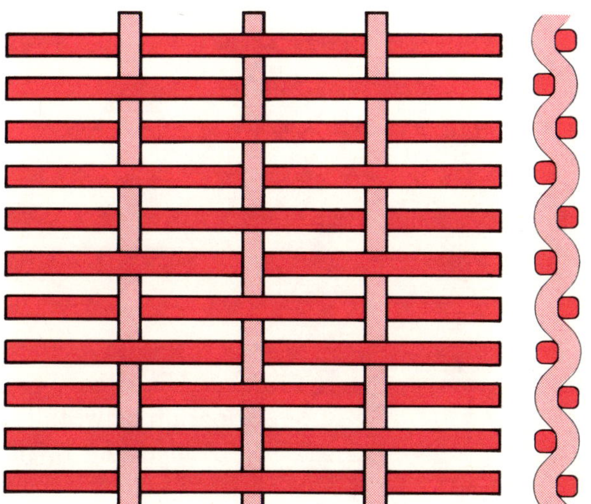

(B) Unidirectional weave.

(C) Square-weave woven rovings.

(D) Photo of (C); fibers are not wound or twisted. (Owens/Corning)

(E) Multifilament wound or twisted roving used in making heavy woven fibrous glass fabrics. (Owens/Corning)

Fig. 6-11. Weave patterns and rovings.

offer lightweight reinforcements, lower costs, better produce finish, and cause little molding damage to equipment.

Crystal "whisker" fibers of aluminum oxide, beryllium oxide, magnesium oxide, and silicon carbide are very expensive to manufacture with pres-

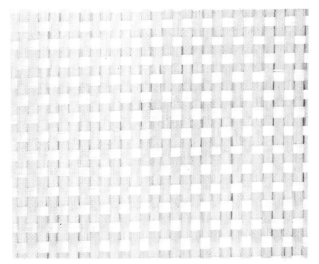

(A) Woven roving.

(B) Another style of woven roving.

(C) Fine strand mat.

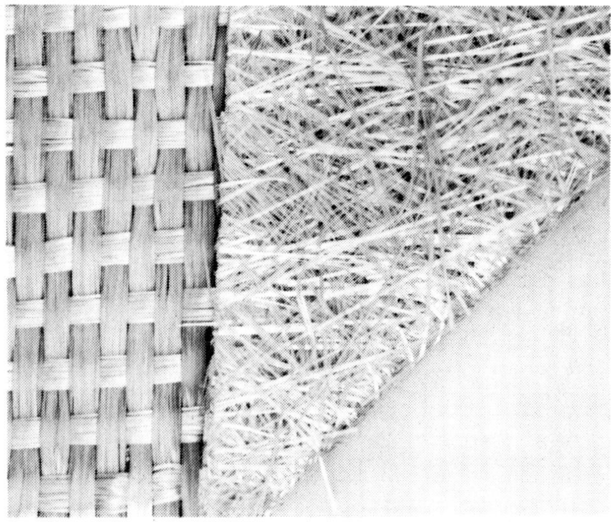

(D) Combination woven roving and mat.

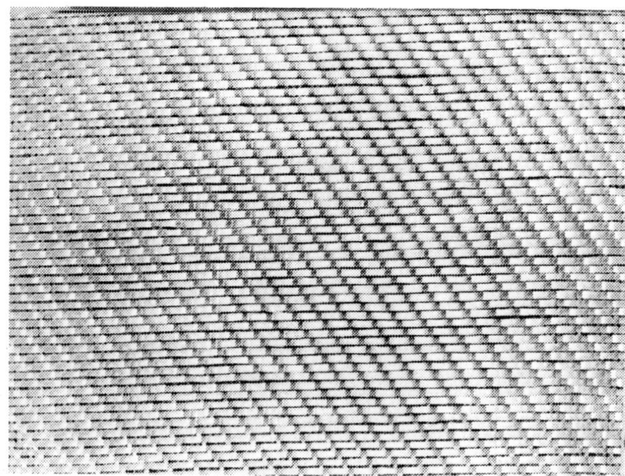

(E) Zero-twist weave.

Fig. 6-12. A few of the many forms of fibrous glass reinforcements. (FERRO)

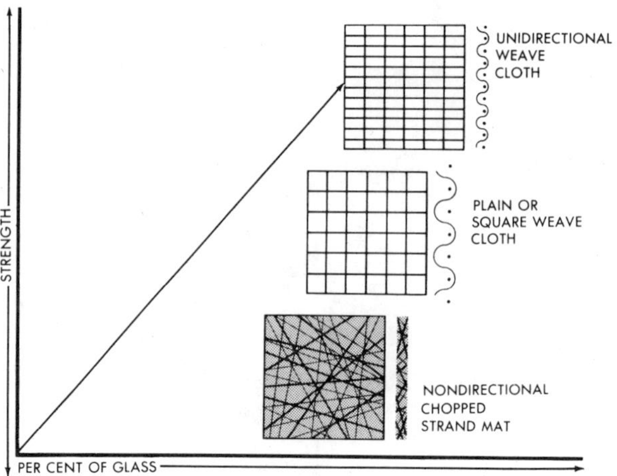

Fig. 6-13. Strength and percent glass relationship.

ent technologies; however, they display tensile strengths greater than six million pounds per square inch. Research in utilizing whisker reinforcements in dental plastics fillings, turbine compressor blades, and special deep water equipment has shown encouraging results. Fig. 6-14 shows a tail rotor shaft of reinforced boron/epoxy for a helicopter.

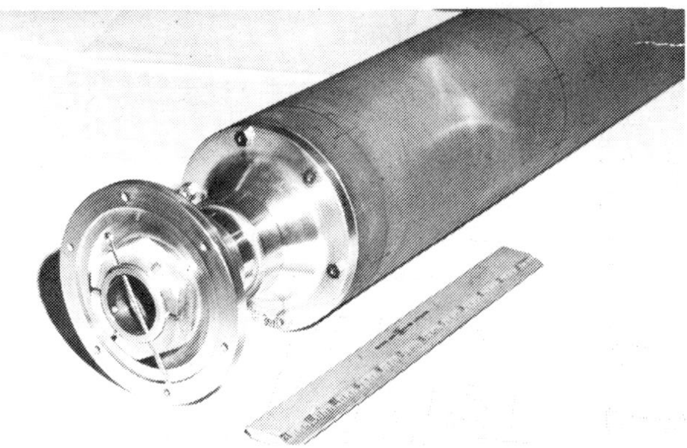

Fig. 6-14. Rotor driveshaft with end fittings and bearing supplements. (Wittaker Corp.)

Carbon and graphite fibers are comparable to glass in strength. They are finding numerous applications as self-lubricating materials, heat resistant re-entry bodies, blades for turbines and helicopters, and as valve packing compounds. (See Fig. 6-15.)

Fig. 6-15. Carbon fiber reinforced Kayak is lighter yet more rigid than glass reinforced one. *(Popular Mechanics)*

Ceramic fibers have high tensile strengths and low thermal expansion. Some fibers may reach a tensile strength of two million psi. Present applications for ceramic fibers include dental fillings, special electronics, and spacecraft research.

Metal fibers (Fig. 6-16) are finding a growing market as filament reinforcements in plastics. Steel,

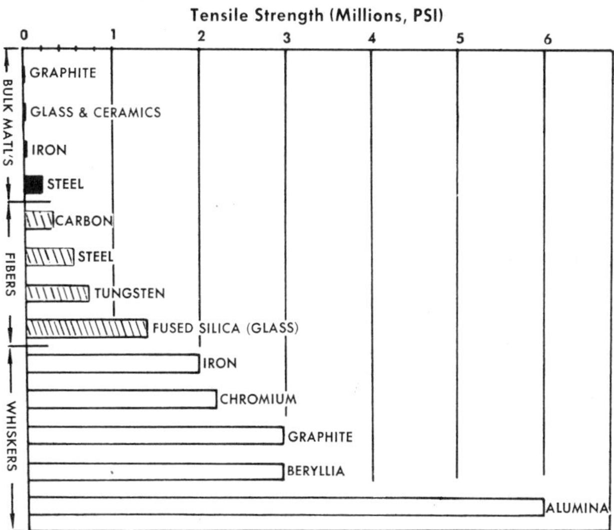

Fig. 6-16. Measured maximum tensile strengths of various materials in bulk, fiber, and whisker form.

aluminum, copper, nickel, and others are draw formed through dies, which improves the strength of the metallic fiber. Metal reinforcing fibers are used in epoxy to provide greater strength and heat transfer. These fibers can also be used for exotic electron control or electrical conductivity.

Solvents, Lubricants, Plasticizers, and Stabilizers

Solvents are added to resins for several reasons. Many natural resins are very viscous and hard; consequently, they are commonly diluted or caused to be dissolved by the solvent before processing. Resinous varnish and paints must be sufficiently thinned with solvents to ensure proper application.

Solvent molding is possible because the solvent holds the resin in solution while it is being applied to the mold. The solvent is rapidly evaporated, leaving a layer of plastics film on the mold surface.

Because solvents may dissolve many thermoplastics, they are used for identification and cementing purposes.

Solvents are particularly useful for cleaning resins from tools and instruments.

Lubricants are an important ingredient of plastics. During the manufacture of polymers, lubricants are added for three basic reasons. (1) They help eliminate some of the friction between the resin and the manufacturing equipment. (2) These internal lubricants aid in emulsifying other ingredients and provide internal lubrication for the resin. (3) Some of these agents prevent the plastics from adhering to the mold surface during processing. As the products are withdrawn from the mold, lubricants may exude from the plastics and prevent the products from adhering to each other or may actually provide a nonsticking or slip quality to the plastics surface.

There are numerous lubricants used as ingredients in plastics. Waxes and metallic soaps such as montan, carnauba, paraffin, stearic acid, and stearates of lead, cadmium, barium, calcium, and zinc are typical examples. Most of the lubricant is lost during the process of manufacturing the resin. If there is an excess of lubricant, polymerization may be slower in occurring or a "lubrication bloom" may appear. A lubrication bloom appears as an irregular, cloudy patch on the plastic surface (Fig. 6-17).

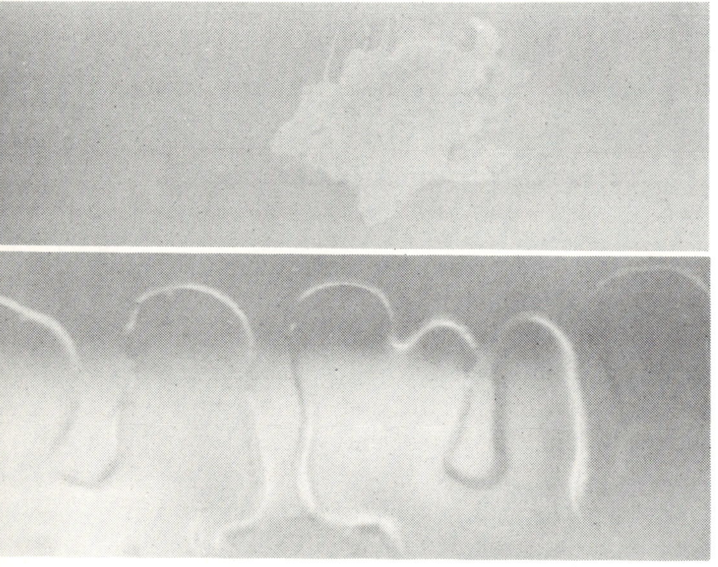

Fig. 6-17. Examples of lubrication bloom.

Because some plastics families exhibit the property of nonstick and self-lubrication, they are sometimes used as lubricants in other polymers. Fluorocarbons, polyamides, polyethylene, and the silicone plastics possess these properties.

Plasticity refers to the ability of a material to flow or become fluid under a force. A *plasticizer* is a chemical agent added to plastics to: (1) increase flexibility, (2) reduce melt temperature, and (3) lower viscosity. All of these properties aid in processing and molding.

Plasticizers act much as solvents by lowering the resin viscosity and like lubricants by allowing slip to occur between molecules. You will recall that van der Waals' bonds are only physical attractions and not chemical reactions. Plasticizers help neutralize most of these forces. Plasticizers thereby produce a more flexible polymer. Although these agents are much like solvents, they are not designed to volatilize or evaporate from the polymer during normal service life.

Camphor, the material Hyatt added to cellulose nitrate, may be considered as a plasticizer. It makes the plastics more moldable, flexible, and less explosive.

Over 500 different plasticizers are formulated and may be used to modify polymers. Plasticizers are important ingredients in plastics coatings, extrusions, moldings, adhesives and films.

A selected list of plasticizers is shown below in Table 6-4.

Table 6-4. Selected Plasticizers and Compatability With Selected Resins.

Plasticizer ↓ / Resin →	Polyvinyl Acetate	Polyvinyl Chloride	Polyvinyl Butyral	Polystyrene	Cellulose Nitrate	Cellulose Acetate	Cellulose Acetate Butyrate	Ethyl Cellulose	Acrylic	Epoxy	Urethane	Polyamide
Butyl benzyl phthalate	C	C	C	C	C	P	C	C	C	C	C	C
Butyl cyclohexyl phthalate	C	C	C	C	C	P	C	C	C	C	C	C
Butyl decyl phthalate	I	C	C	C	C	I	C	C	C	P	C	P
Butyl octyl phthalate	I	C	P	C	C	I	C	C	C	P	C	C
Dioctyl phthalate	I	C	P	C	C	I	C	C	C	I	C	C
Diphenyl phthalate	C	C	C	C	C	I	P	C	C	P	P	P
Cresyl diphenyl phosphate	C	C	C	P	C	C	C	C	C	C	C	C
Methyl phthalyl ethyl glycollate	C	P	C	C	C	C	C	C	C	C	C	I
N-Cyclohexyl-*p*-toluenesulfonamide	C	I	C	P	C	P	C	C	C	P	P	C
N-Ethyl-*o,p*-toluenesulfonamide	C	I	C	P	C	C	C	C	C	C	P	C
o,p-toluenesulfonamide	C	I	C	P	C	C	C	C	C	C	P	C
Chlorinated biphenyls	C	P	C	C	C	I	C	C	P	C	C	C
Chlorinated paraffins	C	P	P	C	P	I	P	C	P	P	C	C
Didecyl adipate	I	C	I	C	C	I	C	C	I	I	P	C
Dioctyl adipate	I	C	C	C	C	I	C	C	I	I	P	C
Dioctyl azelate	I	C	P	C	C	I	C	C	P	I	P	C
Dioctyl sebacate	I	C	P	C	C	I	P	C	I	I	P	C

C = Compatible I = Incompatible
P = Partially compatible

It is important that plastics products maintain their physical and chemical properties during normal service life. There are a number of factors, however, which may alter these properties. Degradation may be caused by many forms of energy: heat, light, oxidation, or mechanical shear. If sufficient energy is applied, the chemical and physical bonds of the plastics may be destroyed. To help prevent this deterioration of resins and plastics, *stabilizers* are sometimes added to the resin during manufacture. These chemical stabilizers help to prevent discoloration and decomposition during the storage (shelf life) and service life of the polymer.

Solar radiation on polymers may result in crazing, chalking, color changes, or loss in physical, electrical, and chemical properties. This weathering damage is caused when the polymer absorbs photons (small particles of light energy) which may possess enough energy to break chemical bonds between atoms. Ultraviolet light is the most destructive of the solar radiation striking plastic products.

Liquid or powder stabilizers are added to the resin to absorb energy, transfer energy to other molecules, or actually screen out the harmful ultraviolet light waves. If chemical bonds are broken in the polymer, further reactions such as crosslinking and oxidation may occur.

Nearly all formulations of plastics must contain a small proportion of stabilizer in order for conventional processing application to be applied. When plastics is "plasticized" or caused to become fluid by heat and pressure, thermal degradation may be arrested effectively with proper stabilizers.

The storage or shelf life of many resins may be considerably shortened when they are exposed to heat, light, or other forms of energy. Stabilizers are added to inhibit this action and increase shelf life of the resin.

There are several stabilizer types to consider when formulating the basic resin. When liquid or powder stabilizer types are used, they must not be toxic if used in contact with food products. Regulatory agencies such as the United States Food and Drug Administration, the National Sanitation Foundation, the Meat Inspection Division of the United States Department of Agriculture, and recent food additive laws help to impose standards for plastics packaging of food products.

Barium, cadmium, zinc phosphates, lead sulfate, lead carbonate, tin mercaptides, and benzophenones are only a few of the many stabilizing compounds used in resin formulation.

Colorants

Plastics possess a wide range of colors—a property that other materials cannot duplicate. One reason plastics have enjoyed a wide variety of product use is because they are produced in a multitude of colors.

Basically, there are four types of *colorants* used in the formulation of plastics: dyes, organic pigments, inorganic pigments, and special effect pigments.

One of the basic differences between pigments and dyes is in solubility. Dyes are more complex and color the material by forming chemical linkages with molecules. Dyes have excellent clarity and optical properties, but they have poor thermal and light stability. Thousands of dyes are synthesized from coal-tar chemicals for use in plastics.

Pigments are not soluble in common solvents or the resin. They must be mixed and evenly dispersed in the resin. Migration of these colorants or "bleeding" during processing or during the life of the product may cause serious surface color defects. Color bleeding may stain or cause allergic conditions if it comes in contact with the skin.

Organic pigments may provide the most brilliant and brightest opaque colors available. Translucent and transparent colors are not as brilliant as those produced with dyes, but are better than those achieved with inorganic pigments.

Inorganic pigments are often quite simple chemicals such as: carbon (black), iron oxide (red), cobalt oxide (blue), cadmium sulfide (yellow), and lead sulfate (white). These metallic oxides and sulfides are easily dispersed in the resin but do not produce as brilliant a color as organic pigments and dyes. Because of their inorganic structure, they are more light and thermal resistant than other colorants. Most of the inorganic pigments are used in high concentrations to produce opaque colored plastics. Low concentrations of iron oxide pigment will produce a translucent shade.

Special effect pigments may be organic or inorganic compounds. Colored glass is used, in a fine powdered form, as heat and light stable pigments for plastics. This colored glass powder is particularly effective in exterior applications because of its color stability and chemical resistance.

Metallic flakes of aluminum, brass, copper, and even gold may be used to produce a striking metallic sheen. Iridescent plastics are used by the automotive industry in producing the "metallic" finish on many of their products. When metallic powders are mixed with a quantity of colored finish and applied to the automobile, a finish varying in highlights and reflective hues is produced.

Pearl essence, both natural and synthetic, may be used to produce a brilliance in finishing jewelry, toys, toilet seats, fishing tackle and other articles where a pearl luster is desired. Natural pearl essence is obtained from the fatty skin of several species of fish in the form of *quanine crystals*. Synthetic, multifaceted quanine crystals are easily produced and are less expensive than natural ones.

When energy is absorbed by a material, a portion of that energy may be released in the form of light. This light is radiated when the molecules and atoms have their electrons excited to such a state that they begin to lose energy in the form of *photons*, or light particles. If heat causes these electrons to release photons of light energy, the radiation is called *incandescence*.

When chemical, electrical, or light energy excites electrons, the radiation of light is called *luminescence*. Luminescent materials are often added to plastics for special effects. Luminescence is categorized into *fluorescence and phosphorescence* (Fig. 6-18).

(A) Illuminated signs.

(B) Nonilluminated signs.

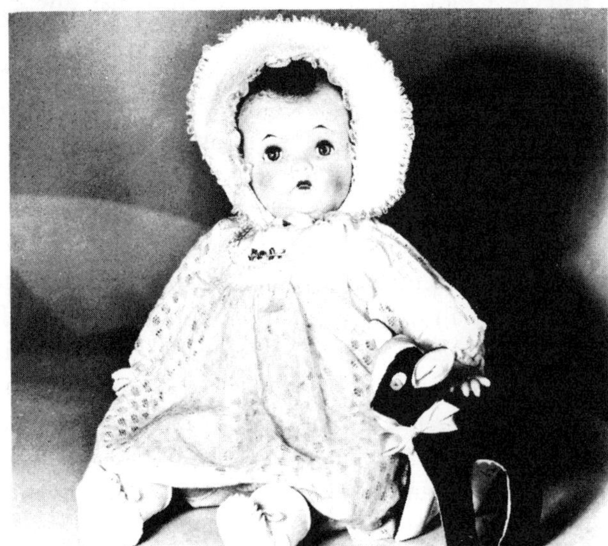

(C) Illuminated doll.

(D) Nonilluminated doll.

Fig. 6-18. Phosphorescent pigments glowing in the dark after exposure to light. (U.S. Radium Corp.)

Fluorescent materials emit light only when the electrons are being excited. This pigment ceases to emit light when the exciting energy source is removed. Many gems are fluorescent and illuminate in natural light. Fluorescent materials are made from sulfides of zinc, calcium, and magnesium. Fluorescent paint, applied to the instrument panels of many airplane cockpits, allows the pilot to read his instruments with little visible light being emitted. Other applications for fluorescent materials include hunting jackets, hard hats and gloves, lifesavers, rain slickers, bicycle strips, road warning signs, and so on.

Phosphorescent pigments possess an "afterglow" or they continue to emit light for a limited amount of time after the exciting force has been removed. The most commonly viewed phosphorescence is seen on television picture tubes when electrical energy excites the phosphorescent materials. Phosphorescent pigments used in plastics and paints are made from calcium sulphide or strontium sulphide.

Mesothorium and radium compounds are radioactive materials sometimes used for special applications requiring luminescence. There may be harmful effects, however, from prolonged exposure to radioactive materials.

Because it is often necessary to mix colorants to achieve the desired shade or color hue, a color chart of primary and secondary colors (Fig. 6-19) should be of value.

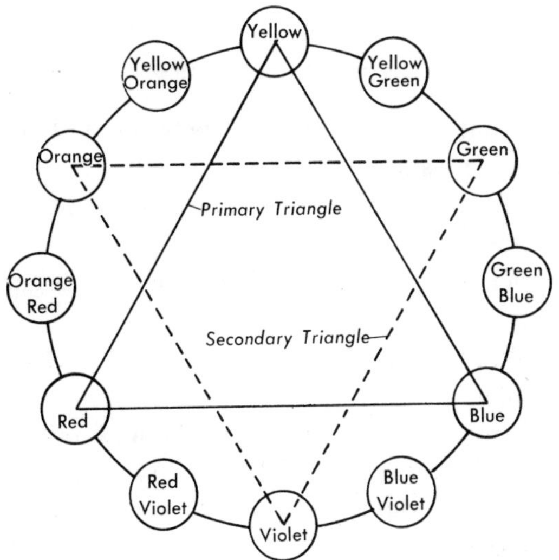

Fig. 6-19. Pigment primary and secondary colors on color chart.

By mixing the primary colors, yellow and blue, the secondary color green is produced. Yellow and green will yield a yellow-green color.

Catalysts and Promoters

Catalysts are used to initiate or accelerate the rate of polymerization. Two commonly used catalysts are organic peroxides and hydroperoxides. These catalysts (or more correctly "initiators") cause the ends of the monomers to join and form the long polymer chains and possible crosslinkage. You may recall that stabilizers (inhibitors) are frequently used in resins to prolong storage life by blocking polymerization.

When catalysts are added, the polymerization begins and is influenced little by the inhibitors in the resin. As organic peroxides are added to polyester resin, the polymerization reaction begins and produces exothermic heat. This formation of heat further accelerates the crosslinking and polymerization.

Promoters are additives which react just the opposite of inhibitors. They are often added to resins to aid in polymerization. They react only when the catalyst is added. This reaction, which causes the polymerization, produces varying amounts of heat energy. A common promoter used with methyl-ethyl-ketone-peroxide catalyst is cobalt naphthanate. All promoters and peroxides should be handled with caution. Peroxides may cause skin irritation or acid burns. If promoters and catalysts are added directly together, a violent reaction may occur. *Always* mix in thoroughly the promoter and then add the desired amount of catalyst to the resin.

Resins which have not been prepromoted generally have a longer shelf life. Remember: other forms of energy will cause polymerization. Heat, light or electrical energy may initiate this reaction.

Flame Retardant Additives

Flame retardant additives are often added to the basic resin. Most commercial flame resistant chemicals are based on combinations of bromine, chlorine, antimony, boron, and phosphorus.

Many of the retardants emit a fire extinguishing gas when heated. Others react by swelling or foaming and thus forming an insulation barrier against heat and flame (Fig. 6-20).

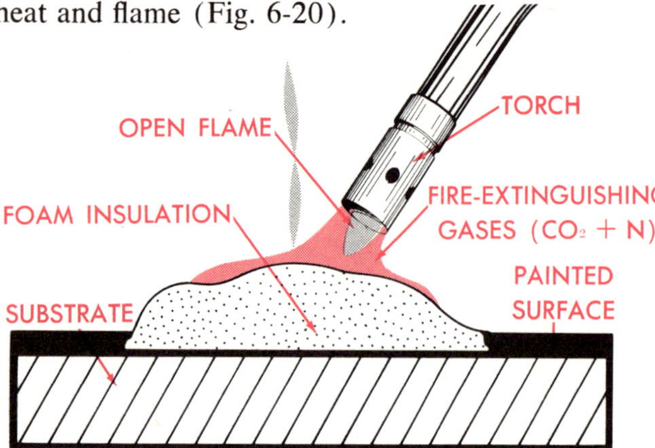

Fig. 6-20. This protective finish swells when heated, forming a foam insulation barrier. It also emits a fire-extinguishing gas to retard the flame and burning.

According to the National Fire Protection Association, nearly two billion dollars worth of damage is caused annually by fires. Although the fire record of plastics used for construction has been good, continued expansion and growth will depend on reduction of flammability of materials and well defined fire test standards and building codes.

Some families of plastics possess natural flame retardants. During decomposition by flame, they emit flame extinguishing gases.

Glossary of New Terms

Colorants—Dyes or pigments which impart color to plastics.

Exothermic—Evolving heat during reaction (cure).

Filler—An inexpensive, inert substance added to a plastic to make it less costly. Fillers may improve physical properties and are usually small, in contrast to reinforcements.

Fluorescence—A property of a substance, such as fluorite, to produce light while it is being acted on by radiant energy such as ultraviolet light or X-rays.

Inhibitors—A substance that slows down chemical reaction. Inhibitors are sometimes used in certain types of monomers and resins to prolong storage life.

Initiators—Agents necessary to cause polymerization, especially in emulsion polymerization processes.

Lubrication bloom—An irregular, cloudy, greasy film on a plastic surface.

Luminescence—Light emission by the radiation of photons after initial activation. Luminescent pigments are activated by ultraviolet radiation, producing very strong luminescence.

Photon—A quantum of light energy, analogous to the electron.

Phosphorescence—Luminescence which lasts for a period after excitation.

Plasticizer—Chemical agent added to plastic compositions to make them softer and more flexible.

Primary colors—The fundamental colors from which all others are made.

Promoters—A chemical, itself a feeble catalyst, that greatly increases the activity of a given catalyst.

Secondary colors—Color obtained by mixing two or more primary colors.

Stabilizers—An inexpedient used in the formnulation of some plastics, especially elastomers, to assist in maintaining the physical and chemical properties of the compounded materials at their initial values throughout the processing and service life of the material.

Thixotropy—State of materials that are gel-like at rest but fluid when agitated. Liquids containing suspended solids are apt to be thixotropic.

Review Questions

6-1. What is the difference between polymers and plastics?

6-2. What functions do fillers provide in plastics?

6-3. What are the disadvantages of organic fillers?

6-4. What is the major purpose of a thixotropic filler?

6-5. What are the major differences between reinforcements and fillers?

6-6. Why is glass fiber selected as a reinforcement?

6-7. What are rovings?

6-8. Why are solvents added to resins?

6-9. Why are lubricants used in the manufacture of plastics?

6-10. What causes lubrication bloom?

6-11. Why are plasticizers used in resins and plastics?

6-12. Of what value are stabilizers? What properties are affected?

6-13. Why is solar radiation particularly harmful to plastics?

6-14. What are the four types of colorants used in plastics?

6-15. What is the difference between luminescence and incandescence? Between fluorescence and phosphorescence?

7

Various Forms of Plastics in Use

Many people are aware that metals can be purchased in a variety of forms. Plastics, like metal, may be obtained as structural shapes, thin fibers, powders, and several other forms. It is because of these available forms that plastics are used for such a broad range of product applications.

The eight basic forms of plastics discussed on the following pages are not "forms" in the physical sense. In order to be a "plastics," the material must be solid in the finished state; consequently, the names of many available forms of plastics are really a description of their use or product application. These forms may begin as monomer resins, partially polymerized resins, or fully polymerized solid plastics.

It is not too difficult to visualize cutting, sawing, and welding tubes, "I" beams, and channel shapes, or forming flat sheets into various contours of the finished product. Craft classes have been cutting, sanding, buffing, and forming plastics for some time. This "fluff and buff" method, however, is not typical of industry. In industrial plastics the resins and polymers are often formed in a die or mold to produce the completely finished product.

Molding Compounds

Many of the available forms of plastics begin as molding compounds. In fact, large volumes of this form of plastics are molded into other familiar forms. Profile shapes, fibers, and films may originate as a basic molding compound.

It should be evident that some of the available forms of plastics are never seen by the consumer. Some plastics forms, like casting resins, are intermediate to the production of other plastics forms. Rods, sheets, coatings, and adhesives may begin in resinous form.

Combinations of plastics forms may be utilized in the production of some laminates. These laminate forms may be composed of sheets of reinforced plastics bonded to a core of cellular plastic by adhesive compounds.

Molding compounds are formulated into different solid forms to facilitate molding. *Powdered* forms are used in rotational molding and in various coating applications.

The powder grain size seldom exceeds 0.005 of an inch in diameter.

Many thermosetting powders are pressed into convenient *preform* shapes. These preforms are often referred to as pills, tablets, biscuits, or premolds. Although many of these preforms are tablet or pill shaped, they may be made into any desired shape or size to fit the molding needs.

Specially designed machines weigh and compact the plastics powder into the desired preform shape. Preforms are often used in compression molding thermosetting plastics.

To simplify mold construction and shorten the molding cycle, the preform is preheated prior to molding operation. This preheating helps to remove moisture and shorten molding time in both thermosetting and thermoplastic compounds.

This preheating is accomplished by several methods including the use of hot plates, ovens, infrared radiation, and high-frequency heating. The latter method is in wide use because the preform is very rapidly and uniformly heated. (See Fig. 7-1.)

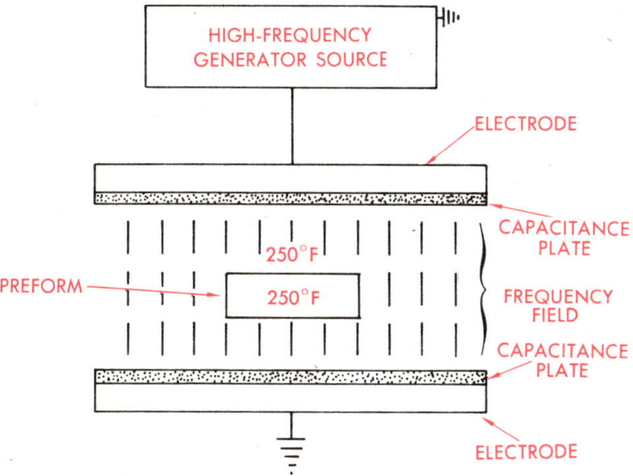

Fig. 7-1. Preheating preform by high-frequency heating.

Fig. 7-2 will help to illustrate how the impregnated preform is produced. These preforms are easily handled, produce less waste, and result in shorter molding cycles. These products have a greater resin-to-reinforcement ratio than many reinforced forms. Consequently their maximum strength is less.

Large products such as automobile bodies and boat hulls may be produced by this method. The matched molded preforms will produce finished surfaces on both sides of the product.

Molding compounds formulated in a "putty" form are referred to as a *premix*. This putty form is sometimes referred to as "gunk" or "slurry molding." These compounds are usually a mixture of monomer resin, short fibrous reinforcements, and fillers. When the reinforcing material is combined with the full complement of resin before molding, the mixture is called a *prepreg* (Fig. 7-3). The premix may be molded by several processes or may be used for filling cracks and voids in various materials. Some are polymerized by adding a catalyst or heat, while others consist of collodial suspensions and harden when the solvents evaporate.

Polyester or epoxy resins are typically mixed with the following percentages:

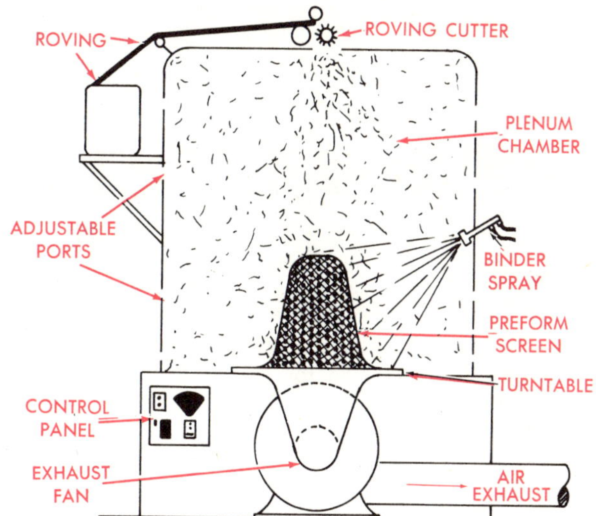

(A) Plenum chamber preform process.

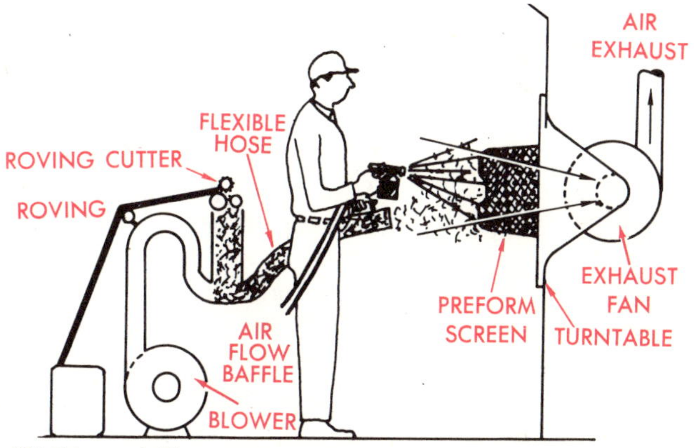

(B) Directed fiber preform process.

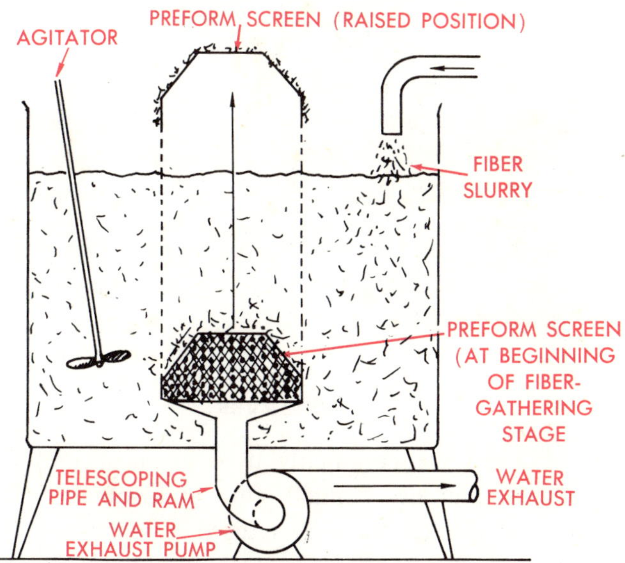

(C) Water slurry or "Aqua-glass" preform process.

Fig. 7-2. Three ways how the preform is produced. (Owens Corning Fiberglas Corp.)

Fig. 7-3. Diallyl phthalate prepregs are built up for a large aircraft radome in this internally heated kettlelike mold. (FMC Corporation)

Premix "Putty"

Resin	38%
Chopped fibers	18%
China clay (filler)	44%
	100%

Pellet forms are convenient to dispense to equipment and molding machines. Granular pellets seldom exceed 0.250 of an inch in diameter (see the illustration in Fig. 7-4).

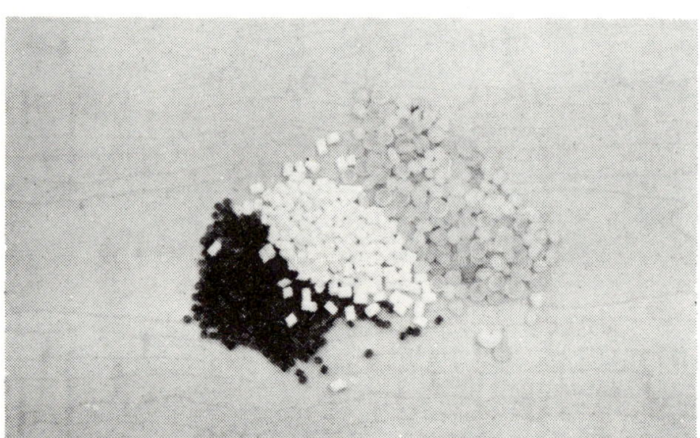

Fig. 7-4. Types of pellet forms.

Injection and extrusion molding processes utilize these pellet forms. The molding processes will be discussed in Chap. 15.

Adhesives

Plastics are commonly used as substitutes for other materials. Animal glue and some of the natural plastics have been used as adhesives for centuries. Wax and shellac were used to seal letters and important documents while many ancient civilizations used pitch to seal cracks in boats and rafts. Archeologists have evidence that nearly 3300 years ago the Egyptians used adhesives to adhere gold leaf to wooden coffins and crypts.

Today, plastics adhesives have replaced most of the natural adhesives and provided man with strong, durable bonding materials. (see Fig. 7-5).

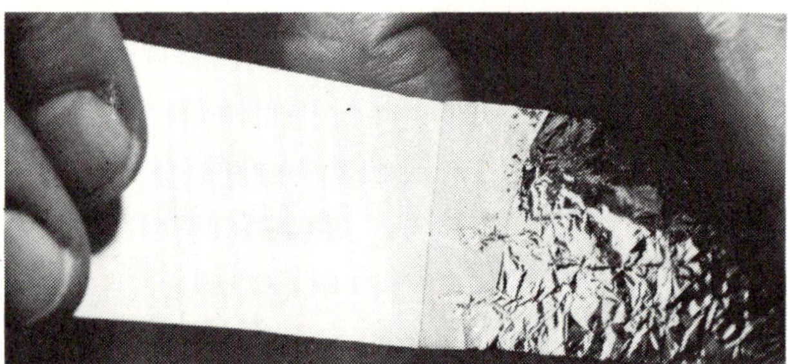

Fig. 7-5. Adhesive properties are illustrated by failure of the paper rather than the hot-melt adhesive in this aluminum-foil-to-paper laminate. (General Mills, Inc.)

Adhesive bonding has replaced many of the more conventional methods of bonding materials together. Nails, bolts, rivets, solder, welding, and other assembly methods are giving way to adhesive bonding in the production of countless consumer and industrial products.

Because adhesives are used for so many different purposes and in different ways, a precise definition of "adhesives" may appear hazy. Adhesives are a broad class of substances which "adhere" materials together by surface attachment.

At one time the word "glue" referred to an adhesive obtained from hides, tendons, cartilage, bones, and blood of animals. Today, this term is synonymous with "adhesive." Modern adhesives include both natural and synthetic plastics.

Cement and liquid glue adhesives are general terms used interchangeably. The term "cement," when used in the bonding of plastics, is generally considered to mean a liquid adhesive employing a solvent base of the synthetic or resin variety. The bond is chemical in nature and is classed as cohesive bonding. The term "glue" is probably used more often when referring to the bonding of wood. Building blocks are commonly held together with a mortar sometimes called "cement." This cement has a true adhesive action in holding the blocks together; however, the word "cement" is also used to describe many of the plastics adhesives.

Adhesives are commonly used in gluing wood, repairing broken china, and as bonding agents on tape. When selecting adhesives, consider the materials to be joined, the surfaces, the method of application, and the service conditions expected of the bonded product.

There are two basic forces which attribute to adhesion. (1) *Mechanical* or *physical adhesion* has been defined as "adhesion between surfaces in which the adhesive holds the parts together by interlocking action." Although adhesives are sold in various available forms, they must be in a liquid or semiliquid state during the bonding operation to ensure sufficiently close contact with the adherends. Physical adhesion usually involves secondary or van der Waals' forces and simple mechanical interlocking on the surface. This type bond does not attack the surface or result in any material flow between adherends. In organic materials there may also be some primary forces at work. Metal-to-metal bonding forces are due largely to secondary and physical forces.

Types of Bonding Action

wood to wood ....physical bonds
metal to metal
physical bonds: cohesive when welded
ceramic to ceramic
physical bonds: cohesive when "fired"
paper to paper ....physical bonds
leather to leather ....physical bonds
plastics to plastics
physical bonds: cohesive when welded or solvent cemented

(2) *Chemical* or *specific adhesion* has been defined as "adhesion between surfaces which are held together by valence forces of the same types as those which give rise to cohesion." The forces which hold the molecules of all materials together are referred to as cohesive forces. These forces include the strong primary valuence bonds and the weaker secondary bonds.

In chemical adhesives, there is a strong valence attraction between the materials as the molecules flow together. During the welding of two metals, molten metal flows and there occurs a chemical cohesive action between the pieces.

There are cohesive forces involved "internally" in all adhesives as they are pulled apart. Both cohesion and adhesion is employed when "contact cement," a popular polymer adhesive, is used. This adhesive is applied to both surfaces and allowed to set. The two surfaces are then pressed together, resulting in a strong cohesive bond between the adhesive surfaces and an adhesive bond to the adherends (substrates).

It should be apparent that only by causing a softening or flow in the two materials can cohesive or chemical bonding occur. If the surfaces of the adherends are not caused to flow, physical forces are holding the pieces together.

In bonding plastics or metals, either bonding method may be employed. In the cohesive bonding of metals, heat must be applied to cause the surface flow and molecules to intermingle. In cohesively bonding plastics, solvents may be used on most of the thermoplastic polymers. These solvents soften the two surfaces being joined, allowing strong valence bonding between molecules. Solvent cementing will be discussed in greater detail in Chapter 22.

A *solvent cement* is one which dissolves the plastic being joined, forming strong intermolecular bonds as it evaporates.

A *monomeric cement* is a monomer of at least one of the plastics to be joined, and is catalyzed so that a bond is produced by polymerization in the joint. Cohesive or adhesive bonding will occur with monomer cements depending on the chemical composition of the substrates.

If the surfaces are properly prepared, nearly all materials may be bonded with an adhesive.

Synthetic adhesives come in a number of physical forms: powders, films, dispersions, hot melts, pastes, liquids, and two-part components (see Table 7-1). The adhesives on tape are generally designed to remain tacky throughout service life.

Casein and urea formaldehyde (amino resins) are used principally in the woodworking industries. Some of the urea resins are sold in liquid forms for use in the manufacture of plywood, particle boards, and hardboards such as Masonite.

Shell-molding is an important process used by foundrymen in casting metal parts. Phenolic and amino resins are used to bond the sand mold together. Fig. 7-6 shows how resins are used to bond sand and mold together.

Table 7-1. Selected List of Plastics Adhesives.

Plastics Adhesives	Available Forms
Thermosetting	
Casein	Po, F
Epoxy	Pa, D, F
Melamine formaldehyde	Po, F
Phenol formaldehyde	Po, F
Polyester	Po, F
Polyurethane	D, L, Po, F
Resorcinol formaldehyde	D, L, Po, F
Silicone	L, Po, Pa
Urea formaldehyde	D, Po, F
Thermoplastic	
Cellulose acetate	L, H, Po, F
Cellulose butyrate	L, Po, F
Cellulose/carboxymethyl	Po, L
Cellulose/ethyl	H, L
Cellulose/hydroxyethyl	Po, L
Cellulose/methyl	Po, L
Cellulose/nitrate	Po, L
Polyamide	H, F
Polyethylene	H
Polymethylmethacrylate	L
Polystyrene	Po, H
Polyvinyl acetate	Pt, D, L
Polyvinyl alcohol	Po, D, L
Polyvinyl chloride	Pa, Po, L

Po=Powder; F=Film; D=Dispersion; L=Liquid; Pa=Paste; H=Hot Melt; Pt=Permanently tacky

Phenol-formaldehyde (phenolic) resins are sold in liquid, powder, and film forms. Films of these adhesives are made about 0.001 of an inch thick and are placed between the materials to be bonded. Moisture in the material itself or external steam is applied, causing the adhesive film to flow and liquify. The curing reaction is accomplished at temperatures in the 250°F to 300°F range. A reinforced film is often used for many applications. These films are usually very thin tissuelike papers, which are saturated with the resin adhesive. They are used in the same manner as unreinforced films but are easier to handle and apply.

During the manufacture of exterior plywood and tempered or smooth-finished hardboard, large quantities of this adhesive are utilized.

Resorcinol-formaldehyde, another phenolic based adhesive, usually comes as a liquid, to be mixed with a powdered catalyst at the time of use. This phenolic resin has the advantage of curing at room temperature as well as being water and heat resistant. High grades of exterior and marine plywoods are bonded with resorcinol adhesives.

(A) After resin-sand mixture sets, it is stripped off the matchplate. It will be used to make welding rods.

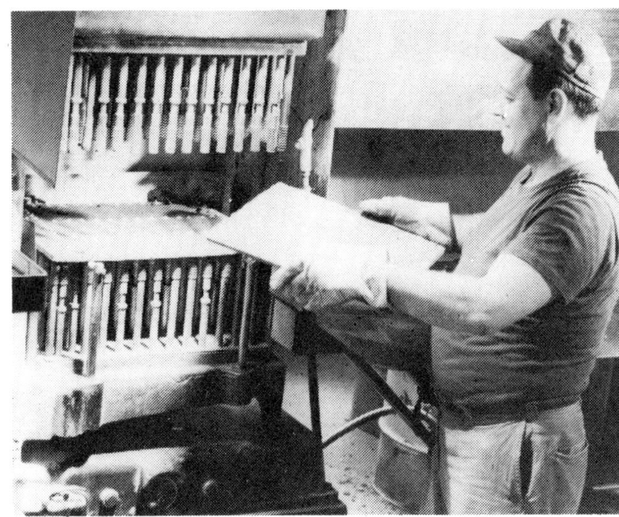

(B) Resin-sand molds are joined by heat sealing with small spots of resin. Later the mold will be sent to the pouring floor.

Fig. 7-6. Resins are used to bond the sand and molds together. (Stellite Div., Cabot Corp.)

High-frequency heating of adhesives greatly speeds the curing or polymerization time of many plastics used as adhesives. The high-frequency field excites the molecules of the moist adhesive, causing heat and rapid polymerization. Wood joints are readily assembled by this method (Fig. 7-7).

Resin-bonded grinding wheels and sandpapers are made from abrasive grains and a plastics bonding agent. The grinding wheels are made from abrasive grains, powdered resin, and a liquid resin by a cold molding process (discussed in Chap. 15).

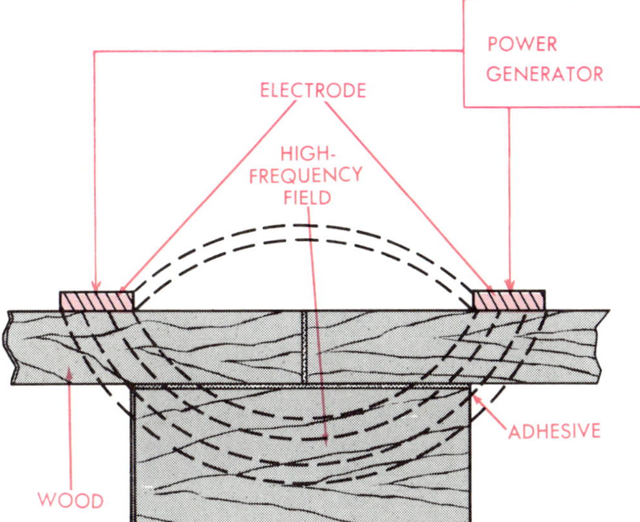

Fig. 7-7. High-frequency (radio-frequency) heating of adhesives.

Fig. 7-8 shows some typical sandpapers and grinding wheels.

The "glue pots" associated with the older animal derived glues have disappeared in favor of the more modern and easily used adhesives. A very popular one is the "white" polyvinyl acetate adhesive. This adhesive comes ready to use, in a fast setting liquid form. This familiar white adhesive is a dispersion of polyvinyl acetate in a solvent. Often the solvent is water, and, consequently, it must be kept from freezing. Carpenters, artists, secretaries, and many others utilize the adhesive properties of this low cost, widely compatible material.

Hot melt adhesives are gaining popularity because they are easy to use, somewhat flexible, and obtain maximum adhesive qualities when cooled. Several thermoplastic plastics are used, including polyethylene, polystyrene, and polyvinyl acetate.

Fig. 7-8. Resin-banded grinding wheels and sandpapers.

Small sticks or rods of these plastics are heated in an electric gun. The hot plastics is forced from the gun nozzle onto the gluing surface. Fig. 7-9 shows a leather strap being assembled using hot-melt plastics in an electrically heated applicator gun.

Small, rapidly assembled articles may be bonded with this method. Probably the most serious disadvantage is the difficulty of making large area glue joints. The adhesive simply cools too rapidly to ensure adhesion on large bonding surfaces.

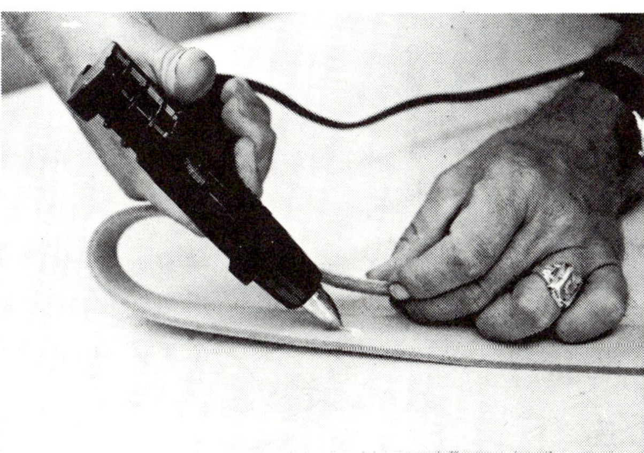

Fig. 7-9. Hot melt "glue" gun.

The shoe industry is presently utilizing this adhesive as an effective means in the assembly of leather goods.

Epoxy resin adhesives are thermosetting plastics which are available in two-part paste components. Epoxy resin and either a powder or a resinous catalyst are mixed to polymerize the resin. Heat is sometimes used to aid or speed this hardening process. Specially formulated one-part epoxy resins may be polymerized by the application of heat alone.

Epoxy adhesives have excellent adhesion to nearly all materials if each surface is properly prepared. However, polyethylene, silicones, and fluorocarbons are among the most difficult materials to bond. The excellent adhesive properties of epoxies are utilized to mend broken china, bond copper to phenolic laminates in printed circuits, and can function as bonding agents in sandwich or skin type structures.

(A) Various plastics adhesives.

(B) Familiar two-part epoxy adhesives.

Fig. 7-10. Epoxy, casein, urea, and resorcinol adhesives are used extensively in industry, while "white" polyvinyl adhesives are commonly used in the home.

Vinyl adhesives may be used to bond framing members and interior reinforcements to several automotive parts.

Adhesive faced tapes of all kinds are important members of the adhesive family. The adhesive is generally activated by one of the following methods: (1) solvents—water on gummed labels and stamps, (2) heat—iron-on fabric patches and some heat sealing packages (both plastics and paper), and (3) pressure—masking tapes, labels, and many other applications.

Caulking, sealing, patching, glazing, and putty compounds may be classified with other plastics adhesives because they depend to a certain extent on their ability to bond effectively with the substrate to which they are applied. Their primary purpose is usually to eliminate the infiltration of moisture, air, or other materials from penetrating cracks or small openings. These compounds include many elastomers as well as plastics.

Fig. 7-11. Only a few of the many cartridge-type packaged caulkings, sealants, or glazings are shown.

The term "caulking" is probably derived from the practice of forcing tar soaked wicking into the seams of wooden boat hulls. Though there are few wooden boats being "caulked" today, the word "caulking" implies that a compound is being forced into a crack.

The word "sealant" is a modern term describing compounds which fill cracks. The words "putty," "patching," and "glazing compounds" may mean sealant compounds but are usually descriptive of a special application. Putties and patching compounds, for example, are used to fill large cracks or openings and are applied from bulk containers with spatula or putty knife. These compounds contain a large amount of filler; consequently, they are less expensive and shrink less than many compounds. Sealing large cracks in wood, steel, and concrete are among the many typical uses.

Glazing compounds are most often associated with sealing openings around windows. Because a great variety of construction materials are joined and because each material has a different expansion rate, sealants must remain flexible, nondrying, and adhesive (Fig. 7-12). Aluminum, for example, will expand about 2½ times more than glass.

A modern and efficient method of applying sealants is with the use of compressible tapes or extruder guns. Compressible tapes are simply rolls of sealant in ribbon form. They are effectively used in the automotive industry for sealing two metal joints and as adhesive sealants in window construction. See Fig. 7-13.

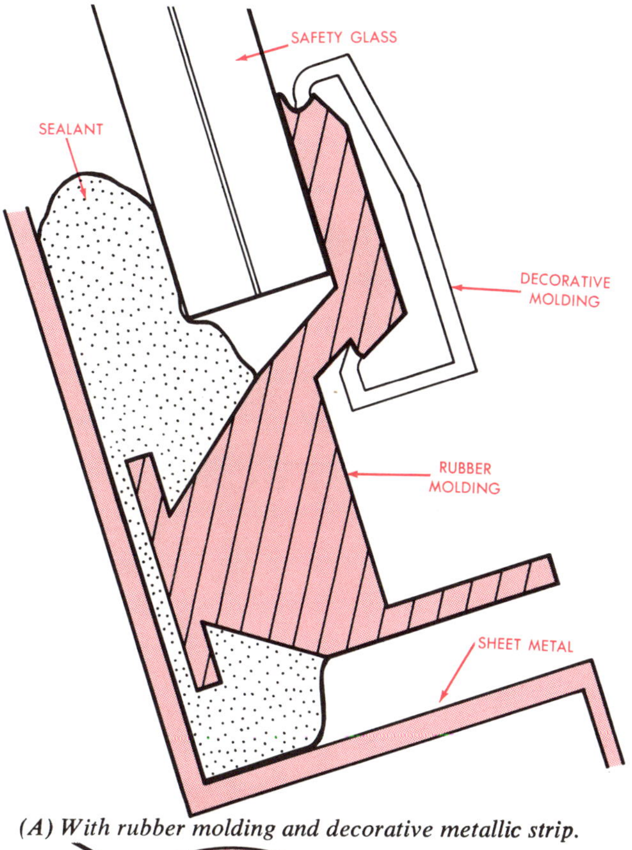

(A) With rubber molding and decorative metallic strip.

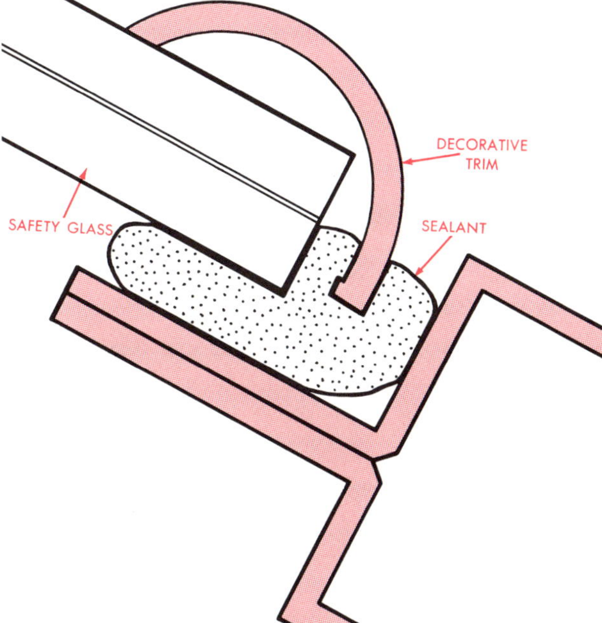

(B) With molding strip in automobile.

Fig. 7-12. Sealing safety windshield.

Extruder guns are convenient applicators which use disposable or refillable containers of sealing compound. These applicator "guns" push the sealant out a nozzle opening during application. (See Figs. 7-14 and 7-15.)

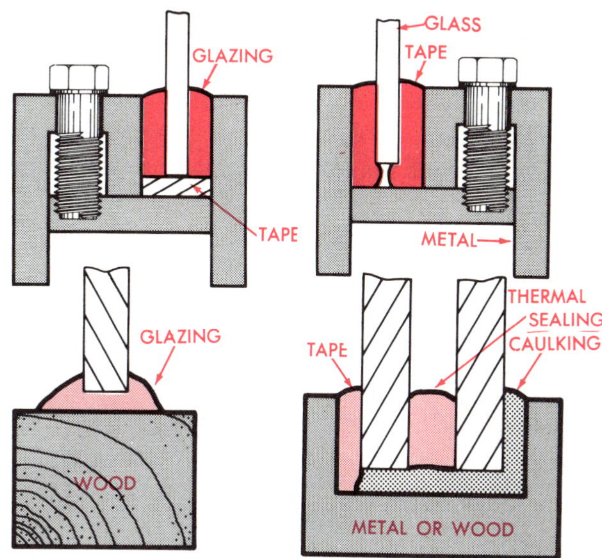

Fig. 7-13. Other sealing methods.

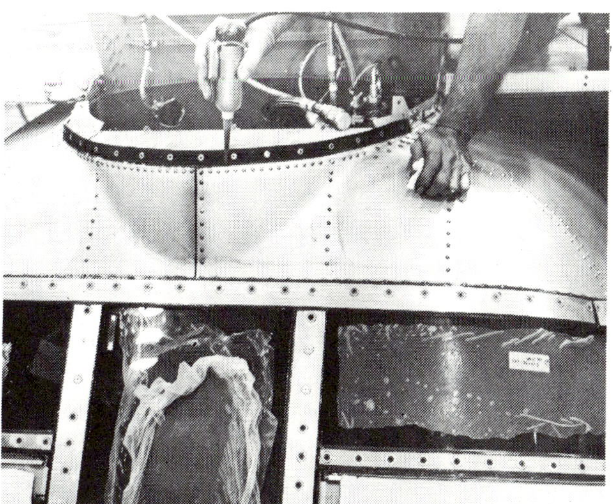

Fig. 7-14. A special applicator gun is applying sealant around motor housing of an aircraft section. (Thiokal Chemical Corp.)

Fig. 7-15. This overlap angle joint, used in aerospace vehicles, is receiving an application of polymeric sealant. (Thiokal Chemical Corp.)

Some of the more common sealants include polysulfides, acrylics, polyurethanes, silicones, and numerous natural and synthetic rubber compounds.

The polymer polysulfide is a very effective sealant with a wide range of applications in the aircraft, electrical, and building trade industries today. (See Fig. 7-16.)

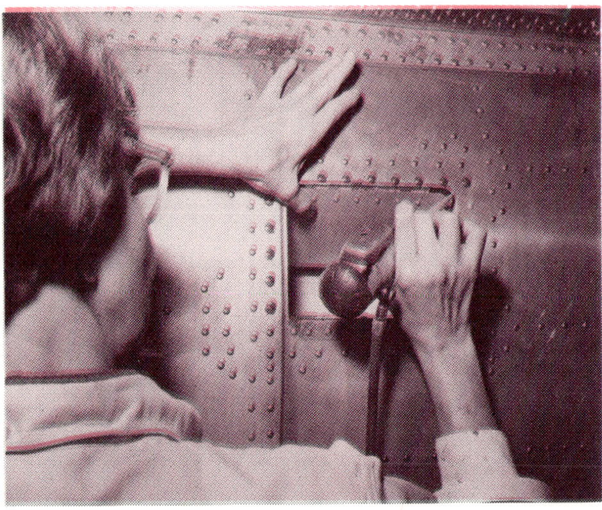

Fig. 7-16. Polysulfide polymer is used to seal many parts used in the aircraft and aerospace industries. (Thiokal Chemical Corp.)

Glazing compounds and putties may be made of acrylic compounds which are easily applied, will not crack, sag, or deteriorate.

Special formulations of epoxy and silicon compounds find applications as sealants around lavatories and bathtubs.

Silicone construction sealant is relatively expensive but is easily used, cures tack free in one hour, and remains permanently flexible.

Dimensional or Profile Forms

Most of the dimensional or profile forms of plastics are produced by the extrusion process. This familiar form is available in various structural shapes including rods, tubes, bars, and sheets. These forms are produced in continuous lengths but are cut to standard lengths to aid in handling and shipping. A few of the typical dimensional or profile shapes are shown in Fig. 7-17. Plastic tubing, pipe, and sheets are common examples of this available form.

Fig. 7-17. Assorted Extruded Profiles form of plastics. (Fellows Gear Shaper Co.)

Not all dimensional or profile shapes are made by the extrusion process. The casting process may be utilized to produce tubes, rods, bars, and other shapes.

Because filaments and films have different product applications and are normally produced with small cross-sectional areas, they will be discussed separately. Films and filaments, for example, seldom exceed 0.010 of an inch in thickness. The term "sheet" refers to a plastics form greater than 0.010 of an inch in thickness.

Artists and hobbycraft people use this familiar sheet form because it is easily formed and shaped with simple hand tools. Sheet may be veneered or clad to metal and wood as durable materials for countertops, wall, and exterior siding. Fig. 7-18 shows vinyl siding for use on homes.

Sheets are thermoformed into various products including picnic coolers, car bodies, toys, outdoor signs, luggage, refrigerator liners and other products. Sheet plastics are commonly produced by forcing the melted plastics through an adjustable die opening and onto cooling or takeup rollers (Fig. 7-19A).

Quenching tanks may be used in place of the cooling rollers (Fig. 7-19B). Calendering is yet another method of producing sheet forms.

Fig. 7-18. This wood grain effect siding is made of solid vinyl for the exterior of homes. (Bird and Son, Inc.)

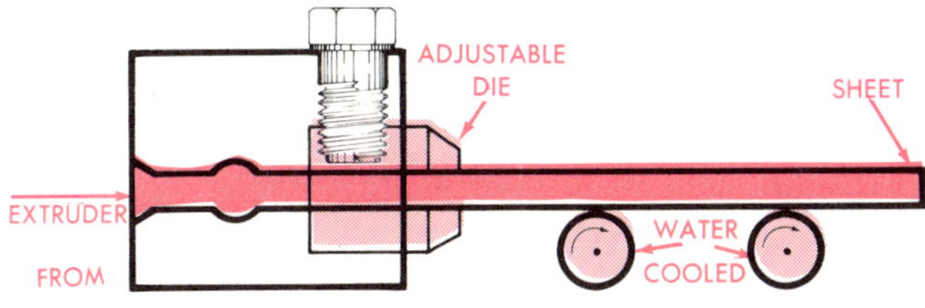

(A) Adjustable sheet die with chilled takeup rollers.

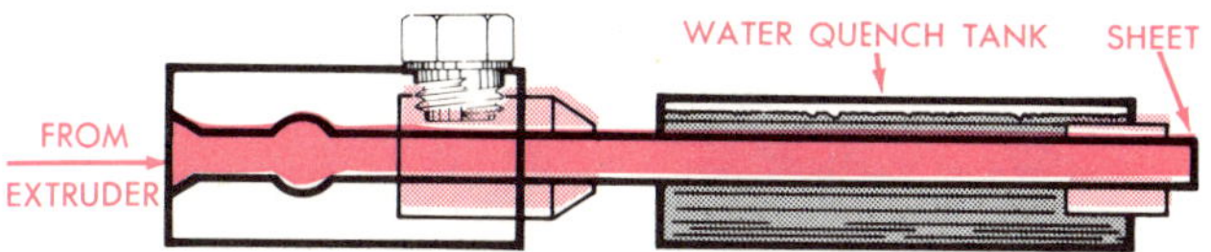

(B) Adjustable sheet die with water quench bath.

Fig. 7-19. Production of sheet plastics.

A number of thermoplastic and thermosetting plastics may be cast into sheet form (Fig. 7-20). Acrylic sheets are commonly produced by pouring a catalyzed monomer or partially polymerized resin between two parallel plates of glass. The glass is usually sealed with a gasket material to prevent leakage and help control the thickness of the cast sheet.

After the resin has fully polymerized in an oven or autoclave, the acrylic sheet is separated from the glass plates. The sheets are further processed by reheating to relieve stresses caused during the casting process. The faces are usually covered with

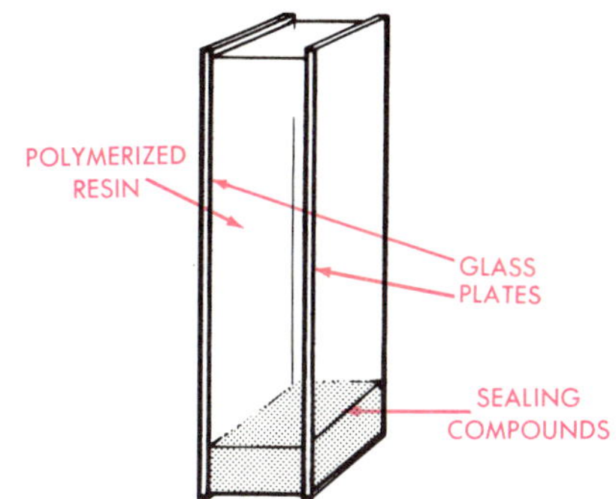

Fig. 7-20. Casting plastics sheets.

masking paper to protect the sheet during shipment, handling, and fabrication. Sheets may be purchased in untrimmed state, with the sealing material still adhering to the edges (Fig. 7-21).

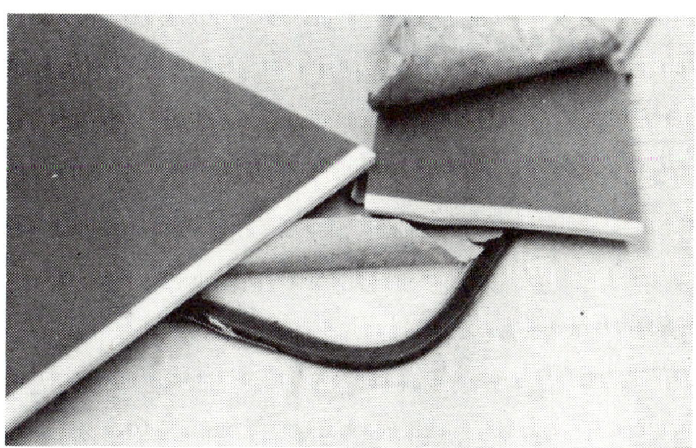

Fig. 7-21. The sealing materials may be seen on the edges of these untrimmed sheets of acrylic plastics.

Films

The American Society of Testing Materials (ASTM) has defined *film* as plastics sheeting less than 0.010 of an inch in thickness. The term *sheet* refers to plastics sheeting 0.010 of an inch or thicker. The word *sheeting* refers to flat pieces of plastic that are manufactured in a long continuous section, while the term "sheet" is used to refer to relatively small pieces of plastics sheeting cut to dimensional sizes.

The consumer may find virtually every thermoplastic family represented in the versatile form of films. Common methods of manufacturing film include extrusion (both lip die and blow), calendering, and casting.

Plastics films have three broad categories for consumption: (1) packaging, (2) industrial, and (3) soft goods.

The largest use of film is for *packaging* applications. Packaging meats, breads, produce, toys, frozen foods, housewares, garments, and trash are only a small sample of the many established uses for plastics films. Boilable and sterilizable bags are growing in popularity.

Packaging meats in transparent polyvinyl chloride film permits the consumer to see the product and helps preserve the freshness and red color of the meat. Several films are semipermeable. They selectively allow certain materials to permeate or enter through the film walls. Oxygen must permeate the film package to keep the appealing red color in meats.

Container type plastics bags have grown in consumer popularity to package bread, cover garments, and contain trash.

Heat sealing and vacuum packaging is commonplace in most supermarkets. Cookies, hardware, meats and other products are packaged in plastics films that have been formed or shaped to help hold and display the product. The "blister" packaging of hardware and other products are familiar examples (Fig. 7-22).

Fig. 7-22. These toy premiums are in "blister" packages for a stand-up, stand-out promotion. (Celanese Plastics Co.)

Because polyvinyl alcohol film has a high water permeability, it is used as packaging dispensers for soaps and other chemicals. These products conveniently release their contents when placed in water.

Plastics in film form are used for numerous *industrial applications*. Film sheeting may be used as protective covering for plants, silage, and other agricultural applications. Fig. 7-23 shows degrad-

(A) Film applicator closeup.

(B) Rows of film mulch.

(C) Zucchini and eggplant in rows of mulch.

Fig. 7-23. Applying degradable plastics mulch on soil. (Princeton Chemical Research, Inc.)

able plastics mulch being placed on the prepared soil using a film applicator which is attached to a tractor. The front discs open up a shallow trough into which the edges of the film are pressed by the first pair of wheels. The spades cover the edges with soil and the rear wheels press the soil down. One farmer can apply about 32,000 sq ft (one acre coverage for 5-ft center rows) in 20 minutes.

Sheeting films may provide the necessary moisture barrier in various residential and commercial constructions.

The industrial market utilizes film for magnetic recording and adhesive tapes. Because of the excellent electrical resistance and dielectric properties of plastics, they are used as protective sheathing on electrical conductors.

Plastics are used in photographic films and as laminates on paper and cardboard. Films are veneered or clad to metal and wood as durable materials for countertops, walls, and exterior sidings of buildings.

Soft goods are produced from a variety of films. They have a fabriclike hand (feel) and drape like other woven textiles. These "fabrics" or soft goods are either unsupported or supported films. The supported film may have a woven or bonded backing. Polyvinyl chloride is a very popular film used for raincoats, dresses, handbags, shoes, shower curtains, and wall coverings. These films can be embossed to resemble woven fabric, leather, or other materials.

Foam backed or supported films are used by the automotive industry for upholstery. Balloonlike furniture and other types of home furnishings utilize these plastics films.

The leather industry has probably felt the greatest impact of plastics soft goods which have replaced much of the genuine leather used in garments, handbags, luggage, and shoes. Naugahyde, a tradename for a fabric supported polyvinyl film, is a soft but tough upholstery material. Corfam is also a tradename for a leathery polyurethane material reinforced with polyester. This material was manufactured by du Pont until 1970 in any thickness with more than a million micropores per square inch. Because of its porous structure, a broad line of materials are made of this type film including gaskets, seals, and leatherlike products.

The largest proportion of film is produced by a variation of the extrusion process called *blown film extrusion* (Fig. 7-24). The hot plastics is forced out a circular die opening, forming a tube.

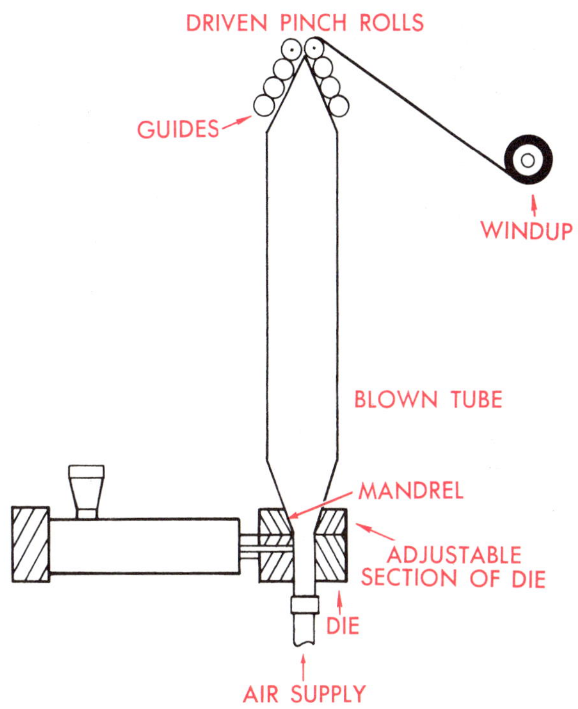

Fig. 7-24. Sketch of a blown film extrusion process. *(Modern Plastics Encyclopedia)*

Air is forced into this extruded tube, expanding it to the desired diameter. This blow or air expanding helps regulate the film thickness after it leaves the adjustable extruder die.

From a die opening of approximately twelve inches in diameter, emerges a thin walled tube about twelve feet in diameter. This air expanded tube may be wound onto takeup rollers or split and wound as flat film.

Another rapid method of producing film is *calendering*. Calendering is an old, widely established process used by many industries. It is a process where hot mixtures of plastics are pressed between a series of rollers. These rollers regulate the thickness of the film and may emboss a textured finish upon the stock (Fig. 7-25).

Much of the calendered film is used by the textile industry. The embossed or textured film is used to produce leatherlike apparel, handbags, shoes, luggage, etc.

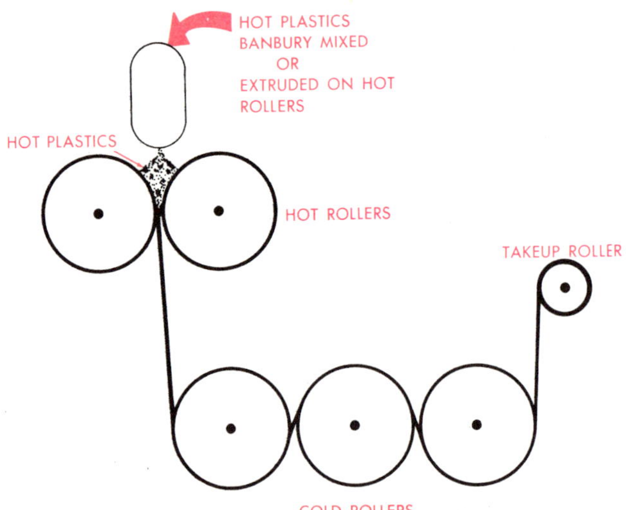

Fig. 7-25. Sketch of calendering process.

Some film is laminated to a layer of foam and fabric to produce new fashions with numerous applications for the sportsman, police, and other groups. There are two basic methods for producing bonded textiles. (See Fig. 7-26.)

The casting of film requires solvents rather than heat to cause the plastics to become liquid. Plastics are dissolved along with other additives and poured onto a polished surface of stainless steel. After the solvents have evaporated, the film is stripped from the polished surface and wound on a roll. The film may be cast directly on fabric or paper for special product applications. Fig. 7-27 shows two methods of solvent casting film.

Early photographic film and plates were produced by solvent casting of cellulose nitrate on smooth glass plates. When the solvents evaporated, the film could be stripped from the glass and further processed.

In order to be economically feasible, a solvent recovery system is necessary in solvent casting of film.

Solvent casting of film offers several advantages over other heat melt processes. Additives for heat stabilization and lubrication are not necessary. Films uniform in thickness, optically clear, and without orientation or stress are possible with this

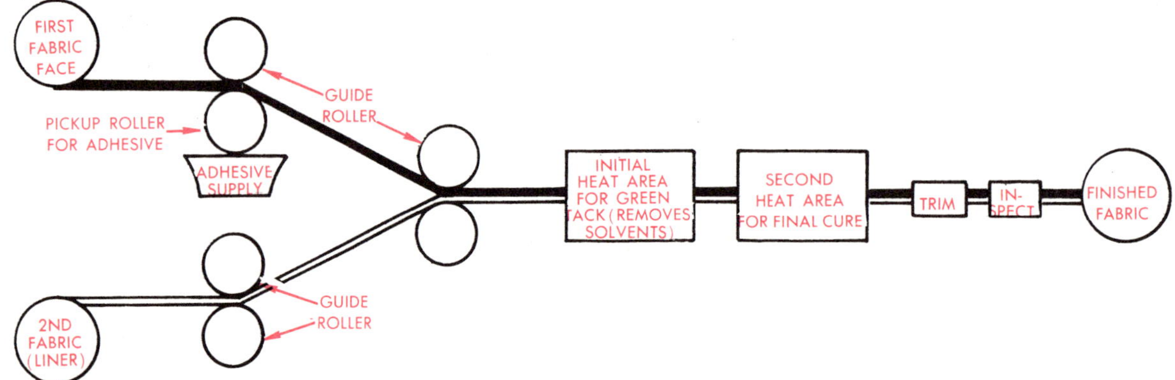

(A) Wet adhesive bonding.

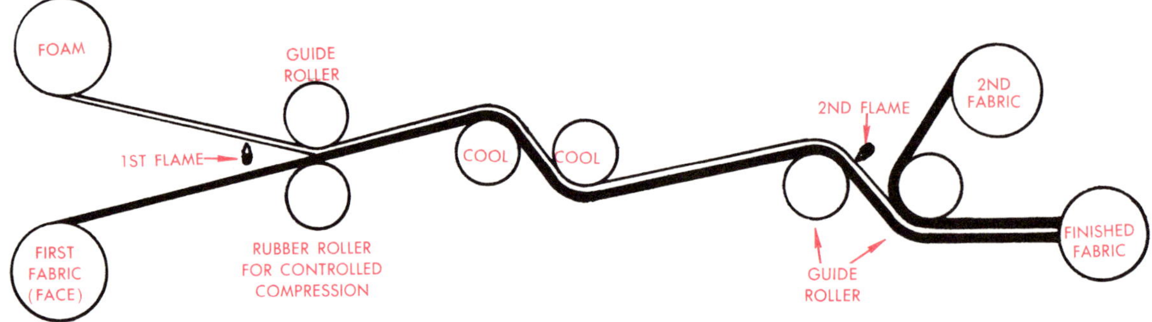

(B) Foam flame bonding.

Fig. 7-26. Two methods for producing bonded textiles. (*American Fabrics*, No. 68. Summer, 1965)

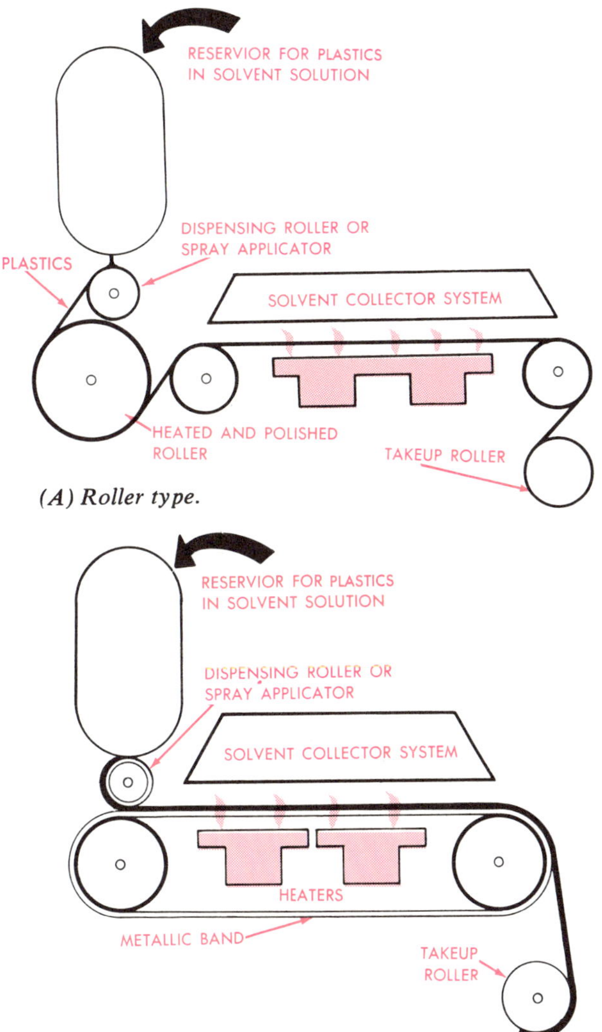

(A) Roller type.

(B) Band type.

Fig. 7-27. Two methods of solvent casting of film.

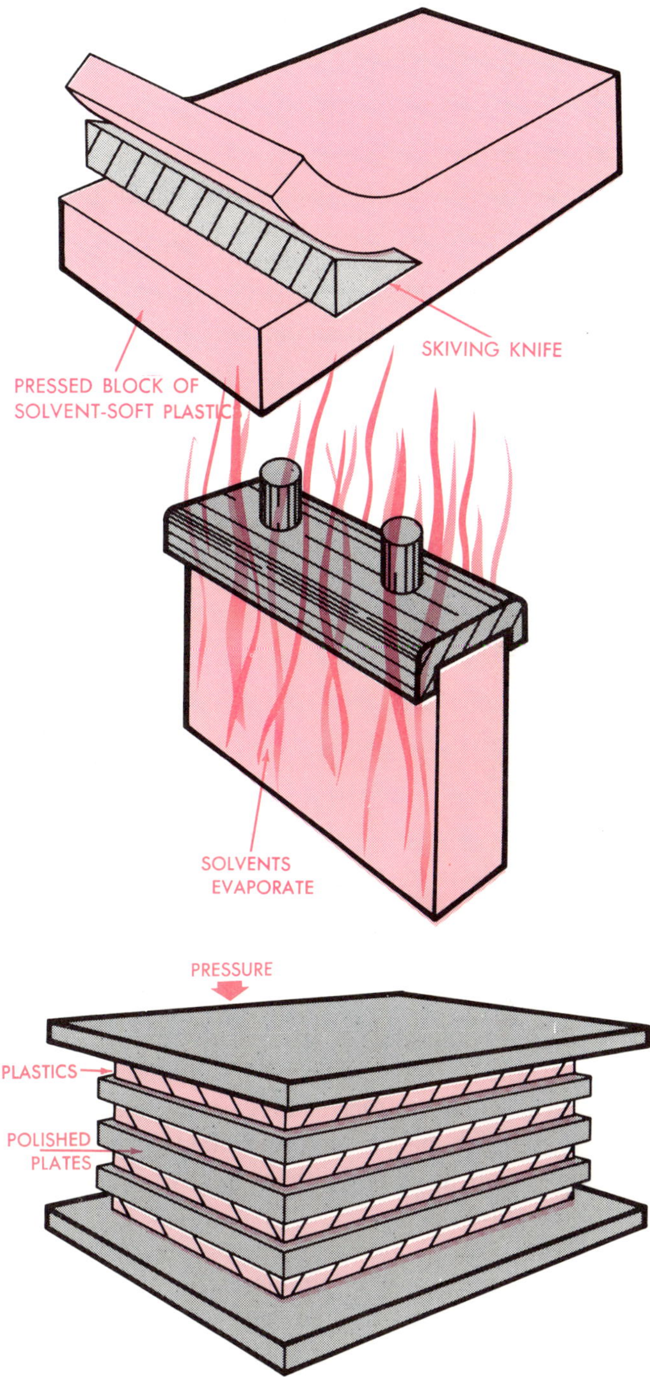

Fig. 7-28. Skiving sheets of plastics.

method. Solvent casting has a maximum allowable thickness of approximately 0.010 of an inch.

When heat processing or solvent casting cannot produce film or sheet stocks of required thickness, another process, called *skiving,* is used. Cellulose nitrate and other plastics may be sliced or skived from blocks of solvent-softened plastics. After the residual solvents have had time to evaporate, the skived piece is pressed between polished plates to improve the surface finish. (See Fig. 7-28.)

Filaments and Fibers

Everyone is an ultimate consumer of plastics filaments and fibers. Large volumes of these special plastics forms are used in the textile industry.

A *filament* is a single, long slender shaft of plastic. The fisherman is probably most familiar with "monofilament" fishing line. This single filament of plastics may be manufactured in any length desired.

Yarns may be composed of either a monofilament or multifilament shaft of plastics.

The term *fiber* is used to indicate all types of filaments, both natural and plastics, monofilament or multifilament. In other words, these fibers are first spun or twisted into yarns and then made into the finished fabrics ready for consumer use.

Many people are unaware of the significant role that plastics play in the textile industry. Natural fibers such as wool, cotton, flax, and silk have been used for years. The synthetic fiber industry is relatively new; in fact, all commercial fibers used before the twentieth century were of natural origins. Fig. 7-29 shows a popular application of plastics fiber.

Fig. 7-29. These ladies' wigs are made of 100 percent plastics fiber.

Because thousands of articles are produced from fibers, the textile industry is important to the world's economy. One source states that nearly thirty million people in the United States alone receive their livelihood either directly or indirectly from textiles.

There are a number of reasons that synthetic textiles have grown in popularity. (1) They may be produced year around with uniform quality. (2) They may be tailor made as to size, shape, luster, length, and other properties. (3) They may be made to simulate leather and fur. (4) They may be made to stretch or hold their shape (stretch fabrics and permanent press). (5) They may be used for carpeting both indoor and out. (6) They are mothproof, nonallergic, mildew resistant, and are free from odor when wet. All of these qualities and others make plastics fibers an important area for study.

Because of the apparent importance of the textile industry, more young men and women may choose to include the subject of textiles in their high school curriculum and consider it as a possible vocation. See Fig. 7-30.

Fig. 7-30. Look at the labels next time you visit a fabric center and notice how many plastics filament, yarn, and woven fabrics there are!

Not all the filaments are used by the textile industry. Some monofilaments are used for bristles in brooms, toothbrushes, and paint brushes. The relative shapes of these filaments vary as shown in Fig. 7-31.

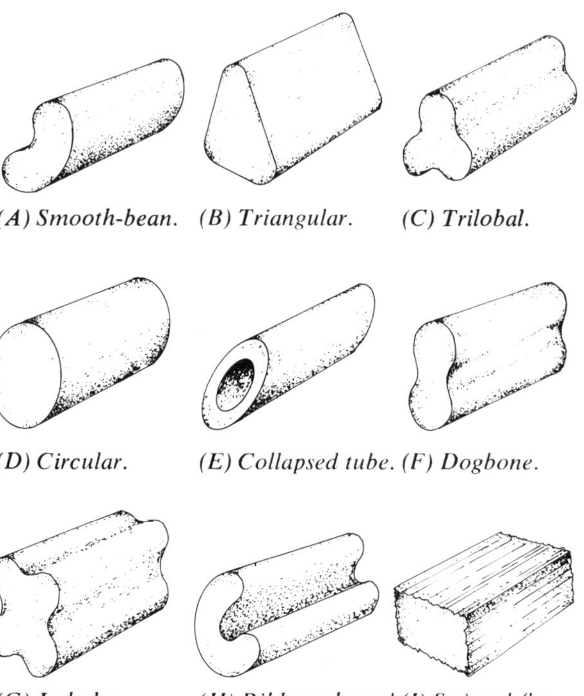

Fig. 7-31. Cross-sectional shapes of selected fibers.

The mass per length or fineness of a fiber is measured in a unit called a *denier*. One denier equals 4,464,528 yards of fiber per pound or the weight of 9000 meters of fiber in grams. For example, 9000 meters of a 10-denier yarn weigh 10 grams. The name of the unit may have evolved from the name of a weight of a 16th century French coin used as a standard for measuring the fineness of silk fibers.

Plastics fibers vary in denier because of the difference in density. Two filaments, for example, may have the same denier but one may have a larger diameter because of a lower specific gravity.

To figure the denier of filaments in yarns, divide the yarn denier by the number of filaments:

$$\frac{\text{80-denier yarn}}{\text{40 filaments}} = \text{2 denier for each filament}$$

REMEMBER: 1 denier of 9000 meters weighs one gram.
2 denier of 9000 meters weigh two grams.

If the fiber is to be pliable and soft, a fine filament is needed. If the fiber is to withstand crushing and be stiff, a thicker filament is used. Clothes fibers, for example, range from two to ten denier per filament. Fibers for carpeting range from 15 to 30 denier in fineness per filament.

The cross-sectional shape of the fiber helps determine the texture of the finished product. Triangular and trilobal shapes impart many of the properties of the natural fiber, silk. Many of the ribbon- and bean-shaped filaments resemble cotton fibers.

Filament shapes are made by forcing plastics through small orifices or openings. This special extrusion process is referred to as *spinning* because the plastics is shaped by the opening in the die or *spinneret*.

These spinnerets are often made of platinum or other metals that will resist acids and orifice wear. These orifices are often much finer than the diameter of human hair. This process may have been named "spinning" from the primitive method of spinning natural fibers. The small opening under the jaw of the silkworm is also called the "spinneret." In order to be forced or extruded out these small openings, the plastics must be made fluid.

Acrylic fiber is produced in the spinning process shown in Fig. 7-32. A thick chemical solution is extruded into a coagulation bath through tiny holes in a spinneret. In the bath the solution coagulates (becomes a solid) and this is Acrilan acrylic fiber. The fiber is washed, dried, crimped, cut into staple lengths, and baled for shipment to textile mills, where it will be converted into carpeting, wearing apparel, and many other products.

Fig. 7-32. The production of acrylic fiber.

There are three basic methods of spinning the fibers. (1) Polyethylene, polypropylene, polyvinyl, polyamide, and thermoplastic polyesters are melted and forced out the spinneret. As the filaments hit the air, they solidify and are passed through other conditioners. This process is called *melt spinning* (Fig. 7-33A). (2) In *solvent spinning,* plastics such as acrylics, cellulose acetate, and polyvinyl chloride are dissolved by selected solvents and forced through the spinnerets (Fig. 7-33B). The filament then passes through a stream of hot air which aids in evaporating the solvents from the slender fibers. To be economical these solvents must be recovered and reused. (3) The first step of *wet spinning* (Fig. 7-33C) is similar to solvent spinning. The plastics is dissolved in selected chemical solvents. This fluid solution is then forced out the spinnerets and into a coagulating chemical bath. The chemical bath makes the plastics "gel" into a solid filament form. Selected members of the cellulosic, acrylic, and polyvinyl plastics families may be processed by wet spinning.

All three of the processes begin by forcing fluid plastics through a spinneret and end by solidifying

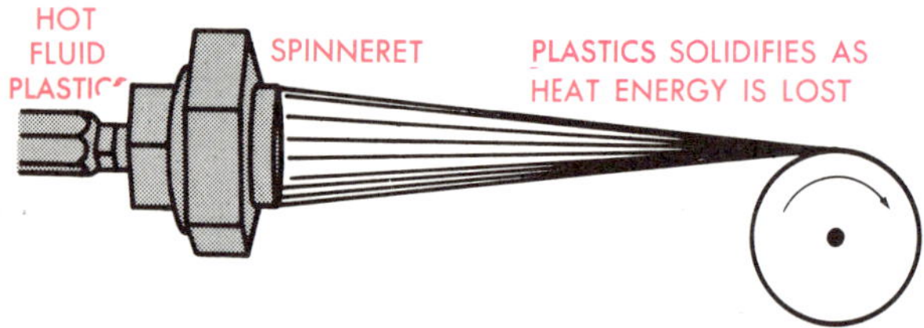

(A) Melt spinning.

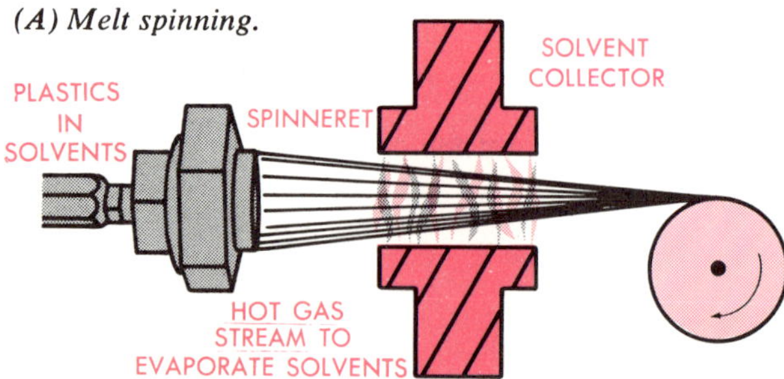

(B) Solvent Spinning.

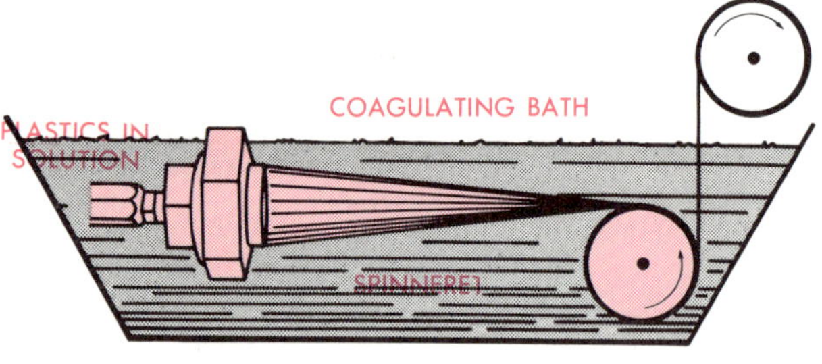

(C) Wet spinning.

Fig. 7-33. Three basic methods of spinning plastics fibers.

the filament by methods of cooling, evaporation, or coagulation.

The strength of the individual filaments may be determined by several factors. (1) Most of the selected filament fibers are linear and crystalline in composition. When groups of molecules lay together in long chains and these long molecular chains are parallel, there are additional strong bonding sites available. (2) By forcing the liquid plastics out the spinnerets many of the molecular chains are forced closer together and parallel to the filament axis. This packing, arranging and drawing provides increased strength throughout the filament. (3) By mechanically working the filament, further molecular orientation and packing can be accomplished. This mechanical process is called *drawing*. The drawing of noncrystalline plastics also helps to orient molecular chains and thus improve strength.

Drawing or stretching of the plastics is accomplished by running the filaments through a series of variable-speed rollers. The drawing of crystalline plastics is continuous. As the fiber is sent through the drawing process, each roller is rotated at a faster speed. The speed of these rollers determines the amount of stretching or drawing. Fig. 7-34 will help to depict this action.

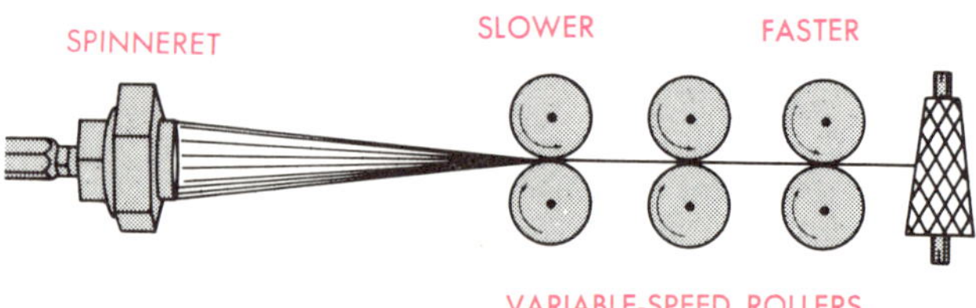

Fig. 7-34. Drawing plastics filaments.

High-bulk fibers and yarns may be produced from films. In this process a composite film which has been coextruded or laminated is mechanically drawn and cut (fibrillated) into fine fibers. The fibrillation is accomplished during the stretching or drawing process as the film passes between serrating rollers of different circumferential speeds. The "teeth" of these rollers cut the film into fibrous form (Fig. 7-35). The composite film develops internal stresses as it is extruded, drawn, and fibrillated. This unequal stress orientation in the film layers induces the fibers to curl and exhibit properties much like natural fibers.

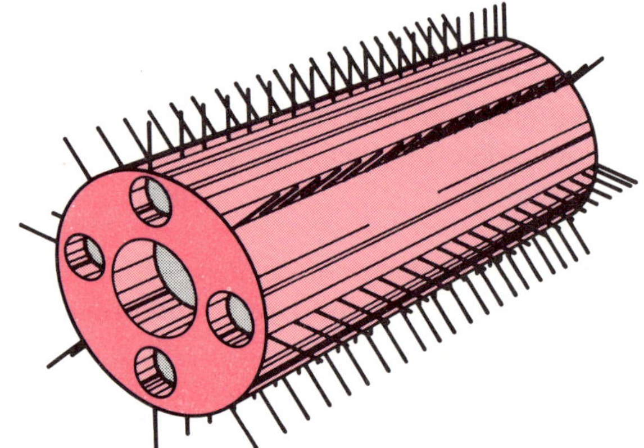

Fig. 7-35. This fibrillating device has rows of pins in roll form and produces fibrillated yarn of very fine denier.

The future for plastics fibers will undoubtedly expand the many known uses and properties of plastics. Besides the manufacture of textiles, these fibers are used in the manufacture of automobile upholstery, marine nets, ropes, screening, carpets, draperies, and flock.

Some of the outstanding characteristics of plastics filaments may be exhibited by the performance of fibers used by the carpeting industry. Carpets are used in homes, schools, hospitals, churches, hotels, and other areas of severe traffic wear. These fibers are often exposed to wet soil, harsh sunlight, extreme atmospheric conditions, and insects, and yet retain their color, resilience, and durability. They are nonallergenic, come in any color or texture, and are densely woven.

There are two complaints lodged against plastics fibers: flammability and low moisture absorbency. Fabrics made of wool and fiber glass, for example, are "fire retardant." This means that they will burn slowly. Cotton and cellulosic plastics fibers are "flammable or inflammable"; both terms indicate the fabric will burn. Cigarette sparks and glowing embers will melt holes in most synthetic fabrics. Very sheer or napped fabrics are more likely to burn than tightly woven ones.

Because most of the plastics fibers absorb only small amounts of moisture, they have an important bearing on health and comfort. Wool and cotton absorb moisture readily. By selective construction of the fiber weave and by chemically treating the fibers, the absorptive capacity of most synthetic fibers can be increased.

Casting Compounds

Casting compounds are another form of plastics familiar to many. These compounds may come as syrupy liquids in monomer form, as dispersions of plastics in a solvent, or as solid plastics forms which are melted and cast much like candle wax.

Because the casting of plastics is one of the simplest of all the molding techniques and requires no molding pressure, it has been a very popular hobbycraft and artcraft casting material. Castings of unusual or complicated shapes are possible with this technique because glass, rubber, plaster, or metal molds may be used. Many of these molds are used only once.

Although many plastics may be obtained in syrupy resin form, polyester, epoxy, acrylic, and polystyrene casting resins are among the most important. Polymerization of these resins is carried out by the addition of catalysts and/or heat.

Phenolic castings were part of the early development of the plastics industry when Baekeland introduced numerous cast articles. Today, cast phenolic resins are used to produce most of the world's billiard balls. Cast phenolics have replaced many ivory and horn items.

Large amounts of fillers and reinforcements are generally used in casting polyesters. "Cultured marble" is a product which contains marble dust as the filler and polyester plastics as the binding material. This product is reportedly superior to genuine marble articles and is used to produce paperweights, lamp bases, table tops, exterior veneers, statues, and other marble products (Fig. 7-36). Literally thousands of items are fabricated from clear polyester resins.

Fig. 7-36. This one-piece sink is made of filled plastics and cast to resemble marble.

Acrylic plastics are seen as cast rods, sheets and tubing. A monomer of partially polymerized syrupy resin is used for this purpose.

A major use of all casting resins and plastics has been in encapsulation, embedment, and potting. Polyester and acrylic resins are used to encase various objects in clear, transparent plastics for the purposes of preservation, display, and study. In the biological sciences, animal and plant specimens are embedded to help preserve the specimen and allow handling of the most fragile sample. Embedments are always placed in a transparent plastics (Fig. 7-37).

Fig. 7-37. These objects are embedded in transparent plastics.

Polyester and epoxy compounds are sometimes used for potting and encapsulation.

Potting is used to protect electrical and electronic components from poor environments and possible damage. The potting process completely encases the desired component in plastics (Fig. 7-38). In order that all voids are filled with resin, vacuum, pressure, or centrifugal forces are frequently applied.

Fig. 7-38. Potting an electronic unit with silicone plastics (Dow Chemical Corp.)

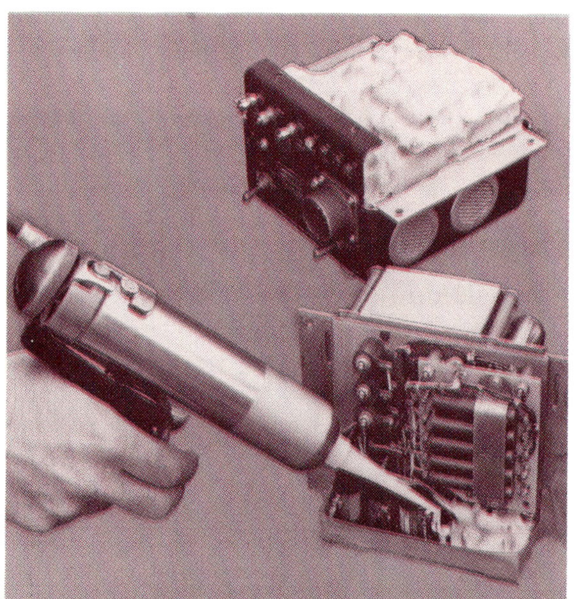

Encapsulation is similar to potting, but encapsulation is only a covering on electrical components (Fig. 7-39) and does not fill all the voids. This envelope of plastics is usually applied by immersing the object in the casting resin. After potting, many components are also encapsulated.

(A) Of transformer with silicone elastomer.

(B) Of electronic components into convenient modules.

Fig. 7-39. Examples of encapsulation. (Dow Corning Corporation)

Potting and encapsulation sometimes employ a second form of casting compounds. These are referred to as *hot-melt* plastics. They are already polymerized and in the solid plastics state. Many of the hot-melt adhesive plastics may be utilized. Cellulose acetate, ethyl cellulose, polyolefins, polyamides, and other plastics are heated to a syrupy state and forced around the electrical components.

Another variation using fully polymerized casting compounds is *rotational casting*. Small pellets or powders of plastics are placed in hollow metal molds and rotated in two planes simultaneously. As the metal mold begins to heat, the plastics melts on the mold walls. The rotation causes the plastics to flow over the surface of the mold. Seamless, hollow objects are possible with this process (see Fig. 7-40).

These pellets and powders may be cast in flat sheet form by placing the plastics powders on flat pieces of glass or metal and applying heat.

(A) Doll parts.

(B) Toy balls.

(C) One-piece modern chair.

Fig. 7-40. Examples of rotational casting.

The last group of casting compounds are usually dispersions of plastics in selected solvents, plasticizers, and other additives. Centrifugal, rotational, slush, and solvent castings are often employed utilizing these compounds. Organosols and plastisols are plastics compounds most commonly associated with these processes.

Organosols are vinyl or polyamide dispersions in organic solvents and plasticizers (Fig. 7-41). A plastisol may be converted to an organosol by the addition of selected solvents.

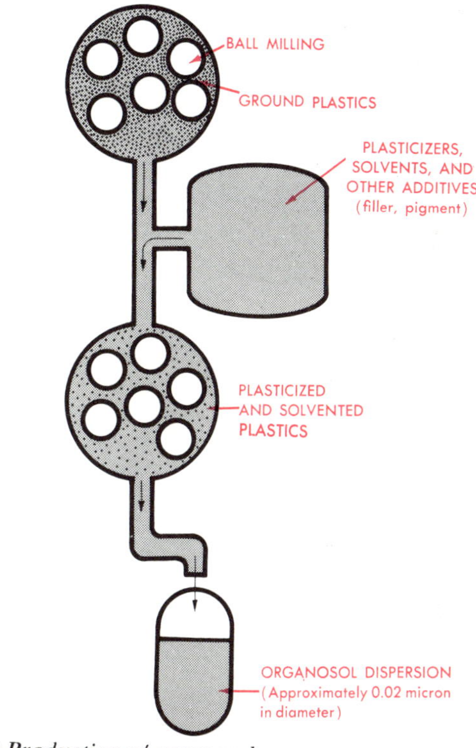

(A) Production of organosols.

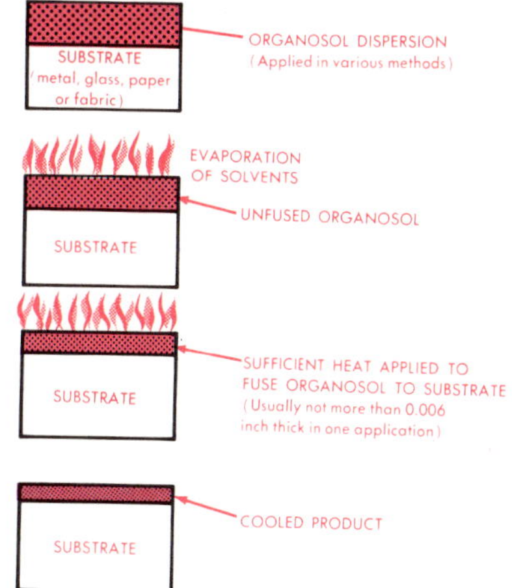

(B) Fusion of organosols on substrate.

Fig. 7-41. Organosols.

Plastisols (Fig. 7-42) may be nearly 100 percent solids because they contain only plasticizers. Organosols may be 50 to 90 percent solids because they contain both plasticizers and varying amounts of solvents. Both compounds may contain thixotropic materials. When these agents are added, a paste form called *plastigel* or *organogel* is formed. Plastigels and organogels are easily spread. They are probably associated more often with the plastics coatings on paper and fabrics.

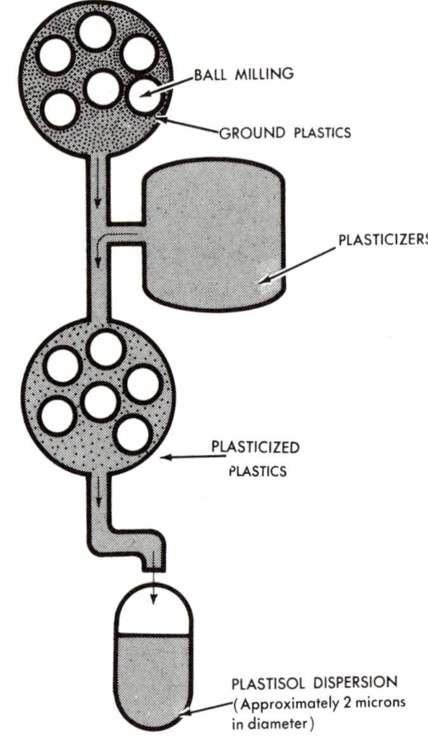

(A) Production of plastisols.

SUBSTRATE (metal, glass, paper)
PLASTISOL DISPERSION (applied in various methods)
SUBSTRATE
SUFFICIENT HEAT APPLIED TO FUSE PLASTISOL TO SUBSTRATE
(Usually not more than 0.250 inches thick in one application)
SUBSTRATE
COOLED PRODUCT

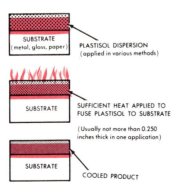

(B) Fusion of plastisols to substrate.

Fig. 7-42. Plastisols.

Plastisols may be cast in hollow, open metal molds or in closed rotational molds. Plastisols are converted to the final product by the addition of heat.

Organosols are cast and solvents allowed to escape. The dry, unfused plastics is left on the substrate. Heat is then applied to fuse the plastics remaining on the mold surface. Many modified casting procedures are used by the plastics industry today.

Dip casting is a simple process where a heated mold is dipped into liquid dispersions of plastics. The plastics melts and adheres to the hot metal surface as the mold is withdrawn. Additional curing is required in ovens to ensure proper fusion of the plastics particles. After this curing operation, the plastics is peeled from the mold and the process repeated.

Coatings

Many plastics come in the form of *coatings*. The main goal in the formulation of coatings is to form a film over the substrate to which it is applied. Sometimes coating and casting processes are overlapping and confusing. Casting compounds are applied and then *removed* from the molds when polymerized or cooled. Coatings are normally *thin protective coatings* on a substrate and are *not removed*.

Every polymer and plastics has been considered for use as coatings. The paint, varnish, and lacquer industries utilize large quantities of plastics compounds. Plastics based paint, varnish, and lacquers are generally far superior to their natural counterparts. One-coat house paints, tough transparent varnishes, both stain and water resistant, and beautiful quick-drying finishes are all advents of the plastics age.

Plastisols and organosols are used as coating materials. They may be applied by several processes. A simple dipping or submerging of articles in plastics dispersions is easily done and requires little equipment.

Numerous products are coated in this manner and include coated fabric gloves, dish drainers, tool handles, bobby pins, and many other special applications.

Plastics coating of paper and fabric is applied by several methods (Fig. 7-43). The "knife" and roller method is commonly used for these coating applications.

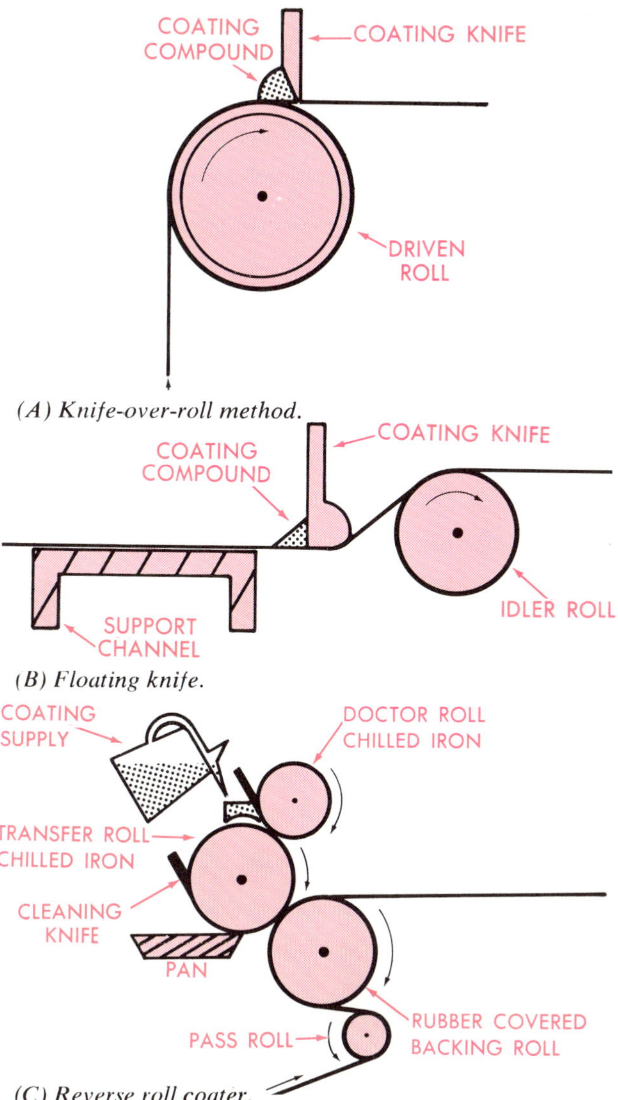

(A) Knife-over-roll method.

(B) Floating knife.

(C) Reverse roll coater.

Fig. 7-43. Plastics coating on paper and fabric. (John Waldron Corp.)

Thermoplastic compounds are sometimes applied to metal surfaces by the *fluidized* bed and *spray* processes. In the fluidized bed process, preheated objects are placed into dry plastics powders. As the powders strike the hot surface of the object, they melt and fuse. With the spray process, hot parts are simply sprayed with fine, dry plastics powders. Electrostatic spray methods are often employed to eliminate overspray, ensure even coatings, and produce coatings without preheating the part. Hundreds of products are coated with powder coatings, including outdoor fencing, chemical tanks, plating racks, and numerous appliance parts for dishwashers, refrigerators and washers.

To ensure that replacement parts arrive or may be stored in any part of the world, stripable coatings are placed on gears, guns, and other hardware. Any time machined or polished surfaces need protecting during fabrication, shipping, handling, or storage, stripable coatings may be used. They are formulated with good cohesion but relatively poor adhesion so that they may be stripped or peeled from the part when needed. Stripable coatings are sometimes used as masking films in electroplating or applying paints.

Cellular or Foamed Rubber

Cellular or foamed rubber has been in existence for a number of years. Most thermoplastic and thermosetting plastics may be processed into a cellular form.

There are two general types of foamed plastics relative to cellular structures. (1) *Open-celled* foams are like sponges. The cells are interconnected and capable of holding liquids. (2) If the cells are completely closed by thin walls of plastic, the structure is referred to as *closed-celled* foam.

Fig. 7-44 shows both open and closed cellular plastics. In the closeup, the cellular plastics on the left is closed-celled, while the two on the right are open-celled.

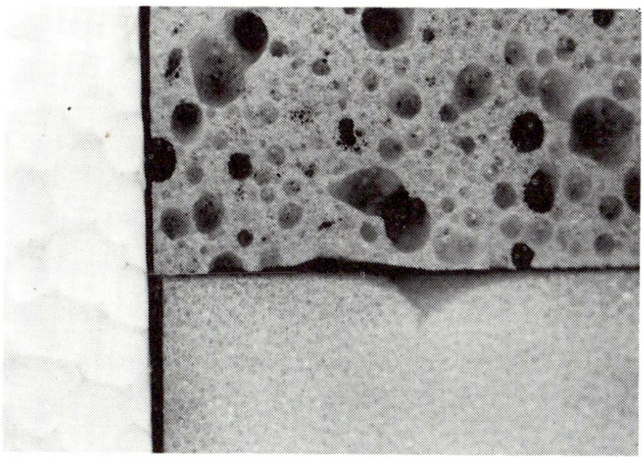

Fig. 7-44. Open-celled (right) and closed-celled (left) cellular plastics.

There are basically three broad methods of foaming resins and plastics: mechanical, physical, and chemical. (1) Air may be *mechanically* whipped into many resins just prior to polymerization. Whipping air into egg whites and baking is a simplified analogy. (2) A chemical with a low boiling point, usually below 110 degrees Celsius, may be added to the resin or plastics prior to processing. This chemical, either liquid or solid, may change *physical* state when its boiling point is reached and cause gaseous bubbles to form in the resin (Fig. 7-45). (3) Chemical foaming agents are compounds which yield or liberate a gas by decomposition at room temperature or with the addition of heat. Both the chemical and physical foaming agents are referred to as *blowing agents* because a gas is released by them to form the cellular structures.

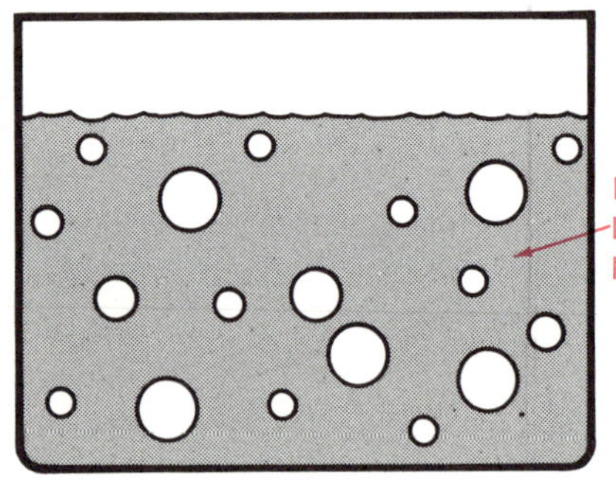

Fig. 7-45. Gaseous bubbles are formed as chemicals change physical states.

Fig. 7-46 shows two applications of foamed polystyrene, a popular cellular plastics.

A special fourth class of foamed plastics is sometimes included. This special type of foam utilizes "microballons" or hollow spheres of glass or plastics with an average diameter of 0.0013 inch, to form the cellular structure. Various microballons and fibrous fillers have been developed for use in *syntactic foams*. These syntactic foams are produced by blending these fillers with a basic resin such as polyester, phenolic, or epoxy. (See Fig. 7-47.)

More than ten families of foamed plastics are commercially manufactured in the United States. Many of these plastics foams may be obtained in rigid, semirigid, or flexible foams. Many may be made flame retardant.

(A) Cellular for foamed polystyrene has excellent thermal insulating qualities.

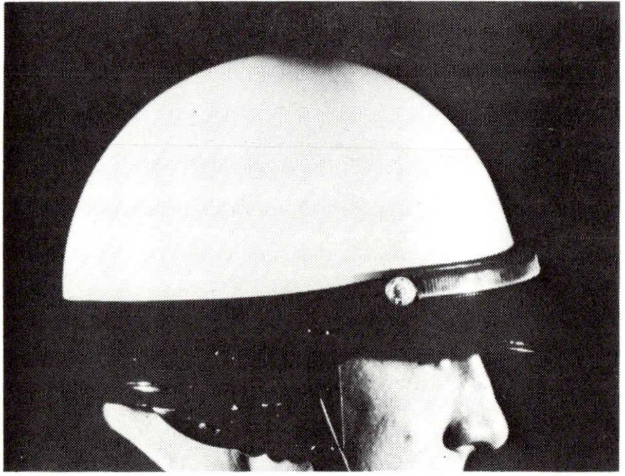

(B) The inner liner of this crash helmet is molded of expanded polystyrene.

Fig. 7-46. Applications of polystyrene. (Sinclair-Koppers)

Fig. 7-47. Syntactic foam.

Foamed polystyrene and the tradename Styrofoam are probably the most familiar of the foams. Because nearly all the foams are waterproof and have low densities, they are used for all types of thermal insulation, acoustical insulation, cores for sandwich structures, electrical insulation, lightweight toys, and containers.

Reinforced and Laminated Plastics

Reinforced and laminated plastics are fulfilling a number of functions. The automotive, boating, architectural, and aerospace industries utilize these forms because they may be formulated to meet varying needs and because they offer physical, chemical, and electrical properties superior to other materials. Nearly all of the families of plastics may be economically reinforced or used for laminating.

The major confusion in using the term "reinforced plastics" is that the term is sometimes used to discuss laminated plastics and molded products in which the reinforcement is not in laminated form. In other words, if the reinforcing is placed in layers it may be called a *reinforced plastics laminate*. If the reinforcement is not deposited in layers or piles, the resulting form should be called a *reinforced plastics*. The word "reinforced" is used to indicate that an additive or special ingredient (a reinforcement) was added to the molding compound. It does not refer to a specific molding or forming process. The words "reinforcing" and "laminating" should be used to describe a process, while "reinforced" and "laminated" should be used to describe a plastics form. Reinforced plastics products are produced by several processing operations, including low-pressure and high-pressure lamination compression, transfer, injection, and rotational molding.

Although fibrous glass is by far the most used reinforcing agent, asbestos, paper, and numerous other filaments are also used in large quantities. Fig. 7-48 shows some examples. Diallyl phthalate resin, a thermosetting material widely used to mold electrical and electronic components in military and industrial applications, has entered a nonelectrical consumer market. A stable, white asbestos-filled resin is now being used to mold the rear handle cover of this household iron (Fig. 7-48A). The asbestos improves strength and heat resistance.

The fibrous glass reinforced polyacetal parts in Fig. 7-48B have improved tensile strength and deflection temperatures. Mold shrinkage, thermal expansion, and creep are also improved.

The front plates of a portable writing recorder (Fig. 7-48C, left) are transfer-molded from a glass reinforced diallyl phthalate resin-based compound to reduce production costs and improve instrument performance. The plastics front plate redesign (right) replaces a machined aluminum part.

In Fig. 7-48D the pipe and duct fittings have been developed to afford advantages of polypropylene with the strength of a fibrous glass reinforcement. This material has been developed to afford the pipe and duct industry with a construction combining the advantages of polypropylene with the reinforcement of glass fibers dispersed through the wall of the polypropylene (Fig. 7-48D). This combination results in a highly chemical resistant, strong, lightweight pipe and duct designed to meet the growing demands for corrosion control and air pollution control products.

The term "laminate" may appear confusing at first but it is used when two or more layers of materials are bonded together by either cohesive or adhesive action. These layers of material may function as reinforcing agents, much like the wire screen and steel rods used to strengthen concrete, or as decorative surfaces on furniture or table tops.

Another reason for the confusion is that the plastics industry is using the term to refer to other nonplastics laminates. Laminated wooden beams and plywood are typical "wood" laminates in which the wood is the reinforcing agent and the plastics the bonding agent (Fig. 7-49).

The plastics "resin" does not necessarily have to impregnate the materials to be considered a laminate. Laminates may be tubes, rods, sheets, or other molded shapes. Sheets of aluminum, steel, plastics, wood, or fabric are commonly laminated to produce special parts with improved properties. Fig. 7-50 shows illustrations of commercial, military, and private radar domes (radomes) of composite laminates of plastics and other materials.

Composite laminates of plastics and other materials are used in aircraft structures, helicopter blades, and as structural materials for automobiles, furniture, bridges, and homes (Fig. 7-51).

Thin metallic layers of foil may be clad or bonded to layers of paper or fabric. These lam-

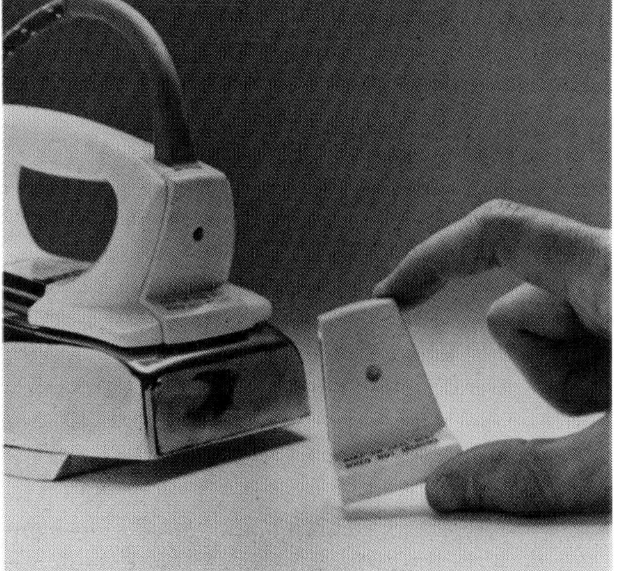

(A) Rear handle cover in asbestos-filled resin. (FMC Corporation)

(B) Fibrous glass reinforced polyacetal. (E. I. du Pont de Nemours & Co., Inc.)

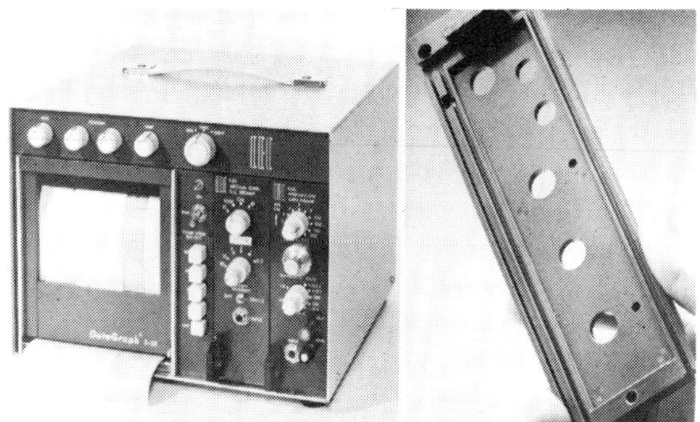

(C) Glass reinforced diallyl phthalate front plates. (FMC Corporation)

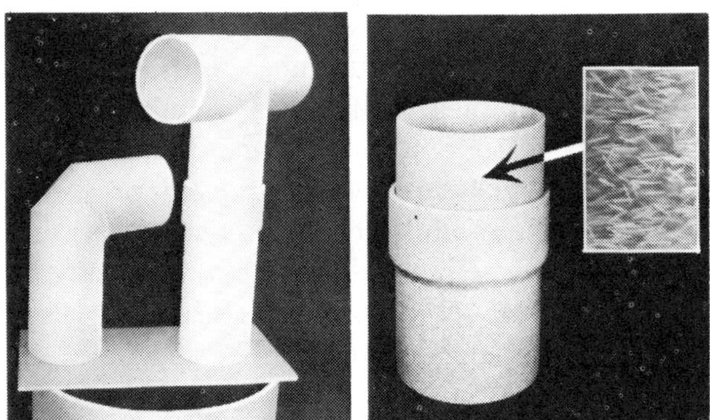

(D) Polypropylene copolymer and homopolymer in pipes and pipe fittings (left), with closeup at right. (Druid Plastics, Inc.)

Fig. 7-48. Various reinforced plastics.

Fig. 7-49. In this plant large laminated beams and arches are being fabricated. (AITC photo)

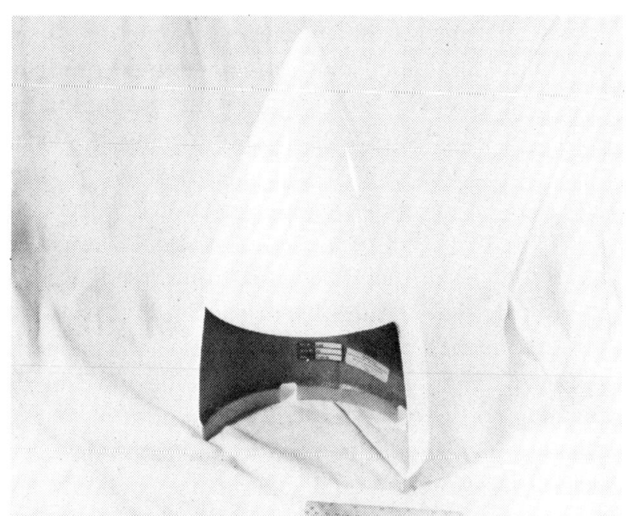

(A) Aircraft-missle countermeasure radome. (McMillan Radiation Labs, Inc.)

(B) Plastics radome (left) protects the radar antenna. (FMC Corporation)

Fig. 7-50. Radomes of composite laminates.

inated "boards" produce strong, corrosion- and electrical-resistant materials for printed circuits used by the electronics industry (Fig. 7-51B).

(A) Over 60 million pounds of plastics will be used in aircraft this year. The strong lightweight laminates account for much of this use.

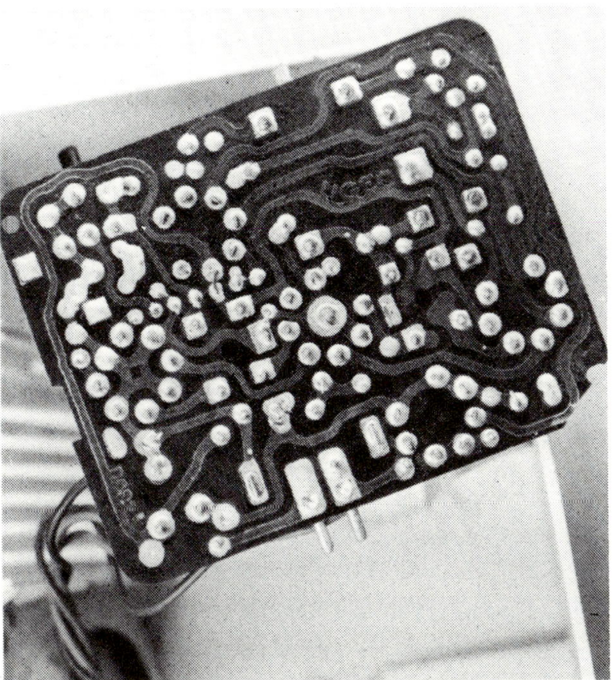

(B) High-pressure laminate is used as electronic circuit board in radio.

Fig. 7-51. Some of the versatility of laminates.

There are numerous applications of laminates in the textile industry, where layers of cloth, plastics film, and foam are bonded together to produce special purpose materials.

Coextruded composite films of polyethylene and vinyl acetate produce a tough, durable two-layer film which may be heat sealed. The two or more plastics layers are brought together in the molten state and extruded out a single die opening to produce the multilayered laminate film (Fig. 7-52).

Plastics films are often combined with paper, cloth, and metal foil for unlimited varieties of properties.

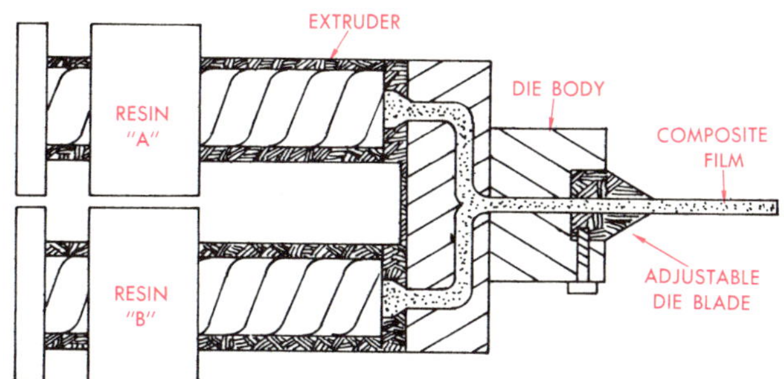

Fig. 7-52. Production of coextruded film.

Many laminates are composed of layers of resin-impregnated materials where the pores of interstices of the material are filled with resin (Fig. 7-53). Once the resin is polymerized, there is a strong cohesive bond formed between layers of resin. The resin has completely impregnated the reinforcing material.

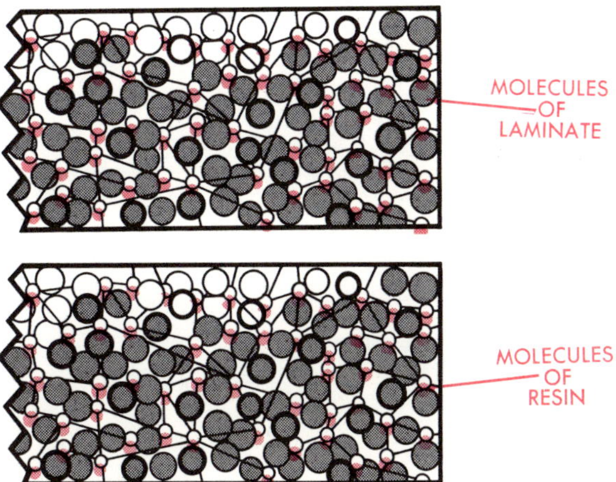

Fig. 7-53. Resin impregnated material.

In order for an impregnated material to be classed as a laminate, it must be composed of two or more resin soaked layers. "Particle and fiberboard" materials are not laminates. They are resin impregnated wood or cellulose fibers, pressed into a board or sheet form.

Formica, Micarta, and Texolite are tradenames for a familiar laminate form. These laminates are used as countertops, wall panels, flooring, and for other applications where a decorative and durable surface is needed. Most of these laminates are composed of several layers of resin impregnated paper. The surface may be textured or patterned by several methods or it may contain metallic flakes, abrasives, or be embossed to simulate wood grain, fabric, or stone.

Sandwich and cellular laminates hold great promise for the future in residential home fabrication. Fig. 7-54 shows the roof of a prototype four-bedroom home which is mechanically fastened and requires no nails. Panels, shown during installation, have rigid polyurethane foam core tapering from 4½ to 3 inches to give a 1-in-12-inch pitch to the otherwise flat roof surface.

(A) Installing panels.

(B) Sealing panels.

Fig. 7-54. Panelled polyurethane-foam roof requires no nails.

Although conventional in appearance, the home in Fig. 7-55 was made with construction techniques which were completely unconventional. The 5000 square feet of panels for walls and roof have aluminum exterior and polyurethane foam core. They were produced in a plant, trucked to the site, and assembled with screws.

(A) Installing walls.

(B) Construction of roof.

(C) Front view of home.

Fig. 7-55. Home with polyurethane-foam and aluminum panelled walls and roof.

Many small "prefabricated" structures are presently being produced with a "honeycomb" sandwich or reinforced plastics (Fig. 7-56).

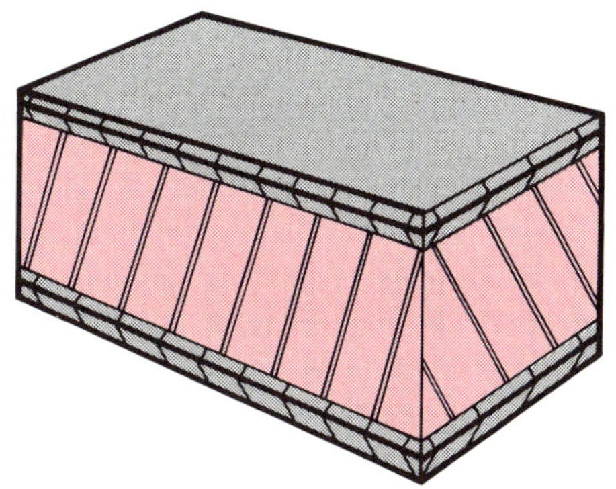

(A) Basic design.

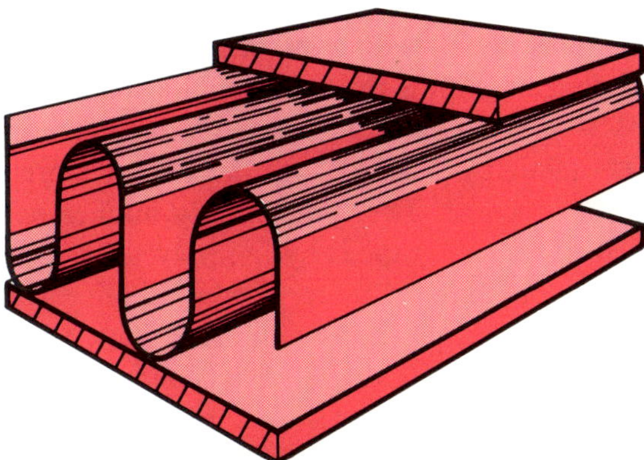

(B) Corrugated paper core.

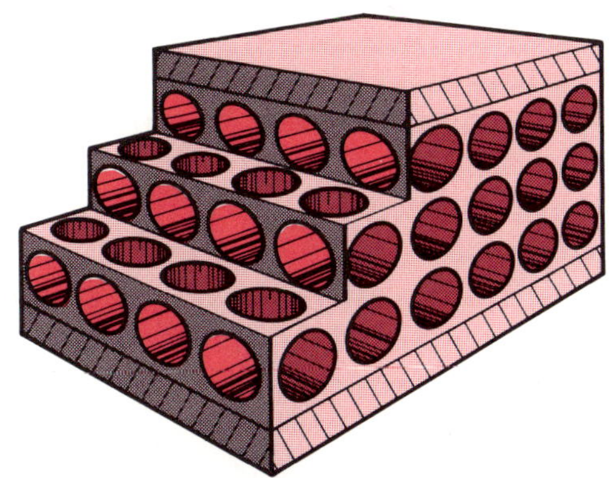

(C) Cellular plastics core.

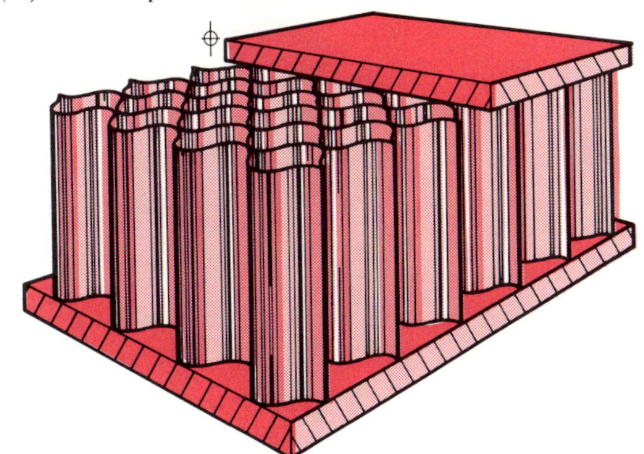

(D) Honeycomb core.

Fig. 7-56. Types of core constructions.

Cellular and sandwich construction is nothing new but it is beginning to become attractive to the consumer. Trailerhomes, automobiles, and the aircraft industry utilize these sandwich structures because they are extremely strong for their weight. Fig. 7-57 shows a paper honeycomb sandwich material beefed up by fiber glass reinforced polyester. This sandwich material cuts fabrication costs, maintenance, and reduces weight by over 20 percent compared to previously used materials. It is now making inroads into three markets currently dominated by aluminum, steel, wood, and other materials. It essentially consists of fiberglass reinforced polyester skins around a phenolic impregnated paper honeycomb interior.

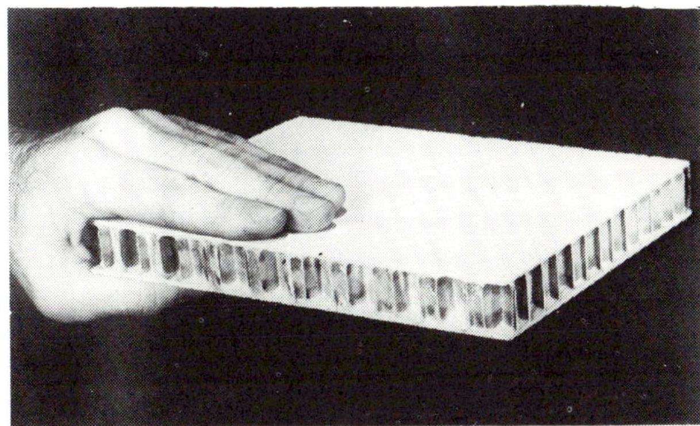

Fig. 7-57. New honeycomb material enters truck trailer, machine container, and modular housing field. (Panel-Comb Industries Corp.)

Glossary of New Terms

Adhesion—The state in which two surfaces are held together by interfacial forces which may consist of interlocking action (mechanical means).

Adhesives—A substance capable of holding materials together by surface attachment.

Calendering—To produce film or sheets of material by pressure between two or more counter-rotating rolls.

Cellular or foamed—Resins in sponge form. The sponge may be flexible or rigid, the cells closed or interconnected, the density anything from that of the solid parent resin down to 2 lbs per cubic foot. The terms *cellular, expanded,* and *foamed plastic* are used synonymously; however, "cellular" is most descriptive of the product.

Cement—To bond together as to adhere with a liquid adhesive; a liquid adhesive employing a solvent base of the synthetic elastomer or resin variety.

Closed-cell—Describing the condition of individual cells which make up cellular or foamed plastics; cells not interconnected.

Coating—A form of plastics or process describing various technologies for applying a thin layer of plastics to a substrate.

Cohesion—The propensity to adhere to itself; the internal attraction of molecular particles toward each other; the ability to resist partition from the mass.

Denier—The weight in grams of 9000 meters of synthetic fiber in the form of a single continuous filament.

Drawing—The process of stretching a thermoplastic sheet, rod, film, or filaments to reduce its cross-sectional area and change its physical characteristics.

Embedment—Enclosing an object in a closed envelope of transparent plastics by immersing the object in a casting resin and allowing the resin to polymerize.

Encapsulation—Enclosing an article, usually an electronic component, in a closed envelope of plastics by immersing the object in a casting resin and allowing the resin to polymerize or, if hot, to cool.

Fibrillating—The process of cutting or forming films into filaments.

Fibers—This term usually refers to relatively short lengths of very small cross sections of various materials. Fibers can be made by chopping filaments.

Filament—A variety of fiber characterized by extreme length, which has little or no twist and is usually produced without the spinning operation required for fibers.

Foamed—*See* Cellular.

Foaming agents—Chemicals added to plastics and rubbers that generate inert gases on heating, causing the resin to assume a cellular structure.

Glue—An adhesive prepared from hides, tendons, cartilage, bones, etc., of animals by heating with water. Through general usage, the term is now synonymous with the term "adhesive."

Honeycomb—Manufactured product of metal, paper, plastics, etc., which is resin impregnated and has been formed into hexagonal shaped cells. Used as a core material for sandwich type construction.

Hot melt—A general term referring to thermoplastic synthetic resins used as 100 percent solid adhesives at temperatures between 250°F and 400°F.

Impregnated—The process (impregnation) of thoroughly soaking a material such as wood, paper, fabric, etc., with synthetic resin so that the resin gets within the body of the material.

Laminates—(Synthetic resin bonded laminate.) A product made by bonding together two or more layers of material. The term usually applies to preformed layers joined by adhesives or by heat and pressure. The term also applies to composites of plastics films with other films, foil, and paper even though they have been made by spread coating or by extrusion coating. A reinforced laminate refers mainly to superimposed layers of resin impregnated or resin coated fabrics or fibrous reinforcements which have been bonded, especially by heat and pressure. When the bonding pressure is at least 1000 psi, the product is called a *high-pressure laminate*. Products pressed at pressures under 1000 psi are called *low-pressure laminates*. Products produced with little or no pressure, such as hand lay-ups, filament wound structures, and spray-ups, are sometimes called *contact-pressure* laminates.

Molding compounds—A group of plastics or resin materials in varying stages of formulation (powder, granular, preform, etc.) comprising resin, filler, pigments, plasticizers, or other ingredients ready for use in the molding operation.

Organosol—A dispersion (usually vinyl or polyamide) in liquid phase of which contains one or more organic solvents.

Orifice—The opening in the extruder die or spinneret formed by the orifice bushing ring and mandrel.

Open-celled—Referring to the interconnecting of cells in cellular or foamed plastics.

Pellets—One of the many formulations of molding compounds.

Plastisol—Mixtures of resins and plasticizers which can be molded, cast, or converted to continuous films by the application of heat. If the mixtures contain volatile thinners also, they are known as *organosols*.

Potting—Similar to encapsulating except that steps are taken to ensure that the polymeric material covers only the object to be protected and not the surrounding area.

Reinforcing—A process where reinforcements are used to improve selected properties of plastics parts.

Reinforcement—A strong inert material bound into a plastic to improve its strength, stiffness, and impact strength. Reinforcements are usually long fibers of glass, sisal, cotton, etc., in woven or nonwoven form. To be effective, the reinforcing material must form a strong adhesive bond with the resin.

Sandwich—Referring to a laminated construction, consisting of thin facings, bonded to relatively thick lightweight core, resulting in a rigid lightweight panel or product.

Skiving—A process of cutting a thin slice or split of plastics material from a thicker sheet or block of material.

Spinneret—A type of extrusion die, usually a metal plate with many tiny holes, through which a plastics melt is forced to make fine fibers and filaments. Filaments may be hardened by cooling in air, water, etc., or by chemical action.

Syntactic foams—Cellular plastics produced by blending hollow glass or plastics microballoons in a base resin such as polyester, phenolic, or epoxy.

Yarns—A single or multifilament which may be twisted and woven into fabric.

Review Questions

7-1. Name eight available forms of plastics and give a product example of each.

7-2. What is the difference between cohesion and adhesion? Give an example of each.

7-3. What adhesives are used to bond plywood?

7-4. What is the difference between films and sheets?

7-5. Why must some materials be solvent cast into sheets?

7-6. What is the major difference between melt spinning and wet spinning and solvent spinning? Which materials are used with each? Why?

7-7. What is the major difference between the processes of potting, embedding, and encapsulation?

7-8. What is the difference between organosols and plastisols? Name a product application of each.

7-9. What is the difference between a casting and a coating?

7-10. Name ten products that have plastics coatings.

7-11. How does open-celled cellular plastics differ from closed-celled cellular plastics?

7-12. Name a product that could utilize syntactic foams. Why?

7-13. Why are plastics made into cellular products? List five products made of cellular plastics.

7-14. Why are reinforcements sometimes listed separately from laminated forms of plastics?

7-15. Name eight laminating materials used with resins.

7-16. What importance does denier have in relation to filaments?

7-17. What is dispersion?

7-18. What is made by spinnerets?

7-19. What makes a plastics foam expand?

7-20. Where are some major foams used in day-to-day living?

7-21. What does a forming agent do in the forming of a foam?

7-22. What role does the resin in a foam play?

7-23. Where can we use rigid urethane foam?

7-24. Where is flexible urethane foam primarily used?

8

Identification of Plastics

Because plastics are complex materials and because they are being copolymerized with other polymers and ingredients, identification is difficult. Accurate identification requires complex equipment and techniques. The addition of fillers and the insolubility of some plastics adds to identification problems.

It may be necessary for the student and consumer to identify plastics so that repairs or fabrication of plastics parts can be accomplished. There are four broad methods for identification: (1) appearance, (2) effects of heat, (3) effects of solvents, and (4) specific gravity. There are also numerous laboratory tests which serve to indicate the possible ingredients of the unknown material. The methods presented here are intended for easy identification of basic polymer types and do not include complex instrument identification procedures. Infrared spectroscopy analysis is the only accurate method to obtain the quantitative identification of unknown polymers. Highly complex and costly X-ray diffraction equipment is used for identification of solid crystalline compounds. The elements nitrogen, sulfur, chlorine, and fluorine (halogens) are found in many plastics. Testing for these elements serves to identify many. Casein, cellulose nitrate, polyamide, melamine-formaldehyde, and polyurethane will indicate the presence of the element nitrogen. A simple test (the Beilstein test) for determining the presence of a halogen is to heat a clean copper wire in a Bunsen flame until it glows. Quickly touch the hot wire to the test sample and then return the wire to the flame. A green flame color indicates the presence of a halogen (chlorine, fluorine, bromine, and iodine). Plastics containing chlorine are polychlorotrifluoroethylene, polyvinyl chloride, polyvinylidene chloride, and other chlorinated materials. Polychlorotrifluoroethylene, polytetrafluoroethylene, and polyvinyl fluoride may be similarly tested and give positive results to tests for fluorine. If the tests are negative the polymer may be composed of only carbon, hydrogen, oxygen, or silicon.

For further chemical analysis, the Lassaigne procedure of sodium fusion may be used. In this test, 0.5 gram of the test polymer is placed in an ignition tube with about 0.1 gram of clean sodium. The tube is heated until the polymer is decomposed and the glass is red hot. While the tube is still red hot, plunge it into a container of distilled water. (This test is dangerous and should be carefully controlled. Sodium is highly reactive and the open end of the ignition tube should be kept away from the operator at all times.) The ignition tube and contents are carefully ground while under the distilled water and the mixture is then heated to boiling. While the resulting mixture is hot, the carbon and fragments of glass may be filtered from this mixture. The resulting filtrate may be divided into four equal portions to complete the four following tests:

(1) Nitrogen-containing polymers may include: cellulose nitrate, polyamide, casein (some sulphur may be present), polyurethane, urea-formaldehyde,

acrylonitrilebutadiene-styrene, melamine-formaldehyde, epoxides, and others. To test for nitrogen, place about four drops of freshly prepared aqueous ferrous sulphate in a portion of the previously prepared filtrate. After this solution is brought to a boil and cooled, add about five drops of dilute sulphuric acid and two drops of aqueous ferric chloride. The solution should be slightly acid at this point and the presence of a blue precipitate of ferric ferrocyanide (Prussian blue) indicates the polymer contained nitrogen.

(2) Chlorine-containing polymers may include: polyvinyl chloride, polyvinylidene chloride, polychlorotrifluoroethylene, and others. To test for the presence of chlorine or bromine, add about five drops of dilute nitric acid to the test solution and bring to a boil for two or three minutes. A few drops of aqueous silver nitrate is then added. A white, ammonia-soluble precipitate indicates the presence of chlorine. A yellow precipitate, insoluble in ammonia, indicates the presence of bromine. The halogens iodine and bromine are presently not found in commercial plastics and fluorine does not give a precipitate.

(3) Fluorine-containing polymers may include: polychlorotrifluoroethylene (with chlorine), polytetrafluoroethylene, polyvinyl fluoride, and others. To test for the presence of fluorine, prepare an aqueous (0.1%) solution of zirconium nitrate and alizarin red S. A filter paper is then impregnated with the solution and allowed to dry. The dried filter paper is then moistened with a 50-percent aqueous acetic acid solution. The test sample solution is neutralized with a few drops of hydrochloric acid. One drop of the solution to be tested is then placed on the moistened filter paper. If the red color of the filter paper turns yellow in that spot, fluorine is present.

(4) There are few polymers containing sulphur. Among these polymers are chlorosulphonated polyethylene (with small amounts of chlorine), casein (with small amounts of nitrogen), and various elastomers.

If a negative result is obtained in each of the four tests, the unknown polymer may not contain nitrogen, sulphur, or halogens, and may be composed of only the elements carbon, hydrogen, oxygen, or silicon.

Appearance

Many physical or visual clues are used to help identify plastics materials. While in the raw, uncompounded, or in the pellet stage, plastics are more difficult to identify. Thermoplastics are generally produced in powder, granular, or pellet forms. Thermosetting materials are normally in the form of powders, preforms, or resins.

The method of fabrication and the product application are good clues to identification. Thermoplastic materials are commonly extruded, injection formed, calendered, blow molded, and vacuum molded, while thermosetting plastics are usually compression molded, transfer molded, or cast. Polyethylene, polystyrene, and cellulosic are used extensively in the container and packaging industry. Harsh chemicals and solvents are likely to be stored in polyethylene containers. Polyethylene, polytetrafluoroethylene, polyacetals, and polyamides have a waxy feel not present in other polymers. Some plastics are not available in transparent colors, while others are generally reinforced or heavily filled. Some identifiable characteristics of selected plastics are given in Table 8-1.

Effects of Heat

By heating plastics specimens in test tubes, many odors may be distinctly identified and associated with selected plastics. The actual burning of the sample in an open flame may provide further clues. Polystyrene and copolymers burn with black (carbon) smoke. Polyolefins burn with a clear, flaming drip when molten. (See Table 8-2.)

The actual melting point may provide further clues to identification. Thermosetting materials don't melt. Several thermoplastics melt at less than 200°F. An electric soldering gun can be pressed on the surface of the plastics material. If the material softens and the hot tip sinks into the plastics, it is a thermoplastic; but if it stays hard and only chars the material, it is a thermoset.

The melting or softening point may be observed by placing a small piece of the unknown thermoplastic on an electrically heated platen or in an oven. The temperature must be carefully controlled and recorded. The temperature should be

Table 8-1. Identification of Selected Plastics.

Plastics	Specific Gravity @ 23°C	Physical-Visual Characteristics
ABS	1.02-1.25	Styrenelike, tough, metallike ring when struck, translucent
Acetal	1.40-1.45	Tough, hard, metallike ring when struck, translucent, low coefficient of friction, waxy feel
Acrylics	1.17-1.20	Brittle, hard, transparent
Allyl	1.30-1.40	Hard, filled, reinforced, transparent to opaque
Aminos	1.47-1.65	Hard, brittle, opaque but some translucent
Celullosics	1.15-1.40	(Varies) tough, transparent
Chlorinated Polyesters	1.4	Tough, similar to polyethylene or polystyrene, translucent or opaque
Epoxies	1.11-1.8	Hard, mostly filled, reinforced, transparent
Fluorocarbons	2.1 -2.2	Tough, waxy feel, low coefficient of friction, translucent
Ionomers	0.93-0.96	Tough, impact resistant, similar to polyolefins, transparent
Phenolics	1.25-1.55	Hard, brittle, filled, reinforced, transparent
Phenylene oxide	1.06-1.10	Tough, hard, often filled, reinforced opaque
Polyamides	1.09-1.14	Tough, waxy feel, low coefficient of friction, translucent
Polycarbonate	1.2 -1.52	Styrenelike, tough, metallike ring when struck, translucent
Polyesters	1.3	Hard, brittle, filled, reinforced, transparent
Polyolefins	0.91-0.97	Waxy feel, tough, soft, translucent
Polystyrene	0.98-1.1	Brittle, white bend marks, metallike ring when struck, transparent
Silicones	1.6 -2.0	Tough, hard, filled, reinforced, some flexible, opaque
Urethanes	1.15-1.20	Tough castings, mostly foams, flexible, opaque
Vinyls	1.2 -1.55	Tough, some flexible, transparent
Polysulfone	1.24	Rigid, similar to polycarbonate, transparent to opaque

increased at a rate of one degree Celsius per minute when the specimen is within a few degrees of the suspected melting point.

A standard method of testing polymers is designated in ASTM D-2117-64. For polymers which have no definite melting point such as polyethylene, polystyrene, acrylics, and celluosics or those which melt with a broad transition temperature, the *Vicat softening point* may be of some aid in identification (ASTM D-1525-65T).

Melting point values of selected plastics are shown on Table 8-2.

Effects of Solvents

The solubility and insolubility of plastics are easily performed identification testing methods. With the exception of polyolefins, acetals, polyamides, and fluoroplastics, thermoplastic materials can be considered soluble at room temperature. Thermosetting plastics may be considered "solvent resistant."

In making solubility tests, it must be remembered that some of the solutions are inflammable, give off toxic fumes, are dangerously absorbed through the skin, or all three. Appropriate safety precautions should be taken.

Before proceeding with a solubility test, a chemical solvent must be selected. In order to help identify polymers and solvents which may react molecularly with each other, a solubility parameter number is assigned selected polymers and solvents in Table 8-3. In principle, the solubility parameters indicate that a polymer will dissolve in a solvent with a solubility parameter within the solubility parameter of the polymer. In other words, if the solubility parameters of the plastics and solvent are similar, the plastics may dissolve. Because of high-energy hydrogen bonding or other forms of molecular interaction including crystallization, some polymers may be soluble in solvents of somewhat different solubility parameters.

When making solubility tests, use a ratio of one volume of plastics sample to twenty volumes of boiling or room temperature solvent. A water-cooled reflux condenser may be used to collect or minimize solvent loss when heating solvents. In Table 8-4 selected solvents are indicated with selected plastics.

Specific Gravity

The presence of fillers or other additives and the degree of polymerization make the identification of plastics by specific gravity tests difficult. The presence of these materials can cause the specific gravity to differ greatly from that of the plastics itself. Polyolefins, ionomers, and low-density polystyrene will float in water, which has a specific gravity of 1.00. For comparison of specific gravities of various selected materials see Table 8-5.

The specific gravity of other plastics materials is shown in Table 8-1.

The degree of polymerization (DP) and density of selected plastics may be measured in solutions of known specific gravity contained in the individual cylinders shown in Fig. 8-1. In industry, a *gradient density column* is used to measure densities of small samples of plastics.

Salt solutions of distilled water and calcium nitrate are mixed and measured with technical grade hydrometers until a desired specific gravity is obtained. For densities less than that of water (1.00), isopropyl alcohol may be used. Undiluted isopropyl alcohol has a specific gravity of approximately 0.92. With the addition of small amounts of distilled water, this value may be raised.

It must be remembered that the density of plastics may vary depending on DP, amount of fillers, reinforcements, plasticizers, and other additives present in the plastics specimen. It is best to test only known specimens. Saw marks, dirt, or grease should be removed in order to avoid entrapped air when immersing the specimen.

Utilizing the specific gravity of solutions is a convenient and rapid method of measuring density of plastics.

Table 8-2. Identification Test for Selected Plastics.

Plastics	Flame-Burn	Odor	Melting Point (°F)	Melting Point (°C)
Acetal	Blue flame, no smoke, drippings may burn	Formaldehyde	347	175
Acrylic	Blue Flame, yellow top	Fruit, florallike	375	190
ABS	Yellow flame, black smoke, drips	Rubber smell	160	71
Casein	Yellow Flame	Burnt milk		
Cellulose acetate	Yellow flame, sparks drippings may burn	Acetic acid	446	230
Cellulose acetate butyrate	Blue flame, yellow top, sparks, drippings may burn	Rancid butter, camphor smell	356	180
Epoxy	Yellow flame, some soot	Phenolic-phenol		
Ethylcellulose	Pale yellow with blue-green	Burning wood		
Fluorocarbons PEP	Deforms, chars, won't burn	Slight acid or burned hair	554 621	290 327
Melamine formaldehyde	Difficult to burn, chars, yellow flame with blue-green base	Fishlike		
Phenolic	Cracks, deforms, difficult to burn, yellow flame	Phenolic-phenol		
Polyamides 6/6	Blue flame, yellow tip, melts and drips, self-extinguishing	Burned wool or hair	351 489	177 254
Polycarbonate	Decomposes, chars, self-extinguishing	Characteristic, use controls	430	221
Polychlorotrifluoro ethylene	Yellow	Slight, acrid, acidic fumes	350	177
Polyethylene	Blue flame, yellow top, drippings may burn, transparent hot area	Paraffin	221 (low density) 248 (high density)	105 120
Polytetrafluoroethylene	Yellow	None	550	288
Polypropylene	Blue flame, drips, transparent hot area	Heavy sweet	250	121
Polystyrene	Yellow flame, dense smoke, clumps of carbon in air	Illuminating gas, sweet, marigold, floral	374	190
Polyurethane	Yellow with blue base	Acrid		
Polyvinyl alcohol	Yellow, smoky	Unpleasant, sweet		
Polyvinyl fluoride	Pale yellow	Acrid	300	149
Polyvinylidene chloride	Yellow with green base, smoky	Pungent	160	71
Silicone	Bright yellow-white	None		
Vinyl acetate	Yellow flame, smoke, some soot, green on copper wire test	Acetic acid	140	60
Vinyl chloride	Yellow flame, green at edges, white smoke, self-extinguishing	Hydrochloric acid	302	150

Table 8-3. Solubility Parameters of Selected Solvents and Plastics.

Solvent	Solubility Parameter
Water	23.4
Methyl alcohol	14.5
Ethyl alcohol	12.7
Isopropyl alcohol	11.5
Phenol	14.5
n-Butyl alcohol	11.4
Ethyl acetate	9.1
Chloroform	9.3
Trichloroethylene	9.3
Methylene chloride	9.7
Ethylene dichloride	9.8
Cyclohexanone	9.9
Acetone	10.0
Isopropyl acetate	8.4
Carbon tetrachloride	8.6
Toluene	9.0
Xylene	8.9
Methyl isopropyl ketone	8.4
Cyclohexane	8.2
Turpentine	8.1
Methyl amyl acetate	8.0
Methyl cyclohexane	7.8
Heptane	7.5
Plastics	**Solubility Parameter**
Polytetrafluoroethylene	6.2
Polyethylene	7.9-8.1
Polypropylene	7.9
Polystyrene	8.5-9.7
Polyvinyl acetate	9.4
Polymethyl methacrylate	9.0-9.5
Polyvinyl chloride	9.38-9.5
Bisphenol (a polycarbonate)	9.5
Polyvinylidene chloride	9.8
Polyethylene terephthalate	10.7
Cellulose nitrate	10.56-10.48
Cellulose acetate	11.35
Epoxide	11.0
Polyacetal	11.1
Polyamide (66)	13.6
Coumarone idene	8.0-10.6
Alkyd	7.0-11.2

Table 8-5. Specific Gravity of Selected Materials.

Substance	Specific Gravity
(Woods)	
Ash	0.73
Birch	0.65
Fir	0.57
Hemlock	0.39
Red Oak	0.74
Walnut	0.63
(Liquids)	
Acid, Muriatic	1.20
Acid, Nitric	1.217
Benzine	0.71
Kerosene	0.80
Turpentine	0.87
Water, 39.2°F	1.00
(Gases)	
Air	1.00
Acetylene	0.898
Carbon dioxide	1.529
Ethylene	0.967
Hydrogen	0.069
Nitrogen	0.97
Oxygen	1.105
(Metals)	
Aluminum	2.67
Brass	8.5
Copper	8.85
Iron, cast	7.20
Iron, wrought	7.7
Steel	7.85
(Plastics)	
Acetal	1.425
Acrylic	1.17-1.20
Casein	1.35
Fluorocarbon	2.12-2.2
Phenolic	1.25-1.55
Polyester	1.01-1.46
Silicones	1.05-1.23

If a plastics floats in a solution with a specific gravity of 0.94, it may be a medium- or low-density polyethylene plastics. If the specimen floats in a

Table 8-4. Identification of Selected Plastics by Solvent Test Method.

Plastics	Acetone	Benzene	Furfuryl Alcohol	Toluene	Special Solvents
ABS	Insoluble	Partially soluble	Insoluble	Soluble	Ethylene dichloride
Acrylic	Soluble	Soluble	Partially soluble	Soluble	Ethylene dichloride
Cellulose acetate	Soluble	Partially soluble	Soluble	Partially soluble	Acetic acid
Cellulose acetate butyrate	Soluble	Partially soluble	Soluble	Partially soluble	Ethyl acetate
Fluorocarbon	Insoluble (most)	Insoluble	Insoluble	Insoluble	Dimethyacetamide (not FEP-TFE)
Polyamide	Insoluble	Insoluble	Insoluble	Insoluble	Hot aqueous ethanol
Polycarbonate	Partially soluble	Partially soluble	Insoluble	Partially soluble	Hot benzene-toluene
Polyethylene	Insoluble	Insoluble	Insoluble	Insoluble	Hot benzene-toluene
Polypropylene	Insoluble	Insoluble	Insoluble	Insoluble	Hot benzene-toluene
Polystyrene	Soluble	Soluble	Partially soluble	Soluble	Methylene dichloride
Vinyl acetate	Soluble	Soluble		Soluble	Cyclohexanol
Vinyl chloride		Insoluble			Cyclohexanol

solution of 0.92, it must be either low-density polyethylene or polypropylene. If the specimen sinks in all solutions below a specific gravity of 2.00 the specimen is either a fluorocarbon or a silicone plastics.

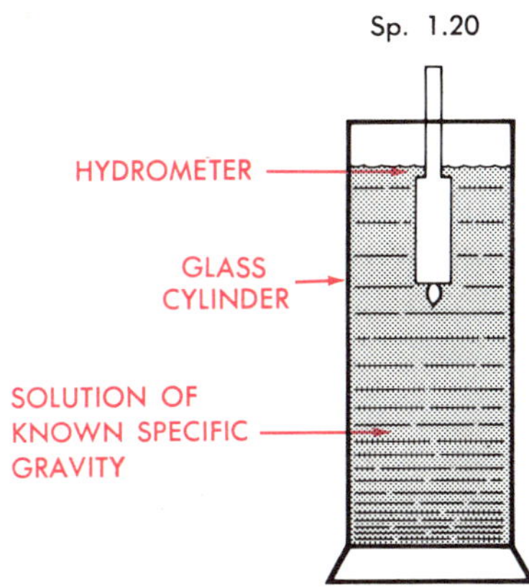

Fig. 8-1. Setup for measuring density and DP of selected plastics.

All of these solutions may be stored in clean containers and reused but the specific gravity of the solutions should be checked during the testing procedure. Factors such as temperature and evaporation may radically change the specific gravity value.

Another method to determine specific gravity of plastics specimens is to weigh the specimen in air and in water. A fine wire may be used to suspend the plastics specimen in the water from a laboratory balance as shown in Fig. 8-2. You may calculate the specific gravity of the plastics by the following formula:

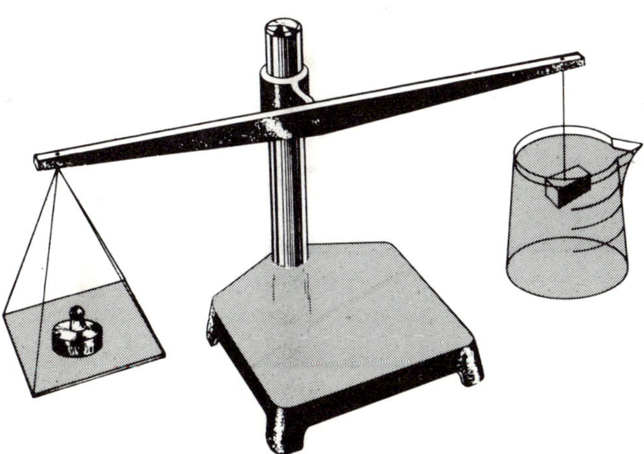

Fig. 8-2. Analytical balance for determining specific gravity of plastics specimens.

$$G_s = \frac{A - B}{A - B + C - D}$$

G_s = specific gravity at 20°C,
A = weight of specimen and wire in air,
B = weight of wire in air,
C = weight of wire with end immersed in water,
D = weight of wire and specimen immersed in water.

Glossary of New Terms

Density—Weight per unit volume of a substance, expressed in grams per cubic centimeter, pounds per cubic foot, etc.

Gradient density column—A means to measure densities of small plastics samples conveniently. A vertical, glass gradient tube is filled with a heterogeneous mixture of two or more liquids, the density of the mixture varying linearly or in other known fashion with the height. The specimen is placed in the gradient tube and falls to a position of equilibrium which indicates its density by comparison with positions of known standard samples.

Halogens—The elements fluorine, chlorine, bromine, and iodine.

Ionomers—Polymers which have ethylene as their major component but containing both covalent and ionic bonds. These resins have high transparency, resilience, tenacity, and many of the characteristics of polyethylene.

Parameters—A term used loosely to denote a specified range of variables, characteristics, or properties relating to the subject being discussed; or an arbitrary constant.

Specific gravity—The density (mass per unit volume) of any material divided by that of water at a standard temperature, usually 20° or 23°C. Since water's density is nearly 1.00 gram per cubic centimeter, density in grams per cubic centimeter and specific gravity are numerically nearly equal.

Solvent resistance—The ability of a plastics material to withstand exposure to a solvent.

Vicat softening point—The temperature at which a flat-ended needle of 1 square millimeter circular or square cross section will penetrate a thermoplastic specimen to a depth of 1 mm under a specified load using a uniform rate of temperature rise. (Definition taken from ASTM D-1525-64T.)

Review Questions

8-1. What are four broad methods for identification of plastics?

8-2. Name three polymers which contain the element chlorine.

8-3. Describe the Beilstein test. What does it test for?

8-4. Why is it necessary to identify plastics?

8-5. Describe the basic procedures for the Lassaigne test.

8-6. Name three fluorine-containing plastics.

8-7. What are the dangers when working with solvents?

8-8. Which "variables" should be considered in the identification of plastics?

8-9. Which plastics materials have a specific gravity less than that of water?

9

Selected Properties and Tests of Plastics

In order to describe meaningfully and understand the characteristics of various plastics materials, it is necessary to become acquainted with the nomenclature and testing methods used by the plastics industry.

Test data are valuable for identification, ensuring uniformity, and indicating probable performance of plastics products.

The plastics material manufacturer, who produces the basic plastics from chemicals, conducts tests to help describe his product and serve as reference points in quality control.

The processor, who converts the basic plastics into solid shapes, performs standard tests that will measure factors which affect handling and processability.

The fabricator and finisher, who further fashions and decorates the plastics forms, must be able to select the best plastics material to suit a particular application.

Most of these tests are performed on specified specimens under controlled laboratory conditions. Consequently, these tests should be used only as indicators for product service and design. These testing limitations should be kept in mind when making judgments about plastics. The true test comes when the plastics product is used in actual service conditions.

In order to ensure product success, the service requirements, design, and properties of the plastics product must be carefully considered.

There are several agencies which conduct and publish testing specifications on plastics. The United States of America Standards Institute, the United States military, and the American Society of Testing and Materials are currently involved in testing of plastics.

The American Society for Testing and Materials (ASTM) is an international, nonprofit, technical society devoted to ". . . the promotion of knowledge of the materials of engineering, and the standardization of specifications and methods of testing." The standardization and procedures for testing plastics is under the jurisdiction of the ASTM Committee D on Plastics. The *Book of ASTM Standards* is published annually with the plastics section in a two-volume edition.

Mechanical Properties

Mechanical properties are associated with the reaction that results when a force or load is applied to a material. These properties are sometimes referred to as "physical properties." The term "mechanical properties," however, is much to be preferred. Fig. 9-1 is a typical graph showing certain mechanical properties of selected plastics (relation of crystallinity and molecular weight).

There are three direct kinds of stress (force) which may be applied to materials: *compression, tension,* and *shear.* (See Fig. 9-2.) The rate (time) at which all three types of stress are applied is usually specified in testing procedures. Extremely rapid or slow applied forces may yield false values.

Compressive strength is a value which indicates how much force (maximum weight or load) is re-

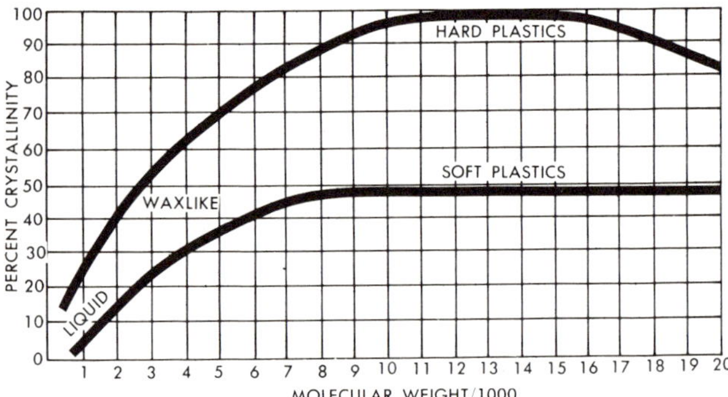

Fig. 9-1. Mechanical properties of selected plastics to show relation of crystallinity to molecular weight.

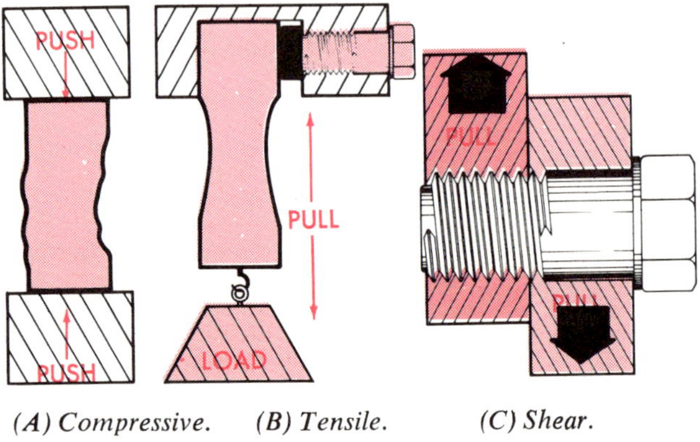

(A) Compressive. *(B) Tensile.* *(C) Shear.*

Fig. 9-2. Three types of stress.

quired to rupture or crush a material. The values are usually expressed in thousands of pounds per square inch (10^3 psi). Compressive strength is calculated by dividing the maximum load (force) in pounds by the area of the specimen in square inches:

$$\text{compressive strength (psi)} = \frac{\text{weight in pounds of force}}{\text{cross-sectional area in square inches}}$$

For example,

$$\text{compressive strength} = \frac{100 \text{ pounds}}{\frac{1}{2} \text{ sq in}} = 200 \text{ psi}$$

if 100 lbs is required to rupture a ½-inch plastics bar.

These values may be useful in distinguishing between grades of plastics and in comparing plastics to other materials. Compressive strength is especially significant in testing cellular or foamed plastics.

If stress is applied to a material in a pulling fashion until it breaks, *tensile strength* may be calculated. Tensile strength is calculated by dividing the maximum load (force) by the original cross-sectional area:

$$\text{tensile strength (psi)} = \frac{\text{pulling force in pounds}}{\text{cross-sectional area in square inches}}$$

The pulling stress usually causes the material to deform by stretching in length. This change in dimension in relation to its original dimension is called *strain* (Fig. 9-3). Strain is measured in inches per inch of length, in percent of elongation:

$$\text{strain (psi)} = \frac{\text{deformation in inches}}{\text{original dimension in inches}}$$

$$\text{strain (\%)} = \frac{\text{final length less original length}}{\text{original length}} \times 100$$

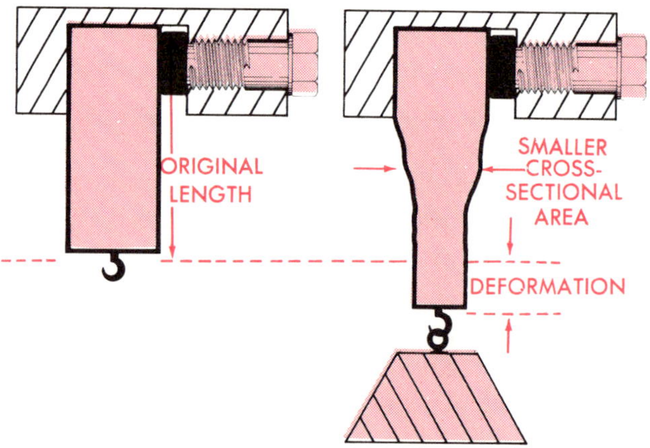

Fig. 9-3. Deformation due to pulling stress is called *strain*.

The significance of strain becomes evident in plastics because most plastics which are not reinforced deform in cross-sectional area before breaking (Fig. 9-5). Consequently, the breaking strength may be less than the greatest tensile strength, when strain has reduced the cross-sectional area of the specimen (See Fig. 9-6B.)

Stress-strain diagrams (Fig. 9-6) are convenient means of expressing and plotting the strength of plastics.

A certain amount of strain is called *elastic strain* and indicates that the molecules will return

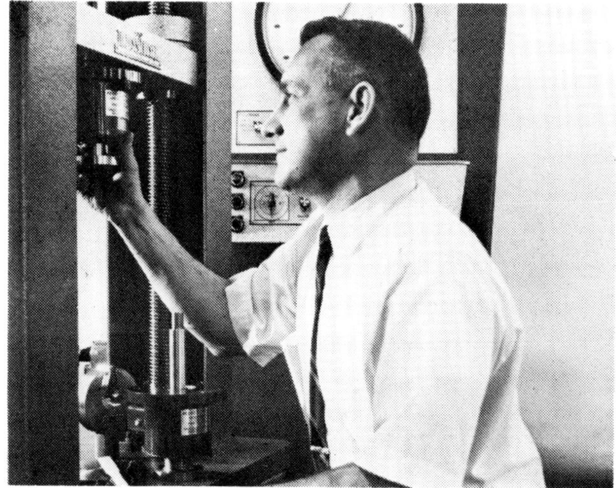

Fig. 9-4. Tensile strength testing machine for plastics specimens. (Tinius Olsen Testing Machine Co., Inc.)

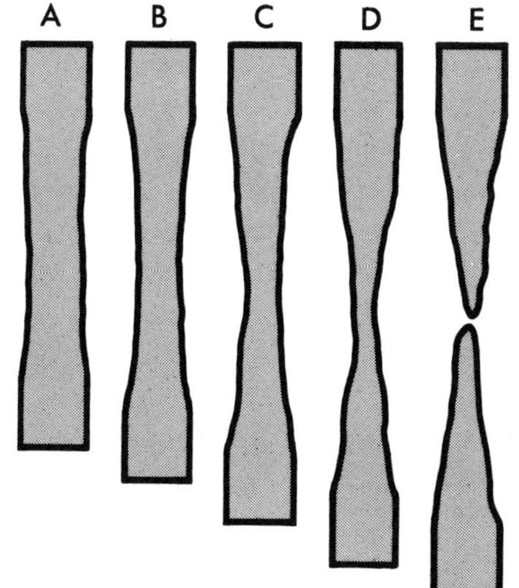

Fig. 9-5. Stages of deformation of unreinforced plastics.

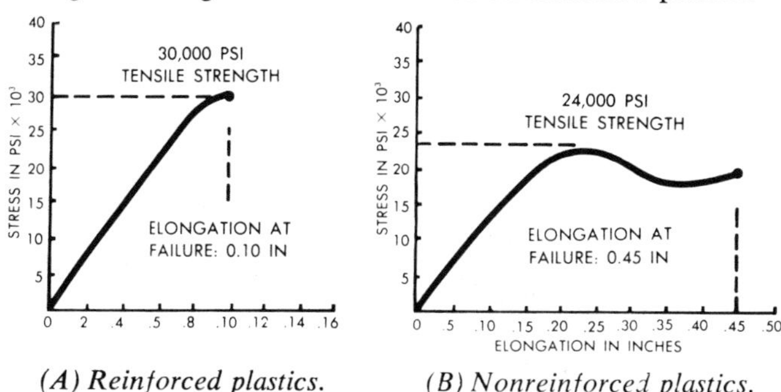

(*A*) *Reinforced plastics.* (*B*) *Nonreinforced plastics.*

Fig. 9-6. Typical stress-strain curves.

to their original position once the stress is removed. *Plastic strain* is the region where molecules begin to slip and slide past each other in permanent deformation. The point where the elastic strain ends and the plastic strain begins is referred to as *proportional limit* (Fig. 9-7).

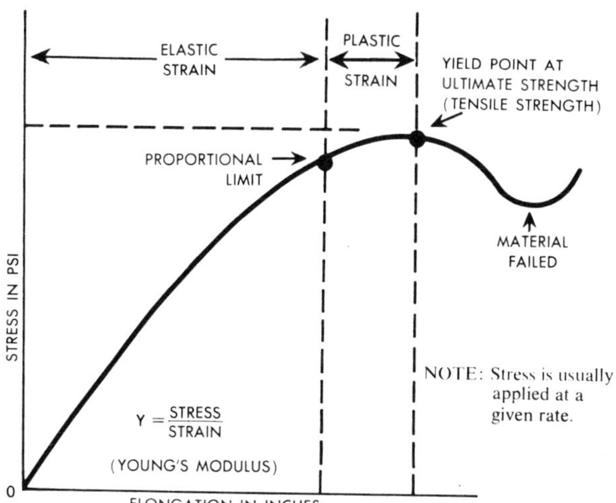

Fig. 9-7. Elastic strain, plastic strain, and proportional limit.

Young's modulus or the modulus of elasticity (tensile modulus) is a ratio between stress applied and the strain, within the elastic range. It is calculated by dividing the stress (load) in pounds per square inch by the strain in pounds per square inch:

$$\text{Young's modulus of elasticity} = \frac{\text{stress}}{\text{strain}}$$

Young's modulus has no meaning above the proportional limit. Plastic strain and permanent deformation cannot be allowed in the design of products. The ratio of tensile force to elongation is useful in determining how long a plastics specimen will get under a predetermined load. A large tensile modulus would indicate that the plastics is rigid and resistant to stretch or elongation.

In the three stress-strain curves represented in Fig. 9-8, the area under the curve is used to represent the energy required to break a plastics specimen. This area is an approximate measure of *toughness*. The toughest specimen is depicted by the largest area under the stress-strain curve (see Fig. 9-8C).

Although the term has not been clearly defined, toughness can be associated with *impact strength*. Impact strength is *not* a measure of stress required to break a specimen but a measure of the energy

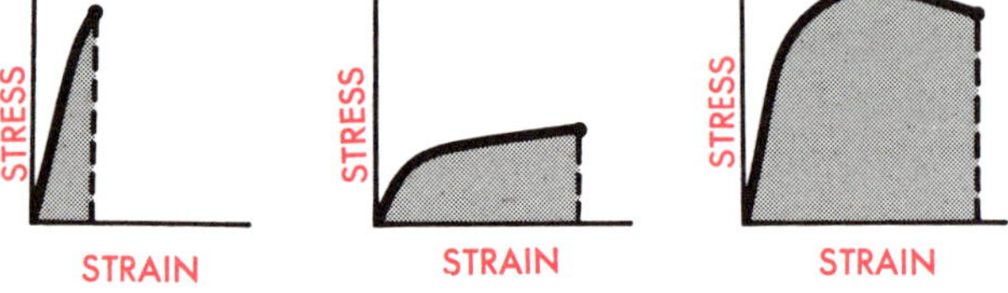

(A) Brittle plastics. *(B) Soft and weak plastics.* *(C) Hard and tough plastics.*

Fig. 9-8. Toughness is a measure of energy required to break a material.

or work required or absorbed in breaking the specimen. There are two basic methods for testing impact strength: (1) falling weight tests, and (2) pendulum tests.

There are numerous modifications to impact testing to accommodate various product shapes and designs.

The falling weight method of testing plastics is performed by dropping a ball-shaped weight from a predetermined height onto the plastics surface. Plastics containers, dinnerware, and helmets are commonly tested in this manner (see Fig. 9-9A and Fig. 9-9D).

A blunt shaped "dart" is commonly used for testing films up to 0.010 inch in thickness. (See Figs. 9-9B and 9-9E.)

Sometimes the plastics specimen itself is allowed to slide down a trough and strike a metal anvil. This test may be repeated from various heights and the specimen examined for impact damage (cracks, chips, or other fractures). See Fig. 9-9C.

The pendulum test method utilizes the energy of a swinging pendulum hammer to strike the plastics sample. The energy is recorded on the machine as a measure of energy or work absorbed by the specimen during fracture. The values are given in foot-pounds. The hammers of most plastics testing machines have a kinetic energy of 2 to 16 foot-pounds (Fig. 9-10).

In the *Charpy* pendulum method (simple beam), the test piece is supported at both ends without being held down and struck in the center of the sample by the pendulum hammer. (See Figs. 9-10A and 9-10B.)

When the test specimen is rigidly supported at one end and struck by the pendulum hammer (cantilever beam), the test is called *Izod* method (Figs. 9-10C and 9-10D). The test specimens may be notched or unnotched in Izod testing. In both

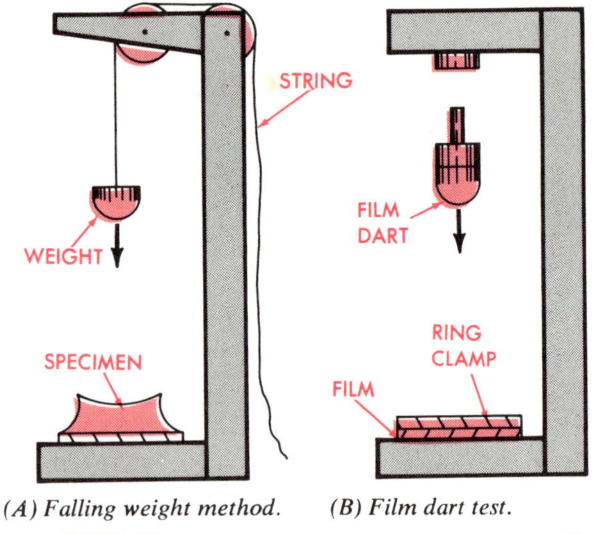

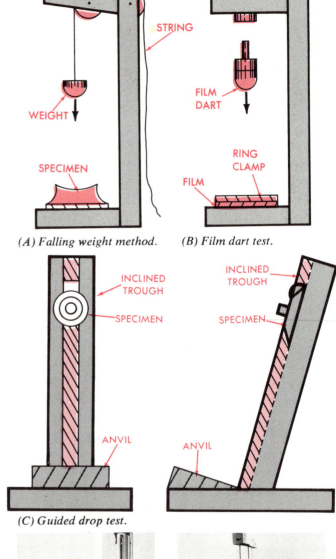

(A) Falling weight method. *(B) Film dart test.*

(C) Guided drop test.

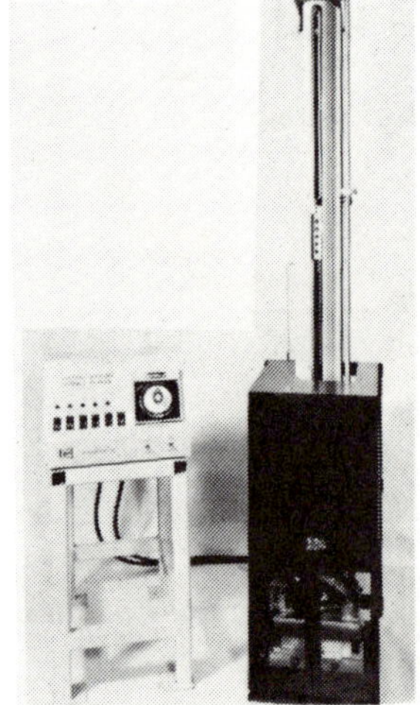

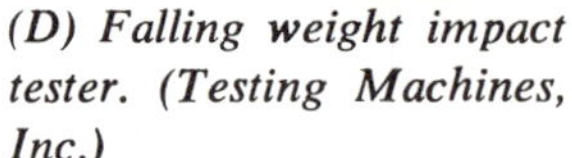

(D) Falling weight impact tester. (Testing Machines, Inc.)

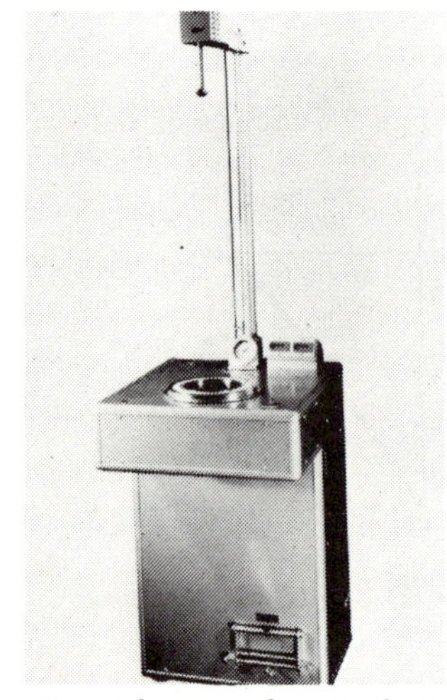

(E) Photograph of dart tester for films. (Testing Machines, Inc.)

Fig. 9-9. Methods of testing impact strength of plastics.

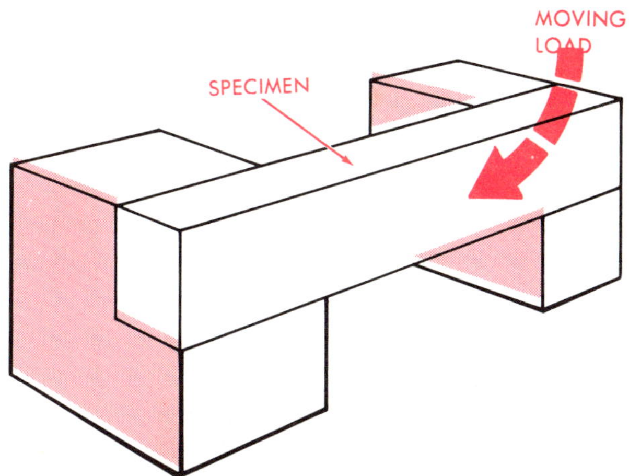

(A) Charpy pendulum method.

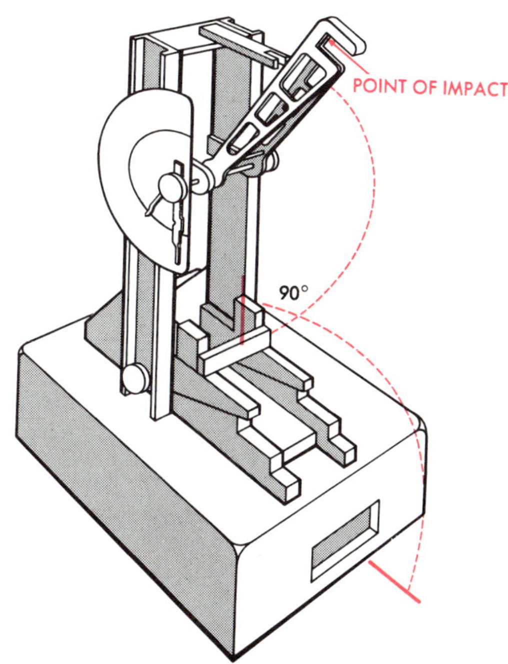

(B) Charpy-type simple-beam impact machine. (Tinius Olsen Testing Machine Co., Inc.)

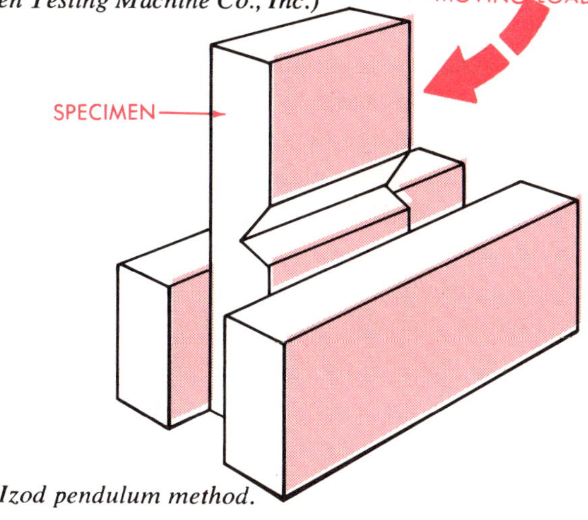

(C) Izod pendulum method.

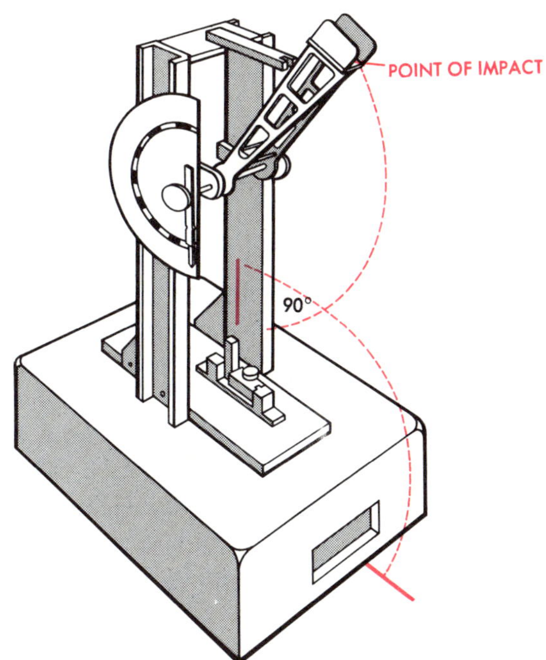

(D) Izod-type cantilever-beam impact machine. (Tinius Olsen Testing Machine Co., Inc.)

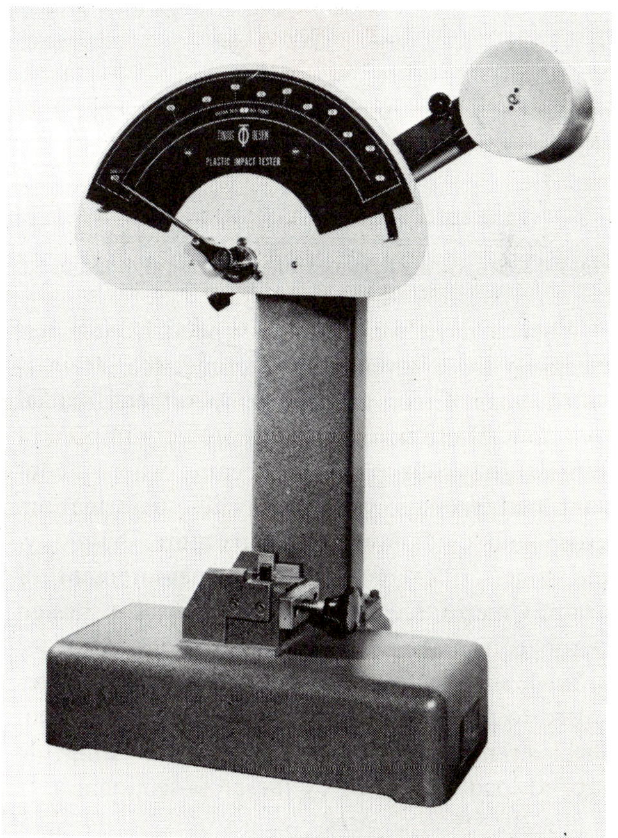

(E) Impact tester for Charpy and Izod testing. (Tinius Olsen Testing Machine Co., Inc.)

Fig. 9-10. Impact testers for Charpy and Izod testing.

testing methods, impact values are usually expressed in terms of the energy required to break the test specimen.

The ability to resist sharp blows is an important property of many plastics. Some possess extremely high impact strength and surpass the toughness of steel.

Flexural strength is a measure of how much stress (load) can be applied to a specimen before it breaks. The specimens are supported on test blocks four inches apart and the load is applied in the center (Fig. 9-11).

Because most specimens do not break when deflected, the flexural strength cannot be calculated. Instead, most thermoplastics and elastomers are referenced when five percent strain occurs in the specimen. The flexural property is found by measuring the load in pounds per square inch which caused the surface of the test piece to stretch five percent. Both tensile and compressive stresses are involved in bending the specimen.

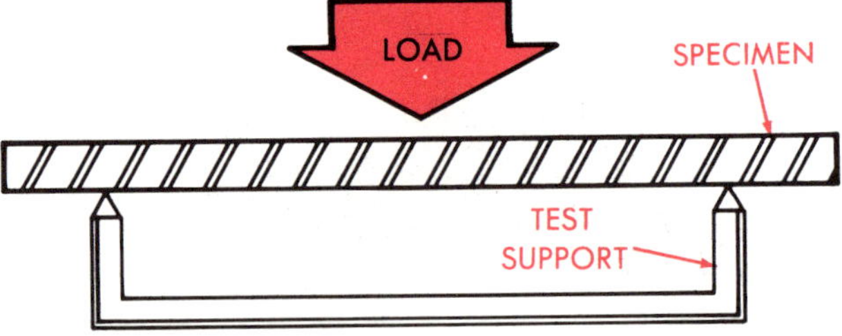

Fig. 9-11. Measuring flexual strength (flexural modulus).

When a dead weight is suspended from a test specimen for a given lapse of time, the strain is called *creep*. Creep at room temperature is called *cold flow*. These two properties are very important in pressure vessels, pipes, and beams, where a constant load (pressure-stress) may cause deformation. Creep and cold flow are temperature, humidity, and time critical for accurate measurement of strain. Creep forces must be considered in design factors of rapidly rotating rotors, blades, or parts.

Shear strength is the maximum load (stress) required to produce a fracture by a shearing action. Shear strength may be calculated by dividing the applied load in pounds by the cross-sectional area of the specimen sheared:

$$\text{shear strength (psi)} = \frac{\text{load in pounds}}{\text{cross-sectional area in square inches}}$$

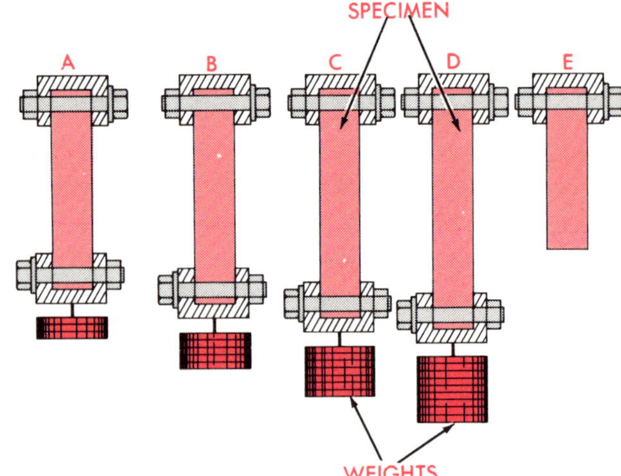

Fig. 9-12. Stages of creep and cold flow.

There are several methods used in testing shear strength. Fig. 9-13 shows three.

Shear strength is an important factor in plastics adhesives and in designing film and sheet products which may be exposed to shear stress.

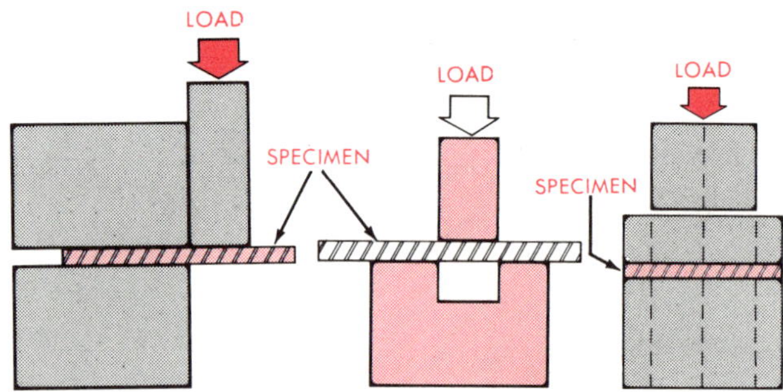

Fig. 9-13. Various methods for testing shear strength of plastics.

Because plastics weigh considerably less than metals and most other materials, plastics occupy a proportionately large volume. *Strength-to-weight ratio* is sometimes used to express the ratio of the tensile strength to the density of materials. A reinforced plastics with a density of 0.02 pound per cubic inch and exhibiting a tensile strength of 100,000 psi would have a strength-to-weight ratio of $50 \times 100{,}000 = 5.0 \times 10^6$. A piece of steel with a tensile strength of 280,000 psi and a density of 0.28 pounds per cubic inch would have a ratio of $280{,}000/0.28 = 1.0 \times 10^6$. These ratios are sometimes used in design criteria.

Fatigue is a general term used to express the number of cycles through which a specimen can be deformed before fracture. Fatigue fractures are dependent on frequency, amplitude, temperature, stress, and mode of stressing.

If the load (stress) does not exceed the elastic limit, many plastics may be stressed for an infinite number of cycles without failure. In the production of plastics integral hinges and one-piece box-and-lid containers, the fatigue characteristics of the plastic materials must be considered (Fig. 9-14).

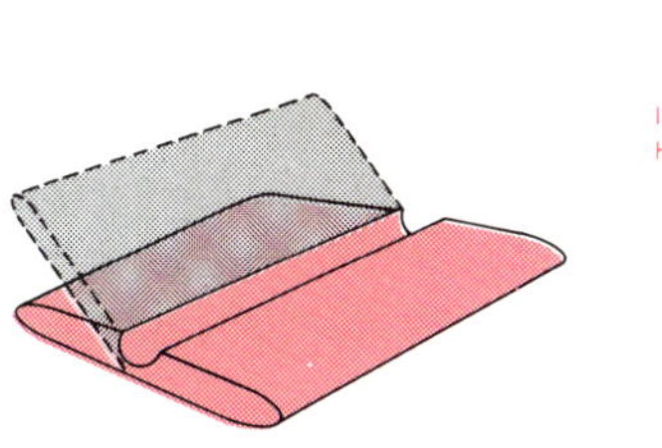

(A) Flexible hinge.

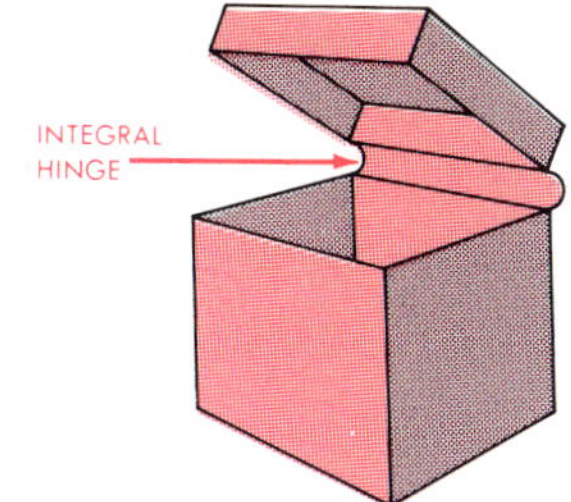

(B) One-piece box and lid.

(C) This folding endurance tester records the number of flexes on a dial before failure occurs in the plastics specimen. (Tinius Olsen Testing Machine Co., Inc.)

Fig. 9-14. Fatigue characteristics of plastics articles are important.

Plastics are capable of absorbing or dissipating vibrational energy. This resistance to vibration transmission is a property called *damping*. On an average, plastics possess ten times more damping capacity than steel. Plastics gears, bearings, appliance housings, and even some architectural applications make effective use of this vibration-reducing property.

The term *hardness* does not describe a definite or single mechanical property of plastics. Scratch, mar, and abrasion resistance as well as strength are closely related to hardness. Surface wearing of floor tile and marring of optical lenses are affected by these properties.

A widely accepted definition of hardness, as it applies to plastics, is "resistance to penetration or indentation by another body." Stated in another way, the hardness of a plastics is "the resistance of the plastics material to compression, indentation, and scratching."

There are several types of instruments used to measure hardness. When a hardness value is given, the type of instrument or scale is also indicated (Fig. 9-15A).

The Mohs scale of hardness is used largely by geologists and mineralogists, and it has been used to test the hardness of various plastics specimens. This scale is based on the fact that harder materials scratch softer ones.

Another simple test is based upon the hardness ranges of sharpened pencils. The hardness index is the first numbered pencil to scratch the test surface.

A nondestructive test may be performed with a *scleroscope* (Fig. 9-16). The scleroscope is used to measure the rebound height of a freely falling hammer called a *tup*. There are limitations in the rebound method when used in testing plastics; however, meaningful values of hardness may be made.

For more sophisticated quantitative measurements, indentation type instruments are used. Rockwell, Wilson, Barcol, Brinell, and Shore are well-known measuring instruments. Table 9-1 will help you formulate the basic differences in hardness tests and scales. In these tests, either the depth or area of indentation by the indentor is a measure of hardness.

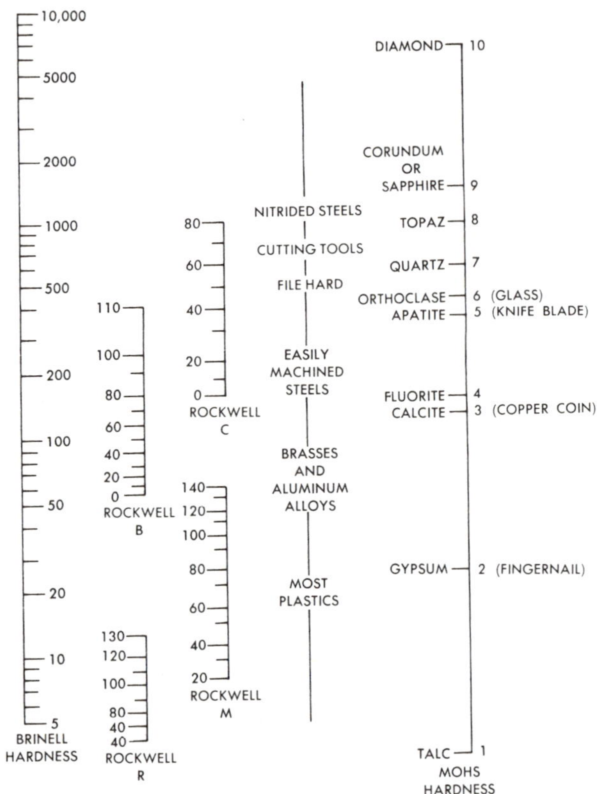

(A) Comparison of various hardness scales.

(B) Recording results of Rockwell hardness test on ABS bar specimen. (Wilson Instrument Div. of ACCO)

(C) Molded bar of ABS is positioned under indenter of Rockwell hardness tester. (Wilson Instrument Div. of ACCO)

Fig. 9-15. Hardness scales and tests.

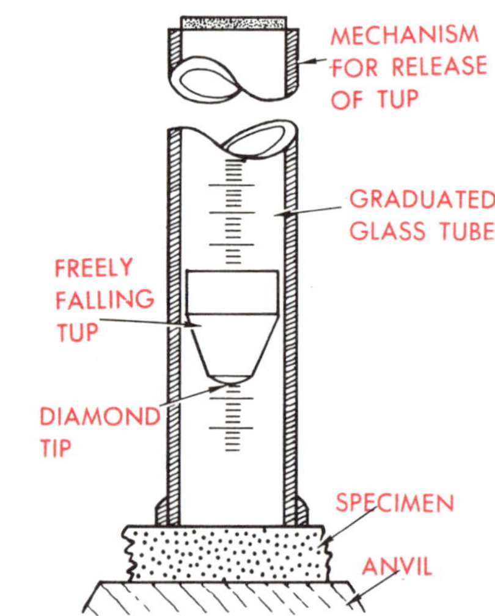

Fig. 9-16. Testing hardness with a scleroscope.

The Brinell test relates hardness to the area of the indentation. Fig. 9-17 shows a Brinell hardness tester. The Rockwell instrument relates hard-

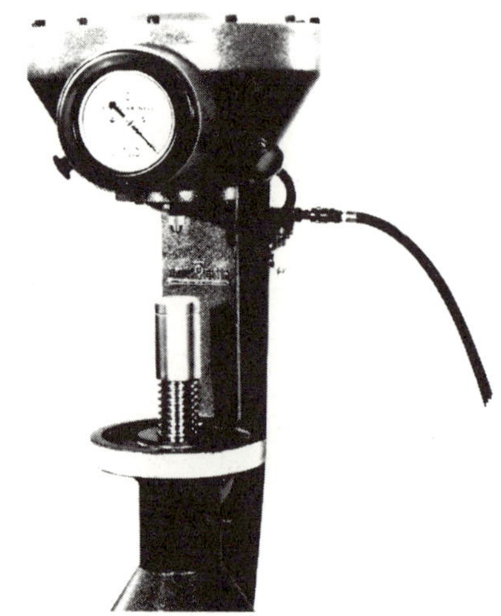

Fig. 9-17. This Brinell hardness tester is air operated. (Tinius Olsen Testing Machine Co., Inc.)

Table 9-1. Comparison of Selected Hardness Tests.

Instrument	Indentor	Load	Comments
Brinnell	Ball, 10-mm diameter	500 kg 3000 kg	Averages out hardness differences in material. Load applied for 15-30 seconds. View through Brinell microscope shows and measures diameter (value) of impression. Not for materials with high creep factors
Barcol	Sharp-point Rod 26° 0.157-mm flat tip	Spring loaded. Push against specimen with hand (10-15 lbs)	Portable. Readings taken after 1 or 10 seconds
Rockwell C	Diamond cone	Minor 10 kg / Major 150 kg	Hardest materials, steel. Table model
Rockwell B	Ball 1/16 in	Minor 10 kg / Major 100 kg	Soft metals and filled plastics.
Rockwell R	Ball ½ in	Minor 10 kg / Major 60 kg	Within 10 seconds after applying minor load, apply major load. Remove major load 15 seconds after application. Read hardness scale 15 seconds after removing major load or Apply minor load and zero within 10 seconds. Apply major load **immediately** after zero adjust. Read number of divisions pointer passed during 15 seconds of major load
Rockwell L	Ball ¼ in	Minor 10 kg / Major 60 kg	
Rockwell M	Ball ¼ in	Minor 10 kg / Major 100 kg	
Rockwell E	Ball ⅛ in	Minor 10 kg / Major 100 kg	
Shore A	Rod, 1.40-mm diameter. Sharpened to 35° 0.79-mm	Spring loaded. Push against specimen with hand pressure	Portable. Readings taken in soft plastics after 1 or 10 seconds
Shore D	Rod, 1.40-mm diameter. Sharpened to 30° point with 0.100-mm radius	Same as above	Same as above

ness to the difference in the depth of penetration of the minor (usually 10 kilograms) and major (from 60-150 kilograms) loads applied to a ball shaped indentor (Figs. 9-18 and 9-19).

The Barcol apparatus uses a sharp-pointed indentor to measure hardness (Fig. 9-20).

For soft or flexible type plastics, the Shore durometer instrument may be used. There are two ranges of durometer hardness. Type A utilizes a blunt rod-shaped indentor to test soft plastics. Type D utilizes a pointed, rod-shaped indentor to measure harder materials. The reading or value is taken after one or ten seconds of applying "hand" pressure.

Fillers and reinforcements may cause considerable local variations in hardness values taken with any of the hardness instruments. Numerous readings should be taken and expressed as an arithmetic mean (average).

The term *dimensional stability* refers to a broad number of properties affecting dimensional changes in plastics products. Changes in dimensions may be caused by:

(a) cold flow and creep,
(b) swelling or shrinking,
(c) internal stresses,
(d) temperature changes,
(e) weathering and aging.

Observe the important fact that the depth measurement does not employ the surface of the specimen as the zero reference point and so largely eliminates surface condition as a factor.

Dial now reads B-C plus a constant amount due to the added spring of the machine under major load, but which value disappears from dial reading, when major load is withdrawn.

NOTE—The scale of the dial is reversed so that a deep impression gives a low reading and a shallow impression a high reading; so that a high number means a hard material.

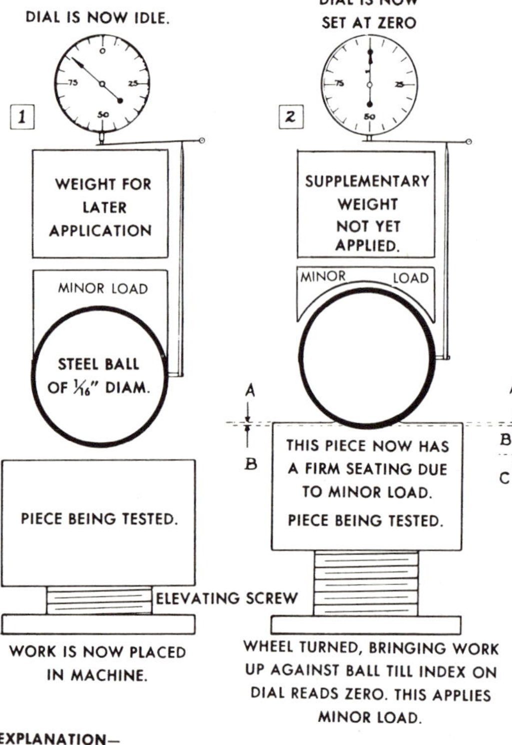
DIAL IS NOW IDLE.
1
WEIGHT FOR LATER APPLICATION
MINOR LOAD
STEEL BALL OF 1/16" DIAM.
PIECE BEING TESTED.
ELEVATING SCREW
WORK IS NOW PLACED IN MACHINE.
DIAL IS NOW SET AT ZERO
2
SUPPLEMENTARY WEIGHT NOT YET APPLIED.
MINOR LOAD
A
B
THIS PIECE NOW HAS A FIRM SEATING DUE TO MINOR LOAD.
PIECE BEING TESTED.
WHEEL TURNED, BRINGING WORK UP AGAINST BALL TILL INDEX ON DIAL READS ZERO. THIS APPLIES MINOR LOAD.

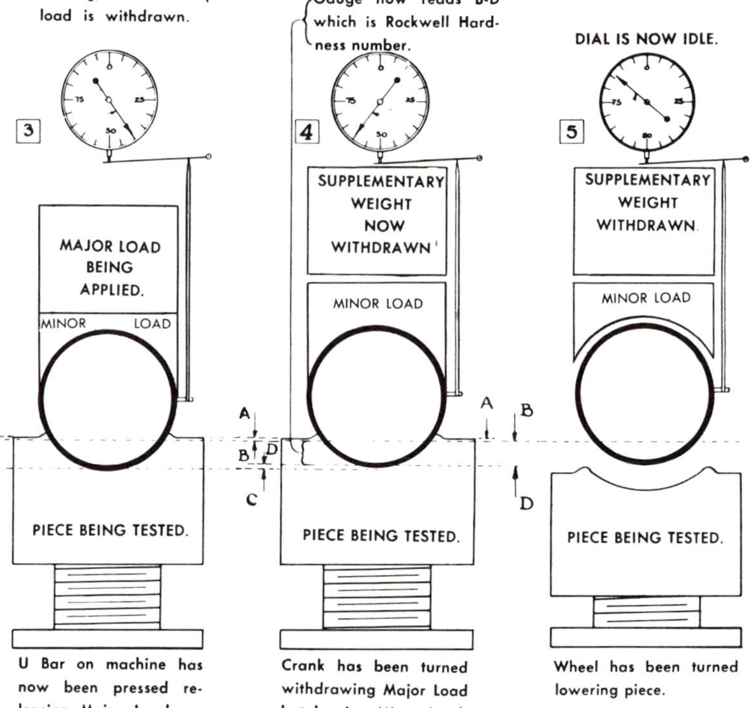
3
MAJOR LOAD BEING APPLIED.
MINOR LOAD
A
B
C
PIECE BEING TESTED.
U Bar on machine has now been pressed releasing Major Load.
Gauge now reads B-D which is Rockwell Hardness number.
4
SUPPLEMENTARY WEIGHT NOW WITHDRAWN
MINOR LOAD
A
B
D
C
PIECE BEING TESTED.
Crank has been turned withdrawing Major Load but leaving Minor Load.
DIAL IS NOW IDLE.
5
SUPPLEMENTARY WEIGHT WITHDRAWN.
MINOR LOAD
A
B
D
PIECE BEING TESTED.
Wheel has been turned lowering piece.

EXPLANATION—
Diagrammatically the cycle of operation of the Rockwell Direct-Reading Hardness Tester is here shown. To illustrate the principle and show the action of the ball under application and release of minor and major loads, the size of the 1/16" ball has been enormously exaggerated.

A-B = Depth of hole made by Minor Load
A-C = Depth of hole made by Major Load
D-C = Recovery of metal upon reduction of Major to Minor Load. This is an index of the elasticity of metal under test, and does not enter the hardness reading.
B-D = Difference in depth of holes made = Rockwell Hardness number.

Fig. 9-18. The Rockwell hardness tester accomplishes a test which in principle is represented by this series of sketches. (Wilson Instrument Div. of ACCO)

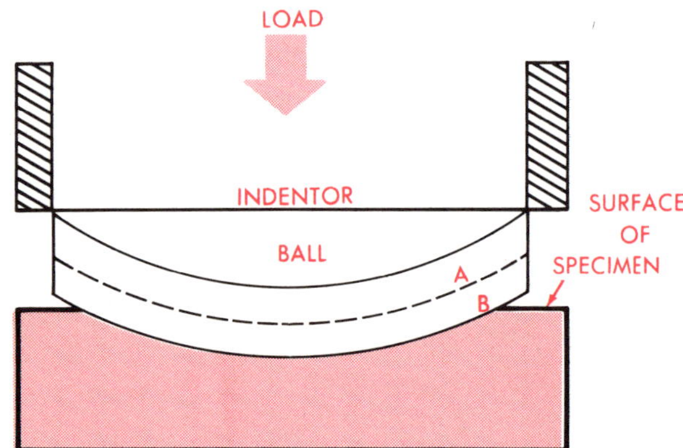
LOAD
INDENTOR
BALL
A
B
SURFACE OF SPECIMEN

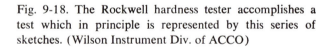

Fig. 9-19. Line *A* shows depth of minor load. Line *B* shows depth of major load. Distance between *A* and *B* is basis of Rockwell readings.

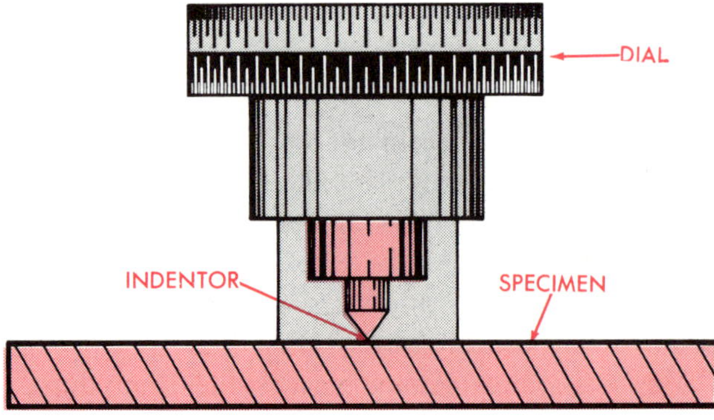
DIAL
INDENTOR
SPECIMEN

Fig. 9-20. Barcol instrument uses sharp-pointed indentor.

Fillers and other additives are used to help prevent much of this instability.

As previously stated, creep is the strain in inches that a material changes in a period. Creep that occurs at room temperature is called "cold flow."

Both of these properties are important property considerations in pressure vessels, pipes, beams, rotors, blades, or other parts where a constant pressure or stress may cause dimensional changes. Fig. 9-21 shows a section of pipe being tested for its burst strength. It is important to know how much stress and strain the pipe will sustain. Rupture or burst strength is often carried out under water as pressure is increased in the test specimen. This specimen ruptured while under 853 pounds per square inch (see arrow and circled area).

Metals also flow and deform if sufficient force is applied.

Fig. 9-21. Testing the burst strength of pipe. (Schloemann-Fellows)

Although plastics materials may be processed in a variety of physical states, internal stresses may be locked into the finished product. This locked-in stress may cause warpage, crazing, or cracking under varying environmental conditions. Solvents are also more likely to craze or crack parts held under stresses (see Chap. 10).

Plastics are temperature sensitive and change dimensionally with changes in temperature. Thermal expansion is commonly expressed with a coefficient of expansion in inches per inch of length per degree Fahrenheit.

Loss of various properties and dimensional changes may be a result of the all-inclusive term "weathering." The combined effects of temperature, light radiation, moisture, gases, and other chemical environments all affect dimensional and physical changes in plastics. Table 9-2 gives water absorption characteristics of various plastics.

Table 9-2. Water Absorption.

Material (24-Hour Immersion)	Percent Absorbed
Polychlorotrifluoroethylene	0.00
Polyethylene	0.01
Polystyrene	0.04
Epoxy	0.10
Polycarbonate	0.30
Polyamide	1.50
Cellulose acetate	3.80

Optical Properties

Optical properties are closely linked with molecular structuring such as chemical bonding and crystallinity. In view of this molecular structuring, the electrical, thermal, and optical properties are interrelated.

Gloss, luster, haze, transparency, color, clarity, and refractive index are only a few of the many optical properties of importance to plastics.

The light reflective quality of a plastics surface is referred to as *gloss* or *luster*. A cloudy or milky appearance in plastics is known as *haze*. The plastics commonly termed *transparent* is one that absorbs very little light in the visible spectrum. The uneven or selective absorption of light results in the material being colored. In general, materials with free electrons are opaque because the electrons absorb light energy. It is therefore not likely with present technology for a transparent material to be a good electrical conductor. *Clarity* is a measure of distortion seen when viewing an object through a transparent plastics. Fillers, chemical bonds, or crystalline planes all distort or interfere with the passage of light.

When light enters a transparent material, part of that light is reflected and part is refracted (Fig. 9-22). The index of refraction n may be expressed

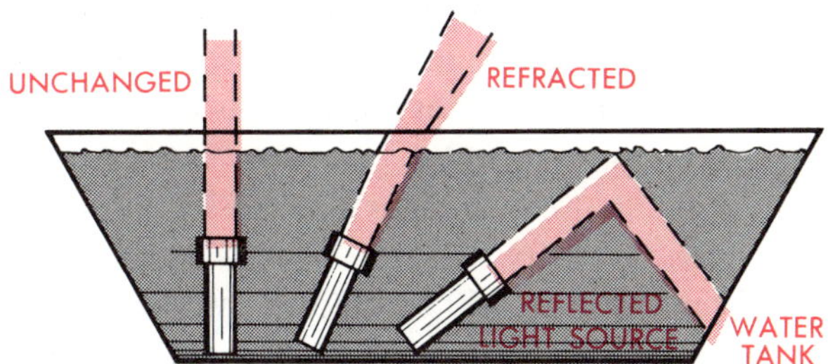

Fig. 9-22. Refraction and reflection of light.

in terms of the angle of incidence *i* and the angle of refraction *r*:

$$n = \frac{\sin i}{\sin r}$$

where *i* and *r* are taken relative to the perpendicular to the surface at the point of contact. The index of refraction for most transparent plastics is in the vicinity of 1.5, not greatly different from most building glass. For selected plastics it may be 1.35. Table 9-3 gives indexes of refraction for some plastics.

Table 9-3. Optical Properties of Plastics.

Material	Refractive Index	Light Transmission (%)
Methyl methacrylate	<1.49	94
Cellulose acetate	1.49	87
Polyvinyl chloride acetate	1.52	83
Polycarbonate	1.59	90
Polystyrene	1.60	90

Viscosity is a property of a liquid which describes its internal resistance to flow. Thus the more sluggish the liquid, the greater its viscosity. Viscosity is measured in units called *poises* or fractions of poises: centipoises and millipoises (see Table 9-4).

Table 9-4. Viscosity of Selected Materials.

Material	Viscosity in Centipoises
Water	1
Kerosene	10
Motor oil	10-100
Glycerine	1000
Corn syrup	10,000
Molasses	100,000
Resins	$<100\text{-}>10^6$
Plastics (hot, viscoelastic state)	$<10^5\text{-}>10^{10}$

Viscosity is an important factor in transporting resins, injecting plastics in the liquid state, and obtaining critical dimensions of extruded shapes. It must be remembered that fillers, solvents, plasticizers, thixotropic agents, degree of polymerization, and density all affect changes in viscosity measurements. The viscosity for resins such as polyester range from 1000-10,000 centipoises, while the hot melt of polyethylene may be as high as 300,000 centipoises.

Electrical Properties

Most plastics may be considered as electrical insulators. The electrical behavior of plastics which makes them suitable for various electrical or electronic products can be described in terms of five fundamental properties: arc resistance, insulation resistance, dielectric strength, dielectric constant, and dissipation (power) factor. The predominantly covalent bonds of polymers limit their electical conductivity.

Arc resistance is a measure of time for a given electrical current to render the surface of plastics conductive because of carbonization. The units are reported in seconds, and the higher the value the more resistive the plastics is to arcing. This breakdown may be a result of corrosive chemicals such as ozone and nitric oxides or due to an accumulation of moisture or dust. With the addition of fillers such as graphite or metals, plastics can be made conductive or semiconductive.

Insulation resistance is the resistance between two conductors of a circuit or between a conductor and the earth when they are separated by an insulator. The insulation resistance is equal to the product of the resistivity of the plastics and the quotient of its length divided by its area:

$$\text{insulation resistance} = \text{resistivity} \times \frac{\text{length}}{\text{area}}$$

Resistivity is expressed in ohm-centimeters. Table 9-5 gives resistivities for selected plastics.

Dielectric strength is a measurement of electrical voltage required to break down or "arc" through a plastics material. The units are reported as volts per mil (0.001 inch) of thickness. This electrical property gives an indication of the ability of a plastics to act as an electrical insulator. See Fig. 9-23 and Table 9-5.

The *dielectric constant* of a plastics is a measure of the ability of the plastics to store electrical energy. Plastics are used as dielectrics in the production of capacitors used in radios and other electronic equipment. The dielectric constant is usually designated by the letter *K*, and is based upon air, which has a value of 1.0. Plastics with a dielectric constant of 5 will have five times the electricity-storing ability as air or a vacuum.

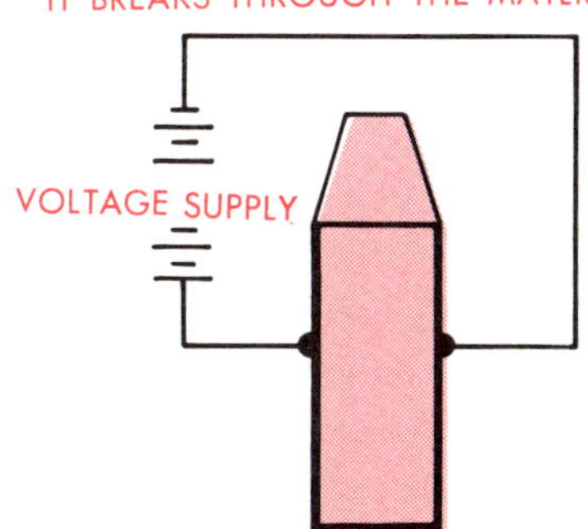
HOW GREAT CAN THE VOLTAGE BE BEFORE IT BREAKS THROUGH THE MATERIAL?
VOLTAGE SUPPLY

(A) Testing dielectric strength.

(B) Insulating pieces molded of premium plastics are used by General Electric in high-capacity power transformers such as this one on test. (FMC Corporation).

Fig. 9-23. Dielectric strength is an important characteristic of plastics used as insulators.

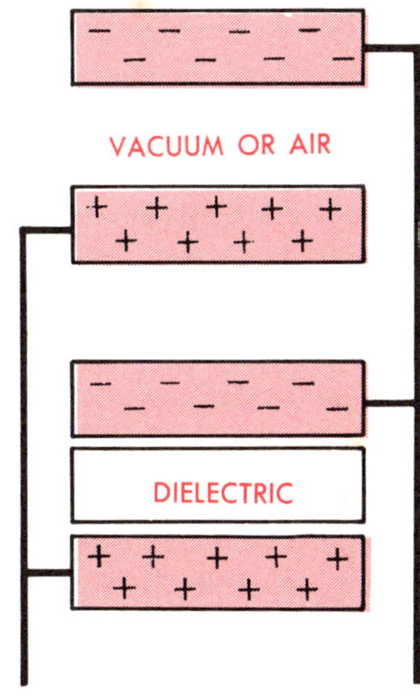
VACUUM OR AIR
DIELECTRIC
+

Fig. 9-24. The dielectric constant is the quantity of electricity stored across an insulating material divided by the quantity stored across air or vacuum.

Nearly all electrical properties of plastics are time-, temperature-, or frequency-critical. For example, the values may vary as frequency or cycles per second (hertz) are increased (see Table 9-5, "Dielectric Constant" and "Dissipation Factor.")

Dissipation (power) factor or *loss tangent* also varies with frequency. It is a measure of the power (watts) lost in the plastics insulator. A test procedure similar to that used in determining the value of the dielectric constant is used to measure this power loss. The measurements are usually made at million hertz frequencies and are indicators of the percent of alternating current lost as heat within the dielectric material. Plastics with low dissipation factors waste little energy and do not become overheated. For some plastics this is a disadvantage because they cannot be preheated or heat sealed by high-frequency methods of heating (see Table 9-5 for various dissipation factors).

The exact relation between heat, current, and resistance is shown in the power equation:

$$P = I^2R$$

Table 9-5. Electrical Properties of Selected Plastics.

Plastics	Resistivity (ohm-cm*)	Dielectric Strength (V/mil)	Dielectric Constant (K) 60 Hz	Dielectric Constant (K) 10^6 Hz	Dissipation (Power) Factor 60 Hz	Dissipation (Power) Factor 10^6 Hz
Acrylic	10^{16}	400-500	3.0-4.0	2.2-3.2	0.04-0.06	0.02-0.03
Cellulosic	10^{15}	200-600	3.0-7.5	2.8-7.0	0.005-0.12	0.01-0.10
Fluoroplastics	10^{18}	260-600	2.1-8.4	2.1-6.43	0.0002-0.04	0.0003-0.17
Polyamides	10^{15}	300-840	3.7-5.5	3.2-4.7	0.020-0.014	0.02-0.04
Polycarbonate	10^{16}	350-500	2.97-3.17	2.96	0.0006-0.0009	0.009-0.010
Polyethylene	10^{16}	445-1000	2.25-4.88	2.25-2.35	<0.0005	<0.0005
Polystyrene	10^{16}	300-600	2.45-2.75	2.4-3.8	0.0001-0.003	0.0001-0.003
Silicones	10^{15}	200-550	2.75-3.05	2.6-2.7	0.007-0.001	0.001-0.002

* Divide by 2.54 to get ohm-inches.

The power (P) used to perform wasted work is lost or dissipated power. In this formula, the amount of power can be decreased by lowering either the current (I) or the resistance (R). In electrical appliances designed to produce heat, a low dissipation factor is not considered desirable.

Thermal Properties

The important thermal properties of plastics are: thermal conductivity, specific heat, coefficient of expansion, heat distortion, resistance to cold, and flame resistance (see Fig. 9-25).

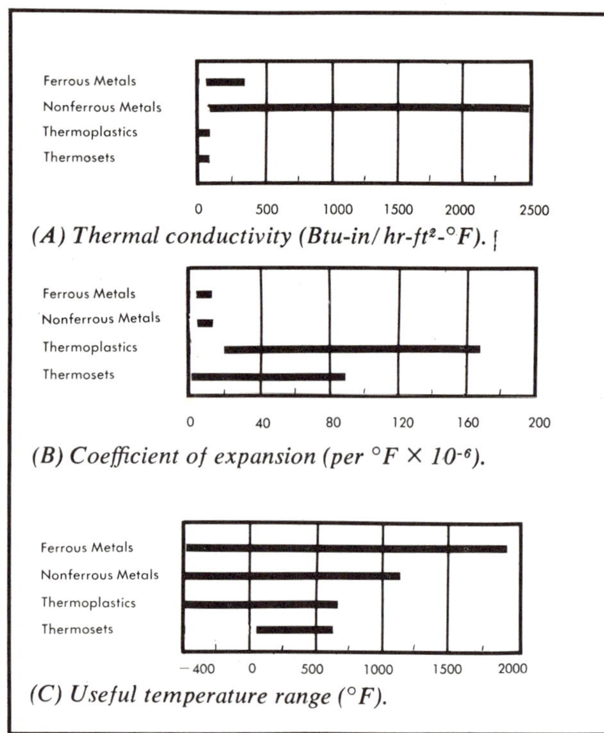

Fig. 9-25. Three selected thermal properties of various materials.

As plastics are heated, molecules and atoms within the material begin to oscillate and the molecular chains become longer. Additional heat may cause slippage between the weaker van der Waals' forces as the material forms a viscous liquid. In thermosetting plastics, bonds are not allowed to become essentially free but must be broken or decomposed.

Because calculations are more easily made and increasing numbers of industries utilize the Celsius (centigrade) scale, students should be aware of the following relations:

$$\text{degrees Fahrenheit} = 1.8(°\text{C}) + 32°$$

$$\text{degrees Celsius} = \frac{5}{9}(°\text{F} - 32°)$$

For example, 10°C is 18 + 32 = 50°F, and 50°F is (5/9) (50 − 32) = 10°C.

Thermal conductivity represents the transmission of heat energy from one molecule to another (Fig. 9-26). For the same (molecular) reasons that plastics are electrical insulators, they are also thermal insulators.

Fig. 9-26. The ice melts rapidly due to a high thermal conductivity of the metal bar.

Thermal conductivity is expressed in British thermal units (Btu) or calories. The amount of heat energy needed to raise the temperature of one pound of water one degree Fahrenheit is one Btu. One Btu equals 252 calories. A calorie is a unit of heat energy required to raise the temperature of one gram of water one degree Celsius.

The coefficient of thermal conductivity is commonly called the k factor. It should not be confused with the capital letter K designating dielectric constant. The amount of heat in Btu transferred in one hour through one square foot of a material that is one inch thick and has a temperature difference between its surfaces of one degree Fahrenheit is called the k factor. Materials with k values greater than 1 are good thermal conductors. Aluminum has a k factor of 1416.00, while foamed or cellular plastics have k values of less than 0.11.

Specific heat is the number of Btu of heat required to raise the temperature of one pound of material one degree Fahrenheit. Water has a specific heat of 1.0. The values of most plastics indicate that the material requires a greater amount of input heat energy to raise the temperature than does water. The amount of heat may be expressed in calories per gram per degree Celsius (see the problem in Fig. 9-27).

Specific heats are important in calculating processing and forming costs.

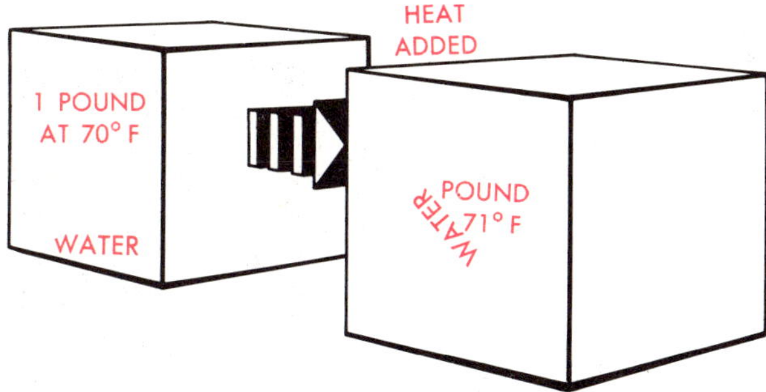

Fig. 9-27. How much heat was added?

The *coefficient of thermal expansion* in unfilled plastics is great. This number is used to represent thermal expansion in length, area, or volume per unit of temperature rise.

Table 9-6 gives the coefficient of thermal expansion for several materials. From these values a practical estimation of changes in length may be obtained.

Table 9-6. Thermal Expansion of Selected Materials.

Substance	Coefficient of Linear Expansion in Inches per Inch of Length per Degree Fahrenheit
Aluminum	0.00001200
Brass	0.00001014
Brick	0.00000306
Concrete	0.00000550 to 0.00000780
Copper	0.00000961
Glass	0.00000399 to 0.00000521
Granite	0.00000460
Iron, Cast	0.00000587
Marble	0.00000400
Steel	0.00000599 to 0.00000720
Wood (pine)	0.00000276
(Plastics)	
Melamine-formaldehyde	0.00001500
Phenol-formaldehyde	0.00004000
Polyamide	0.00005500
Polyethylene	0.00010000
Polystyrene	0.00003500
Polyvinylidene chloride	0.00010500
Polytetrafluoroethylene	0.00005500
Polymethyl methacrylate	0.00005000
Silicones	0.00016666

Example—A steel rod 60 feet in length is heated from 20 degrees Fahrenheit below zero to 100 degrees Fahrenheit. It will expand 120 $\times$ 0.00000599 $\times$ 60 $\times$ 12 = 0.5175 inch in length.

High thermal expansion is often a major disadvantage for many plastics applications. Metals with low coefficients of expansion are used where parts of close tolerances are required. Fillers may be used in plastics for added stability (Fig. 9-28).

Fig. 9-28. This miniature complex part, about the size of a dime, is used in a guidance gyro; dimensional stability from −273°F to +750°F must be maintained.

Heat distortion temperature indicates the maximum continuous operating temperature that the material will withstand. The product is usually tested in an oven and the temperature elevated until the product chars, blisters, distorts, or loses appreciable strength. The boiling point of water is a particularly important temperature level since it is often encountered in practice. Heat distortion temperature may be thought of as the maximum useful temperature range of the plastics product.

Under stress and high heat the thermoplastic polyester in Fig. 9-29 stands up under heat and load. From left to right, test bars of glass-reinforced polycarbonate, polysulfone, and thermoplastic polyester are locked in a lab vise with equal 175-gram weights clipped to their free ends (Fig. 9-29A). After only one minute under 310°F heat from an infrared radiant heater, the polycarbonate test bar began to deflect, and one minute later the polysulfone test bar followed suit (Fig. 9-29B). The thermoplastic polyester test bar survived not only the 310°F temperature, but was still going strong and undeflected after 6 minutes at 365°F.

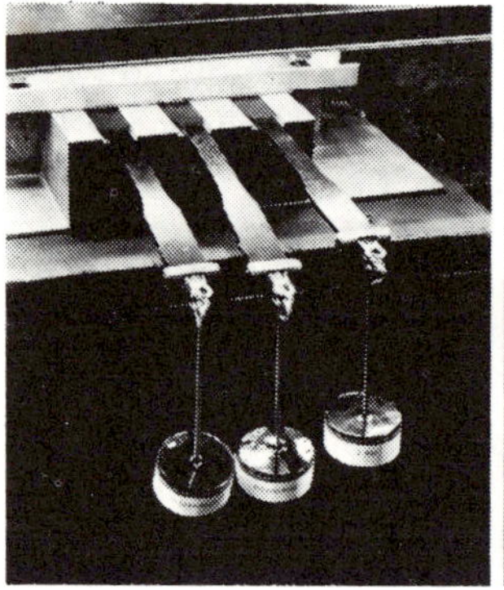

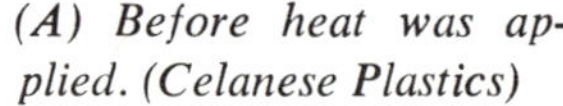

(A) Before heat was applied. (Celanese Plastics)

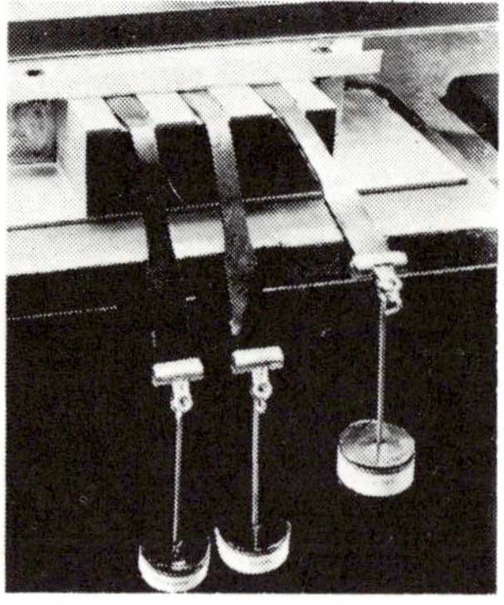

(B) After two minutes of heating. (Celanese Plastics)

(C) Deflection temperature tester determines heat distortion of up to five plastics samples at a time. (Tinius Olsen Testing Machine Co., Inc.)

Fig. 9-29. Testing heat distortion.

Usually plastics are not used where high heat resistance is needed; however, some phenolics have been submitted to temperatures as high as 5000°F, and ablative plastics have been used for missile re-entry. In ablative materials, heat is absorbed through a decomposition process known as *pyrolysis* which takes place in the near surface layer exposed to heat energy. Although much of the plastics is consumed, large amounts of heat energy are also absorbed and sluffed off.

As a rule, plastics have good *resistance to cold.* Polyethylene plastics are used extensively in the food packaging industry at temperatures of −60°F. Some plastics can withstand the extreme temperatures of −320°F with little loss of physical properties.

Flammability or *flame resistance* is a loosely used term indicating a measure of the ability of a material to support combustion. In one test, a plastics strip is started burning and the flame removed. The time and amount of material consumed is measured and the units are expressed in inches per minute. The more combustible plastics including cellulose nitrate will have a high value (see Fig. 9-30).

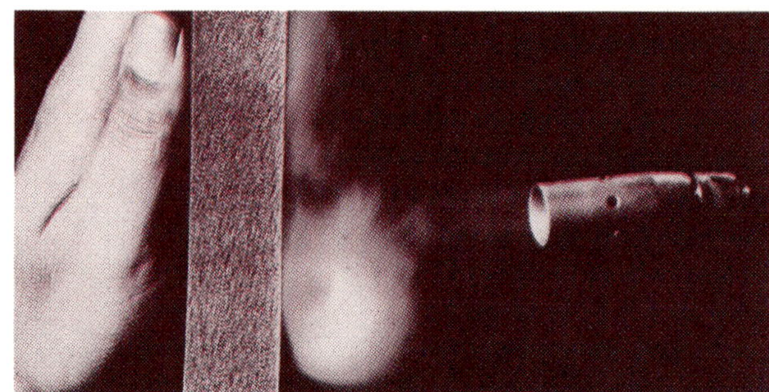

(A) This cellular plastics (polyurethane) demonstrates the thermal insulating ability and flame resistance of specially formulated materials.

(B) Self-extinguishing characteristic is one of several important properties of many plastics. When the flame is removed the plastics will not support combustion and the flame will go out. (General Mills Research Labs.)

(C) Technician ignites fabric to determine minimum amount of oxygen to sustain combustion in a downward pattern. This test essentially provides quantitative measure of flammability. (Gibbs & Soell, Inc.)

Fig. 9-30 Flammability of plastics.

A somewhat loosely used term associated with flammability is *self-extinguishing*. This property indicates that the material will not continue to support combustion once the flame is removed. Nearly all plastics may be made self-extinguishing.

It must be remembered that plastics will combust when exposed to direct flame and at present, there is no universally accepted test for fire retardancy.

Table 9-7 gives further data in the flammability of various materials.

Table 9-7. Ignition Temperatures and Flammability of Various Materials.

Material	Flash-Ignition Temperature (°C)	Self-Ignition Temperature (°C)	Burning Ratio (inches/minutes)
Cotton	230-266	254	SB
Paper, newsprint	230	230	SB
Douglas fir	260		SB
Wool	200		SB
Polyethylene	341	349	0.3 to 1.2
Polypropylene, fiber		570	0.7 to 1.6
Polytetrafluoro-ethylene		530	NB
Polyvinyl chloride	391	454	SE
Polyvinylidene chloride	532	532	SE
Polystyrene	345-360	488-496	0.5 to 2.5
Polymethyl methacrylate	280-300	450-462	0.6 to 1.6
Acrylic, fiber		560	SB
Cellulose nitrate	141	141	Rapid
Cellulose acetate	305	475	0.5 to 2.0
Cellulose triacetate, fiber		540	SE
Ethyl cellulose	291	296	1.1
Polyamide (nylon)	421	424	SE
Nylon 66, fiber		532	SE
Phenolic, glass fiber laminate	520-540	571-580	SE-NB
Melamine, glass fiber laminate	475-500	623-645	SE
Polyester, glass fiber laminate	346-399	483-488	SE
Polyurethane, polyether, rigid foam	310	416	SE
Silicone, glass fiber laminate	490-527	550-564	SE

NB=Non-burning
SE=Self-extinguishing
SB=Slow burning

Radiation Methods

The term "radiation" can refer to energy carried by either waves or particles. The carrier of wave energy is called a *photon*. In radiant energy the photon is wavelike when in motion, but particle-like when absorbed or emitted by an atom or molecule.

The ordinary electric light bulb with an element temperature of approximately 2300°C emits radiations waves that are visible. The sun, with surface temperatures of about 6000°C, emits both visible and invisible radiation. Man can see radiation with wavelengths as small as 3×10^{-5} cm (0.0016 inch) and as long as 7×10^{-5} cm.

Ultraviolet radiations are waves of energy which sunburn or tan the exposed parts of the human body and yet are invisible to the human eye (see Fig. 9-31). Ultraviolet radiation has wavelengths

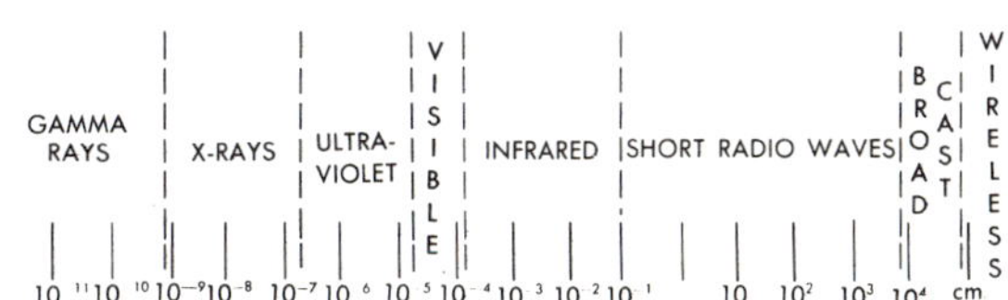

(A) *Wavelengths of radiation.*

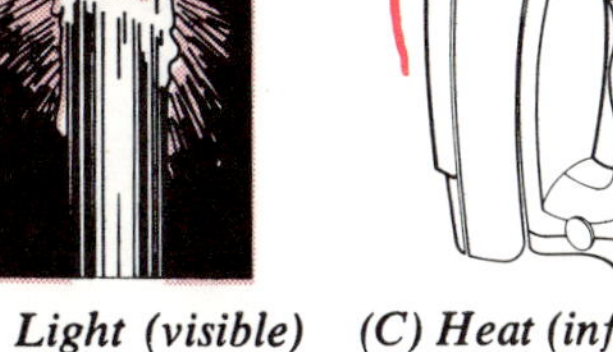

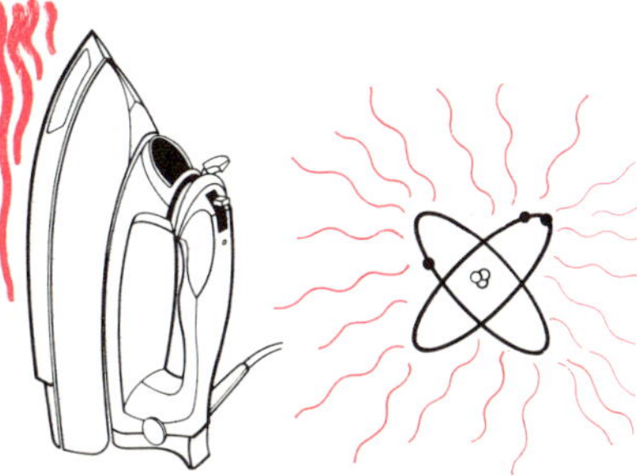

(B) Light (visible) radiation makes things visible. *(C) Heat (infrared) radiation can be felt.* *(D) Radioactive radiation can't be seen or felt.*

Fig. 9-31. Types of radiation.

shorter than 4×10^{-5} cm. Photon wavelengths are measured in units called *angstroms* (one ten-millionth of a millimeter, or 10^{-8} cm).

Radiation from the sun, burning fuels, or radioactive elements are considered *natural sources* of radiation. A few of the more important naturally occurring radioactive elements are uranium, radium, thorium, and actinium. These radioactive materials emit photons of energy and/or particles as the nuclei disintegrate and decrease in weight.

The earth contains small traces of radioactive materials, while the sun and stars are intensely radioactive.

The most important source of controlled radiation is *man-made*. Man has brought the use and control of induced or man-made radiation to a point where it may serve his needs.

Radiations may be produced by nuclear reactors, accelerators, or from natural or man-made radioisotopes.

Most elements have several kinds of atoms. When the number of protons in the nucleus of an atom changes, a different element is formed; however, if the number of neutrons is changed, a new element is not formed and only the mass is different. Different forms of the same element are called *isotopes*. The simple element hydrogen may appear as three distinct isotopes (Fig. 9-32). Most hydrogen atoms have no neutrons and have a mass number (the number of protons and neutrons) of 1. A very small number of naturally occurring hydrogen atoms have one neutron and one proton. These atoms have a mass number of 2. Only when hydrogen is composed of two neutrons and one proton with a mass number of 3, is it radioactive.

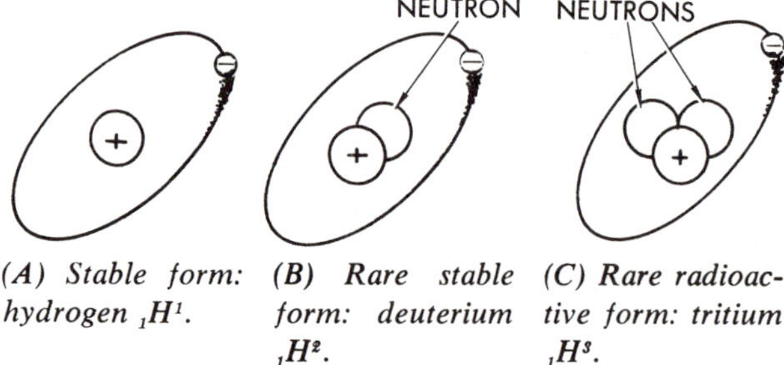

(A) Stable form: hydrogen $_1H^1$. *(B) Rare stable form: deuterium* $_1H^2$. *(C) Rare radioactive form: tritium* $_1H^3$.

Fig. 9-32. Isotopes of hydrogen.

In 1900 the German physicist Max Planck advanced ideas which were developed to show that photons are bundles or packets of electromagnetic energy either absorbed or emitted by atoms or molecules. The unit of energy carried by a single photon is called a *quantum* ("quanta" is plural).

Photons of energy radiation may be classified into two basic groups: electrically neutral and charged radiations.

Alpha (α) particles are heavy, slow moving masses with a double positive charge (two protons and two neutrons). When these particles strike other atoms, the double positive charge results in the removal of one or more electrons. The atom or molecule is then left in a *dissociated* or *ionized* state. Ionization is the process of changing uncharged atoms or molecules into ions. Atoms in the ionized state have either a positive or negative charge.

Electrons being ejected from the nuclei of atoms at very high speed and high energy are called *beta* (β) particles. When a neutron disintegrates, it becomes a proton and an electron. The proton commonly stays in the nucleus while the electron is emitted as a beta particle. Beta particles are electrons (opposite of positrons), with a negative charge. Because beta particles have $\frac{1}{1840}$ the mass of a proton, they are much faster than alpha particles and have greater penetrating power.

Most of the alpha and beta particle energy is lost when they interact with other atomic electrons. As charged particles pass through matter, they gradually lose energy to the nuclei or orbital electrons until all excess energy is transferred or lost. Since beta particles are negative, they can push or repel electrons, leaving the atom with a positive charge, or they may attach to the atom, giving it a negative charge.

Gamma (γ) radiations are short, very high frequency electromagnetic waves with no electrical charge. Gamma rays and x-rays are much alike except for origin and penetrating ability. Gamma photons can penetrate even the most dense materials. Several feet of concrete are required to stop the radiation effect of gamma rays (see Fig. 9-33).

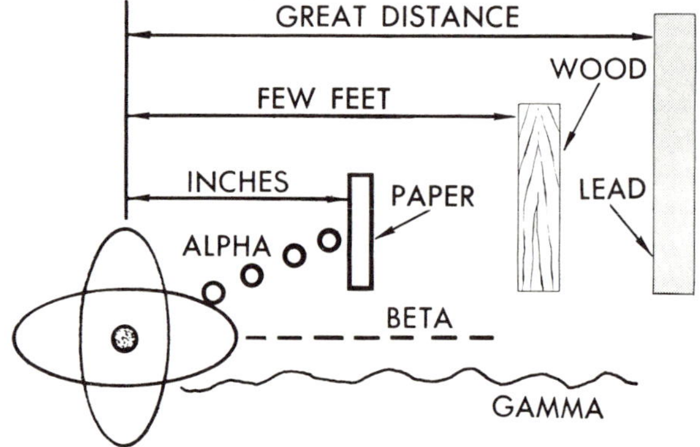

Fig. 9-33 Three types of radiation emitted by unstable atoms or radioisotopes: alpha stopped by paper, beta stopped by wood, and gamma stopped by lead.

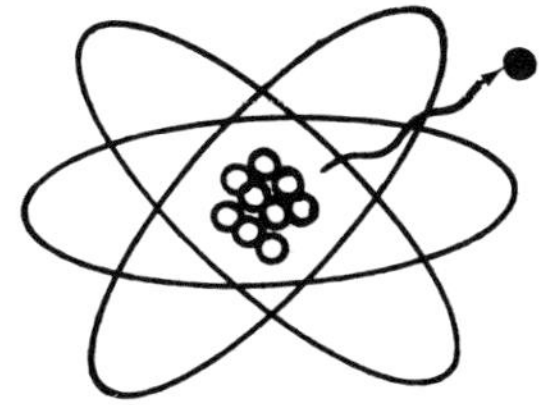

(A) Radioisotope with photon of energy being emitted.

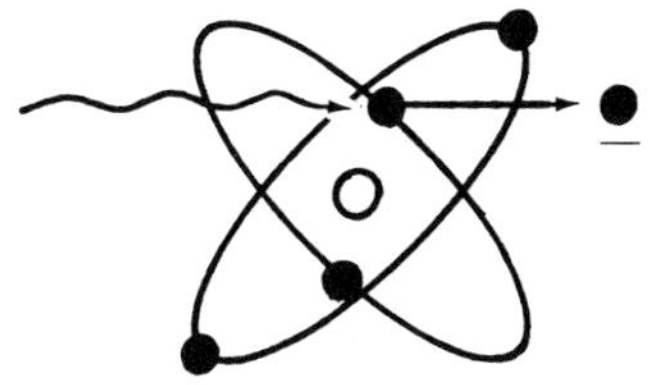

(B) Gamma energy being completely absorbed by forcing the electron from orbit and transferring energy.

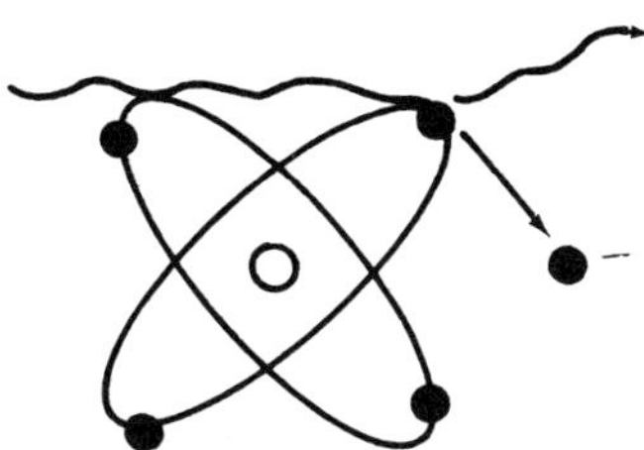

(C) Part of energy continues in new direction and part is used ejecting electron from orbit.

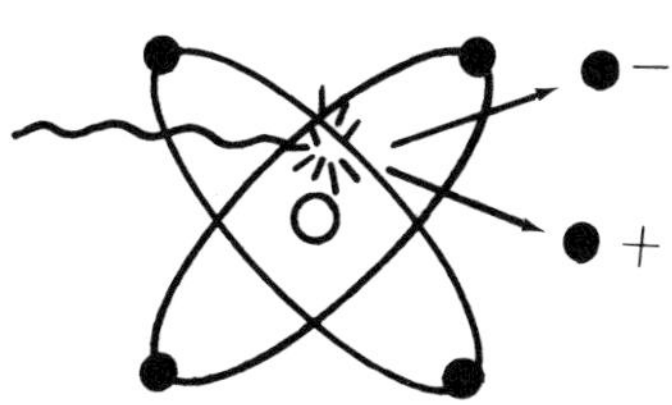

(D) Gamma radiation is annihilated. Electron and position are created and share energy.

Fig. 9-34. Interaction of gamma radiation with matter.

The energy of gamma photons is absorbed or lost in matter in three principal ways (Fig. 9-34). (1) Energy of the gamma photon may be lost or transferred to an electron it strikes, forcing the electron out of orbit. (2) Gamma photon may strike an orbiting electron a glancing blow. When this happens only a portion of the energy is used with the remainder of the energy continuing in a new direction. (3) The gamma ray or photon is completely annihilated when it passes near the powerful electrical field of a nucleus. It separates into two particles of opposite charge, an electron and a positron. The positrons lose their energy very rapidly as they collide with other orbital electrons. The net effect of gamma radiation is like alpha and beta radiation: electrons are knocked from orbit, causing ionization and excitation effects in materials.

Neutrons are uncharged particles which may collide with nuclei, resulting in alpha and gamma radiation as energy is transferred or lost.

Radiation Sources

Cobalt-60, strontium-90, and cesium-137 are three commercially available *radioisotope* sources of radiation. They are used primarily because of their availability, useful characteristics, reasonably long half-life, and cost.

Used or burnt uranium slugs from reactors or fission waste may be a useful source of radiation.

Electron accelerators such as Van de Graaff generators, cyclotrons, synchrotons, resonant transformers, and other machines may also be used for producing radiations.

Ultraviolet radiation sources from plasma arc, tungsten filaments, and carbon arc may produce sufficient radiation penetrating power for film and surface treatments.

All forms of natural radiation must be considered as harmful to polymers because they are not easily controlled.

Irradiation of Polymers

The transfer of energy from the radiation source to materials may assist in breaking bonds and in rearranging atoms into new structures. The various changes in covalent substances directly affect important physical properties.

The effects of radiation of plastics may be divided into four categories: (1) damage by radiation, (2) improvements by radiation, (3) polymerization by radiation, and (4) grafting by radiation.

Damage by Radiation

Breaking of covalent bonds by nuclear radiation is called *scission*. This separation of the carbon-to-carbon bonds may lower the molecular weight of the polymer. In Fig. 9-35 the irradiation of polytetrafluoroethylene is causing the long linear

```
  F F F F F                    F F F F     F F F F F
  | | | | |     RADIATION      | | | |     | | | | |
···C—C—C—C—C··· ~~~~~~~~> ···C—C—C—C—F   C=C—C—C—C···
  | | | | |                    | | | |     |   | | |
  F F F F F                    F F F F     F   F F F
```

Fig. 9-35. Degradation by irradiation.

plastics to break into short segments. As a result of this breaking, the plastic loses strength.

The separation of the carbon-to-carbon bonds may also form free radicals which may lead to crosslinking, branching, polymerization, or the formation of gaseous by-products. In. Fig. 9-36 radiation has caused polymerization and crosslinking of hydrocarbon radicals. In Fig. 9-36B, a gaseous product is formed in the irradiation process. The radical (R) may be H, F, C1, etc. as the gaseous product.

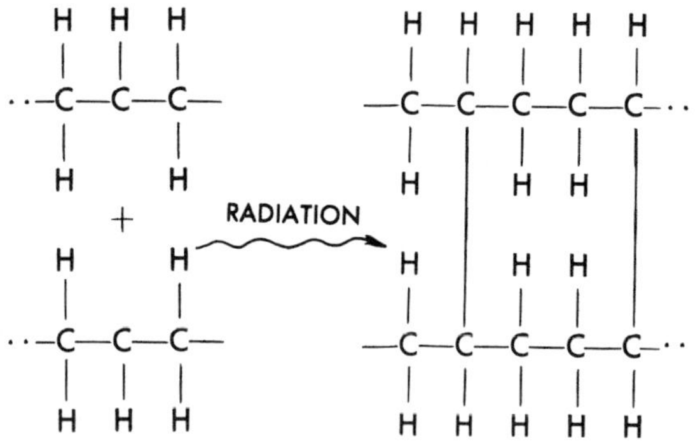

(A) Recombination leading to polymerization are crosslinking of hydrocarbon radicals.

```
  H H                         H H
  | |        RADIATION        | |
—C—C+H ~~~~~~~~~~~~~~~~> —C=C— +HR
  |
  R
```

(B) Gaseous product formed by radiation.

Fig. 9-36. Formation of free radicals due to irradiation.

The irradiation may result in atoms being knocked from the solid material. This disassociation or displacement of an atom produces a defect in the basic structure of the polymer (Fig. 9-37). These vacancies in crystalline structures and other molecular changes result in mechanical, chemical, and electrical property changes in polymers.

Crosslinkage in elastomers may be considered a form of degradation. Natural and synthetic rubbers become hard and brittle with additional crosslinkage or branching. (See Figs. 9-38 and 9-39.)

Vinylidene chloride, methyl methacrylate, and polytetrofluoroethylene have poor radiation resistance. These plastics become brittle and lose much of their desirable optical properties by discoloring and crazing. Fillers and chemical additives may be added to absorb much of the radiation energy, and heavy pigmentation of the plastics may stop deep penetration of these damaging radiations.

Improvements

While some polymers are damaged by radiation, others may actually benefit from controlled amounts. Crosslinkage, grafting, and branching of thermoplastic materials may produce many of the desirable physical properties of the thermosetting plastics. Polyethylene is an example of this phenomenon. Although polyethylene increases in brittleness with loss in elongation and impact strength, it becomes harder and has a higher melting point when irradiated with controlled dosages of nuclear radiation.

Radiation Processing

Radiation processing today is most commonly carried out utilizing electron machines or radioisotope sources such as cobalt-60. This radiation may increase molecular weight by linking polymer molecules together, or decrease molecular weight by degrading others. It is this crosslinking and degradation which accounts for most of the property changes.

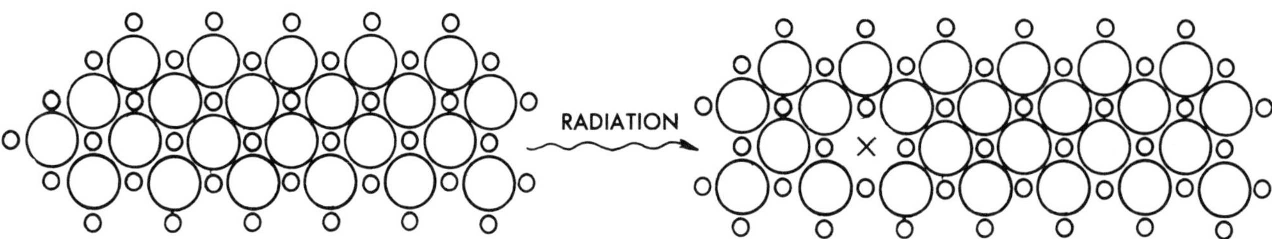

Fig. 9-37. Linear structure of plastics with missing atom. The vacancy in crystalline structure is a potential site for radical attachment.

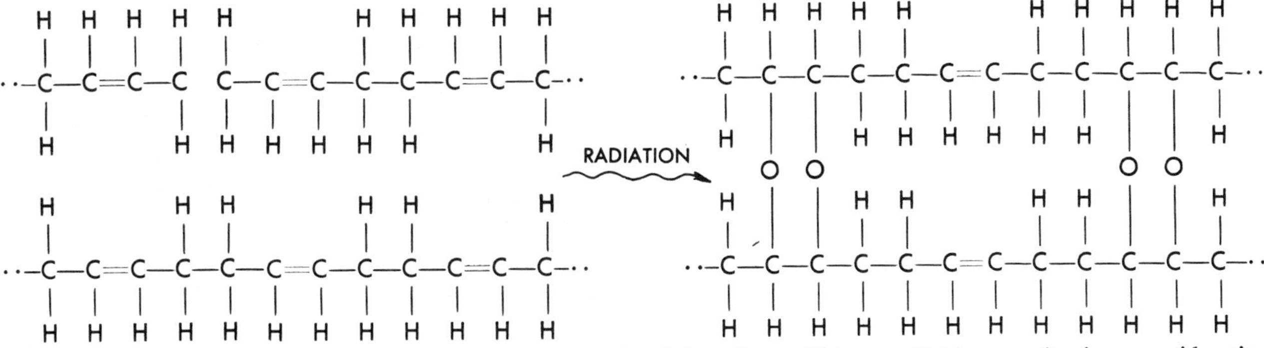

Fig. 9-38. Oxidation (radical attachment of oxygen) of polybutadiene. This crosslinking results in a rapid aging effect with loss of elastic strain.

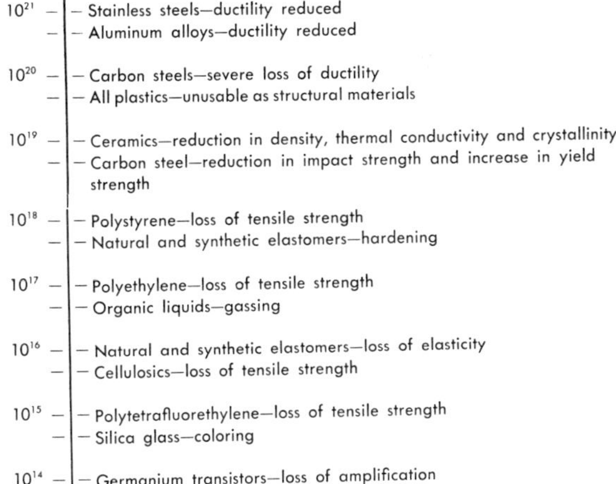

Fig. 9-39. Radiation changes the properties of plastics and other materials. Controlled, it can be beneficial.

The ability of radiation to initiate ionization and free radical formation in polymers under a wide variety of conditions may prove superior to those of other agents, such as heat or chemicals.

The main disadvantage for the industrial application of radiation induced chemical reaction in polymers is high cost of the competitive processing with chemicals. However, with the ability to integrate radiation processing directly into existing processing lines, the cost of radiation systems has been decreasing.

Ultraviolet light treatment may improve surface characteristics such as weather resistance, hardening, dyeability, and neutralizing static electricity.

Polymerization

During the dissociation of a covalent bond by irradiation, a free radical fragment is formed and is immediately available for recombination. The same nuclear energy forces which may cause depolymerization of plastics, may actually initiate "endlinking," crosslinkage, and polymerization of monomer resins (Fig. 9-40).

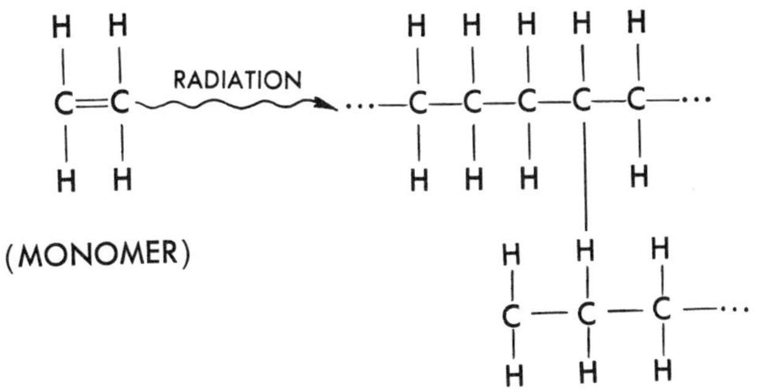

RADIATION

(MONOMER)

Fig. 9-40. Branching of polyethylene.

Grafting

When a given kind of monomer is polymerized and another kind of monomer is polymerized onto the primary backbone chain, a *graft copolymer* results. By irradiating a polymer and adding a different monomer and again irradiating, a graft copolymer is formed (Fig. 9-41).

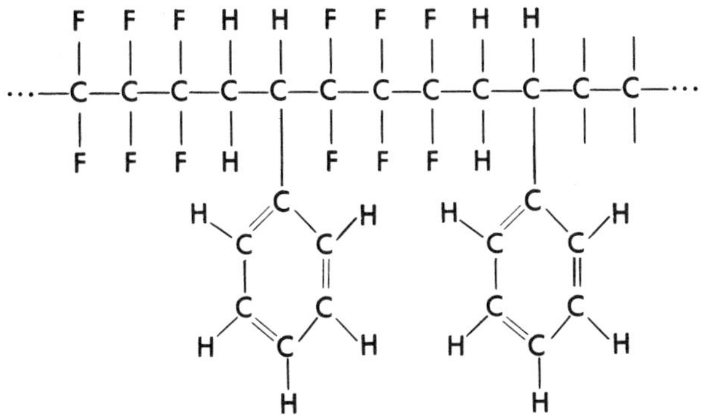

Fig. 9-41. In graft polymerization of monomer of one type (polystyrene) is grafted onto a polymer of another type (polytetrafluoroethylene).

Radiation processing may have numerous broad advantages to offset the main disadvantage.

The first advantage is that reactions can be initiated at generally lower temperatures than in operations employed in chemical processing. A second advantage is good penetration, which allows the reaction to occur inside ordinary equipment and at a uniform rate. Although gamma radiation from cobalt-60 sources can penetrate more than a foot in thickness, the treatment rate is slow and the exposure times are long. Electron radiation sources may very rapidly react with materials less than one half an inch in thickness. For these reasons, over ninety percent of irradiated products are processed by high-energy electron sources.

Because monomers can be polymerized free of chemical catalysts, accelerators, and other components which may leave impurities in the polymer, a third advantage may be listed. Fourthly, radiation induced reactions are little affected by the presence of pigments, fillers, antioxidants, and other ingredients in the resin or polymer. A fifth advantage is that crosslinking and grafting may be performed on previously shaped parts such as films, tubing, coatings, moldings, and other products. Coatings in monomeric form may be applied, eliminating solvents and collection or recovery systems for the solvents. Finally, mixing and storing various chemicals associated with chemical processing may be eliminated.

Applications

In addition to the processing advantages already mentioned, radiation processing may provide unique or other marketable features not achievable by other means (see Fig. 9-42).

Graft and homopolymerization of various monomers on paper and fabrics enhances bulkiness, resilience, acid resistance, and tensile strength. Irradiation of some cellulosic textiles have aided in the development of "dura-press" fabrics. Grafting selected monomers to polyurethane foam, natural fibers, and other plastic textiles improves weather resistance, facilitates ironing, bonding, dyeing, and printing. Small radiation dosages which produce degradation on the surface of some plastics have also proved useful for improving ink adhesion to the surface.

Fig. 9-42. Polyethylene container in center has been exposed to controlled radiation to produce superior heat resistant product. Containers at left and right were not treated and did not hold their shape at 350°F.

Impregnation of monomers in wood, paper, concrete, and selected composites have increased hardness, strength, and dimensional stability after irradiation. The hardness of pine has been increased 700 percent by this method. The novolaks and resols, which are soluble and fusible low molecular weight resins used in the production of prepregs (reinforcement impregnated resins) and impregs (resin impregnated materials), are stages of phenolic resin. The term *A-stage* is used to refer to the novolak and resol resins. Wood, fabric, glass paper, etc. may be supersaturated with A-stage resins while under a high vacuum. This supersaturated prepreg or impreg may then be exposed to cobalt radiation causing the thermosetting, A-stage material to pass through a rubbery stage referred to as the *B-stage*. With continued irradiation, further reaction leads to a rigid, insoluble, infusible, hard product. This last stage of polymerization is known as the *C-stage*. The terms A-, B-, and C-stage resins are also sometimes used to describe analogous states in other thermosetting resins.

A commercially familiar "shrinkable" radiation treated polyethylene film is commonly used for wrapping numerous food items. This irradiated film, crosslinked by radiation, has increased strength and can be stretched more than 200 percent. This film is usually prestressed and when heated to 180°F or above, the film attempts to shrink back to its original dimensions. Radiation is also being used as a heatless sterilization system for packaging of food and surgical supplies.

Radioisotopes are used in numerous measuring applications. The thickness of monomer resins, paints, or other coatings may be measured for thickness without contacting or marking the material surface. The thickness of extruded or "blown" films may be measured for thickness, which can reduce raw material consumption, reduce or eliminate scrap, ensure more uniform thickness, and speed up production (Fig. 9-43).

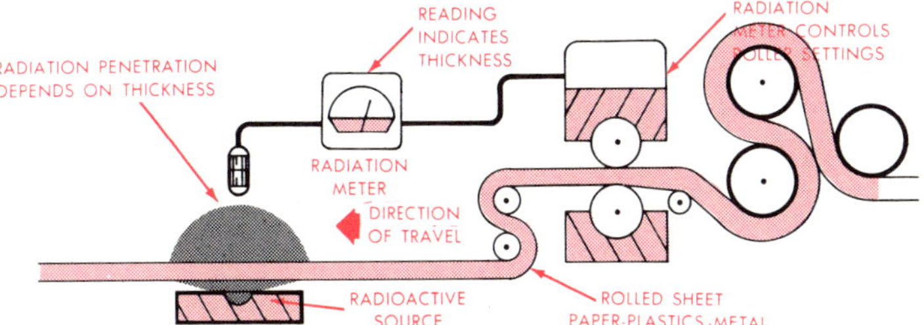

Fig. 9-43. Radioisotopes are useful for gaging thickness on a continuous basis where it is desirable to avoid contacting the item being measured. (AEC)

Chemical Properties

During the polymerization process there is a breaking and recombining of bonds. Chemical deterioration, however, is usually undesirable in plastic products. "The chemical resistance of plastics is as good as its weakest point" is a good rule of thumb statement for all materials. The types of bonds, the distance between bonds, and the energy required to break these bonds are important in considering the chemical reactivity and solubility of plastics.

The chemical resistance of the polyolefins and fluorocarbons is due to the C-C and C-F bonds that are present. These bonds are very stable, resulting in plastics which are exceptionally inert to chemical attack.

The hydroxyl groups (—OH) attached to the carbon backbone of cellulose and polyvinyl alcohol are extremely reactive. Consequently, water and other chemicals may break bonds.

The statement that "most plastics resist weak acids, alkalies, moisture, and household chemicals" must be used only as a rule of thumb. Any statement about the behavior of plastics in chemical environments must be only a generalization. It is best to test each plastics for specific applications and chemicals each plastics material is expected to resist.

Temperature, fillers, plasticizers, stabilizers, colorants, and catalysts can affect the chemical resistance of plastics. In Table 9-8 selected plastics are listed with their chemical resistance indicated at room temperature. A summary of solubility parameters of many plastics is found in Chap. 8. Chemical environments may accelerate cracks in plastics which are under a stress.

Table 9-8. Chemical Resistance of Selected Plastics at Room Temperature.

Plastics	Strong Acids	Strong Alkalies	Organic Solvents
Acetal	Attacked	Resistant	Resistant
Acrylic	Attacked	Slight	Attacked
Cellulose acetate	Affected	Affected	Attacked
Epoxy	Slight	Slight	Slight
Ionomer	Slight	Resistant	Resistant
Melamine	Slight	Slight	Resistant
Phenolic	Resistant	Attacked	Affected
Phenoxy	Resistant	Resistant	Attacked
Pollallomer	Resistant	Resistant	Resistant
Polyamide	Attacked	Slight	Resistant
Polycarbonate	Resistant	Attacked	Attacked
Polychlorotri-fluoroethylene	Resistant	Resistant	Resistant
Polyester	Slight	Affected	Affected
Polyethylene	Resistant	Resistant	Affected
Polyimide	Affected	Attacked	Resistant
Polyphenylene oxide	Resistant	Resistant	Slight
Polypropylene	Resistant	Resistant	Resistant
Polysulfone	Resistant	Resistant	Affected
Polystyrene	Affected	Resistant	Affected
Polytetrafluoro-ethylene	Resistant	Resistant	Resistant
Polyurethane	Resistant	Affected	Slight
Polyvinyl chloride	Resistant	Resistant	Affected
Silicone	Slight	Affected	Slight

In Fig. 9-44 a thermoplastic polyester test piece is being exposed to a stress cracking test. An explosively popping test bar takes its own picture under a spray of acetone that does not affect the similarly stressed test bar of thermoplastic polyester (in foreground) in a typical stress cracking test.

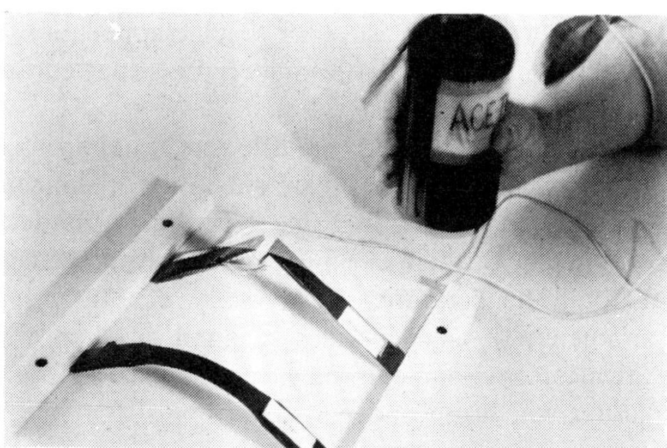

Fig. 9-44. Acetone spray causes glass-reinforced polysulfone test bar to break apart. (Celanese Plastics Co.)

The glass-reinforced polysulfone test bar at the rear cracked apart violently under the spray of acetone to break the connection that triggered the camera taking this picture. Test bars molded of thermoplastic polyester withstand even higher stresses in atmospheres of carbon tetrachloride, methyl ethyl ketone, and other aromatic chemicals and chlorinated hydrocarbons that cause cracking in polysulfone, polycarbonate, and other high-strength engineering thermoplastics.

Glossary of New Terms

Alpha particle—A particle composed of two protons and two neutrons; hence it is identical with the nucleus of a helium atom.

Angstrom—A unit of wavelength equal to 10^{-8} centimeter.

Beta particle—An elementary particle (electron) having 1.602×10^{-19} coulomb of negative charge with the rest mass of an electron, and which is emitted by an atomic nucleus.

Brittleness temperature—The temperature at which plastics and elastomers rupture by impact under specified conditions.

Centipoise—One hundredth of a poise, a unit of viscosity. Water at room temperature has a viscosity of approximately one centipoise. A poise is a metric unit of viscosity, named after the French scientist Poiseuille.

Cold flow—See Creep.

Compressive strength—The maximum load sustained by a test specimen in a compressive test divided by the original area of the specimen.

Creep—A plastics subjected to a load for a period tends to deform more than it would from the same load released immediately after application, and the degree of the deformation is dependent on the load duration. Creep is the permanent deformation resulting from prolonged application as a stress below the elastic limit. Creep at room temperature is sometimes called *cold flow*.

Damping—Variations in properties resulting from dynamic loading conditions (vibrations). It provides a mechanism for dissipating energy without excessive temperature rise, preventing premature brittle fracture, and is important to fatigue performance.

Dimensional stability—The ability of a plastics part to retain the precise shape in which it was molded, fabricated, or cast.

Fatigue (strength)—The maximum cyclic stress a material can withstand for a given number of cycles before failure occurs.

Flexural modulus—The ratio, within the elastic limit, of the applied stress on a test specimen in flexure to the corresponding strain in the outermost fibers of the specimen.

Flexural strength (modulus of rupture)—The maximum stress in the outer fiber at the moment of crack or break. This value is usually higher than the straight tensile strength in the case of plastics.

Gamma-ray—Electromagnetic radiation originating in an atomic nucleus.

Hardness—The resistance of a material to compression, indentation, and scratching.

Haze—The cloudy or turbid aspect of appearance of an otherwise transparent specimen caused by light scattered from within the specimen or from its surfaces.

Impact strength—The ability of a material to withstand shock loading. The work done in fracturing, under shock loading, a specified test specimen in a specified manner.

Index of refraction (refractive index)—The ratio of the velocity of light in a vacuum to its velocity in a transparent specimen. It is expressed as the ratio of the sine of the angle of incidence to the sine of the angle of refraction. The index of refraction of a substance usually varies with the wavelength of the refracted light.

Isotope—One of a group of nuclides having the same atomic number but differing atomic mass numbers.

Photon—The minimum amount of electromagnetic energy which can exist at a particular wavelength.

Plastic strain—The strain permanently given to a material by stresses which exceed the elastic limit.

Proportional limit—The greatest stress which a material is capable of sustaining without deviation from proportionality of stress and strain (Hooke's law). It is expressed in force per unit area, usually pounds per square inch.

Scleroscope—An instrument for measuring impact resilience by dropping a ram with a flattened cone tip from a specified height onto the specimen, then noting the height of rebound.

Stiffness—The capacity of a material to resist a bending force.

Strain—The ratio of the elongation to the gage length of the test specimen; that is, the change in length per unit of original length.

Stress—The force producing or tending to produce deformation in a unit area of substance. The ratio of applied load to the original cross-sectional area.

Toughness—A term with a wide variety of meanings, no single mechanical definition being generally recognized. The energy required to break a material, equal to the area under the stress-strain curve.

Viscosity—A measure of the internal friction resulting when one layer of fluid is caused to move in relationship to another layer. The units of measure are poises.

Review Questions

9-1. What is Young's modulus? How is it used?

9-2. How does the weight of plastics compare with that metal? Wood? Glass?

9-3. How does the heat resistance of plastics compare with that of metal? Wood?

9-4. What properties would you consider in making the following plastics products: ice chest, freezer container, gas tank, raincoat, furniture covers, counter top, fishing rod, eye lenses, and washing machine agitator?

9-5. What does the term "self-extinguishing" mean?

9-6. How do plastics react to acids? Alkalies? Solvents?

9-7. How do most plastics react to freezing temperatures?

9-8. What is creep? Cold flow? Linear expansion?

9-9. Why are plastics chosen as handles for pots, pans, and other appliances?

9-10. What is arc resistance? Dielectric strength? Dissipation (power) factor? Dielectric constant?

9-11. How are plastics grafted to one another?

9-12. How do plastics compare in compressive strength, tensile strength, and impact strength with other materials?

9-13. What is the strength-to-weight ratio?

9-14. What is specific heat?

9-15. As applied to plastics, what does atomic irradiation do to plastics or resins, and how is it used?

9-16. What does the term "damping" mean?

9-17. What are the four main categories of tests of plastics?

9-18. How does radiation affect plastics adversely?

9-19. What are ablative plastics?

9-20. What do the terms "viscosity" and "centipoise" mean?

10

Design Considerations

Plastics are often selected as substitutes for other materials because they have combinations of properties no other materials possess. Plastics, for example, are the only materials which can be simultaneously strong, light, flexible, and transparent (see Table 10-1). During the early de-

Table 10-1. Plastics vs Metals.

Properties of Plastics Which May Be . . .

Favorable

1. Lighter weight.
2. Better chemical and moisture resistance.
3. Better resistance to shock and vibration.
4. Transparent or translucent.
5. Tend to absorb vibration and sound.
6. Higher abrasion and wear resistance.
7. Self-lubricating.
8. Often easier to fabricate.
9. Can have integral color.
10. Cost trend is downward. Today's composite plastics price is approximately 11% lower than five years ago. However, the long-established, high-volume plastics—phenolics, styrenes, vinyls, for example—appear to have reached a price plateau, and prices change only when demand is out of phase with supply.
11. Often cost less per finished part.

Unfavorable

1. Lower strength.
2. Much higher thermal expansion.
3. More susceptible to creep, cold flow, and deformation under load.
4. Lower heat resistance—both to thermal degradation and heat distortion.
5. More subject to embrittlement at low temperature.
6. Softer.
7. Less ductile.
8. Change dimensions through absorption of moisture or solvents.
9. Flammable.
10. Some varieties are degraded by ultraviolet radiation.
11. Most cost more (per cubic inch) than competing metals. Nearly all cost more per pound.

Either Favorable or Unfavorable

1. They are flexible. Even rigid varieties are more resilient than metals.
2. They are electrical nonconductors.
3. They are thermal insulators.
4. They are formed through the application of heat and pressure.

Exceptions

1. Some reinforced plastics (glass-reinforced epoxies, polyesters, and phenolics) are nearly as rigid and strong (particularly in relation to weight) as most steels. They may be even more dimensionally stable.
2. Some oriented films and sheets (oriented polyesters) may have greater strength-to-weight ratios than cold rolled steels.
3. Some plastics are now cheaper than competing metals (nylons vs brass, acetal vs zinc, acrylic vs stainless steel).
4. Some plastics are tougher at low than at normal temperatures (acrylic has no known brittle point).
5. Many plastics-metal combinations extend the range of useful applications of both (metal-vinyl laminates, leaded vinyls, metallized polyesters, and copper-filled TFE).
6. Plastics and metal components may be combined to produce a desired balance of properties (plastics parts with molded-in, threaded metal-inserts; gears with cast-iron hubs and nylon teeth; gear trains with alternate steel and phenolic gears; and rotating bearings with metal shaft and housing and nylon or TFE bearing liner).
7. Metallic fillers in plastics can make them electrically or thermally conductive or magnetic.

(*Machine Design:* Plastics Reference Issue)

velopment of the plastics industry and even today, plastics are selected to fulfill a substitute roll in the production of many products. Some of these products were very successful because much consideration was given to the proper selection of materials, properties, and service life.

Basic Considerations

Before any product is made, consideration must be given to eight basic categories or requirements: (1) environment, (2) appearance, (3) electrical, (4) chemical, (5) mechanical, (6) economics, (7) process of manufacture, and (8) the design limitations.

Environment

When designing a plastics product, the chemical and thermal environments are especially important. The useful temperature range of most plastics seldom exceeds several hundred degrees Fahrenheit. Many plastics parts exposed to radiant and ultraviolet energy soon suffer surface degradation, become brittle, and lose mechanical strength. For products operating above 450°F, the fluorocarbons, silicones, polyimides, and filled plastics must be used. The exotic environments of outer space and the human body are becoming commonplace for plastics materials. The insulation and ablative materials used for space re-entry, the artery reinforcements, monofilament sutures, heart regulators, and valves are only a small portion of these new environments.

Some plastics retain their properties at cryogenic temperatures. Containers, self-lubricating bearings, and flexible tubing must function properly in below-zero temperatures. The cold, hostile environments of space and earth are only two examples. Any time refrigeration and food packaging is considered or where taste and odors are a problem, plastics may be selected. The Federal Food and Drug Administration lists acceptable plastics for packaging of foods.

The Child Protection and Toy Safety Act of 1969 is a federal statute governing the manufacture and distribution of children's toys. Should toys present an electrical, mechanical, toxic, or thermal hazard they may be banned from sale.

In addition to extremes of temperatures, humidity, radiation, abrasives, and other environments, the designer must consider fire resistance. While there are no completely fire resistant plastics, fillers or additives can make the plastics fire retardant or self-extinguishing. The dangers of direct open flames are the most serious objection to the use of plastics in fabrics and architectural structures.

Appearance

The consumer is probably most aware of a product's physical appearance and utility. This includes the design, color, optical properties, and surface finish. Elements of design and appearance encompass several properties at once. Color, texture, shape, and material may influence consumer appeal. The smooth, graceful lines of Danish style furniture with dark woods and satin finish are one example. Changing any one of these elements or properties would drastically change the design and appearance of the furniture.

One of the outstanding characteristics of plastics is that they may be transparent or colored and yet be as smooth as glass or as supple and soft as fur. For many applications, plastics may be the only materials with the desired combination of properties to fulfill functional service needs.

Electrical Characteristics

All plastics have useful electrical insulation characteristics. The selection of plastics is usually based on mechanical, thermal, and chemical properties. Much of the pioneering in plastics was for electrical uses. The electrical insulating problems of high altitudes, space, undersea, and underground are solved by the use of plastics. All-weather radar and underwater sonar could not be possible without the use of plastics to insulate, coat, and protect electronic components from hostile environments.

Chemical Characteristics

The chemical and electrical natures of plastics are closely related because of molecular makeup. There is no general rule for chemical resistance; consequently, plastics must be tested in the chemical environment of their actual use. Fluorocarbons and polyolefins are among the best chemical resistant materials. Some plastics react as semipermeable membranes, allowing selected chemicals or gases to pass while blocking others. The inability of polyethylene plastics to stop permeation of gases is an asset in packaging fresh fruits and meats. Silicones and other plastics allow oxygen and other gases to pass through the thin membrane while stopping water molecules and chemical ions. The selective filtration of minerals from water may also be accomplished with semipermeable plastics membranes.

Mechanical Factors

The fifth design consideration includes the mechanical factors of fatigue, tensile, flexural, impact, and compressive strengths, hardness, damping, cold flow, thermal expansion, and dimensional stability. All of these properties have been discussed in Chap. 9. Products that require dimensional stability will require careful selection of materials. Fillers will improve the dimensional stability of all plastics. A factor sometimes used for evaluation and selection is strength-to-weight ratio. This is the ratio of the tensile strength to the density of the material. Plastics can compare and even surpass steels in strength-to-weight ratio.

Example—Divide the tensile strength of the material by the density.

$$\text{selected plastics} = \frac{100{,}000 \text{ lbs per sq in}}{0.070 \text{ lb per cu in}} = 1.42 \times 10^6$$

$$\text{selected steel} = \frac{240{,}000 \text{ lbs per sq in}}{0.280 \text{ lb per cu in}} = 0.86 \times 10^6$$

Economics

Economics is always a decisive factor in design considerations or material selection. Some plastics cost more per pound than metals or other materials; the strength-to-weight ratio, chemical, electrical, and moisture resistance, however, may overcome this disadvantage, and plastics often cost less per finished part. The apparent density and bulk factor are important with respect to molding and cost analysis in any molding operation.

Apparent density, sometimes called *bulk density,* is the weight per unit volume of a material. It is calculated by placing the test sample in a graduated cylinder and taking measurements. The volume (V) of the sample is the product of its height (H) and cross-sectional area (A): $V = HA$. Thus

$$\text{apparent density} = \frac{W}{V}$$

where

V = volume occupied by the material in the measuring cylinder, in cubic centimeters,
H = height of the material in the cylinder, in centimeters,
A = cross-sectional area of the measuring cylinder, in square centimeters,
W = weight of the material in the cylinder, in grams.

Bulk factor is the ratio of the volume of loose molding powder to the volume of the same weight of resin after molding. Bulk factors may be calculated as follows:

$$\text{bulk factor} = \frac{D_2}{D_1}$$

where

D_2 = average density of the molded or formed specimen,
D_1 = average apparent density of the plastics material prior to forming.

Economics must also include the method of product production and design limitations. One-piece seamless gasoline tanks may be rotationally cast or blow molded. This latter process utilizes more expensive equipment but can produce the products more rapidly, thus reducing costs. Conversely, large storage tanks may be produced more economically by rotational casting than by blow molding. The number of parts to be produced and the initial production costs may be the decisive factor.

Processes of Manufacture

Thermosetting plastics are limited to a few conventional molding processes. They are not easily blow molded, extruded, or injection molded. Selected thermosetting materials, however, may be injection and compression molded. Because of moldability, production rates, and other material properties, seemingly expensive materials thus become inexpensive products.

Design Limitations

Closely related to production is the design of the product and ultimately the design of the mold to produce the product. Production rates, parting lines, dimensional tolerances, undercuts, finish, and material shrinkage are only a few factors that must be considered by the mold maker or tool designer. For the most economical production, products should not have undercuts. Undercuts and inserts slow production rates and require more expensive mold fabrication. Undercuts, inserts, parting lines, and tapers will be discussed later.

The problem of material shrinkage is of equal importance to both the designer of molds and the designer of molded products. The loss of solvents, plasticizers, or moisture during the molding process, together with the chemical reaction of polymerization in some materials, results in shrinkage. The thermal contraction of the material must also be considered a factor. Remember that the thermal expansion values for most plastics are relatively large. This is a necessary asset in removal of molded products from various mold cavities. If close tolerances are needed, material shrinkage and dimensional stability must be considered. Material shrinkage is sometimes used to ensure snug or "shrink" fits of metal inserts.

Often the selection and production considerations to be made when designing products of plastics are divided into two broad categories including (1) functional considerations and (2) production considerations. The problems encountered in producing plastics products often require the selection of the production techniques before the material or functional considerations are discussed.

Functional Considerations

To ensure proper design, there must be close cooperation between the manufacturers, processors, and the fabricators of plastic materials involved in the operation.

Plastics materials must be carefully selected with the final product application in mind. Because the properties of plastics depend more on temperature than do those of other materials and because they are more sensitive to changes in environment, many families of plastics may be limited in application. There is probably no one material that will possess all of the qualities desired. Consequently, undesirable characteristics may be compensated for in the product design.

The ultimate material selection may be based on the most favorable balance of material properties and on the total cost or selling price of the finished article. A comparative price per pound, cubic inch (in^3), or part is often used as a favorable asset for plastics. Selected plastics parts may be produced at a lower cost; the ultimate material selection and application, however, may require a more costly plastics compound. Although these materials may cost more per cubic inch, they may still surpass all property, design, and cost factors of other materials because plastics parts can often be produced with fewer processing and fabrication operations.

Consideration must be given to the design of the plastics part before it is molded, to ensure that the best combination of mechanical, electrical, chemical or thermal properties will be obtained.

There are no hard and fast rules to determine the most practical wall thickness of a molded part. Ribs, bosses, flanges, and beads are common methods of adding additional strength of a plastics part without added wall thickness. Large flat areas should be slightly convex or crowned for additional strength and to prevent warpage from strain. (See Fig. 10-1.)

In Table 10-2 various complexity of parts is shown for several processes. As the part is being molded, it is important that all parts of the mold cavity are easily filled, thus preventing much of the molding stresses in the part. A uniform wall thickness is important in the design to prevent uneven shrinkage of thin and thick sections. If wall thickness is not uniform, the molded part may be distorted, warped, and have internal stresses or cracks. From 0.250 to 0.500 inch in wall thickness may be considered heavy wall thickness in a molded part.

Plastics parts should have liberal fillets and radii to increase strength, assist molten material flow, and reduce possible points of stress concentration. The recommended minimum radius is 0.020 inch while optimum design is obtained with a radius-to-thickness ratio of 0.6.

Table 10-2. Plastics (Processing) Forms—Complexity of Part.

(Processing) Form	Section Thickness (inches) Max.	Min.	Bosses	Undercuts	Inserts	Holes
Blow Molding	> 0.25	0.01	Possible	Yes—but reduce production rate	Yes	Yes
Injection Moldings	> 1.0; normally 0.250	0.015	Yes	Possible—but undesirable; reduce production speed and increase cost	Yes — variety of threaded and non-threaded	Yes—both through and blind
Cut Extrusions	0.50	0.010	Yes	Yes—no difficulty	Yes—no difficulty	Yes — in direction of extrusion only; 0.020-0.040 inch min.
Sheet Moldings (Thermoforming)	3.0	0.00025[b]	Yes	Yes—but reduce production rate	Yes	No
Slush Moldings		0.020	Yes	Yes — flexibility of vinyl allows drastic undercuts	Yes	Yes
Compression Moldings		0.035-0.125	Possible	Possible—but not recommended	Yes—but avoid long, slender, delicate inserts	Yes—both through and blind; but should be round, large, and at right angles to surface of part
Transfer Moldings		0.035-0.125	Possible	Possible—but should be avoided; reduce production rate	Yes — delicate inserts may be used	Yes—should be round, large, and at right angles to surface of part
Reinforced Plastics Moldings	Bag: 1.0; matched die: ¼	Bag: 0.10; matched die; 0.03	Possible	Bag: yes; matched die: no	Bag: yes; matched die: possible	Bag: only large holes; matched die: yes
Castings		⅛-3/16	Yes	Yes—but only with split and cored molds	Yes	Yes

(From *1972 Materials Selector*, Materials Engineering, Reinhold Publishing Corp., Subsidiary of Litton Publications, Inc., Division of Litton Industries.)

Undercuts (either internal or external) in parts should be avoided if possible because tooling costs usually increase. The cost is for providing techniques for molding, part removal, and cooling jigs. A slight undercut may be tolerated with some products when using tough, elastic materials. The molded part may be successfully snapped or stripped out of the cavity while hot if the undercut dimension is less than five percent of the diameter of the part.

Decorating may be considered an important functional consideration for plastics design. If the product is to have instructions, labels, or letters, it must be decorated in such a manner as not to complicate removal from the mold but also provide durable service to the consumer. Letters are commonly engraved, hobbed, or electrochemically etched into mold cavities.

Production Considerations

In any product design the behavior of the material and cost is often reflected in the molding, fabrication, and assembly techniques. The tooling design must consider material shrinkage, dimensional tolerance allowances, inserts, decorations, parting lines, production rates, and other postprocessing operations (Table 10-3).

It should become obvious that any irregularities in wall thickness may create internal stresses in the molded part. Thick sections cool more slowly than thin ones and may create "sink marks." More pressure must be exerted to force material through thin sections, creating further problems. Molds may have to be made to compensate for material shrinkage. Polyethylenes, polyacetals, polyamides, polypropylenes, and some polyvinyls shrink between

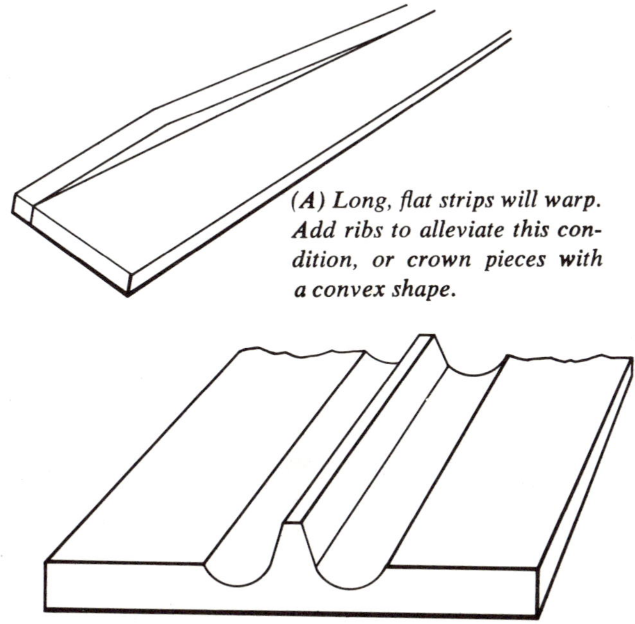

(A) Long, flat strips will warp. Add ribs to alleviate this condition, or crown pieces with a convex shape.

(B) Avoid uneven sections; they cause distortion, warpage, cracks, sinks, etc. because of the differences in shrinkage from section to section.

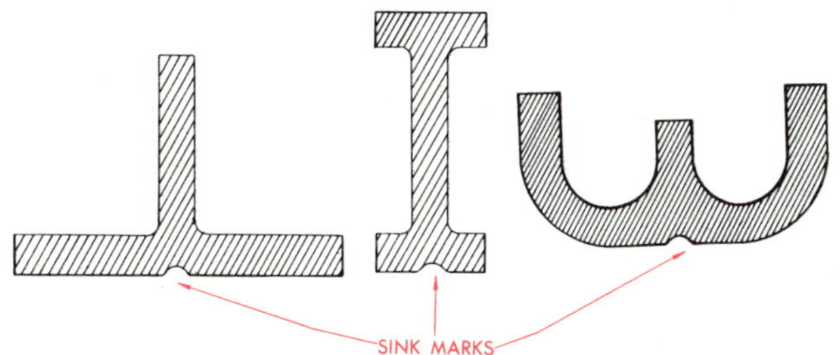

(C) Thickness of walls and ribs in thermoplastic parts should be about 60% of thickness of main walls to reduce the possibility of sink marks.

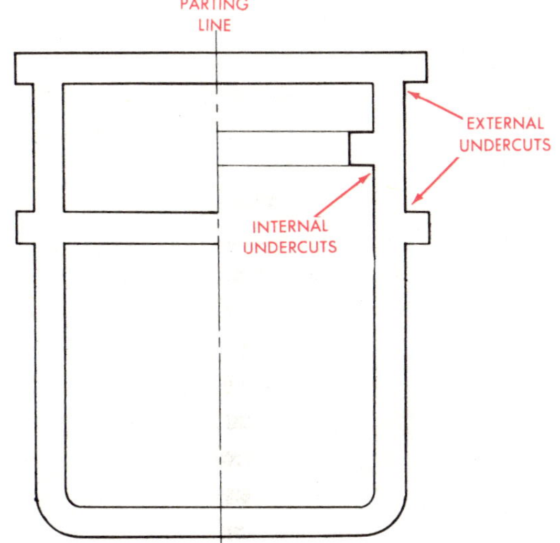

(D) Illustration of a simple molding with external and internal undercuts.

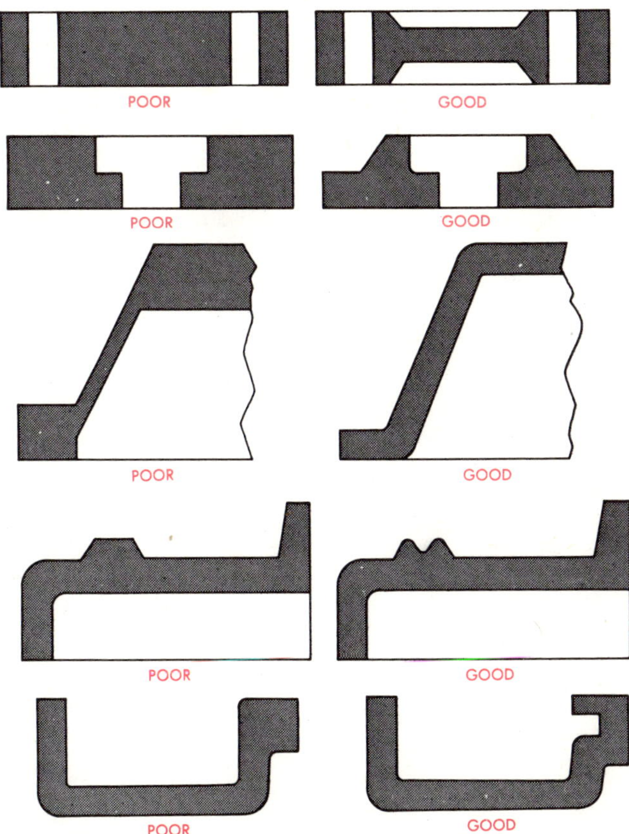

(E) Illustration of importance of uniform sectional thickness of part.

Fig. 10-1. Precautions in plastics production.

0.020 and 0.030 inch after molding. Closely related to shrinkage is dimensional tolerances. Precision tolerance molding of articles requires careful selection of materials, and tooling costs are greater. Dimensional tolerances of single cavity molded articles may be held to ± 0.002 inch per inch or less with selected plastics. As the number of cavities increase in the mold, errors in tooling, variations in shrinkage between multicavity pieces, differences in temperature, loading, and pressure from cavity to cavity all increase the critical dimensional tolerances of multicavity molds. If, for example, the number of cavities is increased to 50, the closest practical tolerance may then be ± 0.010 inch per inch.

Tolerance standards have been established by technical, custom molders and by the Society of the Plastics Industry, Inc. Standards, however, are to be use only as a guide. Each individual plastics material and design of each must be considered in determining dimensions.

Table 10-3. Design Considerations.

Plastics	Approximate Cost(¢/in³)	(Linear) Mold Shrinkage (in/in)	Dimensional Tolerances Practical (in/in) (Single Cavity)			Taper Required (degrees)		
			Fine	Standard	Coarse	Fine	Standard	Coarse
ABS	1.3 - 1.75	.005-.008	.002	.004	.006	¼	½	1
Acetal	3.1 - 4.0	.020-.025	.004	.006	.009	½	¾	1
Acrylic	1.9 - 4.0	.001-.004	.003	.005	.007	¼	¾	1¼
Alkyd (filled)	1.8 - 2.5	.004-.008	.002	.004	.005	¼	½	1
Amino (unfilled)	1.3 - 2.5	.011-.012	.002	.003	.004	⅛	½	1
Cellulosic	1.8 - 2.5	.003-.010	.003	.005	.007	⅛	½	1
Chlorinated polyether	1.7 - 2.6	.004-.006	.004	.006	.009	¼	½	1
Epoxy	2.5 - 6.5	.001-.004	.002	.004	.006	¼	½	1
Fluoroplastic (CTFE)	2.5 - 6.5	.010-.015	.002	.003	.005	¼	½	1
Ionomer	1.7 - 6.5	.003-.02	.003	.004	.006	½	1	2
Polyamide (6/6)	2.93- 3.20	.008-.015	.005	.007	.011	⅛	¼	½
Phenolic (filled)	1.15- 3.01	.004-.009	.0015	.002	.0025	⅛	½	1
Phenytene oxide	2.2 - 3.5	.001-.006	.002	.004	.006	¼	½	1
Pollyallomer	1.8 - 2.2	.01 -.02	.002	.004	.006	¼	½	1
Polycarbonate	2 - 3.5	.005-.007	.003	.006	.008	¼	½	1
Polyester (thermoplastic)	1.7 - 3.5	.003-.018	.002	.004	.006	¼	½	1
Polyethylene (high density)	0.6 - 1.5	.02 -.05	.003	.005	.007	½	¾	1½
Polypropylene	0.8 - 1	.010-.025	.003	.005	.007	1	1½	2
Polystyrene	0.7 - 1.85	.001-.006	.002	.004	.006	¼	½	1
Polysulfone	4.0 - 4.8	.006-.007	.004	.005	.006	¼	½	1
Polyurethane	2.0 - 2.4	.01 -.02	.002	.004	.006	¼	½	1
Polyvinyl (PVC) (rigid)	1.1 - 2.5	.001-.005	.002	.004	.006	¼	½	1
Silicone (cast)	10.5 - 30	.005-.006	.002	.004	.006	⅛	¼	½

There are three classes of allowable dimensional tolerances for molded plastics parts. They are expressed as plus and minus allowable variations in inches per inch. *Fine* tolerance is the narrowest possible limit of variation obtainable under controlled production. *Standard* tolerance is the dimensional control that can be maintained under average conditions of manufacture. *Coarse* tolerance is acceptable on parts where accurate dimensions are not important or critical.

In order that the part may be easily removed from the molding cavity, taper should be provided, both inside and out. The degree of taper may vary according to molding process, depth of part, type of material, and wall thickness. A draft of ¼ degree is sufficient for all shallow, plastics molded parts.

If a part had a depth of 10 inches and a ⅛ degree taper for a fine dimensional tolerance of 0.0022 inch per inch, the total taper of the piece will amount to 0.022 inch per side. See Fig. 10-2.

Parting lines (Fig. 10-3) are normally placed at the greatest radius of the molded part. If the parting line cannot be placed on an inconspicuous edge or concealed, finishing is generally required.

Knockout or ejector pins which push the molded parts from the mold must also be located in hidden or inconspicuous areas. Ejector pins should never be placed on a flat surface unless decorative designs are used to help conceal the marks.

Knockout or ejector pins are usually attached to a master bar or pin plate depending on mold design. Pins are usually drawn flush with the surface of the mold by spring action. Upon opening the die, a rod which is attached to the master bar or pin plate, strikes a stationary stop in the machine, pushing all pins forward. The part is then forced from the cavity by these pins.

Inserts and holes must be carefully designed and placed in the molding cavity or part. A liberal taper should be given long pins and plugs. A general rule for hole depth is to never have the depth more than four times the diameter of the pin or plug. Long pins are often made to meet halfway through the hole, thus gaining twice the hole depth limitation. Long pins are more easily broken and bent by molding pressures of the flowing molding compounds.

The placement of molding gates is important in relation to holes. As the molten material is forced

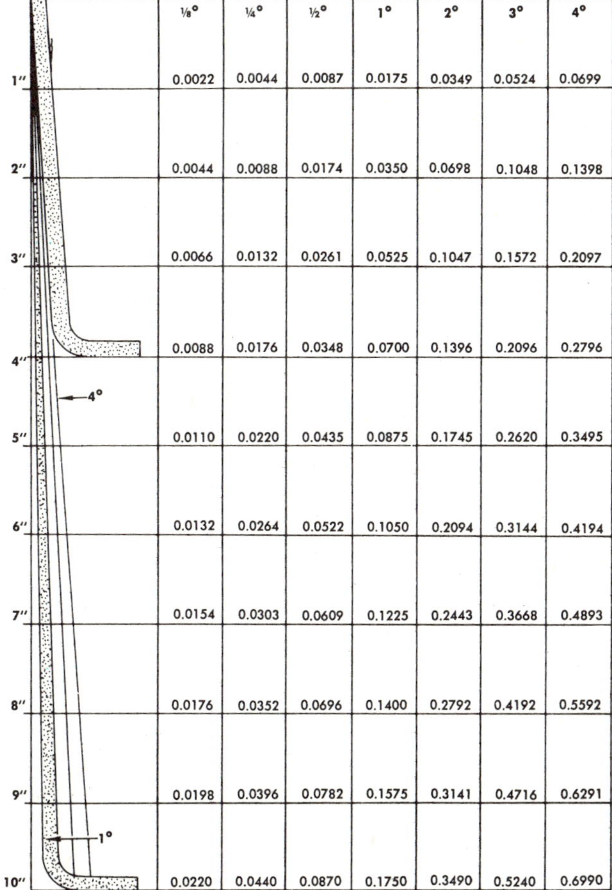

	1/8°	1/4°	1/2°	1°	2°	3°	4°
1"	0.0022	0.0044	0.0087	0.0175	0.0349	0.0524	0.0699
2"	0.0044	0.0088	0.0174	0.0350	0.0698	0.1048	0.1398
3"	0.0066	0.0132	0.0261	0.0525	0.1047	0.1572	0.2097
4"	0.0088	0.0176	0.0348	0.0700	0.1396	0.2096	0.2796
5"	0.0110	0.0220	0.0435	0.0875	0.1745	0.2620	0.3495
6"	0.0132	0.0264	0.0522	0.1050	0.2094	0.3144	0.4194
7"	0.0154	0.0303	0.0609	0.1225	0.2443	0.3668	0.4893
8"	0.0176	0.0352	0.0696	0.1400	0.2792	0.4192	0.5592
9"	0.0198	0.0396	0.0782	0.1575	0.3141	0.4716	0.6291
10"	0.0220	0.0440	0.0870	0.1750	0.3490	0.5240	0.6990

Fig. 10-2. Degree taper per side, in inches. (SPI)

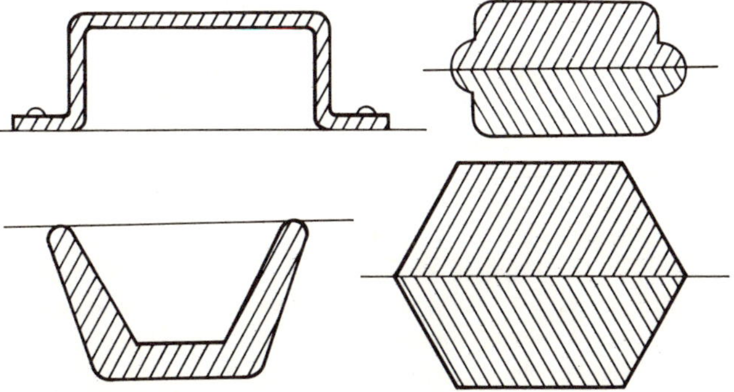

Fig. 10-3. Various locations of parting lines.

into the mold, it must flow around pins which protrude into the mold cavity. These pins are usually withdrawn when the mold opens, leaving the desired hole in the part. If parts are to be assembled on these holes, the material should be made thicker by producing a boss (Fig. 10-4). This adds strength and prevents possible sources of cracks. The placement of the pins which produced the holes often restricts the flow of material and

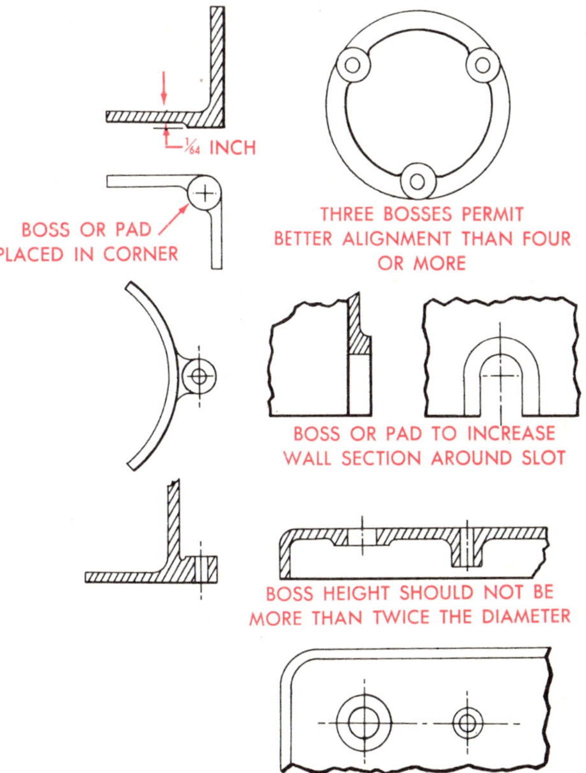

Fig. 10-4. Importance of boss design. (SPI)

causes flow marks or possible cracking because of the molding stress. Flow lines and patterns are shown in Fig. 10-5.

When inserts are used and molded in place, the molten material is forced around the insert and on cooling shrinks around the metal insert. This shrinkage substantially contributes to the holding power of the insert. Inserts may be placed in thermoplastic parts by ultrasonic techniques or posi-

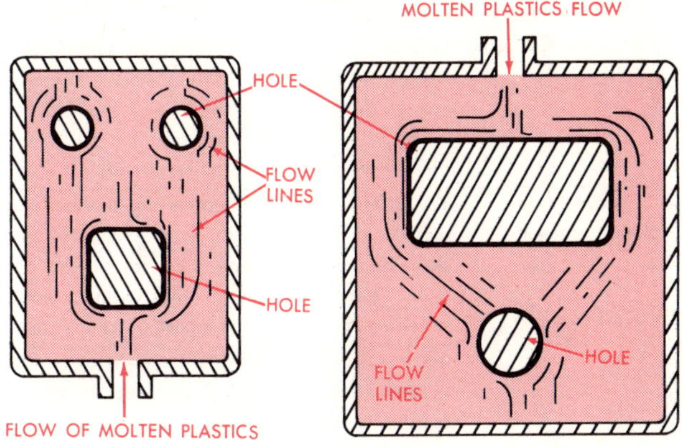

Fig. 10-5. Flow patterns must be kept in mind because flow lines may appear around holes, ribs, and opposite gates.

tioned before molding on small pins in the mold cavity by hand or automation. It is important that sufficient material is provided around all inserts to avoid cracking.

With pinpoint gating on the dies, the pieces produced may not require any gate or sprue cutting. Many dies are designed so that gates are sheared off the molded part automatically each time the press is opened.

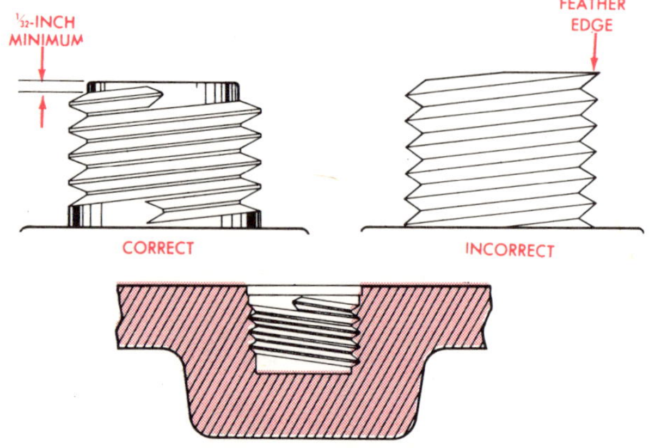

(A) Correct and incorrect threading.

Internal or external threads may be molded in plastic parts. Internal threaded parts may require an unscrewing mechanism for removal from the mold. A 1/32 of an inch clearance should be provided at the end of all threads (Fig. 10-6A). For blind corepin holes, a minimum of 1/64 of an inch should be left as clearance of screws, inserts, or other holding devices (Fig. 10-6B). Table 10-4 shows several process advantages and limitations.

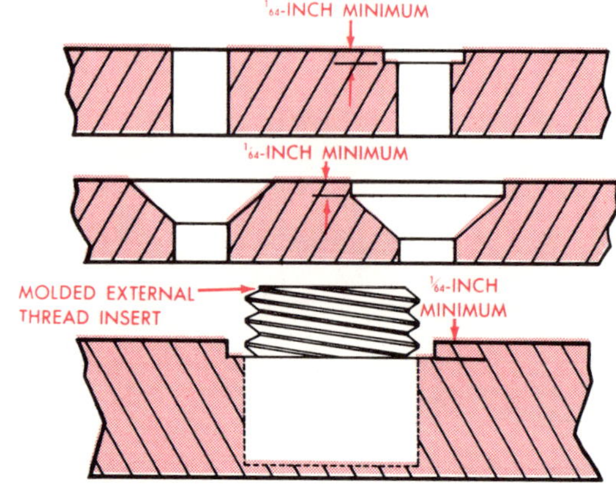

(B) Clearance of corepins.

Fig. 10-6. Putting threads and corepins in plastics.

Table 10-4. Plastics (Processing) Forms—Processes, Advantages, Limitations.

Processing Form ↓ Plastics	The Process	Advantages	Limitations
Injection Moldings	Similar to die casting of metals. A thermoplastic molding compound is heated to plasticity in a cylinder at a controlled temperature and then forced under pressure through sprues, runners and gates into a cool mold; the resin solidifies rapidly, the mold is opened, and the parts ejected; with certain modifications, thermosetting materials can be used for small parts	Extremely rapid production rate and hence low cost per part; little finishing required; excellent surface finish; good dimensional accuracy; ability to produce variety of relatively complex and intricate shapes	High tool and die costs; high scrap loss; limited to relatively small parts; not practical for small runs
Cut Extrusions	Thermoplastic molding powder is fed through a hopper to a chamber where it is heated to plasticity and then driven, usually by a rotating screw, through a die having the desired cross section; extruded lengths are either used as is or cut into sections; with modifications, thermosetting materials can be used	Very low tool cost; material can be placed where needed; great variety of complex shapes possible; rapid production rate	Close tolerances difficult to achieve; openings must be in direction of extrusion; limited to shapes of uniform cross section (along length)
Sheet Moldings (thermoforming)	VACUUM FORMING—Heat-softened sheet is placed over a male or female mold; air is evacuated from between sheet and mold, causing sheet to conform to contour of mold. There are many modifications, including vacuum snapback forming, plug assist, drape forming, etc.	Simple procedure; inexpensive; good dimensional accuracy; ability to produce large parts with thin sections	Limited to parts of low profile
	BLOW OR PRESSURE FORMING—Actually the reverse of vacuum forming in that positive air pressure rather than vacuum is applied to form sheet to mold contour	Ability to produce deep drawn parts; ability to use sheet too thick for vacuum forming; good dimensional accuracy; rapid production rate	Relatively expensive; molds must be highly polished

(**1972 Materials Selector**, Materials Engineering, Reinhold Publishing Corp., Subsidiary of Litton Publications, Inc., Div. of Litton Industries.)

Table 10-4. Plastics (Processing) Forms—Processes, Advantages, Limitations (Continued)

Processing Form ↓	The Process	Advantages	Limitations
Plastics			
	MECHANICAL FORMING—Sheet metal equipment (presses, benders, rollers, creasers, etc.) forms heated sheet by mechanical means. Localized heating is used to bend angles; where several bends are required, heating elements are arranged in series	Ability to form heavy and/ or tough materials; simple; inexpensive; rapid production rate	Limited to relatively simple shapes
Blow moldings	An extruded tube (parison) of heated plastics within the two halves of a female mold and expanded against the sides of the mold by air pressure; the most common method uses injection molding equipment with a special mold	Low tool and die cost; rapid production rate; ability to produce relatively complex hollow shapes in one piece	Limited to hollow or tubular parts; wall thickness difficult to control
Slush, Rotational, Dip Moldings	Powder (polyethylene) or liquid material (usually vinyl plastisol or organosol) is poured into a closed mold, the mold is heated to fuse a specified thickness of material adjacent to mold surface, excess material is poured out, and the semifused part placed in an oven for final curing. A variation, rotationtal molding, provides completely enclosed hollow parts	Low cost molds; relatively high degree of complexity; little shrinkage	Relatively slow production rate; choice of materials limited
Compression Moldings	A partially polymerized thermosetting resin, usually preformed, is placed in a heated mold cavity; mold is closed, heat and pressure applied, and the material flows and fills mold cavity; heat completes polymerization and mold is opened to remove hardened part. Method is sometimes used for themoplastics, e.g., vinyl phonograph records; in this operation, the mold is cooled before it is opened.	Little waste of material and reduced finishing costs due to absence of sprues, runners, gates, etc.; large, bulk parts possible	Extremely intricate parts involving undercuts, side draws, small holes, delicate inserts, etc., not practical; extremely close tolerances difficult to achieve
Transfer Moldings	Also used primarily for thermosetting materials, this method differs from compression molding in that the plastic is 1) first heated to plasticity in a transfer chamber, and 2) fed, by means of a plunger, through sprues, runners and gates into a closed mold	Thin sections and delicate inserts are easily used; flow of material is more easily controlled than in compression molding; good dimensional accuracy; rapid production rate	Molds are more elaborate than compression molds; and hence more expensive; loss of material in cull and sprue; size of parts somewhat limited
Reinforced Plastics Moldings	CONTACT—The lay-up, which consists of a mixture of reinforcement (usually glass cloth or fbers) and resin (usually thermosetting), is placed in mold by hand and allowed to harden without heat or pressure	Low cost; no limitations on size or shape of part	Parts are sometimes erratic in performance and appearance; limited to polyesters, epoxies and some phenolics
	VACUUM BAG—Similar to contact except a flexible polyvinyl alcohol film is placed over layup and a vacuum drawn between film and mold (about 12 psi)	Greater densification allows higher glass contents, resulting in higher strengths	Limited to polyesters, epoxies and some phenolics
	PRESSURE BAG—A variation of vacuum bag in which a rubber blanket (or bag) is placed against film and inflated to apply about 50 psi	Allows greater glass contents	Limited to polyesters, epoxies and some phenolics
	AUTOCLAVE—The vacuum-bag setup is simply placed in an autoclave with hot air at pressures up to 200 psi	Better quality moldings	Slow rate of production
	MATCHED DIE—A variation of conventional compression molding, this process uses two metal molds which have a close-fitting, telescoping area to seal in the resin and trim the reinforcement; the reinforcement, usually mat or preform, is positioned in the mold, a premeasured quantity of resin is poured in, and the mold is closed and heated; pressures generally vary between 150 and 400 psi	Rapid production rates; good quality and excellent reproducibility; excellent surface finish on both sides; elimination of trimming operations; high strength due to very high glass content	High mold and equipment costs; complexity of part is restricted; size of part limited

(1972 Materials Selector, Materials Engineering, Reinhold Publishing Corp., Subsidiary of Litton Publications, Inc., Div. of Litton Industries.)

Table 10-4. Plastics (Processing) Forms—Processes, Advantages, Limitations (Continued)

Processing Form ↓	The Process	Advantages	Limitations
Plastics			
Reinforced Plastics Moldings (Continued)	FILAMENT WOUND—Glass filaments, usually in the form of rovings, are saturated with resin and machine wound onto mandrels having the shape of desired finished part; finished part is cured at either room temperature or in an oven, depending on resin used and size of part	Provides precisely oriented reinforcing filaments; excellent strength-to-weight ratio; good uniformity	Limited to shapes of positive curvature; drilling or cutting reduces strength
	SPRAY MOLDING—Resin systems and chopped fibers are sprayed simultaneously from two guns against a mold: after spraying, layer is rolled flat with a hand roller. Either room temperature or oven cure	Low cost; relatively high production rate; high degree of complexity possible	Requires skilled workers; lack of reproducibility
Castings	Plastics material (usually thermosetting except for the acrylics) is heated to a fluid mass, poured into mold (without pressure), cured, and removed from mold	Low mold cost; ability to produce large parts with thick sections; little finishing required; good surface finish	Limited to relatively simple shapes
Cold Moldings	Method is similar to compression molding in that material is charged into a split, or open, mold; it differs in that it uses no heat—only pressure. After the part is removed from mold, it is placed in an oven to cure to final state	Because of special materials used, parts have excellent electrical insulating properties and resistance to moisture and heat; low cost; rapid production rate	Poor surface finish; poor dimensional accuracy; molds wear rapidly; relatively expensive finishing; materials must be mixed and used immediately

(1972 Materials Selector, Materials Engineering, Reinhold Publishing Corp., Subsidiary of Litton Publications, Inc., Div. of Litton Industries.)

Glossary of New Terms

Apparent density—The weight per unit volume of a material, including the voids inherent in the material.

Bosses—A protuberance on a part designed to add strength, provide assembly, etc.

Bulk factor—The ratio of the volume of any given quantity of loose plastics material to the volume of the same quantity of the material after molding or forming.

Fillets—A rounded filling of the internal angle between two surfaces of a plastic molding.

Gates—In injection and transfer molding, the orifice through which the melt enters the cavity. Sometimes the gate has the same cross section as the runner leading to it.

Parting lines—Marks on a molding or casting where halves of the mold met in closing.

Ribs—A reinforcing member of a fabricated or molded part.

Sprue—In the mold, the channel or channels through which the plastics is led to the mold cavity.

Undercuts—Having a protuberance or indention that impedes withdrawal from a two-piece rigid mold. Flexible materials can be ejected intact with slight undercuts.

Review Questions

10-1. List several favorable and unfavorable properties of plastics.

10-2. What considerations must be given before any product is produced?

10-3. Why is the apparent density or bulk density important in estimating the cost of a plastics product?

10-4. What considerations must be given to functional and production categories?

10-5. What is the importance of external or internal undercuts in product design?

10-6. What is the importance of uniform sectional thickness?

10-7. What are flow lines, and how may they be prevented?

10-8. Where should parting lines be placed?

10-9. How may internal threads be molded?

10-10. What are a few of the advantages and processing limitations for injection, compression, blow, and cast products?

11

Commercial Considerations

The actual production of plastics parts is a competitive business. The material selection, the processing techniques, production rates, and other variables must be considered in the selling price (see Chap. 10).

Resin manufacturers and custom molders are the best sources of information about the performance of a plastics material. In ordering the grade of resin, the quantity and the compounding of the ingredients are important variables to consider. When estimating or planning new articles, confer with resin manufacturers and indicate all specifications: (1) how the part is to be used, (2) grade of resin that is to be used, and (3) what physical requirements the finished part must withstand. Price quotations may vary a great deal. The price per pound for polyethylene may exceed 65 cents a pound in five-pound bags, 40 cents in 100-pound bags and less than 30 cents per pound by the truck or traincar load.

Plastics Molding

There has been much information written about the general properties and forming processes. However, the molding of plastics is difficult and requires a considerable amount of experience to solve many of the production problems involved with the continually changing forming techniques. As a result, only fundamental information and precautions may be given to help in understanding these basic production considerations.

The molding capacity of the equipment may be limited in production by the maximum available pressures of the press. Compression presses, for example, may vary in capacity from less than five tons to more than 2000 tons of molding pressure. Extruder equipment is usually based on the amount of material plasticized per minute, while injection machines are based more on the amount of material plasticized per cycle. In all cases the capacity of the equipment limits the size of the part. Extruder machines may plasticize less than 8 to more than 10,000 pounds per hour. Injection machines may have a maximum of less than one ounce per cycle. It is common to run most equipment at about 75 percent capacity rather than maximum capacity.

Auxiliary Equipment

Because plastics materials are poor conductors of heat and many thermoplatsics are *hygroscopic* (moisture absorptive) preheating auxiliary equipment may be required. Preliminary heating will tend to reduce the polymerization or forming time. Hopper driers are used extensively on injection and extruder machines in order to remove moisture from the molding compounds and obtain reliable molding conditions. Preheating of thermosetting materials may be accomplished by various thermal heating methods including infrared, sonic, or radio-frequency energy. This preheating may increase cure time, cycle time, eliminate streaking or color segregation, reduce molding stresses, reduce part shrinkage, and allow for more even flow of heavily filled molding compounds.

Preform equipment and loaders are important considerations for compression molding.

Injection molding and other molding techniques may require the use of regrinders or granulators to grind sprues, runners, or other cut-off pieces into usable molding materials.

Annealing tanks are sometimes used with thermoplastic products molded at high temperatures to reduce the possibility of sink marks and distortion. Large housings or pieces are often placed over shrink blocks or in jigs or dies to help maintain correct dimension with minimum warpage while the cooling process is completed. Many of the early steering wheels cracked after a length of time in service because of latent shrinkage of the molded part. Today, proper annealing and selection of materials have solved many of these problems.

Molding Temperature Control

One of the most important factors in efficient molding is the control of molding temperatures. The control system in a particular heating or temperature control zone may consist of four basic parts: (1) the thermocouple, (2) the temperature controller, (3) the power output device, and (4) the heaters. This kind of control system is shown in Fig. 11-1.

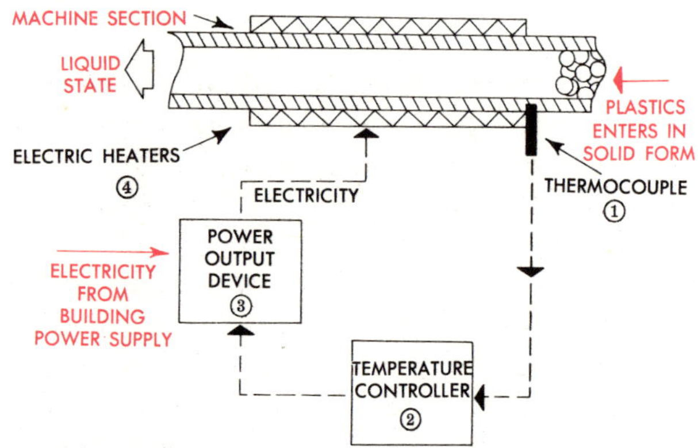

Fig. 11-1. The four basic parts of the temperature control system used by the plastics industry. (West Instrument Division)

The thermocouple is a device made of two dissimilar metals. Combinations of iron-constantan or copper-constantan metals are commonly used for plastics processing. If heat is applied to the junction of the two metals, electrons are freed and the current is measured on a meter calibrated in degrees of temperature. Over 98 percent of all temperature sensors used by the plastics industry today are thermocouples. Resistance temperature detectors (resistance bulbs) have been used for few installations. The West thermocouple installation used to minimize thermal variation is shown in Fig. 11-2.

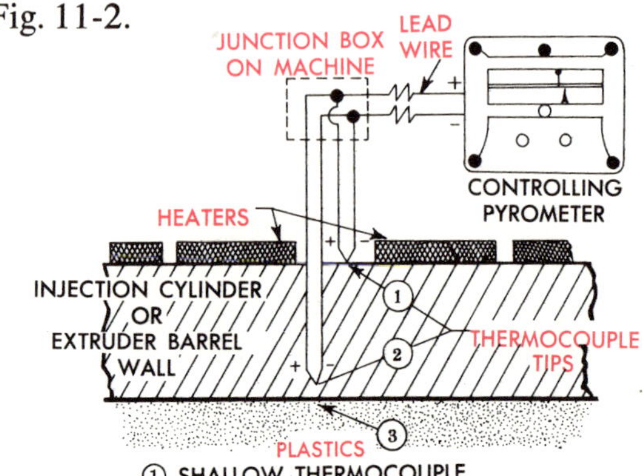

(A) Physical arrangement.

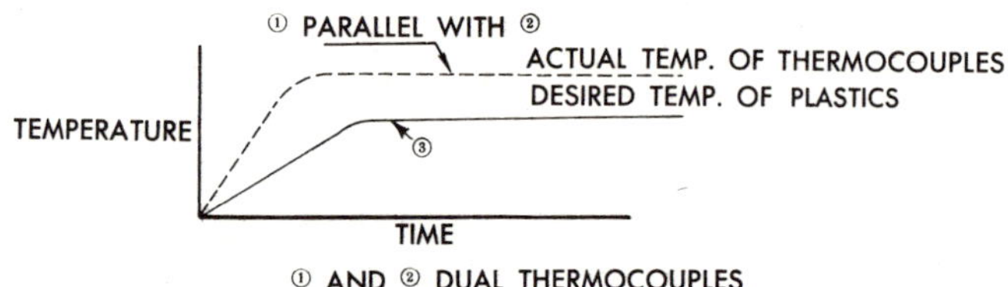

(B) Graph of temperatures.

Fig. 11-2. The physical and electrical arrangements of dual thermocouples are used to minimize thermal variation. (West Instrument Division)

The millivolt and the potentiometric controllers are two basic types of temperature controllers widely used. The potentiometric controller differs from the millivoltmeter type thermocouple controller in that the signal from the thermocouple is electronically compared to a set point temperature. Millivolt and potentiometric controllers may be designed to control the power to the heaters and hold them at a set temperature (stepless control) or designed to turn off the power to the heaters in accordance with a set temperature (proportional control). The power input device used to control the power to the heaters usually consists of a mechanical relay or a solid state control circuit. Instrumentation may also be used to control cooling cycles. Fig. 11-3 shows an instrument panel for a whole operation.

Fig. 11-3. This instrument panel is used to control a large blow molding operation. (Chemplex Co.)

Most modern machines are operated by hydraulic and electrical energy; however, a vacuum, compressed air, hot water, or chilled water may also be required. Cold water is commonly used to cool mold dies in order to reduce the molding cycle time.

Gloves, balances, scales, pyrometers, clocks and various timing devices are important accessories. There are numerous manufacturers of machine and auxiliary equipment for each processing technique used by the plastics industry.

Pneumatics and Hydraulics

Pneumatic and hydraulic actuated accessories and equipment are important considerations for much plastics processing. See Fig. 11-4.

Pneumatics (flow and pressure of air and other gases) are used to activate air-cylinders and provide compact, lightweight, and vibrationless power. Filters, air dryers, regulators, and lubricators are important accessories for pneumatic systems.

Hydraulic (fluid power systems using principles of Pascal's law) power systems may be divided into four basic components: (1) *pumps,* which force the fluid through the system; (2) *motors or*

(A) Balanced piston relief valve.

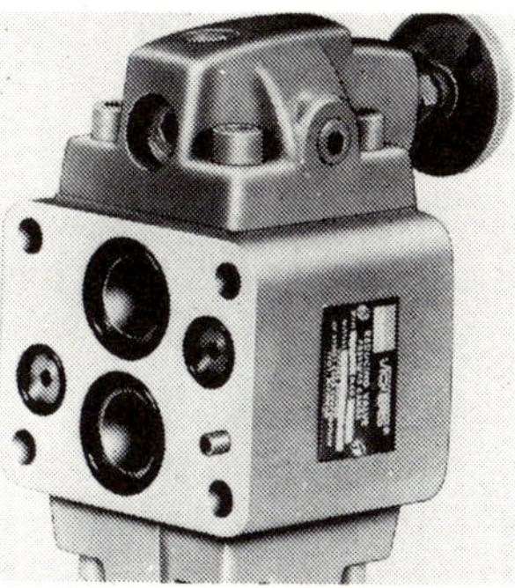

(B) Pressure reducing valve.

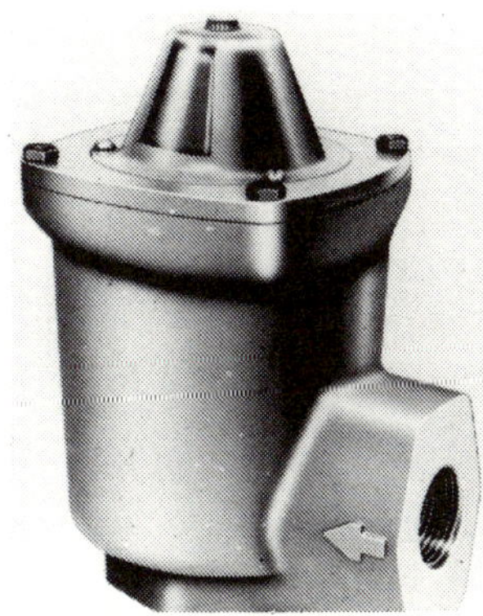

(C) Inlet filter.

(D) High-torque low-speed vane motor.

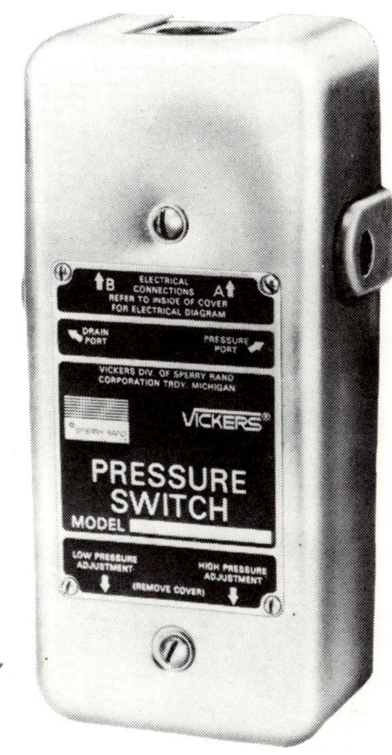

(E) Pressure switch.

Fig. 11-4. Various fluid power components. (Sperry Vickers)

cylinders, which utilize the fluid energy of the fluid pressure and convert it into rotation or extension of a shaft; (3) *control valves,* which control pressure and direction of fluid flow; and (4) *auxiliary components* including piping, fittings, reservoirs, filters, heat exchanges, manifolds, lubricators, and instrumentation.

Graphic symbols are used to show schematically information about the fluid power system. These symbols do not represent pressures, flow, or compound settings. Some devices have their schematic diagrams printed on their faces (Fig. 11-5).

It is often a difficult decision to choose between a hydraulic or pneumatic system for many applications. As a rule, when a great amount of force is required, use hydraulics; and when high speeds or rapid response are required, use pneumatics.

Since all plastics parts are formed by various processes utilizing molds or dies, it is only logical that much consideration be given to them (see Chap. 23).

Price Quotations

All price quotations for molding products should be based on the design and condition of the mold in general. In the long run, the inexpensive die is not the best buy. Compression, transfer, and injection molds are very expensive. It is common for

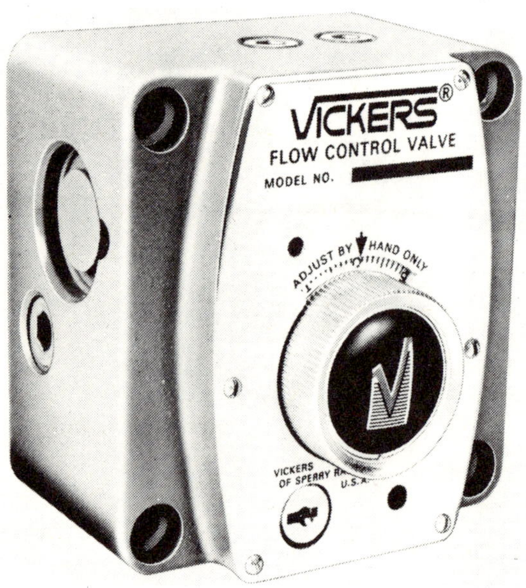

(A) Flow control valve to compensate for fluid pressure and temperature variations.

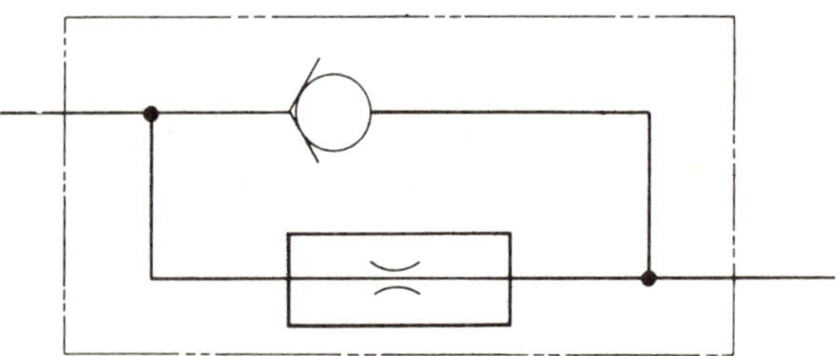

(B) Schematic diagram of fluid power flow control valve shown in (A).

(C) Control panel with schematic on face.

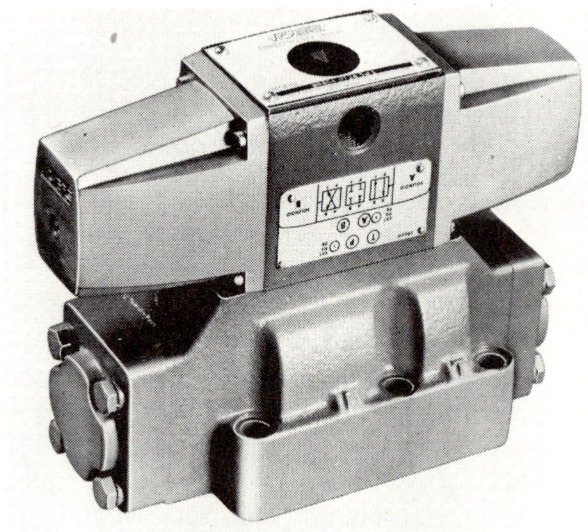

(D) Directional control valve.

Fig. 11-5. Fluid power control devices. (Sperry Vickers)

price quotations to be based on custom molds. These are molds which the customer has given to the molder to produce parts.

The greatest danger in quoting prices referring to custom molds is not taking into consideration the condition of the custom mold. All quotations should be based on the approval of the custom mold.

If the die is owned or made by the molder, it is called a *proprietary mold*. In quotations for proprietary molds, the molder must make enough profit to pay for the mold. Often a portion of the cost of making the mold is calculated into the production of each piece or per thousand pieces until there is complete *amortization* of the mold.

With custom or proprietary molds, corrosion is an enemy. Molds should be placed in storage with all accessories and given a moisture resistant, noncorrosive coating. Water, air, or steam holes should be dry and given a coating of oil.

Plant Locality

Consideration must be given the plant locality in relation to the proximity of raw material, sales offices, and potential market. Freight must be reflected in the cost of each plastic product. A quotation must reflect the tax rates, labor conditions, and wages. If there is an anticipated increase in taxes or wages, the price quotation must include this anticipated increase in production costs.

Shipping

Shipping plastics, resins, and chemicals has come under several special governmental regulations. By departmental ruling of the United States Postal Service, any liquid which gives off flammable vapors at or below a temperature of 20°F is nonmailable under provision of Section 124.2d, Postal Manual. All kinds of poison or matter containing poison are generally regarded as nonmailable by provisions of 124.2a. Caustic or corrosive substances are prohibited in 124.22. These prohibitions and special requirements are specified by the law in Section 1716 of Title 18, U.S. Code.

Acids, alkalies, oxidizing materials, or highly flammable solids, highly flammable liquids, radioactive materials, or articles emitting objectionable odors are considered nonmailable. The Department of Transportation (DOT) regulates the interstate transportation of cellulose nitrate plastics by rail, highway, or water. Special packaging must be used when shipping these materials.

Postmasters and other employees at post offices will not give opinions to the public concerning the mailability of materials. When mailers are in doubt as to whether any material is properly mailable, they should write to the Mailability Division, Office of the General Counsel, Washington, D.C., for instructions.

The Interstate Commerce Commission regulates the packing, marking, labeling, and transportation of dangerous or restricted materials. Many shipping containers are specified and regulations specify the use of labels. Flammable liquids require a red label; flammable solids a yellow label; corrosive liquids a white label. There are also other labels for poisons and radioactive material shipments. It is the responsibility of the REA or other shipping agencies to check the label to make certain it is the correct label and that it is filled in by the shipper.

If there is any doubt about the shipment of flammable or plastics materials, it is best to check local ICC offices. Certain cities have ordinances which prohibit vehicles containing these articles from operating through their tunnels or via bridges. These Interstate Commerce Commission regulations must always be observed.

Glossary of New Terms

Amortization—The gradual repayment of the cost of equipment or molds. It may be done by contribution to a sinking fund at the time of each periodic interest payment.

Annealing—A process of holding a material at a temperature near, but below, its melting point, the objective being to permit stress relaxation without distortion of shape.

Auxiliary equipment — Additional equipment needed to control or form the product. Filters, vents, ovens, takeup reels, etc.

Custom molds—Molds made by the customer and used by the molder.

Estimating—The act of appraising or valuing from statistical examples, experience, or other parameters, the cost of a product or service.

Hydraulics—Referring to the branch of science which deals with the applications of liquids in motion: the transmission, control, or flow of energy.

Hygroscopic—Tending to absorb and retain moisture from the atmosphere.

Pneumatics—A branch of science dealing with the mechanical properties of gases.

Proprietary molds—Molds made and owned by the molder.

Pyrometer—A device used to measure thermal radiation.

Review Questions

11-1. Why must the capacity of the molding equipment be considered when the production of plastics parts is to be considered?

11-2. Name several pieces of auxiliary equipment which may be required for injection, compression, transfer, blow, extrusion, or rotational processing.

11-3. Why is it important that price quotations for molding all plastics parts include the condition of the mold?

11-4. What is the difference between custom and proprietary molds?

11-5. What is the amortization?

11-6. What significance does plant locality, tax rates, labor conditions, and wages have to do with production costs?

11-7. What factors may influence the cost of each plastics product when estimating and planning new articles?

11-8. Where can you find out if a chemical or plastics resin or compound is mailable or can be transported interstate?

11-9. What type of shipping labels are required on transportation of flammable liquids, flammable solids, and corrosive materials?

11-10. When are pneumatics and hydraulic power systems selected, and what are the major differences between the two systems?

12

Health and Safety

As raw chemicals are transformed into useful consumer products, hazardous processes and materials must be used. In going from the raw material to the final product, the hazards are: solvents and intermediates (which are often flammable liquids or gases), monomers, catalysts, dusts, static electricity, processing by heat, operating equipment improperly without guards, and hydraulic, steam, air, or other pressure systems.

Hazards in Manufacture

The greatest hazard of fire comes from the many solvents, diluents, plasticizers, and other ingredients which are essential for producing and processing resins and plastics. Cellulose nitrate (nitrocellulose or pyroxylin) is unique in the plastics field. It possesses the most serious burning characteristics of all plastics and must be handled with special precautions.

Although most plastics can be made self-extinguishing, the plastics labeled slow-burning or self-extinguishing often burn vigorously under actual fire conditions and give off quantities of smoke or toxic fumes.

The National Bureau of Standards, National Fire Prevention Association, Underwriter's Laboratory, and the American Society of Testing Materials have developed tests for the flammability of plastics.

Many processes involve solvent recovery systems and the use of heating elements. Others emit explosive or flammable gases. Static charges may be generated on fast moving plastics films during processing. With flammable blowing agents and solvents which burn rapidly, it is imperative that sources of electric sparks and flame be avoided. "No smoking" signs should be posted in all potentially hazardous areas. All vapors should be quickly recovered or removed, and flammable liquids should be carefully stored.

The NFPA has indicated that there are four classes of fires and that fire extinguishers are of primary value on only one class of fire. For extinguishing plastics related fires, a Class B-C extinguisher may be selected.

NFPA Classes of Fires

Class A: Fires involving ordinary combustible materials (such as wood, cloth, paper, rubber, and many plastics) requiring the heat-absorbing (cooling) effects of water, water solutions, or the coating effects of certain dry chemicals which retard combustion.

Class B: Fires involving flammable or combustible liquids, flammable gases, greases, and similar materials where extinguishment is most readily secured by excluding air (oxygen), inhibiting the release of combustible vapors, or interrupting the combustion chain reaction.

Class C: Fires involving energized electrical equipment where safety to the operator requires the use of electrically nonconductive extinguishing agents. (*Note:* When electrical equipment is de-energized, the use of Class A or B extinguishers may be indicated.)

Class D: Fires involving certain combustible metals, such as magnesium, titanium, zirconium, sodium, potassium, etc., requiring a heat-absorbing extinguishing medium not reactive with the burning metals.

In converting monomer solutions to polymers, organic peroxides or other initiators including radiation are used. Organic peroxides decompose at low temperatures and should be treated as explosive material. Violent decomposition may occur if catalyst and promoters are mixed directly together. Cobalt naphthenate and methyl ethyl ketone peroxide are a common promoter and catalyst used to polymerize polyester resins. Dimethyl phthalate is often used as a diluent or thinner.

Many chemicals and solvents can cause severe irritation to the skin. Every effort should be made to avoid body contact with chemicals and solvents. If color or irritation develops on the skin, seek medical assistance at once.

Protective devices such as goggles, masks, and disposable polyethylene gloves may prevent skin reactions. Allergic-reactive people should be warned of possible bronchial or skin irritants. Danger warnings are dermatitis and irritation to the eyes, nose, or throat. Chemicals such as triethanolamine, osmium compounds, ammonia, and others can impair eyesight or cause blindness. B-naphthylamine crysene and selected polynuclear aromatic compounds may produce various forms of cancer. Chlorinated hydrocarbons, trichloroethane, methyl alcohol, methyl chloride, homologues of benzene, chlorobenzene, and hydrogenated cyclic hydrocarbons are considered highly toxic and must be used with caution. The chlorinated hydrocarbon carbon tetrachloride may cause permanent damage to lungs. Other compounds can cause chronic poisoning of the liver, kidneys, and nervous system.

It is best to keep all materials in properly designed containers. Some are designed with safety "pop-off" valves. Chemicals should be stored at temperatures below 90°F and in a storage area away from other products. Fireproof chests and separate rooms or buildings are ideal.

Dust explosions can occur from the ignition of fine particles of plastics. These particles may be ignited by a spark, flame, or metal surface above 700°F. Wherever regrinding, sanding, flash removal, trimming, or compounding of ingredients are involved, there is the possible danger of dust hazards. They not only present a fire hazard but are dangerous to the respiratory system.

All motors and electrical lights should have combustion covers, and a mechanical-draft duct system for removing dust or corrosive vapors should be used. Static discharge which can rapidly build up on printing equipment and various plastics processing techniques may present a possible vapor or dust ignition hazard. Polyvinyl chloride or ethylene sulfide gives off hydrogen chlorine or sulfur dioxide gases that are corrosive to metals and electrical equipment. Overheated polytetrafluoroethylene may cause a symptom similar to that produced from inhaling zinc fumes. Proper ventilation and protective clothing can effectively control irritation to eyes, nose, and lungs.

Although most chemical polymerizing temperatures seldom exceed 500°F, processing temperatures, which cause the plastics to melt or become fluid enough to be forced into molds, may exceed 650°F.

Serious burns may be experienced by exposure to high-frequency or radioactive radiations. Proper shielding must be provided if high energy outputs or long exposures are possible.

Much of the potential danger of cutting blades, moving shafts, shearing blades, or hot surfaces can be overcome if safety covers, screens, or other locking devices are installed so equipment cannot be misused by the operator. Two-hand controls may be installed to ensure that the operator's hands are not in the moving mechanism.

Equipment and machinery manufacturers are major contributors to the efforts of the plastics industry to make processing safer. Accidents are often caused while maintenance or setup operations are in progress. Operators often circumvent safety devices such as taping down limit switches. All the guards and safety devices are of little value if the operator does not use them or is unaware of their purpose. Various industrial associations offer help in the form of literature, organized programs, and authoritative consulting.

High-pressure steam, air, or hydraulic systems may rupture and present danger to the operator.

The petroleum fluids used in hydraulic systems may present additional fire hazards if the equipment is ruptured or leaking. Fire resistant emulsions or liquids are often substituted in hydraulic pressure systems.

In general, good housekeeping, knowledge of the materials or processes being used, adequate maintenance, and planned safety can prevent many potential hazards.

OSHA

The federally enacted William-Steiger Bill, better known as the Occupational Safety and Health Act (OSHA), which became effective April 28, 1971, has special significance for the plastics industry. Although the act is applicable to all businesses engaged in interstate commerce, schools and institutions of higher learning should also be aware of this important law.

The importance of OSHA cannot be overemphasized. Many states are requiring that all institutions comply with the OSHA standards and are enforcing them at the state level.

In order to become more familiar with the wide-ranging provisions of the act, various "hazard categories" have been selected from Part 1910 of Title 29 of the Code of Federal Regulations as published in Part II, Volume 37, No. 202 of the Federal Register. Table 12-1 is to be used only as a guide for the express purpose of highlighting selected "hazard categories" as outlined by law.

Ecology

The average American has become increasingly aware of the need to take proper care of our air, water, and land. One authority indicates that "Man" is the most endangered species on earth and may become extinct within 100 years.

A special report entitled "Plastics and Solid Waste Disposal" by the Phillips Petroleum Company indicated that pollution can be classified by: (1) air, (2) water, and (3) solid waste.

Air Pollution

Automobile exhaust, industrial smoke, and open dump burning are only a few examples. Plastics,

Table 12-1. Selected Rules and Regulations of OSHA.

Paragraph	Hazard Category
	Walking-Working Surfaces
1910.22-a	All places of employment, passageways, storerooms, and service rooms shall be kept clean and orderly and in a sanitary condition.
1910.36-a	This subpart contains general fundamental requirements essential to providing a safe means of egress from fire and like emergencies.
	Occupational Health and Environmental Control
1910.93	This section lists numerous acceptable ceiling concentrations for possible air contaminants. When ceiling limits or controls are not feasible to achieve to full compliance, protective equipment or other protective measures shall be used to keep the exposure of employees to air contaminants within the limits prescribed.
1910.93-b	Permissible exposure to airborne concentrations of asbestos fibers shall not exceed five fibers, longer than 5 micrometers, per cubic centimeter of air.
1910.94-3a	In reference to abrasive blasting operations: all air inlets and access openings shall be baffled or so arranged that by the combination of inward air flow and baffling the escape of abrasive or dust particles into an adjacent work area . . .
	Grinding, polishing, and buffing operations
1910.94-8b	All power-driven rotatable wheels composed all or in part of textile fabrics, wood, felt, leather, and paper, may be coated with abrasives on the periphery of the wheel for purposes of polishing, buffing, and light grinding. Grinding wheels or discs for horizontal single or double spindle grinders shall be hooded to collect the dust generated by the grinding operation . . .
1910.94	All operations involving the immersion of materials in liquids, or the vapors of such liquids, for the purpose of cleaning, the toxic, flammable, or explosive nature of the vapor, gas, or mist shall be determined and held below accepted limits for workers.
	Occupational Noise Exposure
1910.95-a	Protection against the effects of noise exposure shall be provided when the sound levels exceed those shown below. 8 hours duration ....90 dBA sound level 4 hours duration ....95 dBA sound level 1 hour duration ....105 dBA sound level Impact noise not to exceed 140 dB peak sound pressure level.
	Ionizing Radiation
1910.96-a	Radiation includes alpha, beta, gamma, x-rays, neutrons, high-speed electrons, protons, and other atomic particles. The body should not be exposed to more than 3 Rems during any calendar quarter. All radioactive areas, materials or sources

Table 12-1. Selected Rules and Regulations of OSHA. (Continued)

Paragraph	Hazard Category
	should be labeled with the prescribed radiation caution colors.
	Every employer shall supply appropriate personnel monitoring equipment, such as film badges, pocket chambers, dosimeters, or film rings for the purpose of measuring the radiation dose received.
	Nonionizing radiation
1910.97	This section applies to radiation in the radio-frequency region including microwave and radar frequencies.
	For normal environmental conditions and for incident energy frequencies from 10 MHz to 100 GHz, the radiation protection guide is 10 mW/cm^2 (milliwatt per square centimeter) as averaged over any possible 0.1 hour period. This applies whether the radiation is continuous or intermittent.
	Hazardous Materials
1910.101-a	Each employer shall determine that compressed gas cylinders are in safe condition by visual and other inspections.
1910.102	Applies to installation of gaseous acetylene systems.
1910.103	Applies to the installation of gaseous hydrogen systems.
1910.104	Applies to the installation of oxygen systems.
1910.106	Discusses the storage, ventilation, and fire control of flammable and combustible liquids. Storage of flammable or combustible liquids in drums or other containers including aerosols not exceeding 60 gallons individual capacity and portable tanks not exceeding 660 gallons individual capacity is permitted if placed in proper containers.
	Not more than 60 gallons of flammable or 120 gallons of combustible liquids may be stored in a storage cabinet.
	Every inside storage room shall be provided with either a gravity or mechanical exhaust ventilation system.
	A portable fire extinguisher having a rating of not less than 12-B units shall be located outside of, but not more than 10 feet from, the storage room door.
1910.107	Spray finishing using flammable and combustible materials, including aerated solid powders such as fluidized powder bed, dip, and spray areas.
	All spraying areas shall be provided with mechanical ventilation to remove flammable vapors, mists, or powders.
	No-smoking signs shall be conspicuously posted at all spraying areas.
	Electrostatic apparatus shall be equipped with automatic controls which operate without time delay to disconnect the power supply. Electro-

Table 12-1. Selected Rules and Regulations of OSHA. (Continued)

Paragraph	Hazard Category
	statically charged hand-held spraying equipment shall be designed so as not to produce a spark of sufficient intensity to ignite any vapor-air mixtures or result in appreciable shock hazard upon coming in contact with a grounded object.
1910.108	Dip tanks containing flammable or combustible liquids in which articles are immersed for the purpose of coating, finishing, treating, or similar processing shall have ventilating systems. No open flames, spark-producing devices, or heated surfaces having sufficient temperature to ignite vapors shall be used. Adequate arrangements shall be made to prevent sparks from static electricity.
	Personal Protective Equipment
1910.132-a	Protective equipment, including personal protective equipment for eyes, face, head, and extremities, protective clothing, respiratory devices, and protective shields and barriers shall be provided, used, and maintained in a sanitary and reliable condition wherever it is necessary by reason of hazards of processes or environment. Irritants may cause injury or impair the function of any part of the body through absorption, inhalation, or physical contact.
1910.133	Eye and face protection shall be required where there is a reasonable probability of injury. Suitable eye protectors shall be provided where machines or operations present the hazard of flying objects, glare, or liquids injurious radiation, or a combination of these hazards.
1910.134-a2	Respirators shall be provided by the employer when necessary to protect the health of the employee.
	General Environmental Controls
1910.141	All places of employment, passageways, storerooms, and servicerooms shall be kept clean and orderly and in a sanitary condition. Potable water shall be provided for drinking . . . Adequate toilet facilities which are separate for each sex shall be provided.
1910.144	Safety color codes for marking physical hazards shall be used.
	Red: for fire protection equipment and apparatus.
	Orange: for designating dangerous parts of machines.
	Yellow: for marking physical hazards.
	Green: for designating "Safety" and first aid equipment.
	Blue: for warning against starting or moving equipment under repair.
	Purple: for designating radiation hazards.
	Black/white: designating traffic.

Table 12-1. Selected Rules and Regulations of OSHA. (Continued)

Paragraph	Hazard Category
1910.145	This section includes specifications for the design, application, and use of signs or symbols to define or indicate specific hazards.
	Fire Protection
1910.157	Portable extinguishers shall be maintained in a fully charged and operating condition and kept in their designated places at all times. Extinguishers shall be selected for the specific class or classes of hazards to be protected from.
	Machinery and Machine Guarding
1910.213	Woodworking machinery requirements—each circular hand-fed ripsaw shall be guarded by a hood which completely encloses the blade. It shall be furnished with spreader to prevent material from squeezing the blade and fingers or dogs to prevent kickback. Radial saws shall have a hood that completely encloses the upper portion of the blade down to a point that will include the end of the saw arbor. When used for ripping, nonkickback fingers shall be provided. All portions of the bandsaw blade shall be covered except the working portion between the rollers and the table. Jointers shall have a guard which will cover the section of the cutting head and automatically adjust itself to cover the unused portion of the head. The guard shall remain in contact with the material at all times.
1910.216	Mills and calender in the rubber and plastics industries shall have a safety trip control in front and back of each mill. The triprod, cable, or wire shall extend the length of the face of the rolls.
1910.219	This section indicates that most gears, belts, or flywheels must be covered to protect the worker unless they are very slow moving or not used for mechanical power-transmission.
	Hand and Portable Powered Tools and Other Hand-Held Equipment
1910.242-b	Compressed air shall not be used for cleaning purposes except where reduced to less than 30 psi and then only with effective chip guarding and personal protective equipment.
1910.243	Portable circular saws shall be equipped with guards above and below the base plate or shoe. When the tool is withdrawn from the work, the lower guard shall automatically and instantly return to covering position. Belt sanding machines shall be provided with guards at each nip point where the sanding belt runs into a pulley. All hand-held abrasive wheels shall be provided with safety guards.
	Electrical
1910.308	The National Electrical Code NFPA and ANSI.

specifically polyvinyl chloride and polystyrene, have received the most crticism with respect to air pollution. They both emit smoke, and polyvinyl chloride releases chlorine gas when improperly burned in open dumps. When chlorine combines with moisture in the air, it forms hydrochloric acid (HCl) as does other household and municipal waste such as common salt from food, paper, grass clippings, wood, and leather. Wool, when improperly burned, produces a deadly hydrogen cyanide gas. Other nonplastics components are also responsible for sulfur oxides and other corrosive and polluting agents.

When plastics is burned in open pits or dumps it is indistinguishable from pollution produced by other sources. In the nation, over 80 percent of the solid waste is disposed of by burning in open dumps. The chief advantage of incineration is the reduction of weight and volume of the refuse. Volume may be reduced over 90 percent during proper incineration.

Modern incinerator technology may be the best solution to reduce the volume of waste and prevent pollution of the air.

A study by DeBell & Richardson, Inc., an independent research firm, concludes on the basis of experiments in several countries that when plastics are properly incinerated, carbon dioxide (CO_2) and water (H_2O) are the principal products. It is when they are improperly incinerated (in open dumps) that carbon monoxide, soot, and black smoke results. This report indicated that properly designed incinerators could handle domestic, municipal, or industrial wastes containing ten times the present levels of PVC and still be well within established safety limits of hydrogen chloride emission levels.

The charge that plastics clog and burn incinerator grates is unfounded. In fact, plastics are desirable as fuel in modern incinerators because of their high Btu (British thermal unit) or heat content. Plastics may have three to four times the heat content that other combustible waste has. Plastics in waste may have a 15,700 Btu heating value per pound.

According to Professor Elmer R. Kaiser of New York University, an expert on incinerator processes, "When grates get clogged or burned out under ordinary conditions, it is not the fault of plastics. Plastics volatize at 600°F, and then the volatiles *burn* above the grate. They do not touch the grate, and no incinerator grate will burn out at 600°F."

Utilization of the high Btu content of plastics in waste has been done for years in heat recovery incinerators in Europe. Chicago has an incinerator system in the northwest part of that city that disposes of waste, and the enormous heat energy is used to generate electric power. Several other American cities are exploring the idea of building efficient, nonpolluting incinerators which serve as power plants.

According to the Bureau of Solid Waste Management, 75 percent of the nation's 300 municipal incinerators are substandard and do not burn refuse fully or properly.

Modern pollution control equipment and properly controlled combustion will help solve many air pollution problems.

Water Pollution

The waste materials emitted from industries, municipal sewage disposal plants, commercial shipping, agriculture, and other sources are familiar water pollutants. Many of these waste products are of value. In early petroleum development, gasoline was emptied into rivers or burned as waste. Modern sewage plants are seeking economical methods of collecting methane and other gases for possible polymer use. Many companies and industries are installing adequate waste disposal systems rather than dumping them into sewer lines or rivers. All litter in water or soil is an "eyesore" and potentially hazardous. Plastics, however, do not decompose and release explosive gases, nor do they contaminate water.

Solid Waste

Automobile bodies, grass trimmings, packaging materials, and agriculture wastes are familiar solid waste pollution. The four most used methods for disposal of solid waste are: (1) open dumping, which usually results in burning, (2) sanitary landfill, which potentially pollutes water and releases toxic or explosive gases, (3) incineration, which may make a significant contribution toward cleaner air and more healthy environment, and (4) composting, which consists of shredding or pulverizing wastes into the soil as conditioners.

Open dumping is a health hazard and an eyesore. Many communities have banned this method of refuse disposal. Of the four billion tons of solid waste generated each year in the United States, rubbish, trash, and garbage are now recognized as a major problem. Only one percent of this total refuse is presently composed of plastics.

Unfortunately, open dumping is still the most common method of disposing of refuse. These dumps are breeding grounds for vermin and insects. They threaten disease, cause fires, odors, pollute ground water, and blight the land.

Today, about seven percent of our municipal solid waste goes into *sanitary landfills*. The most desirable material for sanitary landfill is dirt, broken concrete, bricks, and other dense materials. Sanitation authorities welcome plastics in sanitary landfills because they do not decompose or produce polluting odors, gases, or liquids. Other organic materials in landfills leach out chemicals which may pollute ground water supplies.

Degradable landfills are often overrated and may create many secondary problems as they rot and burn. Paper may take more than 60 years to fully degrade.

Incineration of municipal waste and its use in future incinerator-power plants can significantly reduce air pollution and reduce refuse volume.

Plastics are good as bulk factors and mulch; however, most plastics do not decompose. *Composting* of plastics is desirable because this inert material acts as a bulk factor retarding soil compacting and thus allows air to circulate and thus facilitate the decomposition of other organic materials.

There are more biodegradable plastics on the market each year, but they represent only a small proportion of the total used in packaging.

The use of edible and water soluble plastics films in packaging is growing. Examples include casings for meats, drug capsules, and coatings for fruits. Cellulose films are based on alpha-cellulose and have been modified to form hydroxypropyl-methyl cellulose. This film is soluble in both organic solvents and water. This film is a thermoplastic resin and may be extruded, injection molded, and blow molded.

A nondigestible coating material used as a protective coating on poultry and other food products is based upon 30 to 40 percent cellulose acetate butyrate and 60 to 70 percent acetylated monoglycerides. Cellulose acetate butyrate is not a digestible material.

Water-soluble, nonedible films may include polyvinyl alcohol and polyethylene oxide. Polyvinyl alcohol films are used as protective coatings and for packaging all products which are ultimately dissolved in water. Polyethylene oxide films are used for soluble laundry bags and for packaging a variety of industrial and household products. Both water-soluble films are made so that they will go into complete solution at temperatures as low as 40°F without any traces of residue. See Fig. 12-1.

In the United States nearly 400 million tons of refuse is collected each year, at a cost of over five billion dollars. Of this total, only three percent of it is plastics.

Although Fig. 12-2 indicates that three percent of solid waste is attributable to packaging and other plastics refuse, many cities estimate that plastics may compose more than six percent of the total waste volume.

The volume of plastics in solid waste is not large, but is growing. In 1974 the total weight of plastics waste may be estimated at over three million tons with about two-thirds (two million tons) of this share as plastics packaging waste and the remainder (one million tons) as all other plastics waste.

(A) Closeup of plastics film strip.

(B) Showing width and layout of strips.

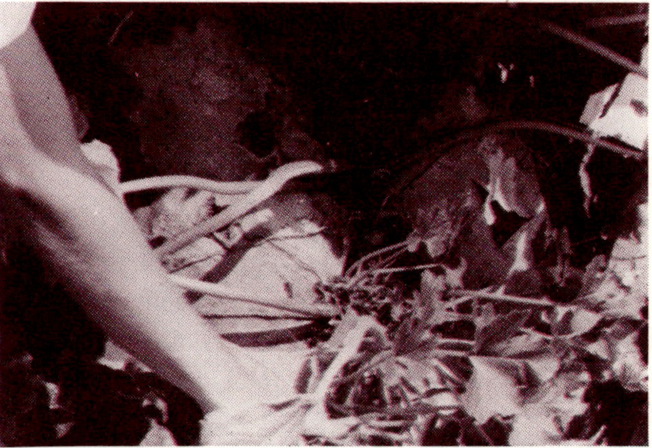

(C) Showing film around plant.

Fig. 12-1. Rows of vegetable crops, transplanted through a degradable plastics film, are growing vigorously (Princeton Chemical Research)

Recycling

The ultimate solution to pollution and solid waste is reclamation or recycling of waste materials. The largest single obstacle to reclamation of plastics is an economical, practical method of separating all materials including plastics. Until plastics are used in a much larger volume, the large volume separation and collection of plastics is not feasible. At the manufacturing level, recycling of scrap plastics is a standard practice. There are new technologies and research projects being developed to make the reuse of plastics economically feasible. The Portland Cement Research Institute is thesting the use of scrap plastics as an ingredient of concrete. "Plastcrete" is lightweight and is as strong as conventional concrete.

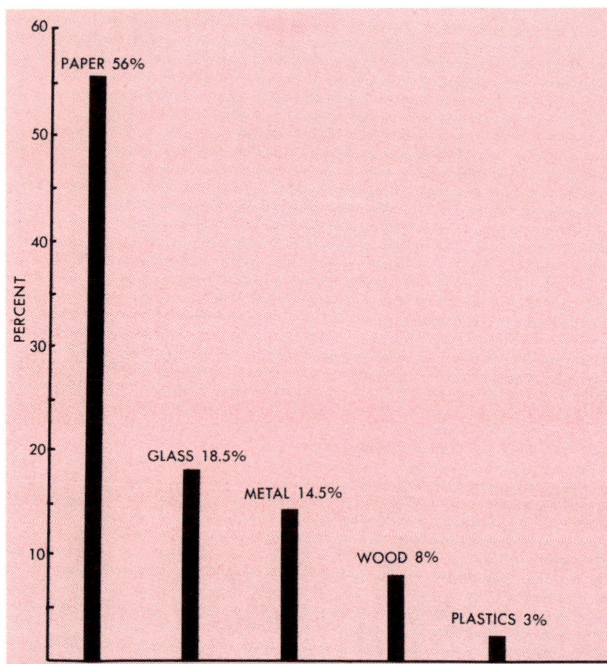

Fig. 12-2. Percent of solid waste attributable to disposable packaging and service containers. (*The Role of Packaging in Solid Wastes Management,* Public Health Service Bulletin No. 1855)

Pyrolysis may provide another method to extract basic plastics chemical compounds from waste plastics. Pyrolysis is a process in which plastic waste is chemically changed by controlled molecular degradation through the application of heat and in the absence of air or oxygen. The resulting compounds from waste plastics include ammonia, oxygen, hydrogen chloride, oil, tar, and a variety of grease and waxlike materials.

The litter and waste disposal problem may be helped if more materials could be economically collected and reprocessed. It has been estimated that more than a billion dollars a year is spent on litter collection along highways, parks, and streets. Litter is a social problem and man is apparently not willing to limit the variety of goods and the packaging conveniences given to him through the use of plastics.

Odor and noise pollution have also been given increased attention as to their effects on man and environment. These effects can involve a remarkable range of problems not only chemical and physical but also social, physiological, and even psychological.

New York State Supreme Court Justice, Saul S. Streit, indicated in his 1971 decision " . . . there is not one shred of evidence presented herein which demonstrates that any form of container, glass, metal, or paperboard, is any more recyclable than plastics containers . . . It thus appears that the discrimination against plastics containers does not rest on any reasonable basis in relation to the objective of promoting the recycling of containers, plastics or otherwise."

Man has the technology for solving his pollution problems. All he has to do is apply that knowledge.

Glossary of New Terms

Air pollution—Filling the atmosphere with undesirable particles some of which may be toxic.

Composting—Grinding trash into small pieces and mixing it with the soil.

Fire resistant—Indicating a material does not easily burn.

Incineration—Burning of municipal waste in a specially designed incinerator.

NFPA—National Fire Prevention Association.

Open dumping—Placing trash or waste in open, uncontrolled areas on the land.

OSHA—Occupational Safety and Health Act.

Pyrolysis—A process used to change waste chemically into usable compounds.

Recycling—Collection and reprocessing of waste materials.

Sanitary landfill—The controlled filling of lowlands or trenches with solid, municipal waste.

Self-extinguishing—A somewhat loosely used term describing the ability of a material to cease burning once the source of flame has been removed.

Solid waste—Refuse that does not rot or decay, such as dirt, concrete, bricks, and many plastics.

Water pollution—Waste products placed into rivers and waterways.

Review Questions

12-1. What class extinguisher may be selected for plastics related fires?

12-2. What chemicals may be toxic when they are inhaled?

12-3. What materials may cause cancerous ulcers on the body?

12-4. What compounds can cause chronic poisoning of the liver, kidneys, and nervous system?

12-5. What materials may cause dermatitis or be absorbed through the skin?

12-6. Where and how should chemicals and plastics be stored?

12-7. What are the dangers in using high-frequency or radioactive radiations?

12-8. What plastics compounds degrade upon high temperature processing and release toxic gases?

12-9. What significance does the Occupational Safety and Health Act of 1971 have to plastics processors? Equipment manufacturers?

12-10. Why is dust, steam, air, and hydraulic pressure dangerous?

12-11. What are the three broad categories in which pollution may be classed?

12-12. Why are plastics desirable in modern incinerators?

12-13. Name four sources of air, water, and solid waste pollution.

12-14. Approximately what proportion of the solid waste refuse is composed of plastics materials?

12-15. What may possibly be the ultimate solution to the problems of pollution and solid waste?

12-16. What compounds are produced from the process called "pyrolysis?"

12-17. What is the difference between open dumping and sanitary land fills?

12-18. How does composting of plastics help soil?

12-19. Name several products where edible and water soluble plastics may be used.

12-20. What are several methods for disposing of plastics?

12-21. Can plastics be made to degrade?

12-22. Do plastics damage incinerators?

12-23. Why are plastics desirable as landfill?

12-24. Can plastics be recycled?

12-25. What does "pollution" mean?

12-26. What are the current methods of disposing of municipal refuse?

12-27. What may you envision as the ultimate long-term solution to the solid-waste disposal problem?

13

Thermoplastics and Their Properties

This chapter covers individual groups of thermoplastics in alphabetical order.

(Poly)Acetal Plastics

A highly reactive gas, formaldehyde (CH_2O), may be polymerized in a number of ways. Formaldehyde of methan*al* is the simplest of the aldehyde group of chemicals. The characteristic ending for the aldehydes is *al*, derived from the first syllable of *al*dehyde.

Because formaldehyde is a water soluble gas with a boiling point of $-21°C$, it is usually shipped as a 40 percent aqueous solution called *formalin.*

Although simple polymers based on formaldehyde have been known since 1859, the first commercially available polyformaldehyde was introduced by du Pont in 1960. The polyformaldehyde or polyacetal polymer is basically a linear, highly crystalline, long molecular structure. The term *acetal* refers to the oxygen atom which joins the repeating units of the polymer structure.

A number of initiators or catalysts are used to polymerize the basic polyacetal resin including acids, bases, metallic compounds, cobalt and nickel.

The polyacetal structure is:

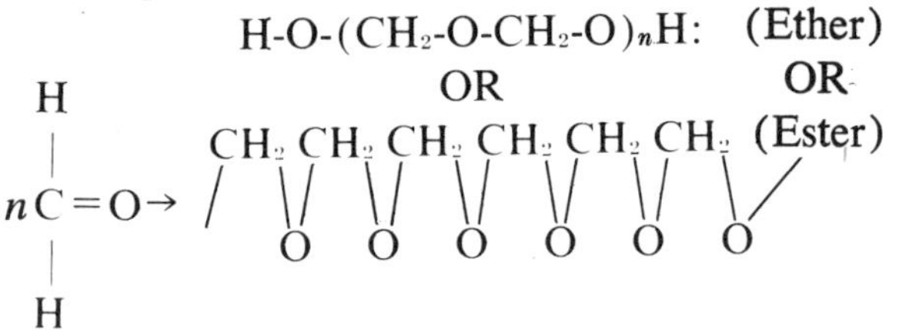

$$H\text{-}O\text{-}(CH_2\text{-}O\text{-}CH_2\text{-}O)_nH: \text{ (Ether)}$$

OR

$$n\,H_2C{=}O \rightarrow \ \diagup\, CH_2\text{-}O\text{-}CH_2\text{-}O\text{-}CH_2\text{-}O\text{-}CH_2\text{-}O\text{-}CH_2\text{-}O\text{-}CH_2\text{-}O \diagup \text{ (Ester)}$$

The best known polyformaldehyde plastics is the oxymethylene linear structure with attached terminal groups. There are, however, numerous miscellaneous aldehyde derived polymers.

Thermal and chemical resistance is increased when esters or ethers are attached to terminal groups. Both esters and ethers are relatively inert toward most chemical reagents and chemically compatible in organic chemical reactions.

H-O-H	R-O-R	R-C(=O)-OH	R-C(=O)-OR
Water	*Ether*	*Carboxylic Acid*	*Ester*

The formulas above indicate some of the structural relationships between water, ethers, carboxylic acid, and esters.

The close packing and short bond lengths of the polyacetal plastics result in a hard, rigid, dimensionally stable material with high resistance to organic chemicals and a wide operating temperature. See Fig. 13-1.

Because they offer easy fabrication and properties not found in metals, polyacetals have become competitive in cost and performance with many nonferrous metals. Polyacetals are similar to polyamides (nylon) in many respects. Acetals, however, may be considered superior in fatigue endurance, creep resistance, stiffness, and water resistance. Although among the strongest and stiffest thermoplastics, they may be filled for

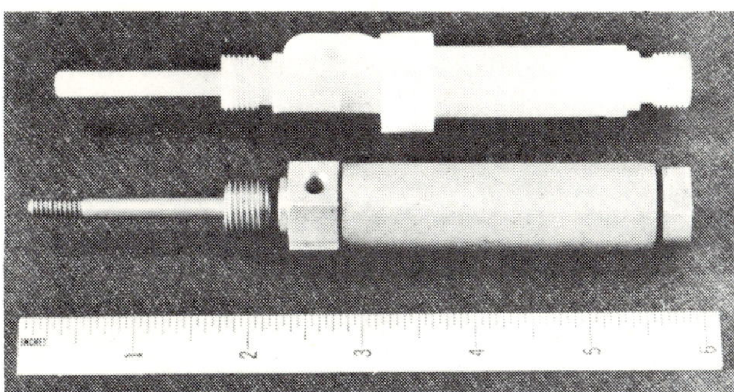

Fig. 13-1. Designed for bursting pressures above 1000 psi and capable of serving within an environment of −150°F to +250°F, this polyacetal cylinder costs less than half its metal counterpart. (Celanese Plastics Co.)

greater strength, dimensional stability, abrasion resistance, plus improved electrical properties (Fig. 13-2).

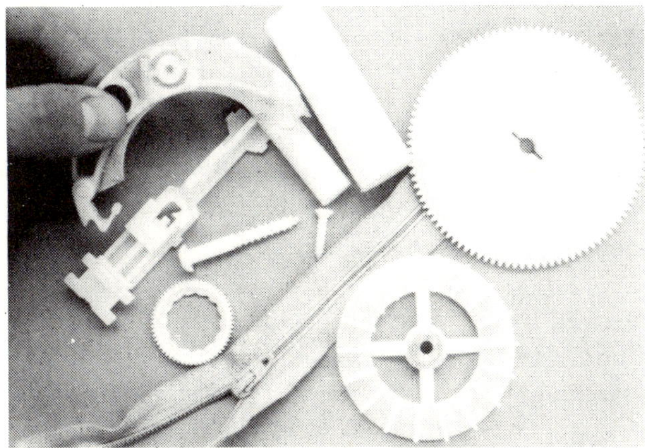

Fig. 13-2. Miscellaneous polyacetal parts.

At room temperatures, acetals are resistant to most chemicals, stains, and organic solvents including tea, beet juice, oils, and household detergents. Hot coffee will usually cause staining. Their resistance to strong acids, strong alkalies, and oxidizing agents is poor; however, copolymerization and filling generally improves chemical resistance.

Exceptional moisture and thermal resistance are among the reasons acetal polymers are used for plumbing fixtures, pump impellers, conveyor belts, aerosol stem valves, and shower heads. See the illustration in Fig. 13-3.

Acetals must be protected from prolonged exposure to ultraviolet light, which causes surface chalking, reduced molecular weight, and gradual

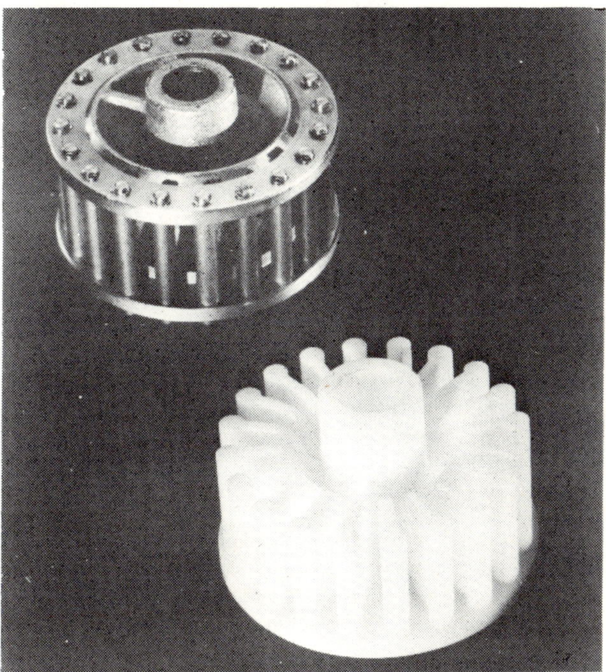

Fig. 13-3. Polyacetal plastics pulley is less than one-third the cost of the metal pulley. (Cellanese Plastics Co.)

degradation. Painting, plating, and filling with carbon black or ultraviolet absorbing chemicals will protect products for outdoor use.

Acetals are available in pellet or powder form and may be processed in conventional injection molding, blow molding, and extrusion equipment. It is not possible to make optically transparent film, because of the highly crystalline structure of polyacetals. Adequate ventilation must be provided the operator when processing polyacetal materials. Upon degrading at high temperature, acetals release a toxic and potentially lethal gas.

Table 13-1 gives some of the most important properties of acetal plastics.

Acrylics

There is evidence that, as early as 1843, acrylic and methacrylic acid and some of their esters were produced as laboratory preparations. By 1900 several investigators, including G. W. Kahlbaum and R. Fittig, reported that these reactive compounds could be polymerized. In 1901 Otto Rohm reported in his doctoral thesis much of the research which later lead to the commercial exploitation of acrylics. Dr. Rohm continued his research in

Table 13-1. Acetal Properties.

Property	Homopolymer	20% Glass Filled
Molding qualities	Excellent	Good to Excellent
Specific gravity (density)	1.42	1.56
Tensile strength (psi)	10,000	8500-11,000
Compressive strength (psi)	18,000 (10% defl.)	18,000 (10% defl.)
Izod, impact (ft-lb/in)	1.4 (Inj.) 2.3 (Ext.)	0.8
Hardness, Rockwell	M94, R120	M75-M90
Thermal expansion (10^{-5}/°C)	8.1	3.6-8.1
Resistance to heat (°F)	195	185-220
Dielectric strength (volts/mil)	380	580
Dielectric constant (60 Hz)	3.7	3.9
Dissipation factor (60 Hz)	3.7	3.9
Arc resistance (seconds)	129	136
Water absorption (24 hrs, %)	0.25	0.25-0.29
Burning rate (in/min)	Slow-1.1	0.8-1.0
Effect of sunlight	Chalks slightly	Chalks slightly
Effect of acids	Resists some	Resists some
Effect of alkalies	Resists some	Resists some
Effect of solvents	Excellent resistance	Excellent resistance
Machining qualities	Excellent	Good to fair
Optical	Translucent-opaque	Opaque

Germany and took an active part in the first commercial development of polyacrylates in 1927. By 1931 there was a Rohm and Haas Company plant operating in the United States. Most of these early materials were used as coatings or as aircraft windshields and "bubble" turrets during World War II.

From these early compounds, an extensive group of monomers has become available and the commercial applications for the polymers have found steady growth.

The principal acid and ester monomers are indicated in Table 13-2.

The term "acrylic" includes acrylic and methacrylic esters, acids, and other derivatives. In order to avoid possible confusion, the basic acrylic formula is shown below with possible side groups of R_1 and R_2:

$$CH_2{=}C\begin{array}{l} / R_1 \\ \backslash COOR_2 \end{array}$$

(A) Basic acrylic formula.

$$CH_2{=}C\begin{array}{l} / H \\ \backslash COOH \end{array}$$

(B) Hydrogen replaces R_1 and R_2 to produce acrylic acid.

$$CH_2{=}C\begin{array}{l} / CH_3 \\ \backslash COOH \end{array}$$

(C) Methyl group replaces R_1 to produce methacrylic acid.

Fig. 13-4. Acrylic formula and two possible radical replacements.

Although there are many monomer possibilities and several ways of preparation, the most important is the commercial preparation of methyl methacrylate from acetone cyanohydrin. These homomonomers and comonomers may be polymerized by one of several commercial methods including bulk, solution, emulsion suspension, and granulation polymerization. In all cases an organic peroxide catalyst is used to initiate polymerization. Many of the molding powders are made by emulsion methods, while bulk polymerization is primarily used for casting of sheets and profile shapes.

The versatility of acrylic monomers in processing, copolymerization, and ultimate properties has

Table 13-2. Principal Acid and Ester Monomers.

Acrylic acid	Methyl acrylate	Ethyl acrylate	*n*-Butyl acrylate	Isobutyl acrylate	2-Ethylhexyl acrylate
$CH_2{=}CHCOOH$	$CH_2{=}CHCOOCH_3$	$CH_2{=}CHCOOC_2H_5$	$CH_2{=}CHCOOC_4H_9$	$CH_2{=}CHCOOCH_2CH(CH_3)_2$	$CH_2{=}CHCOOCH_2CH(C_2H_5)C_4H_9$
Methacrylic acid	**Methyl methacrylate**	**Ethyl methacrylate**	***n*-Butyl methacrylate**	**Isobutyl methacrylate**	**Lauryl methacrylate**
$CH_2{=}CCOOH$ $\vert$ CH_3	$CH_2{=}CCOOCH_3$ $\vert$ CH_3	$CH_2{=}CCOOC_2H_5$ $\vert$ CH_3	$CH_2{=}CCOOC_4H_9$ $\vert$ CH_3	$CH_2{=}CCOOCH_2CH(CH_3)_2$ $\vert$ CH_3	$CH_2{=}CCOO(CH_2)_nCH_3$ $\vert$ CH_3
Stearyl methacrylate	**2-Hydroxyethyl methacrylate**	**Hydroxypropyl methacrylate**	**2-Dimethylaminoethyl methacrylate**	**2-*t*-Butylaminoethyl methacrylate**	
$CH_2{=}CCOO(CH_2)_6CH_3$ $\vert$ CH_3	$CH_2{=}CCOOCH_2CH_2OH$ $\vert$ CH_3	$CH_2{=}CCOO(C_3H_6)OH_4$ $\vert$ CH_3	$CH_2{=}CCOOCH_2CH_2N(CH_3)_2$ $\vert$ CH_3	$CH_2{=}CCOOCH_2CH_2NHC(CH_3)_3$ $\vert$ CH_3	

contributed to their widespread acceptance. Table 13-3 gives some of the basic properties of acrylics.

Polymethyl methacrylate is for the most part an atactic, amorphous, and transparent thermoplastic material. Because of its excellent transparency, with light transmission of about 92 percent, it is used for numerous optical applications including lenses (Fig. 13-5A). Not only is polymethyl methacrylate a good electrical insulator for low frequencies, but it has extremely good resistance to weathering (Fig. 13-5B). The many outdoor advertising signs are probably the most familiar applications of acrylics.

Polymethyl methacrylate is still the standard material used for automobile tail-light lenses and covers (Fig. 13-5C). This material is used extensively on aircraft for windshields, cockpit covers, and "bubble" bodies on helicopters.

Polymethyl methacrylate may be processed by any conventional thermoplastic process and fabricated by solvent cementing. Cast and extruded sheets and profile shapes are very popular forms for the hobby crafts and do-it-yourself fabricators. Sheet forms are growing in popularity for room dividers, skylight domes, and for replacement of glass windows (Fig. 13-5D). There has been widespread acceptance of these plastics in the paint industry in the form of emulsions (Figs. 13-5E and 13-5F). Emulsion acrylics have also gained popularity as a clear, hard, and glossy "wax" coating for floors. Acrylic based adhesives are available with a wide range of application and properties. These adhesives may have the advantage of being transparent and solvent based (air drying), hot melt, or pressure sensitive. Fig. 13-6 shows an acrylic sealant being applied directly to an oily aluminum window frame under water in a tropical fish tank.

Familiar tradenames associated with polymethyl methacrylate are Plexiglas, Lucite, and Acrylite. Can you spot the acrylic objects in Fig. 13-7?

Polyacrylates

Polyacrylates are transparent, chemical, and weather resistant materials. They have a low softening point but find applications as films, adhesives, and surface coatings on paper and textiles. They are usually copolymer compositions. Polyethyl

Table 13-3. Acrylic Properties.

Property	Methyl Methacrylate (Molding)	Acrylic-PVC Copolymer (Molding)	ABS High Impact
Molding qualities	Excellent		Good-excellent
Specific gravity (density)	1.17-1.20	1.30	1.01-1.04
Tensile strength	7000-11,000	5500	4500-7500
Compressive strength	12,000-18,000	6200	4500-8000
Izod, impact (ft-lb/in)	0.3-0.5	15	5.0-8.0 @73°F (1/8 × 1/2 in bar)
Hardness, Rockwell	M85-M105		R75-R105
Thermal expansion (10^{-5}/°C)	5-9	R104	9.5-13.0
Resistance to heat (°F)	140-200		140-210
Dielectric strength (volts/mil)	400-500	400	350-450
Dielectric constant (60 Hz)	3.3-3.9	4	2.4-5.0
Dissipation factor (60 Hz)	0.04-0.06	0.04	0.003-0.008
Arc resistance (seconds)	No track	25	50-85
Water absorption (24 hrs, %)	0.1-0.4	0.13	0.20-0.45
Burning rate (in/min)	Slow 0.6-1.2	Nonburning	Slow-SE
Effect of sunlight	Nil	Nil	Yellows
Effect of acids	Attacked by strong oxidizing acids	Slight	Attacked by strong oxidizing acids
Effect of alkalies	Attacked	None	None
Effect of solvents	Soluble in ketones, esters, aromatic, & chlorinated hydrocarbons	Attacked by ketones, esters, aromatic, & chlorinated hydrocarbons	Soluble in ketones & esters
Machine qualities	Good to excellent	Excellent	Excellent
Optical	Transparent-opaque	Opaque	Translucent

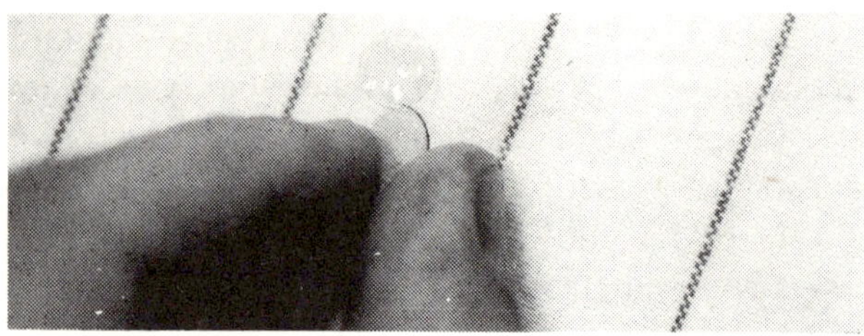

(A) Contact lenses.

(B) Panels on building.

(C) Tail-light lenses.

(D) Room divider, table, and lamp. (Rohm & Haas Co.)

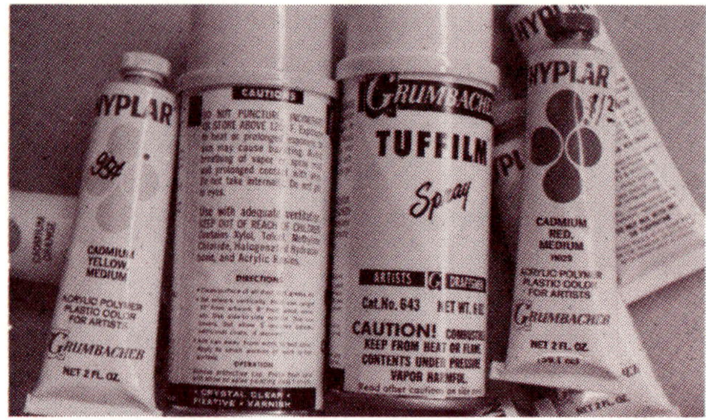

(E) Paints for artists.

(F) Wall paints.

Fig. 13-5. Acrylics have many useful properties.

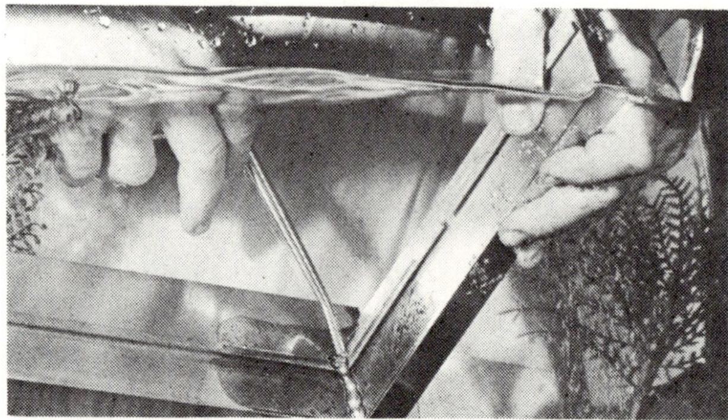

Fig. 13-6. Acrylic sealant being applied under water. (Cabot Corp.)

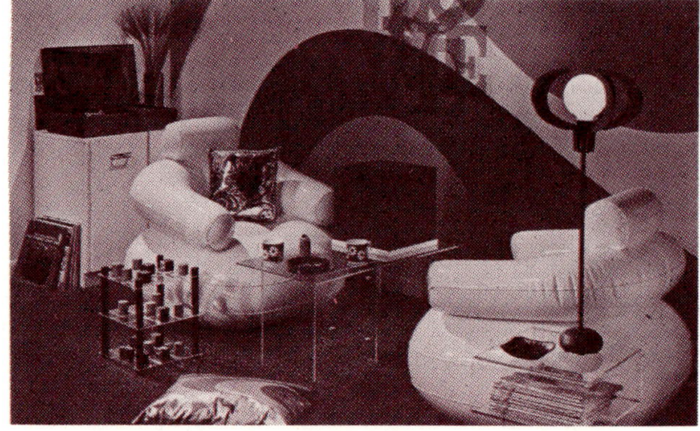

Fig. 13-7. Various uses of acrylics. (Rohm & Haas Co.)

acrylate can be crosslinked to form thermosetting elastomers. Polyacrylate monomers are used as plasticizers for other vinyl polymers.

Acrylic esters may be obtained from the reaction of ethylene cyanhydrin with sulphuric acid and an alcohol.

$$HO \cdot CH_2 \cdot CH_2 \cdot CN \xrightarrow[H_2SO_4]{ROH} CH_2 : CH \cdot CO \cdot O \cdot R,$$

Polyacrylonitrile and Polymethacrylonitrile

The elastomers and fibers which can be produced from these materials were only laboratory curiosities before the Second World War. Since that time, there has been a rapid expansion in the production of acrylonitrile as the main constituent in acrylic fibers. These polymers are copolymerized, stretched to orient the molecular chain, and sold under such familiar tradenames as Orlon, Acrilan, Dynel, and Zefran. Unmodified polyacrylonitrile is only slightly thermoplastic and is difficult to mold. Copolymers of styrene, ethyl acrylates, methacrylates, and other monomers are extruded into the amorphous fiber form. At this point the fiber is too weak to be of value; consequently, it is stretched to produce a degree of crystallization. Tensile strength is greatly increased as a result of this molecular orientation.

Monomers of acrylonitrile and methacrylonitrile may appear as shown below.

$CH_2{=}CHCN$
Acrylonitrile

$CH_2{=}C{-}CN$ (with CH_3 on the central C)
Methacrylonitrile

Acrylonitrile-Butadiene-Styrene (ABS)

ABS polymers are opaque thermoplastic resins formed by the polymerization of acrylonitrile-butadiene-styrene monomers. Because they possess such a diverse combination of properties, many authorities classify them as a family of plastics. They are, however, terpolymers ("ter" meaning three) of three monomers and not a distinct family. Their development resulted from the research efforts on synthetic rubber during and after World War II. The individual proportions of each of the three ingredients may vary, and this accounts for the diverse number of possible properties.

The three ingredients are shown below. (Acrylonitrile is also known as vinyl cyanide and acrylic nitrile.)

Acrylonitrile $CH_2{=}CHCN$

Butadiene $CH_2{=}CH{-}CH{=}CH_2$

Styrene $C_6H_5{-}CH{=}CH_2$

A representative illustration is shown for acrylonitrile-butadiene-styrene.

```
  H   H   H       H       H   H   H       H
  |   |   |       |       |   |   |       |
--C---C---C-------C-------C---C---C-------C--
  |   |   |       |       |   |   |       |
  H  C6H5 C≡N  H--C--H    H  C6H5 C≡N  H--C--H
                  |                       |
                  C--H                    C--H
                  |                       |
                  C--H                    C--H
                  |                       |
               H--C--H                 H--C--H
                  |                       |
                  H                       H
```

Graft polymerization techniques are commonly used to manufacture various grades of this material.

The resins are hygroscopic. Consequently, predrying before molding is advisable. ABS materials can be readily processed on all thermoplastic processing equipment, including blow molding and thermoforming types.

As a group of materials, ABS materials are characterized by their impact, chemical, and heat resistance. They are used as housings for appliances, lightweight luggage, camera housings, pipes, power tool housings, automotive trim, battery cases, tool boxes, and packing crates. Markets have expanded into the furniture industry in the form of radio cases, cabinets, and various furniture components. These materials may be electroplated and used in various automotive, appliance, and housewares applications (Fig. 13-8).

Table 13-3 lists some of the properties of ABS materials.

ALLYLICS

Allylic resins usually involve the esterification of allyl alcohol and a dibasic acid (see Fig. 13-9).

(A) Pipe in construction of apartment complex (Monsanto Company)

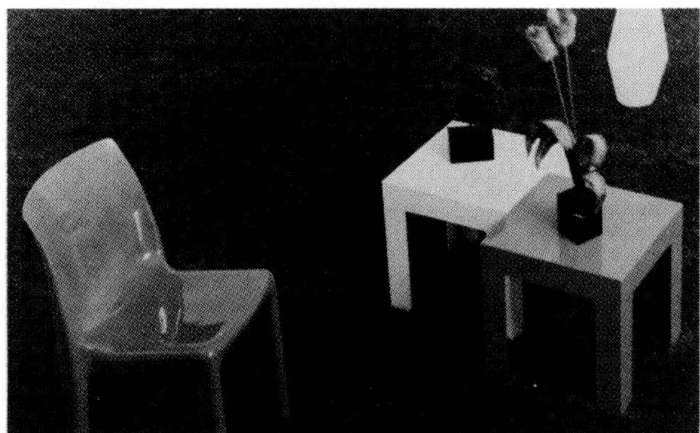

(B) Sturdy lightweight furniture. (Polyform Corp. of America)

(C) Automotive applications. (Monsanto Company)

Fig. 13-8. Applications of ABS materials.

Although these resins did not become commercially important until about 1955, one allyl resin was used in 1941 as a low-pressure laminating resin. The pungent odor of allyl alcohol ($CH_2{=}CHCH_2OH$) has been known to science since 1856. The name *allyl* was coined from the Latin word *allium,* meaning garlic.

Because of the common applications, properties, and historical background of development, allylics are erroneously included in the study of polyesters. Allyl monomers are sometimes used as crosslinking agents in polyesters and thus add to the confusion in nomenclature. Allylics are a distinct family of plastics which are based on monohydric alcohols, while the chemical basis for polyesters is polyhydric alcohols. Allylics are also unique in that they may form prepolymers (partially polymerized resins) and may be homopolymerized or copolymerized.

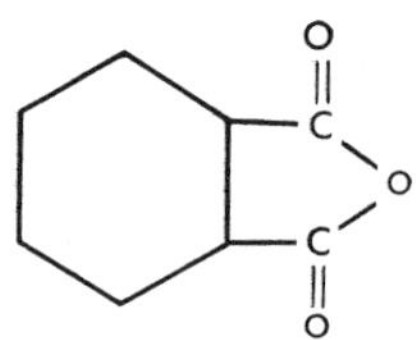

(A) Phthalic anhydride.

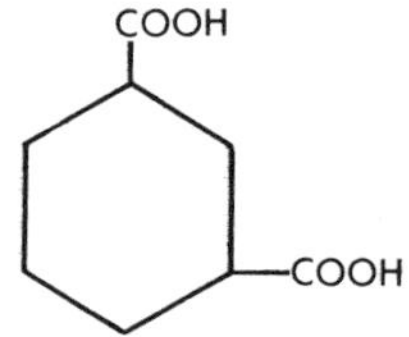

(B) Isophthalic acid.

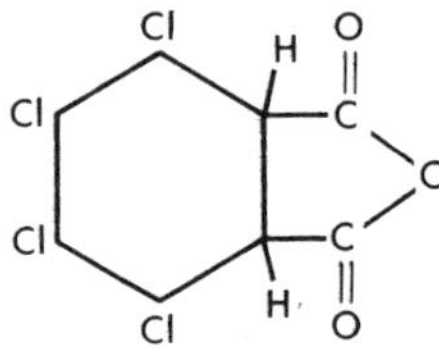

(C) Tetrachlorophthalic acid.

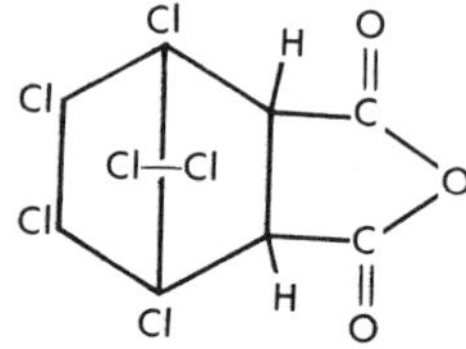

(D) Chlorendic anhydride.

O
||
H—C—C
|| O
H—C—C
||
O

(E) Maleic anhydride.

Fig. 13-9. Dibasic acids used in the manufacture of allylic monomers.

Monoallyl esters may be produced as either thermosetting or thermoplastic resins. The saturated monoallyl esters (thermoplastic) are sometimes used as copolymerizing agents in alkyd and vinyl resins. Unsaturated monoallyl esters have been used to produce simple polymers including allyl acrylate, allyl chloroacrylate, allyl methacrylate, allyl crotonate, allyl cinnamate, allyl cinnamalacetate, allyl furoate, and allyl furfurylacrylate.

Simple polymers have been produced from *di*allyl esters including diallyl maleate (DAM), diallyl oxalate, diallyl succinate, diallyl sebacate, diallyl phthalate (DAP), diallyl carbonate, diethylene glyco bis-allyl carbonate, diallyl isophthalate (DAIP), and others (see Fig. 13-10). The most widely used commercial compounds are diallyl phthalate (DAP: Fig. 13-10A) and diallyl isophthalate (DAIP: Fig. 13-10B). Diallyl resins are usually supplied as monomers or prepolymers. Both forms are converted to the fully polymerized thermosetting plastics by the addition of selected peroxide catalysts. Benzoyl peroxide or tert-butyl perbenzoate are two commonly used catalysts for polymerization of allylic resin compounds. These resins and compounds may be catalyzed and stored for more than a year if kept at low ambient temperatures. When subjected to conventional temperatures of molds, presses, or ovens, complete cure is reached.

The diethylene glycol bis-allyl carbonate resin monomer (Fig. 13-10E) used for laminating and optical castings is illustrated below. This resin could be polymerized with benzoyl peroxide at a temperature of 180°F.

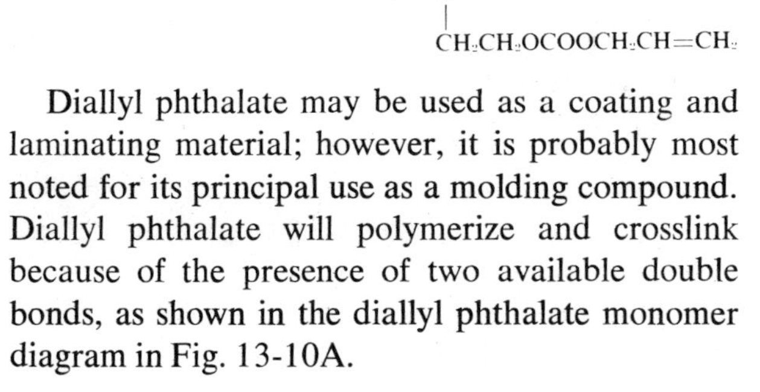

Diethylene Glycol + Allyl Carbonate

Diallyl phthalate may be used as a coating and laminating material; however, it is probably most noted for its principal use as a molding compound. Diallyl phthalate will polymerize and crosslink because of the presence of two available double bonds, as shown in the diallyl phthalate monomer diagram in Fig. 13-10A.

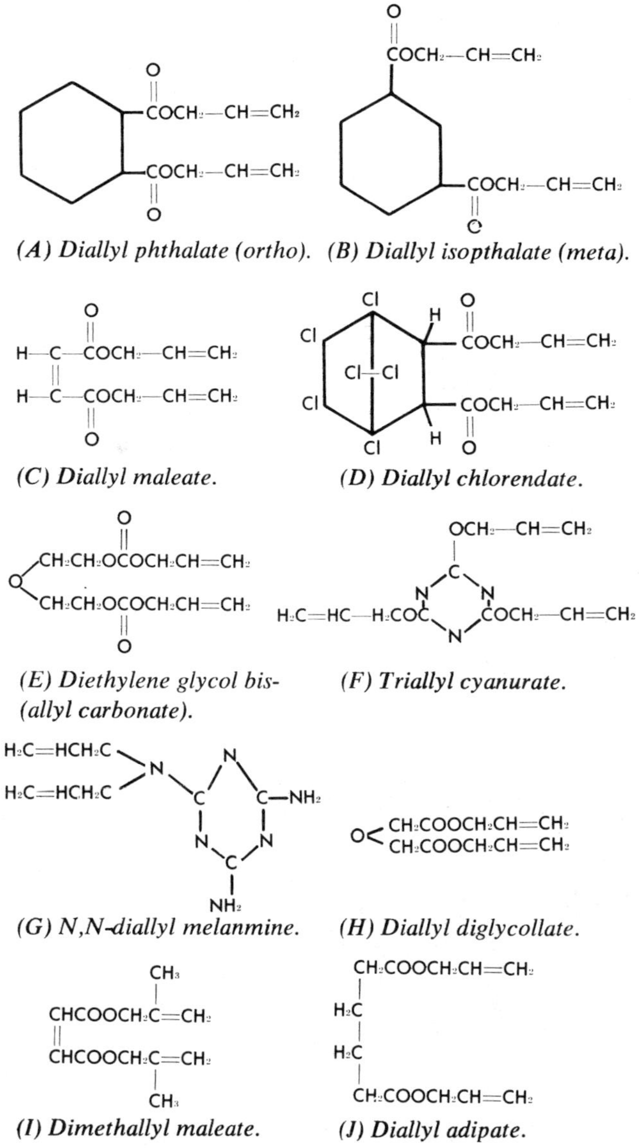

(A) Diallyl phthalate (ortho). *(B) Diallyl isopthalate (meta).*

(C) Diallyl maleate. *(D) Diallyl chlorendate.*

(E) Diethylene glycol bis-(allyl carbonate). *(F) Triallyl cyanurate.*

(G) N,N-diallyl melanmine. *(H) Diallyl diglycollate.*

(I) Dimethallyl maleate. *(J) Diallyl adipate.*

Fig. 13-10. Structural formulas of some commercial allylic monomers.

Good mechanical, chemical, thermal, and electrical properties plus long shelf life and ease of handling are only a few of the attractive attributes of this resin (Fig. 13-11). Radiation and ablation resistances are two reasons why these materials are used in space environments.

Nearly all allylic molding compounds are formulated to include catalysts, fillers, and reinforcements. The bulk of the molding compounds are blended into puttylike premixes. Because of the high cost, allylic molding compounds are limited to applications where the outstanding properties warrant their use. Compounds may be compression

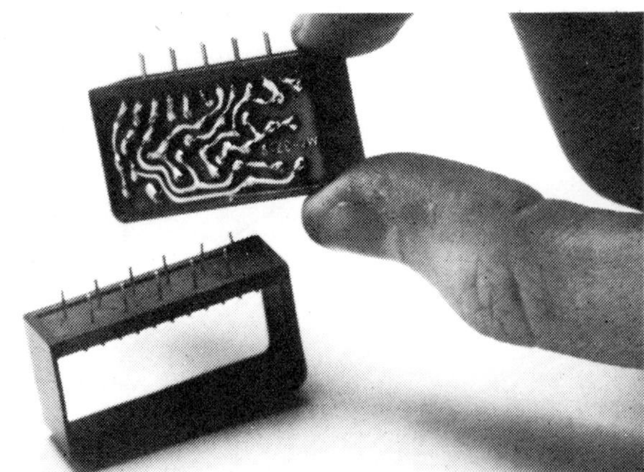

(A) *Delicate electronic circuitry frame with dimensional, temperature, and color stability combined with high volume resistivity.*

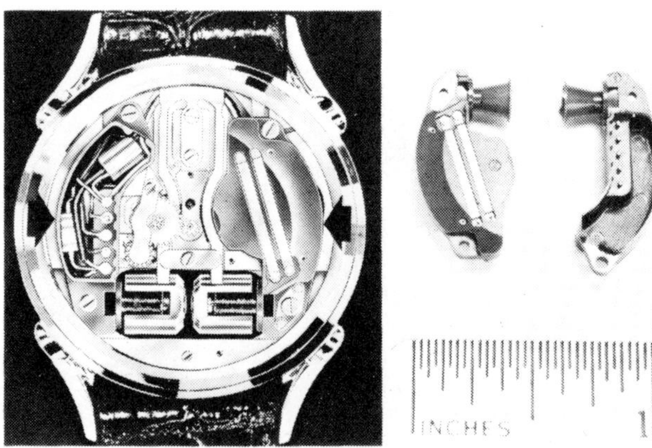

(B) *Two molded parts (right) serve as the chassis for coil assembly (arrows, left) of the Accutron electric timepiece.*

(C) *Shell and head of heavy-duty thermostat.*

Fig. 13-11. Diallyl phthalate resin compounds have wide applicability. (FMC Corporation)

or transfer molded. Some meet the requirements for special high-speed injection machines, low-pressure encapsulation, and extrusion.

Various allylic resin formulations are used in the production of laminates and prepregs (combinations of resins and reinforcements before molding). Monomer resins are readily used to preimpregnate wood, paper, fabrics, or other materials for use in laminate manufacture and to improve various properties in paper and garments. Crease, water, and fade resistance may be improved by the addition of diallyl phthalate monomers to selected fabrics. The furniture and paneling industries utilize preimpregnated decorative laminates or overlays on cores of less expensive materials. Melamine and allylic resin bound laminates have many of the same properties.

Allylic monomers and prepolymers are used in the production of prepregs (wet layup laminates). These prepregs usually consist of fillers, catalysts, and reinforcements. They are usually combined just prior to cure; some, however, are preshaped to facilitate molding. Allylic prepregs may be formulated in advance and stored until ready for use. Premix and prepreg molded parts have good flexural and impact strength. Surface finishes are excellent (See Fig. 13-12).

(A) *Numbers 1-5 designate major parts.*

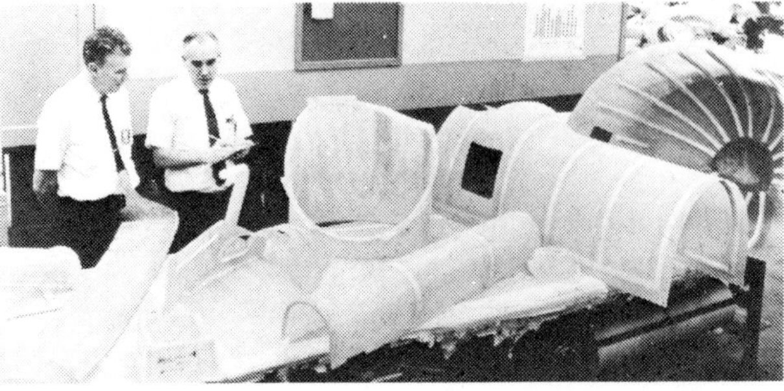

(B) *Various other parts.*

Fig. 13-12. Helicopter parts molded from diallyl phthalate prepregs. (United Aircraft Corp.)

Allylic monomers find use as crosslinking agents for polyesters, alkyds, polyurethane foams, and other unsaturated polymers. They are used because the basic allylic monomer (homopolymer) does not polymerize at room temperatures and may be stored for unusually long periods. At temperatures of 300°F and above, diallyl phthalate monomers may cause polyesters to crosslink and be molded at faster rates than those crosslinked with styrene.

Acrylics (methyl methacrylate) when crosslinked with diallyl phthalate monomers produce plastics with better surface hardness and elasticity.

Allylics have been used to seal the pores in vacuum impregnation of metal castings, ceramics, and other compositions. Other sealing and coating applications include the impregnation of reinforcing tapes used to wrap motor armatures, coating of electrical parts, and the encapsulation of electronic devices.

Table 13-4 gives some of the basic properties of allylic (diallyl phthalate) plastics.

Cellulosics

Cellulose ($C_6H_{10}O_5$) is the material that composes the framework or cell walls of all plants. It is our oldest, most familiar, most useful industrial raw material. Wood, which is composed of about half cellulose, has been man's favorite material. One reason is that cellulose is abundant in all parts of the world in one form or another. Plants are also a very inexpensive raw material from which we may produce shelter, clothing, and food. Cereal straws and grass are composed of nearly 40 percent cellulose, while cotton may be composed of nearly 98 percent cellulose.

These long-chained molecules of repeating glucose units are sometimes referred to as chemically modified natural plastics. It should be obvious why cotton with nearly 98 percent cellulose content and wood with about 50 percent cellulose would be selected as industrial sources of cellulose. The chemical structure of cellulose is shown below. Each cellulose molecule contains three hydroxyl groups (OH) at which different groups may attach for the various cellulosic plastics. Cellulose can also undergo reaction at the ether linkage between the units.

The term "cellulosics" refers to those plastics which are derived from cellulose. This family of plastics consists of many separate and distinct types of plastics. Consequently, each plastics in this family will be discussed briefly. The relation of

Table 13-4. Allylic Properties.

	Diallyl Phthalate Compounds		
Property	Glass Filled	Mineral Filled	Diallyl Isophthalate
Molding qualities	Excellent	Excellent	Excellent
Specific gravity (density)	1.61-1.78	1.65-1.68	1.264
Tensile strength (psi)	6000-11,000	5000-8700	4300
Compressive strength (psi)	25,000-35,000	20,000-32,000	
Izod, impact (ft-lb/in)	0.4-15	0.3-0.45	0.2-0.3
Hardness, Rockwell	80-87 (E scale)	61 (E scale)	238 (Rockwell M)
Thermal expansion (10^{-5}/°C)	1.0-3.6	1.0-4.3	
Resistance to heat (°F)	300-400	300-400	300-400
Dielectric strength (volts/mil)	395-450	395-420	422
Dielectric constant (60 Hz)	4.3-4.6	5.2	3.4
Dissipation factor (60 Hz)	0.01-0.05	0.03-0.06	0.008
Arc resistance (seconds)	125-180	140-190	123-128
Water absorption (24 hrs, %)	0.12-0.35	0.2-0.5	0.1
Burning rate (in/min)	Self-exting. to nonburning	Self-exting. to nonburning	Self-exting. to nonburning
Effect of sunlight	None	None	None
Effect of acids	Slight	Slight	Slight
Effect of alkalies	Slight	Slight	Slight
Effect of solvents	None	None	None
Machining qualities	Fair	Fair	Good
Optical	Opaque	Opaque	Transparent

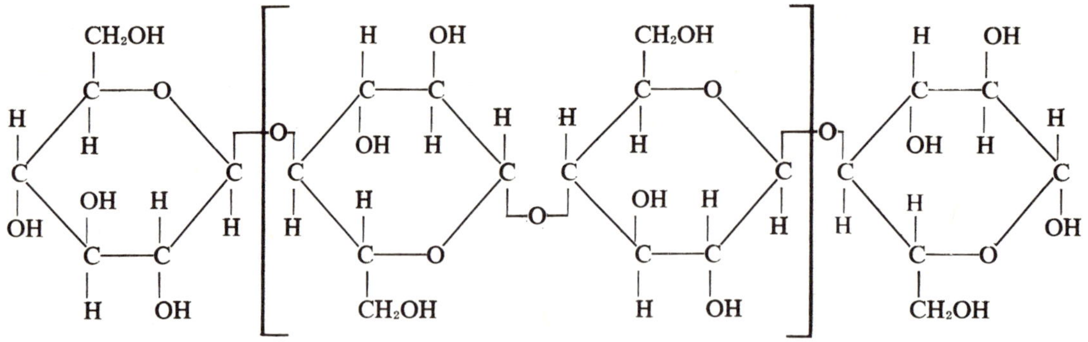

cellulose to numerous plastics and applications is shown in Fig. 13-13.

There are three large categories in which to place cellulose plastics: (1) *regenerated cellulose,* which has first been chemically changed to a soluble soution and then converted by chemical means into its original substance; (2) *cellulose esters,* in which various acids react with the hydroxyl (OH) groups of the cellulose; and (3) *cellulose ethers,* which are compounds derived by the alkylation of cellulose.

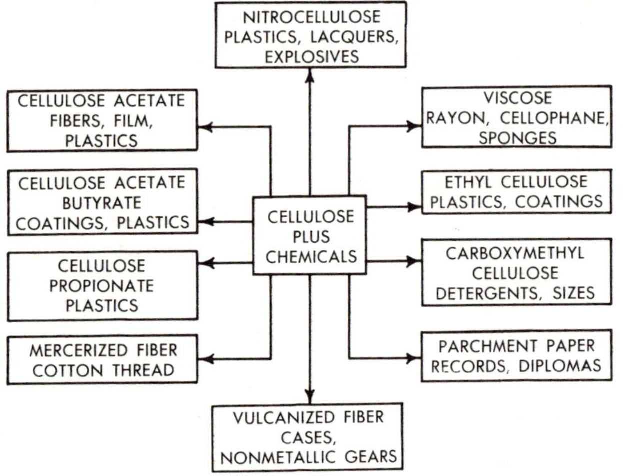

Fig. 13-13. The cellulosic plastics. (E. I. du Pont de Nemours & Co., Inc.)

Regenerated Cellulose

Regenerated cellulose products are cellophane, viscose rayon, and cuprammonium rayon (no longer of commercial importance).

In its natural form, cellulose is insoluble in solvents and incapable of flow by melting. Even in the powdered form it retains a fibrous structure.

There is some evidence that in 1857 cellulose could be dissolved in an ammoniacal solution of copper oxide. By 1897 Germany was commercially producing a fibrous yarn by spinning this solution into an acid or alkaline coagulating bath, thus regenerating the cellulose. Any remaining copper ions were removed by additional acid baths. The process of producing fibers by this method was called *cuprammonium* (copper-ammonia). The fiber was called *cuprammonium rayon.* Because this is an expensive method of manufacture and other synthetic fibers have equally desirable properties, cuprammonium rayon has lost much of its popularity.

In 1892, C. F. Cross and E. J. Bevan of England produced another regenerated cellulose fiber. They treated alkali cellulose (cellulose treated with caustic soda) with carbon disulfide to form a xanthate. Cellulose xanthate is soluble in water to give a viscous solution (hence the name viscose) called *viscose.* Viscose was then extruded through spinneret openings into a coagulating solution of sulfuric acid and sodium sulfate. The regenerated fiber is called *viscose rayon.* Rayon has become an accepted generic name for fibers composed of regenerated cellulose. Rayon continues to be a popular fabric for clothing and finds limited uses as tire cords.

Production patents were granted to J. F. Brandenberger of France in 1908 for the production of an extruded, regenerated cellulose film called *cellophane.* Like viscose rayon, the xanthate solution is regenerated when coagulated in an acid bath. After the cellophane is dried, it is usually given a water resistant coating with materials such as ethyl acetate or cellulose nitrate. These films, coated and uncoated, are used for packaging food products and pharmaceuticals.

Cellulose Esters

Among the esters of cellulose are cellulose nitrate, cellulose acetate, cellulose acetate butyrate, and cellulose acetate propionate.

In this group of plastics, acids are used to react with the hydroxyl (OH) groups to form esters. Professor Henri Bracconot of France carried out the nitration of cellulose as early as 1832. Credit for the discovery of mixing nitric and sulfuric acids with cotton to produce nitrocellulose (or cellulose nitrate) is generally given to C. F. Schonbein of England. This material was an important military explosive but found little commercial value as a plastics. Cellulose nitrate is utilized today as an explosive. Fig. 13-14 shows the nitration of cellulose.

While at the international Exhibition in 1862 Alexander Parks of England was awarded a bronze medal for his exhibit of a new plastics material called "Parkesine." His new material was composed of cellulose nitrate with a plasticizer of castor oil.

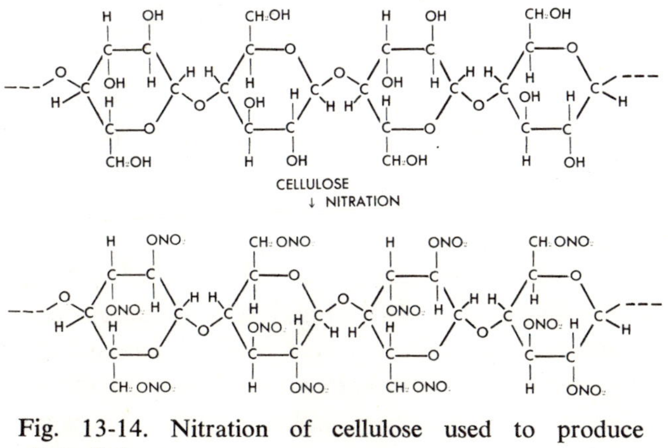

CELLULOSE
↓ NITRATION

Fig. 13-14. Nitration of cellulose used to produce nitrocellulose.

In the United States, the story becomes more familiar. John Wesley Hyatt of Newark, New Jersey, created the same material while looking for a substitute for ivory billiard balls. His original experiments followed the earlier work of the American chemist Maynard. He dissolved cellulose nitrate in ethyl alcohol and ether to form a product used as a bandage for wounds. Maynard gave this solution the name "Collodion." When Collodion solution is spread over the surface of a wound, the solvent evaporates leaving a thin, protective film. One story indicated Hyatt accidently spilled camphor on some pyroxylin (cellulose nitrate) sheets and noticed the improved properties. Another version tells how he had been treating a wound covered by Collodion with a camphor solution and discovered a change in the cellulose nitrate product.

In 1870 Hyatt and his brother patented the process of treating cellulose nitrate with camphor. By 1872 this new plastics, which they called "Celluloid," became a commercial success. Products made of nitrocellulose up to this time were highly explosive, became brittle, and suffered from high shrinkage due to the evaporation of solvents. The addition of camphor as a plasticizer eliminated many disadvantages. Celluloid, made from pyroxylin (nitrated cellulose), camphor, and alcohol, is a highly combustible material but it is not explosive.

At one time cellulose nitrate was used in photographic film, bicycle parts, toys, knife handles, and table-tennis balls (see Fig. 13-15). Today, only a few examples remain because of the difficulty in processing and the high flammability. Cellulose nitrate cannot be injection or compression molded. It is usually extruded or cast into large blocks from which sheets are sliced. Films are produced by continuous casting of a cellulose solution on a smooth surface. As the solvents evaporate, the film is removed from the casting surface and placed on drying rollers. Table-tennis balls and a few novelty items may be made of cellulose nitrate plastics. Sheets and films may be vacuum processed. Cellulose nitrate esters are found in lacquers for metal and wood finishes. They are common ingredients for aerosol paints and fingernail polishes. At one time these lacquers were utilized by the automobile industry, where painting time could be tremendously reduced.

Cellulose acetate is probably the most important of the cellulosic plastics. It was discovered by P. Schutzenberger in 1865. By 1905 George W. Miles and others developed techniques for the acetylation of cellulose. During World War I the British Government employed the aid of Henry and Ca-

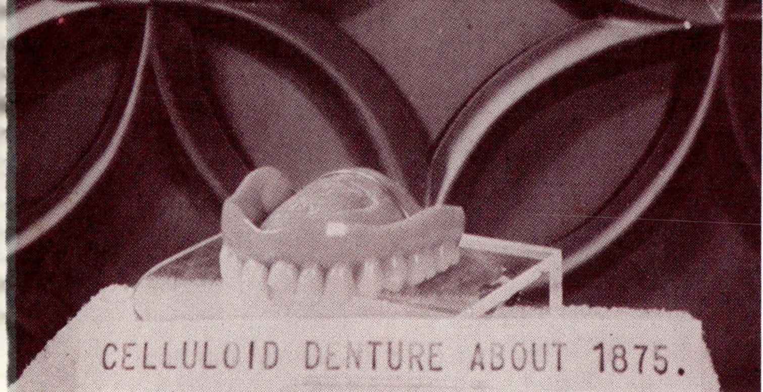

(A) Application in dentistry.

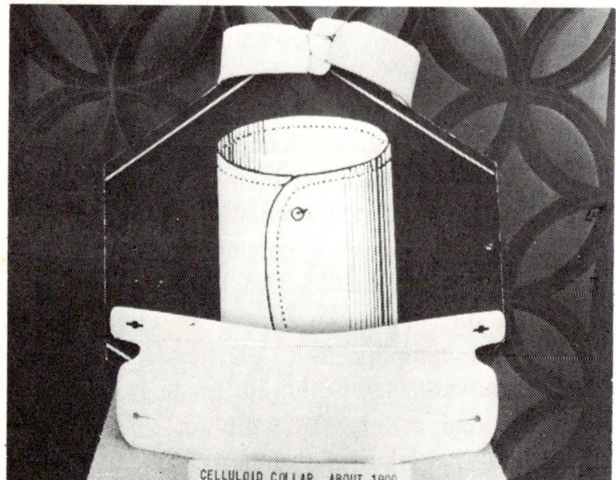

(B) Application in men's clothing.

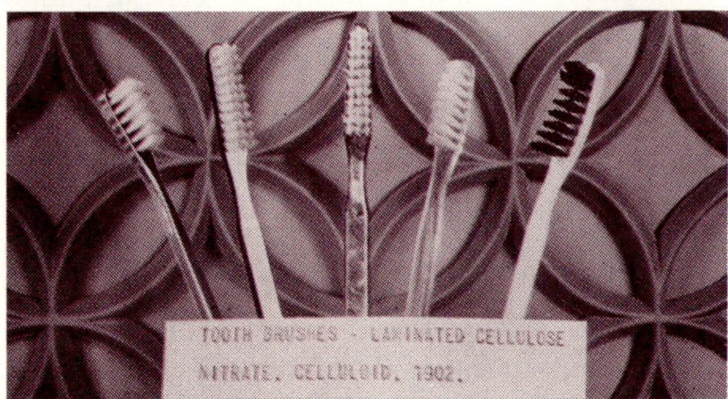

(C) Application in personal hygiene.

(D) Buggy storm front about 1900.

(E) Spanish comb hair ornament about 1910.

(F) Celluloid film led to photography as a hobby.

Fig. 13-15. Some early applications of celluloid (Celanese Plastics Co.)

mille Dreyfus of Switzerland to start large-scale production of cellulose acetate. Cellulose acetate met the need for a fire retardant lacquer for fabric-covered airplanes used at that time. By 1929 the United States produced commercial grades as molding powder, fibers, sheets, and tubes (Fig. 13-16).

Fundamental procedures for making this material is similar to that used in making cellulose nitrate. Acetylation of cellulose is carried out in a mixture of acetic acid and acetic anhydride using sulfuric acid as a catalyst. The acetate or acetyl group CH_3CO is the source of radical chemical reaction with the OH (hydroxyl) groups. The structure of cellulose triacetate is as follows:

```
              H              OCOCH3        CH2OCOCH3
              |              |             |
              C--------------C             C--------------O
       O     /|              |\           /|               \      O
___ /\  \   / OCOCH3         H \ H   H   / H                \   /  \___
       \ C                      C    C                        C
       / \ H                   / \  / \ OCOCH3       H       /
      H   \|                  /   \/   \ |           |      /
            C--------------O      O     C------------C
            |                            |           |
            CH2OCOCH3                    H           OCOCH3
```

Cellulose acetate plastics have poor heat, electrical, weathering, and chemical resistance; however, they are reasonably inexpensive and may be transparent or colored. The main outlets are for films and sheets in the packaging and display industries. They are fabricated by nearly all thermoplastic processes. Brush handles, combs, and spectacle frames are familiar molded applications. Vacuum formed display containers for hardware or food products are seen daily. Films permit the passage of moisture and gases in commercial packaging of fruits and vegetables. Coated films are used in the manufacture of magnetic recording tapes and photographic film. Cellulose acetate plastics are made into fibers for textile manufacturing and as lacquers in the coating industry. Table 13-5 gives some of the properties of cellulose acetate.

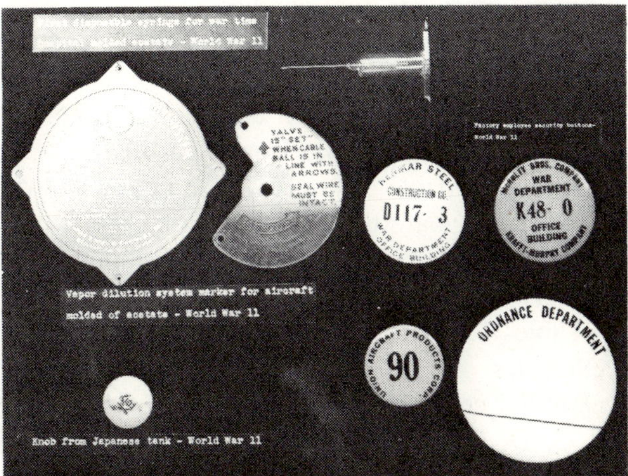

Fig. 13-16. Early applications of cellulose acetate: security buttons, disposable syringe, markers, and knob. (Celanese Plastics Co.)

Cellulose acetate butyrate was developed in the mid 1930's by the Hercules Powder Company and Eastman Chemical. Cellulose acetate butyrate is produced by reacting cellulose with a mixture of sulfuric and acetic acids. Esterification is completed when the cellulose is reacted with butyric acid and acetic anhydride. The reaction is much like the production of cellulose acetate except that butyric acid is also used. The product that results has acetyl groups (CH_3CO) and butyl groups ($CH_3CH_2CH_2CH$) in the repeating cellulose units. This product has better dimensional stability, weatherability, and is more chemical and moisture resistant.

The principal applications of cellulose acetate butyrate are tabulator keys, automobile parts, tool handles, display signs, strippable coatings, steering wheels, tubes, pipes and various packaging components. Probably the most familiar is the application of this material for screwdriver handles (see Fig. 13-17).

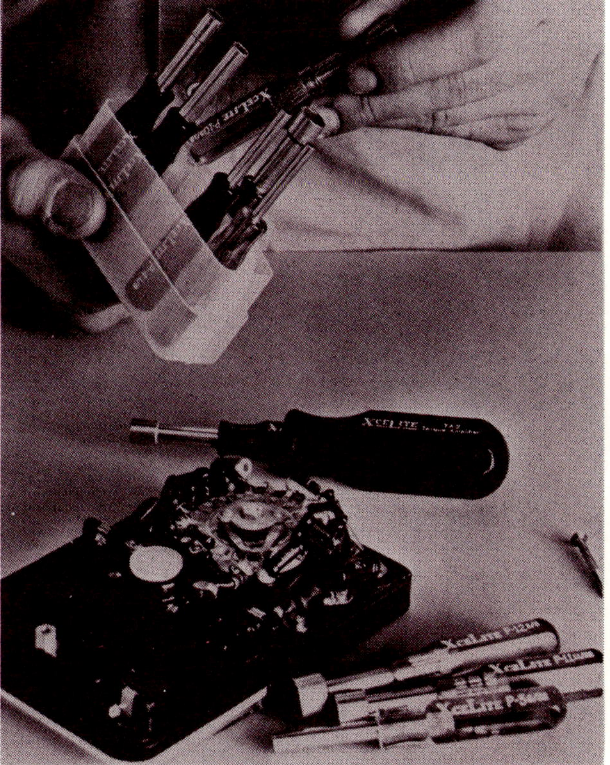

Fig. 13-17. Cellulose butyrate and cellulose acetate handles are used on these tools. (Eastman Chemical Products, Inc.)

Cellulose acetate propionate (also called simply cellulose propionate) was developed by Celanese Plastics Company in 1931 but found little application until the materials shortage of World War II. It is made like other acetates with the addition of propionic acid (CH_3CH_2COOH) in place of acetic anhydride. Its general properties are similar to cellulose acetate butyrate, but it possesses superior heat resistance and has a lower moisture absorption. The main applications of cellulose acetate propionate include pens, automotive parts, brush handles, steering wheels, toys, novelties, and various film packaging displays (see Fig. 13-18).

Fig. 13-18. Cellulose propionate top of portable dishwasher. (Eastman Chemical Products, Inc.)

Cellulose Ethers

Among the ethers of cellulose are ethyl cellulose, methyl cellulose, hydromethyl cellulose, carboxymethyl cellulose, and benzyl cellulose. Their manufacture generally involves the preparation of alkali cellulose and other reactants as indicated in Fig. 13-19.

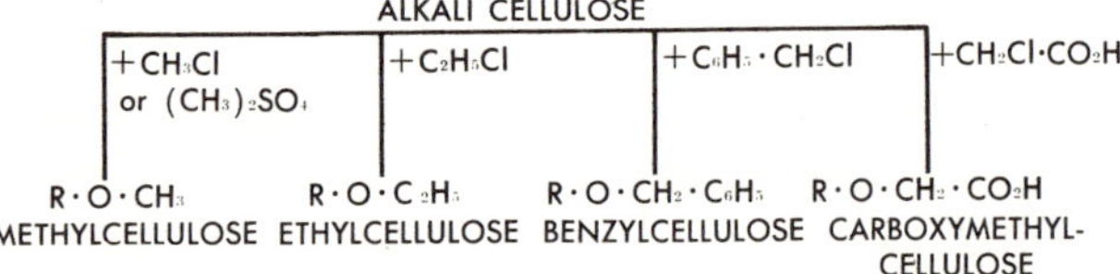

Fig. 13-19. Cellulosic plastics from alkali cellulose and other reactants. The *R* in the formulas represents the cellulose skeleton.

Ethyl cellulose is the most important of the cellulose ethers and the only one used as a plastics material. Basic research and patents were established by Dreyfus in 1912. By 1935 the Hercules Powder Company announced commercial grades in the United States.

Alkali cellulose (sometimes called soda cellulose) is treated with ethyl chloride to form ethyl cellulose. The radical substitution (etherification) of this ethoxy may vary over a wide range of properties. In etherification the hydrogen atoms of the hydroxyl groups are replaced by ethyl (C_2H_5) groups (Fig. 13-20). This cellulose plastics is tough, flexible, and moisture resistant. Table 13-15 gives some of the properties of ethyl cellulose.

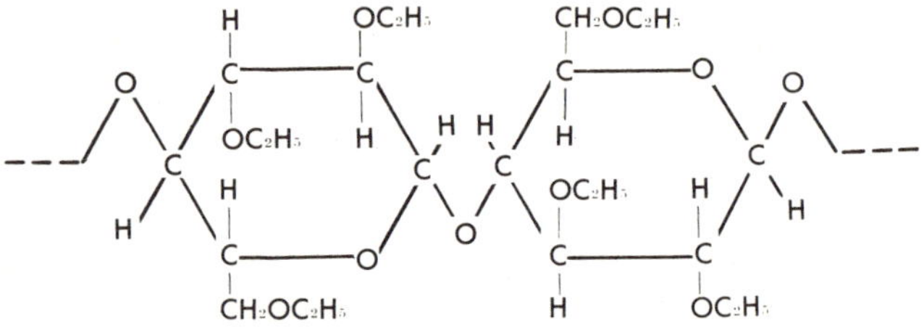

Fig. 13-20. Ethyl cellulose (fully ethylated).

Ethyl cellulose is used in such applications as football helmets, flashlight cases, furniture trim, cosmetic packages, tool handles, and blister packages. It has been used as protective coatings on bowling pins and in the formulations of paint, varnish and lacquers. Ethyl cellulose is a common ingredient in hair sprays. It is often used as a hot melt for strippable coatings. These hot-melt, strippable coatings are used for protecting metal parts against corrosion and marring during shipment and storage.

Table 13-5 lists some of the properties of cellulosics.

Methyl cellulose was developed in England by W. S. Denham and H. Woodhouse in 1914. By 1939 the Dow Chemical Company began commercial production in the form of water soluble flakes. It is prepared in a similar manner as ethyl cellulose, using methyl chloride or methyl sulphate instead of ethyl chloride. In etherification the hydrogen of the OH group is replaced by methyl (CH_3) groups:

$$R(ONa)_{3n} + CH_3Cl \longrightarrow \underset{\textit{Methyl Cellulose}}{R(OCH)_{3n}}$$

Table 13-5. Cellulosic Properties.

Property	Ethyl (Molding)	Acetate (Molding)	Propionate (Molding)
Molding qualities	Excellent	Excellent	Excellent
Specific gravity (density)	1.09-1.17	1.22-1.34	1.17-1.24
Tensile strength (psi)	2000-8000	1900-9000	2000-7800
Compressive strength (psi)	10,000-35,000	2000-36,000	2400-22,000
Izod, impact (ft-lb/in)	2.0-8.5	0.4-5.2	0.5-11.5
Hardness, Rockwell	R50-R115	R34-R125	R10-R122
Thermal expansion (10^{-5}/°C)	10-20	8-18	11-17
Resistance to heat (°F)	115-185	140-220	155-220
Dielectric strength (volts/mil)	350-500	250-600	300-450
Dielectric constant (60 Hz)	3.0-4.2	3.5-7.5	3.7-4.3
Dissipation factor (60 Hz)	0.005-0.020	0.01-0.06	0.01-0.04
Arc resistance (seconds)	60-80	50-310	175-190
Water absorption (24 hrs, %)	0.8-1.8	1.7-6.5	1.2-2.8
Burning rate (in/min)	Slow	Slow to self-exting.	Slow (1.0-1.3)
Effect of sunlight	Slight	Slight	Slight
Effect of acids	Decomposes	Decomposes	Decomposes
Effect of alkalies	Slight	Decomposes	Decomposes
Effect of solvents	Soluble in ketones, esters, chlorinated hydrocarbons, aromatic hydrocarbons	Soluble in ketones, esters, chlorinated hydrocarbons, aromatic hydrocarbons	Soluble in ketones, esters, chlorinated hydrocarbons, aromatic hydrocarbons
Machining qualities	Good	Excellent	Excellent
Optical	Transparent-opaque	Transparent-opaque	Transparent-opaque

Methyl cellulose finds a wide variety of applications because it is water soluble and edible. It is used as a thickening emulsifier and agent in cosmetics and adhesives. Methyl cellulose is a well-known wallpaper adhesive and fabric sizing material. It is useful for thickening and emulsifier agent in water-based paints, salad dressings, ice cream, cake mixes, pie fillings, crackers, and other food products. In pharmaceuticals, it is used to coat pills and as contact lens solutions (see Fig. 13-21).

Hydroxyethyl cellulose is produced by reacting the alkali cellulose with ethylene oxide. It may be used for many of the same applications employed by methyl cellulose. In the schematic equation below, R represents the cellulose skeleton.

$$R(ONa)_{3n} + \overset{\overset{O}{/\ \ \backslash}}{CH_2{-}CH_2} \longrightarrow R(OCH_2CH_2OH)_{3n}$$

Hydroxyethyl Cellulose

At one time benzyl cellulose was marketed for molding and extrusion applications. However, in the United States it is unable to compete with other polymers.

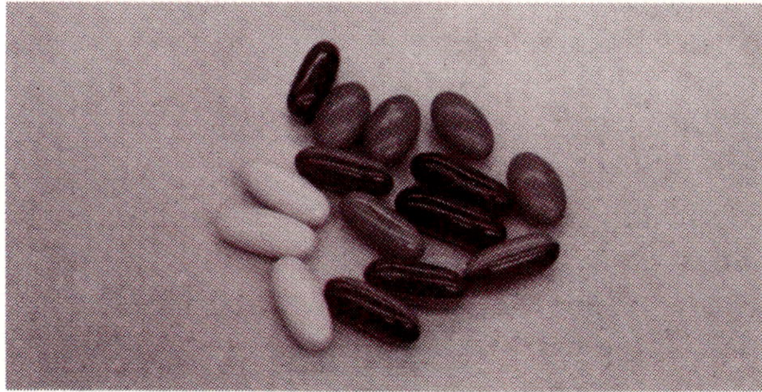

Fig. 13-21. Methyl cellulose may be used to coat pharmaceutical products.

Carboxymethyl cellulose (sometimes called sodium carboxymethyl cellulose) is made from alkali cellulose and sodium chloroacetate. Like methyl cellulose, it is water soluble and used as a sizing, gum, or emulsifying agent. Carboxymethyl cellulose may be found in foods, pharmaceuticals, and coatings. It is an excellent water soluble suspending agent for lotions, jelly bases, ointments, toothpastes, paints, and soaps. It is used to coat pills, paper, and textiles.

$$R(ONa)_{3n} + ClCH_2 \cdot COONa \rightarrow R(OCH_2COONa)_{3n} + naCl$$

Sodium Carboxymethyl Cellulose

Chlorinated Polyethers

In 1959 the Hercules Powder Company introduced chlorinated polyethers with the tradename Penton. In 1972 Hercules announced discontinuation of production and sale of this material. This thermoplastic material was produced by chlorinating pentaerythritol, and the resulting dichloromethyl oxycyclobutane was polymerized into a crystalline, linear product over 45 percent chlorine by weight (Fig. 13-22).

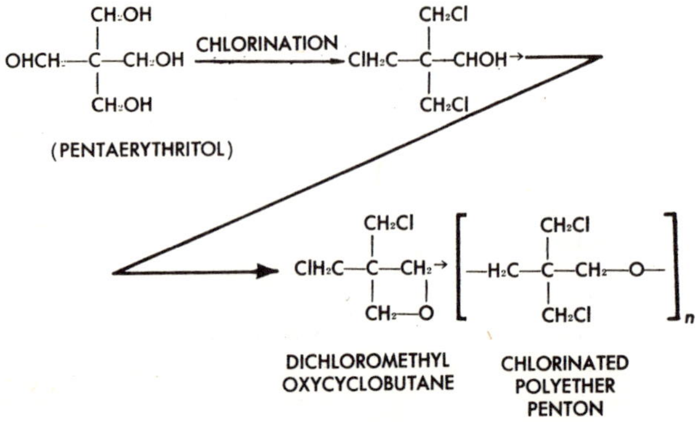

CHLORINATION

(PENTAERYTHRITOL)

DICHLOROMETHYL OXYCYCLOBUTANE

CHLORINATED POLYETHER PENTON

Fig. 13-22. Production of the chlorinated polyether Penthol.

Chlorinated polyethers may be processed on conventional thermoplastic equipment. These materials are high-performance, high-priced plastics (Table 13-6). They have been applied as coating to metal substrates by fluidized bed, flame spraying, or solvent processes. Molded parts possess high strength, heat resistance, excellent electrical and chemical resistance, and low water absorption. Although the high price restricted their wider usage, they did find use as coatings for valves, pumps, and meters. Molded parts did include chemical meter components, pipelining, valves, laboratory equipment, and electrical insulation. At degrading temperatures, lethal chlorine gas is released, however.

To date, there are no plans for another producer to manufacture this plastics with its distinct chemical nature.

Table 13-6. Chlorinated Polyether Properties.

Property	Chlorinated Polyether
Molding qualities	Excellent
Specific gravity (density)	1.4
Tensile strength (psi)	6000
Izod, impact (ft-lb/in)	0.4
Hardness, Rockwell	R100
Thermal expansion (10^{-5}/°C)	8
Resistance to heat (°F)	290
Dielectric strength (volts/mil)	400
Dielectric constant (60 Hz)	3.1
Dissipation factor (60 Hz)	0.01
Water absorption (24 hrs, %)	0.01
Burning rate (in/min)	Self-exting.
Effect of sunlight	Slight
Effect of acids	Attacked by oxidizing acids
Effect of alkalies	None
Effect of solvents	Resistant
Machining qualities	Excellent
Optical	Translucent-opaque

Coumarone-Indene

Coumarone and indene are obtained from the fractionation of coal tar but are seldom separated. These inexpensive products resemble styrene in chemical structure. Coumarone and indene may be polymerized by ionic catalytic action of sulfuric acid. A wide range of properties from sticky resins to brittle plastics may be obtained by varying the coumarone-indene ratio or by copolymerization with other polymers.

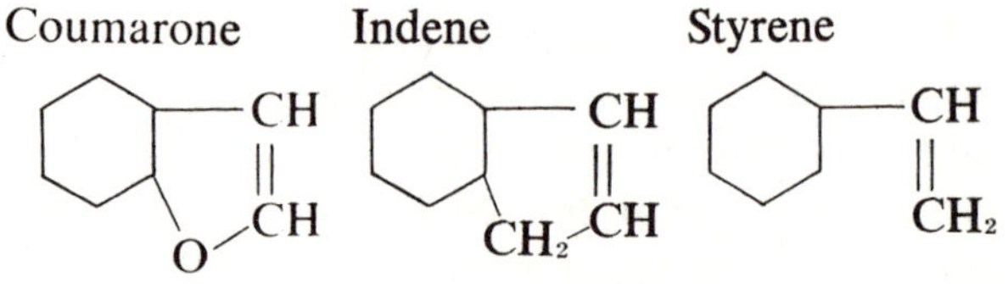

Coumarone Indene Styrene

Although used long before World War II they have found only limited applications. They are not used as molding compounds but find use as binders, modifiers, or extenders for other polymers and compounds. The largest quantities are used in the manufacture of paints or varnishes and as binders in flooring tiles and mats.

The properties of these compounds vary greatly. Coumarone-indene copolymers have good electrical insulation properties but are soluble in hydrocarbons, ketones, and esters. They range from light to dark in color but are inexpensive to manufacture. They are true thermoplastic materials with a softening point of $<100°F$ to $>120°F$. Additional applications include printing inks, coatings for paper, adhesives, encapsulation compounds, some battery boxes, brake linings, caulking compounds, chewing gum, concrete curing compounds, and numerous emulsion binders (see Fig. 13-23).

Fig. 13-23. Floor tile, floor covering, and printing inks may be composed of coumarone-indene polymers.

Fluoroplastics

Elements in the seventh column of the Periodic Table are closely related. These elements, fluorine, chlorine, bromine, iodine, and astatine are called "halogens" from the Greek word which means "salt-producing." Chlorine, for example, is found in common table salt. All of these elements are electronegative (can attract and hold valence electrons) because they have only seven electrons in their outermost shell. Fluorine and chlorine are gases but are not found in a pure or free state. Fluorine is the most reactive element known; consequently, it reacts readily with other elements. Large quantities of fluorine are needed for processes (isolation of uranium metal) connected with the atomic energy program.

Compounds containing fluorine are commonly called *fluorocarbons*. In the strictest definition, the term "fluorocarbon" should be used to refer to compounds containing only fluorine and carbon.

Although the French chemist Moissan isolated pure fluorine in 1886, it remained a laboratory curiosity until 1930. In 1931 the tradename *Freon* was announced. This fluorocarbon is a compound of carbon, chlorine, and fluorine (CCl_2F_2). Freons are used extensively as refrigerants and inert propellents for aerosol cans. This colorless, nontoxic, noninflammable gas is an excellent aerosol propellent for insecticides, hair sprays, paints, suntan lotions, shave creams, etc. In addition to refrigerants and aerosol propellents, fluorocarbons are used in the formation of polymeric materials.

In 1938 the first polyfluorocarbon was discovered by accident in the du Pont research laboratories. It was discovered that tetrafluoroethylene gas (Freon) formed an insoluble, waxy, white powder when stored in steel cylinders. As a result of this chance discovery a number of fluorocarbon polymers have been developed.

The term *fluoroplastic* is used to describe alkene-like structures which have some or all of the hydrogen atoms replaced by fluorine.

```
     MONOMER                     MONOMER
 H  H/ H  H\ H  H          F  F/ F  F\ F  F
 |  | |  |  | |  |         |  | |  |  | |  |
-C--C-+C--C-+-C--C-       -C--C-+C--C-+-C--C-
 |  | |  |  | |  |         |  | |  |  | |  |
 H  H\ H  H/ H  H          F  F\ F  F/ F  F
   Polyethylene            Polytetrafluoroethylene
```

It is the presence of the fluorine atoms which provides the outstanding or unique properties characteristic of the fluoroplastic family. These properties are directly related to the high carbon-to-fluorine bonding energy and to the high electronegativity of the fluorine atoms. Thermal stability and solvent, electrical, and chemical resistances are weakened if fluorine (F) atoms are replaced with hydrogen (H) or chlorine (Cl) atoms. The weaker C-H and C-Cl bonds are more vulnerable to chemical attack and thermal decomposition.

The fluoroplastics of principal commercial importance are shown in Fig. 13-24. There are only two types of fluorocarbon plastics: polytetrafluoroethylene (PTFE) and polytetrafluoropropylene (FEP or fluorinated ethylene propylene). The others must be considered copolymers or fluorine containing polymers.

Polytetrafluoroethylene ($CF_2{=}CF_2$)$_n$ accounts for nearly 90 percent of the fluorinated plastics from the standpoint of volume usage. The monomer tetrafluoroethylene is obtained by pyrolysis of chlorodifluoromethane. Tetrafluoroethylene is polymerized in the presence of water and a peroxide catalyst while under high pressure. PTFE is a high-

```
    F  F
    |  |
  —C—C—
    |  |
   Cl  F
```

(A) Polychlorotrifluoro-ethylene.

```
    F  F
    |  |
  —C—C—
    |  |
    F  F
```

(B) Polyletrafluoroethylene.

```
    H  H
    |  |
  —C—C—
    |  |
    H  F
```

(C) Polyvinyl fluoride.

```
    H  F
    |  |
  —C—C—
    |  |
    H  F
```

(D) Polyvinylidene fluoride.

```
      F       F
      |       |
      C=======C
      |       |
   F—C—F     F
      |
      F
```

(E) Polyhexafluoropropylene.

Fig. 13-24. Monomers of fluoroplastics and of fluorine containing polymers.

ly crystalline, waxy thermoplastic material with a service temperature of −450°F to +550°F. Ablation service temperatures may exceed 1200°F. Because of the high bonding strength and compact interlocking of fluorine atoms about the carbon backbone, this material cannot be processed by conventional thermoplastic techniques. At present it cannot be plasticized to aid processing. Most of the material is made into preforms and sintered. *Sintering* is a special fabricating technique used for metal and plastics. The powdered material is pressed into a mold at a temperature just below its melting or degradation point until the particles are fused (sintered) together. The mass as a whole does not melt in this process. Sintered moldings may be machined. Special formulations may be extruded in the form of rods, tubes, and fibers by using organic dispersions of the polymer which are later vaporized as the product is sintered. Coagulated suspensoids may be utilized in much the same manner. Presintered grades of this material may be extruded through extremely long compacting and sintering zones of special dies. Many films, tapes, and coatings are cast, dipped, or sprayed from PTFE dispersions by a drying and sintering process. Films and tapes may also be cut or sliced from sheet stock.

Teflon is a familiar tradename for homopolymers and copolymers of polytetrafluoroethylene. The antistick or low coefficient of friction property is used to coat various metallic substrates including cookware (Fig. 13-25). There is no known solvent for these materials; however, they may be chemically etched and adhesively bonded with contact

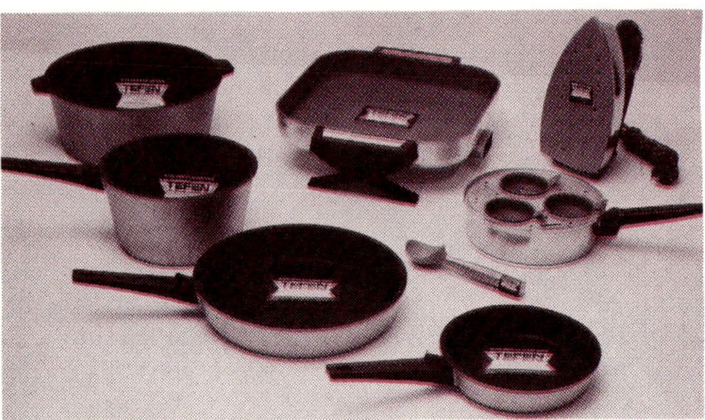

(A) Teflon-II coated cookware and utensils.

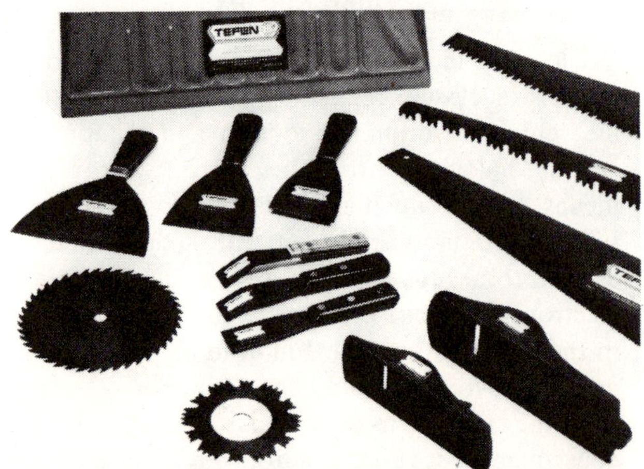

(B) Teflon-S coated tools.

(C) Antistick coating on rolls for dry-copy equipment.

Fig. 13-25. Applications of Teflon. (Chemplast, Inc.)

or epoxy adhesives. Films may be heat sealed together but not to other materials.

Fluorocarbons are heavier than hydrocarbons because the fluorine element is nearly eighteen times heavier than hydrogen. Fluorine has an atomic weight of 18.9984 and hydrogen only 1.00797. As might be expected, fluoroplastics are heavier than other plastics, with densities ranging from 2.0 to 2.3.

Although PTFE requires special fabricating techniques, its chemical inertness, exceptional weathering resistance, excellent electrical insulation characteristics, excellent heat resistance, low coefficient of friction, and nonadhesive properties have led to numerous applications. Parts coated with PTFE have such a low coefficient of friction they release, slide, and require no lubrication. Saw blades, cookware, utensils, snow shovels, bakery equipment, and bearings of various design are common applications. Aerosol spray dispersions of micron-sized particles of polytetrafluoroethylene using a Freon propellent provide an excellent lubricant and antistick agent for metallic, glass, or plastics substrates.

Various profile shapes are used for a variety of chemical, mechanical, and electrical applications. Shrinkable tubing is used to cover rollers, springs, glass and electrical parts (see Fig. 13-26). Skived or extruded tapes and films may be used for seals, packing, and gasket materials. Bridges, pipes, tunnels, and buildings may rest on slip joints, ex-

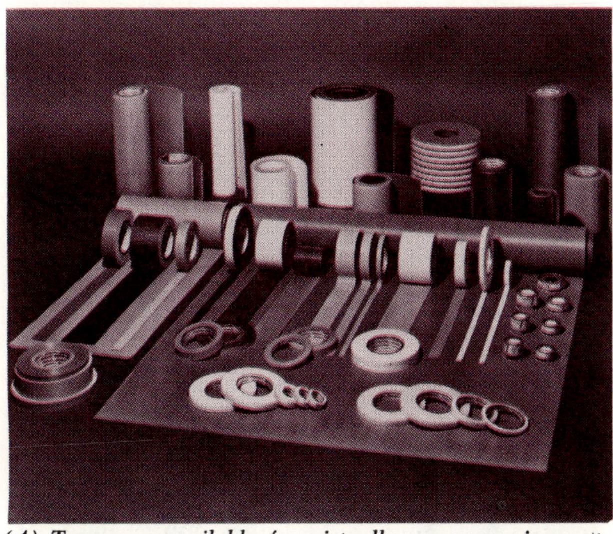

(A) Tapes are available for virtually every requirement. (Chemplast, Inc.)

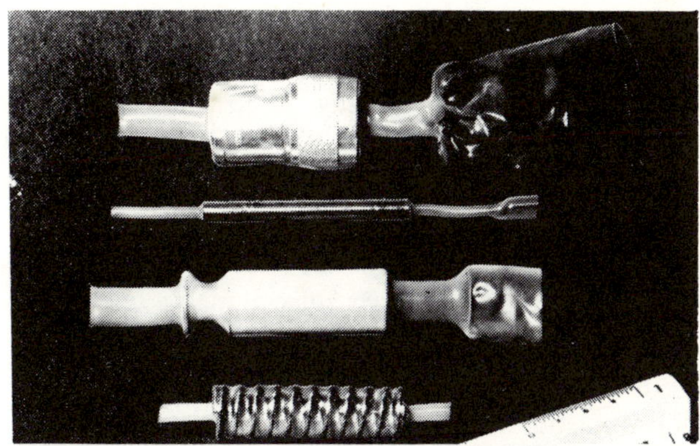

(B) Shrinkable protective tubing. (Chemplast, Inc.)

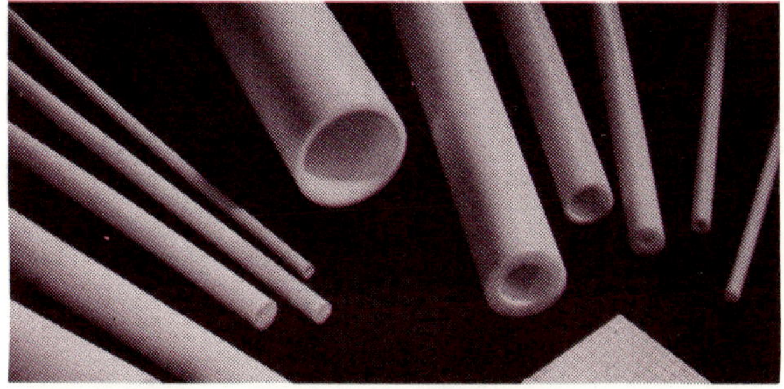

(C) Various rods and tubes. (Chemplast, Inc.)

(D) Envelope gasket. (ICI, Ltd.)

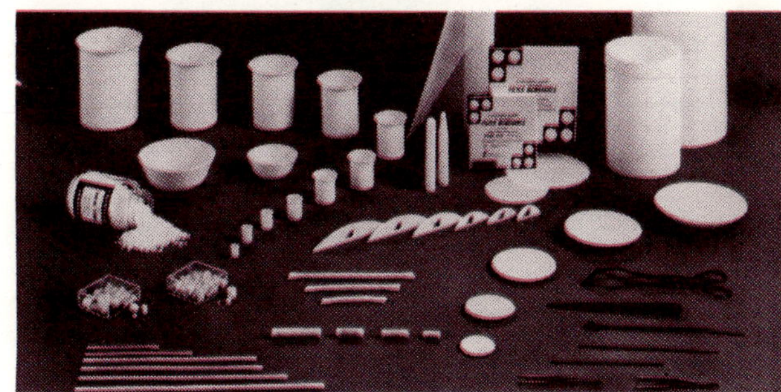

(E) Laboratory items. (Chemplast, Inc.)

(F) Chemical resistant stopcock plugs. (Chemplast, Inc.)

Fig. 13-26. Applications of PTFE.

pansion plates, or bushing or bearing pads of polytetrafluoroethylene (Fig. 13-27).

The excellent electrical insulating and low dissipation factors are used for wire and cable insulation, coaxial wire spacers, laminates for printed circuitry, and many other miscellaneous electrical applications.

In 1956 du Pont announced another Teflon fluoroplastic that is composed wholly of fluorine and carbon atoms. Polyfluoroethylenepropylene (PFEP or FEP) is manufactured by copolymerizing tetrafluoroethylene with hexafluoropropylene (see Fig. 13-28).

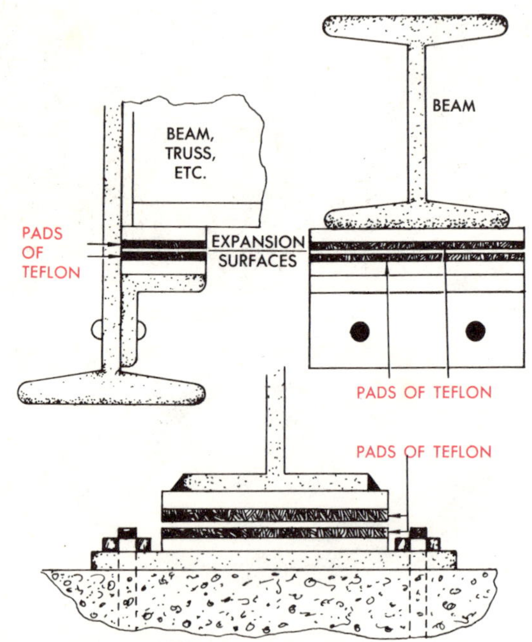

(A) General.

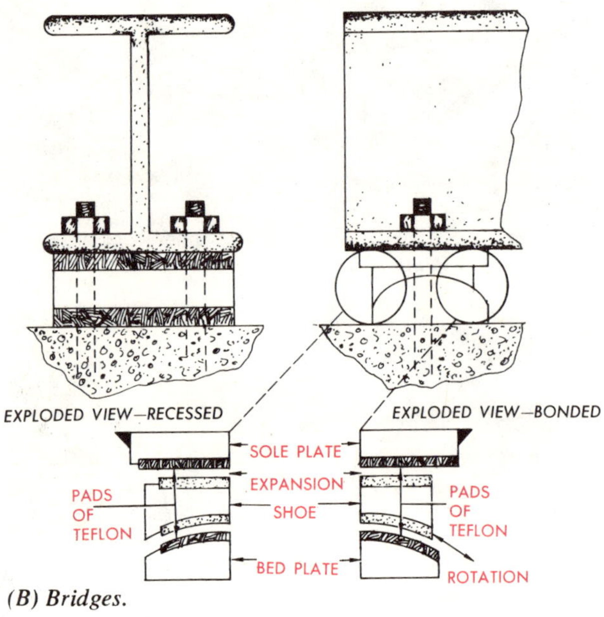

(B) Bridges.

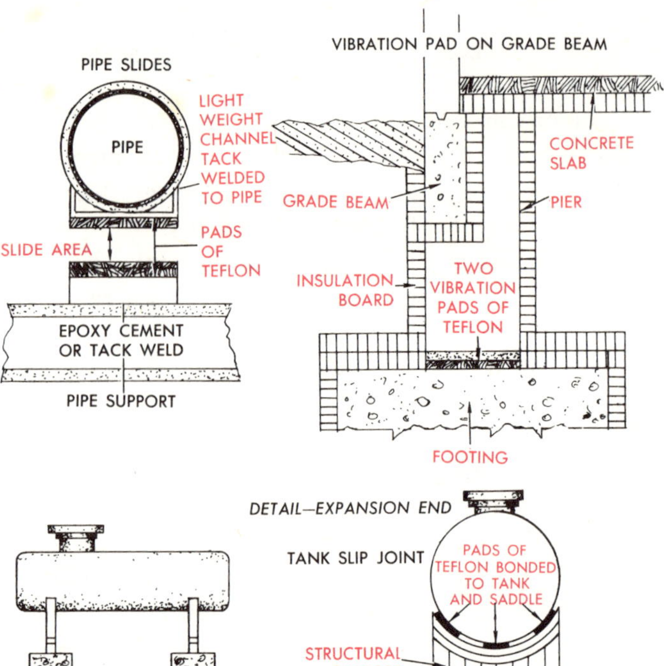

(C) Miscellaneous.

Fig. 13-27. Use of Teflon pads. (E. I. du Pont de Nemours & Co.)

The partial disruption of the polymer chain by the propylenelike groups $CF_3CF{=}CF_2$ results in a reduction in melting point and viscosity for FEP resins. Polyfluoroethylenepropylene may be processed in conventional thermoplastic equipment. Consequently, there is a reduction in production costs of items previously molded of PTFE. Because of pendant CF_3 groups this copolymer is less crystalline, more processable, and transparent in films up to 0.010 inch thickness.

Commercial PFEP plastics possess properties closely resembling those of PTFE. They are chemically inert, have excellent electrical insulation properties, and have a somewhat greater impact strength. Service temperatures may exceed 400°F. Polyfluoroethylenepropylene plastics are used extensively by the military, aircraft, and aerospace industries for electrical insulation and high reliability at elevated or cryogenic temperatures. They are used for lining chutes, pipes, tubes, and coating other objects where a low coefficient of friction

TFE (TETRAFLUOROETHYLENE) + HFP (HEXAFLUOROPROPYLENE) → JOIN TO FORM FEP (FLUORINATED ETHYLENE PROPYLENE)

Fig. 13-28. Manufacture of polyfluoroethylenepropylene.

or nonadhesive characteristics are required. PFEP is molded into numerous parts including gaskets, gears, impellers, printed circuits, pipes, fittings, valves, expansion plates, bearings, and other profile shapes (see Fig. 13-29).

As early as 1933 Germany and the United States were producing a fluoroplastic material used in connection with the development of the atomic bomb and in the handling of uranium fluoride, a compound of uranium with fluorine.

Fig. 13-29. Shrinkable PFEP antistick roll covers on roller. (Chemplast, Inc.)

Polychlorotrifluoroethylene (PCTFE or CTFE) is produced in various formulations. Chlorine atoms are substituted for fluorine in the carbon chain

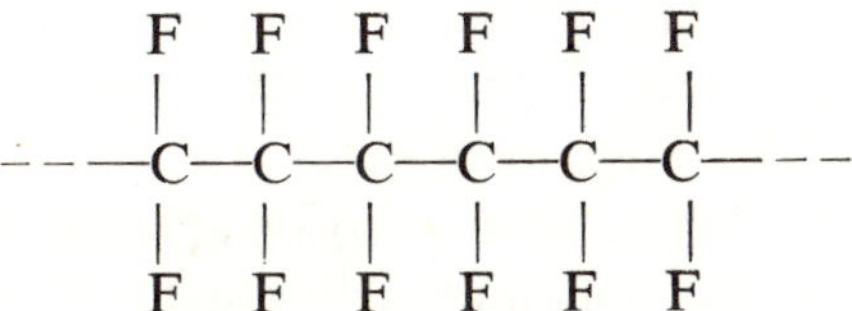

Polytetrafluoroethylene (PTFE)

Monomers are obtained from fluorinating hexachloroethane and then dehalogenating (controlled removal of the halogen, chlorine) with zinc in alcohol:

$$CCl_3CCl_3 \xrightarrow[HF]{Anhydrous} CCl_2FCClF_2 \xrightarrow[\text{boiling ethyl alcohol}]{Zinc} CClF{=}CF_2 + Cl_2$$

Hexachloroethane

The polymerization is similar to PTFE in that it is accomplished in an aqueous emulsion and suspension. During bulk polymerization, peroxide or Ziegler-type catalysts are used.

$$nCF_2{=}CFCl \xrightarrow{polymerize} (\text{—}CF_2CFCl\text{—})_n$$

Chlorotrifluoroethylene   *Polychlorotrifluoroethylene*

The addition of the chlorine atoms to the carbon chain allows this material to be processed by conventional thermoplastic equipment. The chlorine presence also allows selected chemicals to attack and break the partially crystalline polymer chain. PCTFE may be produced optically clear depending on the degree of crystallinity. Copolymerization with vinylidene fluoride or other fluoroplastics provides varying degrees of chemical inertness, thermal stability, and other unique properties.

Polychlorotrifluoroethylene is harder, more flexible, and possesses a higher tensile strength than PTFE. It is more expensive than PTFE and has a service temperature from −400°F to +400°F. Because of the introduction of the chlorine atom, electrical properties are lower while the coefficient of friction is higher. Although it is more expensive than polytetrafluoroethylene, it finds similar applications. Applications for PCTFE include insulation for wire, cable, printed circuit boards, and electronic components. Chemical resistance is utilized in producing transparent windows for chemicals, seals, gaskets, O-rings, and pipe lining as well as pharmaceutical and lubricant packaging (see Fig. 13-30). Various dispersions and films may be used in coating reactors, storage tanks, valve bodies, fittings, and pipes. Films may be sealed by thermal or ultrasonic techniques. Epoxy adhesives may be used on chemically etched surfaces.

(A) Insulation for wires and cables.

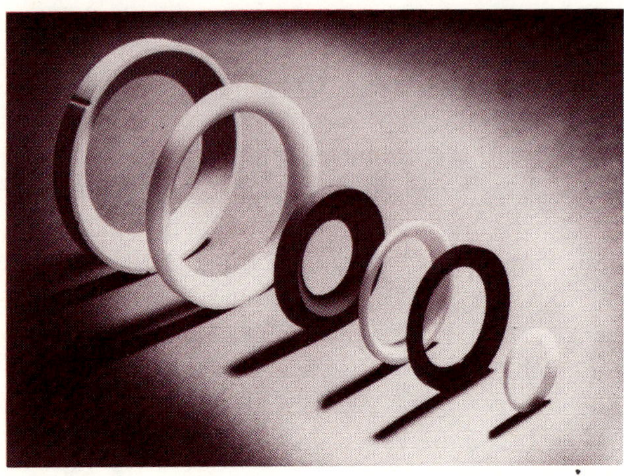

(B) Ball-valve seats.

Fig. 13-30. Two uses of PCTFE. (Chemplast, Inc.)

The early preparation of vinyl fluoride gas in 1900 was considered impossible to polymerize. In 1958, however, du Pont announced the polymerization of vinyl fluoride (PVF). In 1933 monomer resins were prepared in Germany by the reaction of hydrogen fluoride on acetylene in selected catalysts:

$$HF + CH \equiv CH \xrightarrow[\text{catalysts}]{} CH_2{=}CHF$$

Although the monomer has been known to chemists for some time, it is difficult to manufacture or polymerize. Polymerization is carried out using peroxide catalysts in various aqueous solutions at high pressures.

Polyvinyl fluoride may be processed by conventional thermoplastic equipment. Polyvinyl fluoride plastics are strong, tough, flexible, transparent, and have outstanding weather resistance. Polyvinyl fluoride possesses good electrical and chemical resistance with a service temperature near 300°F.

Applications include protective coatings and surfaces for exterior use, finishes for plywoods, sealing tapes, packaging of corrosive chemicals, and numerous electrical insulating applications. Coatings may be applied to automotive parts, lawn mower housings, house shutters, gutters, and metal siding.

Closely resembling polyvinyl fluoride is polyvinylidene fluoride (PVF_2). It became available in 1961 and was produced by the Pennwalt Chemical Corporation under the tradename Kynar. Polyvinylidene fluoride is polymerized thermally by the dehydrohalogenation (removal of selected hydrogen and chlorine atoms) of chlorodifluoroethane under pressure:

$$CH_3CClF_2 \xrightarrow{500\text{-}1700°C} CH_2{=}CF_2 + HCl$$

Like polyvinyl fluoride, it does not have the chemical resistance of PTFE or PCTFE. The alternating CH_2 and CF_2 groups in its backbone contribute to the tough, flexible characteristics. The presence of the hydrogen atoms reduces the chemical resistance, allowing solvent cementing and degradation. PVF_2 materials are processed by conventional thermoplastic equipment and are ultrasonically and thermally sealed. PVF_2 finds wide application in the form of films and coatings because of its toughness, abrasion resistance, chemical resistance, optical properties, and resistance to ultraviolet radiation. Service temperatures range from −80°F to +300°F. A familiar exterior application is the coating seen on aluminum siding and roofs. Molded parts may include such items as valves, impellers, chemical tubing, ducting, and electronic components (Fig. 13-31).

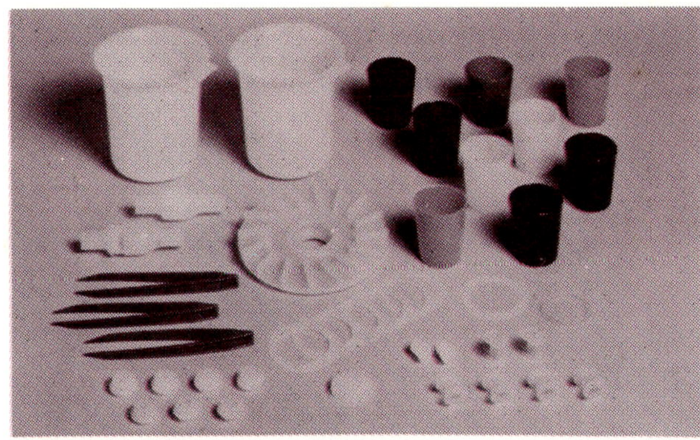

(A) Lab equipment, spools, impellers, containers, and gaskets.

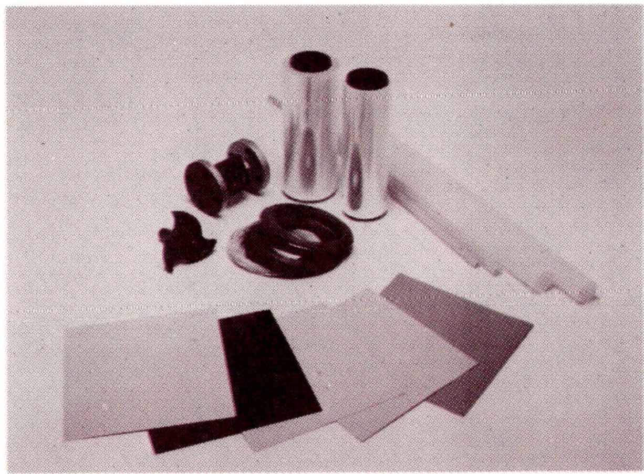

(B) Film, sheet, rods, moldings, and coatings.

Fig. 13-31. Uses of polyvinylidene fluoride. (KREHA)

There are numerous other fluorine containing polymers and copolymers; most, however, are elastomers.

Chlorotrifluoroethylene/vinylidene fluoride is used in fabricating O-rings and gaskets:

$$\left[\begin{array}{cccc} F & H & F & F \\ | & | & | & | \\ -C- & C- & C- & C- \\ | & | & | & | \\ F & H & Cl & F \end{array}\right]_n$$

Hexafluoropropylene/vinylidene fluoride is an outstanding oil and grease resistant elastomer for O-rings, seals, and gaskets.

The chemical structure of two fluoroacrylate elastomers is shown in Fig. 13-32.

Polytrifluoronitrosomethane, silicone, polyesters, and other fluorine containing polymers are also being produced.

$$(CH_2-CH)_n \text{ with } -C{=}O,\ -O-,\ CH_2,\ C_3F_7 \quad\quad (CH_2-CH)_n \text{ with } -C{=}O,\ -O-,\ CH_2,\ CF_2-CF_2-O-CF_3$$

Fig. 13-32. Two fluoroacrylate elastomers.

At degrading temperatures, toxic fluorine gas is released. (See Chap. 12.)

Table 13-7 gives some of the properties of the basic fluoroplastics.

Ionomers

In September of 1964 the du Pont Company coined a new term to describe a material that had characteristics of both thermoplastics and thermosetting plastics. These new materials were known as *ionomers*. Ionic bonding is seldom found in plastics; however, *iono*mers possess polyethylene type chains with ion crosslinks of sodium, potassium, or other similar ions (Fig. 13-33). In this material both organic and inorganic compounds are linked together. Because the crosslinking is predominately ionic the weaker bonds are easily broken on heating and it may be processed as thermoplastics. At atmospheric temperatures the plastics has properties normally associated with linked polymers.

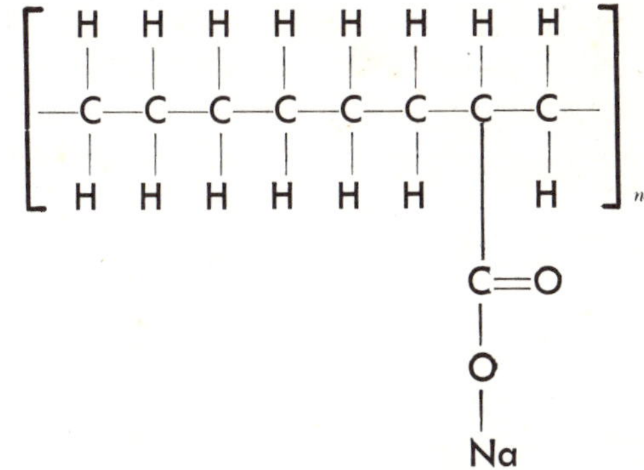

Fig. 13-33. Example of ionomer structure.

Although the basic ionomer chain is produced from polymerization of ethylene and methacrylic acid, other inexpensive polymer chains may be developed with similar crosslinks.

Table 13-7. Fluoroplastic Properties.

Property	Polychloro-Trifluoroethylene (CTFE)	Polytetra-fluoroethylene (TFE)	Polyvinylidene Fluoride (PVF)	Fluorinated Ethylene Propylene (FEP)	Polyvinylidene Fluoride (PVF)
Molding qualities	Excellent	Excellent	Excellent	Excellent	Excellent
Specific gravity (density)	2.1-2.2	2.14-2.2	1.75-1.78	2.12-2.17	1.75-1.78
Tensile strength (psi)	4500-6000	2000-5000	5500-7400	2700-3100	5500-7400
Compressive strength (psi)	4600-7400	1700-2000	8680	....	8680
Izod, impact (ft-lb/in)	2.5-2.7	8	3.6-4	No break	3.6-4.0
Hardness, Rockwell	R75-R95	Shore D50-D65	Shore D80	R25	D80 (Shore)
Thermal expansion ($10^{-5}/^{\circ}C$)	4.5-7	10	8.5	8.3-10.5	8.5
Resistance to heat (°F)	350-390	550	300	400	300
Dielectric strength (volts/mil)	500-600	480	260	500-600	260
Dielectric constant (60 Hz)	2.24-2.8	2.1	8.4	2.1	8.4
Dissipation factor (60 Hz)	0.0012	0.0002	0.049	$<$ 0.0003	0.049
Arc resistance (seconds)	360	300	50-70	165 +	50-70
Water absorption (24 hrs, %)	0.00	0.00	0.04	0.01	0.04
Burning rate (in/min)	None	None	Self-exting.	None	Self-exting.
Effect of sunlight	None	None	Slight	None	Slight bleach
Effect of acids	None	None	Attacked by fuming sulfuric	None	Attacked fuming sulfuric
Effect of alkalies	None	None	None	None	None
Effect of solvents	None	None	Resists most	None	Resists most
Machining qualities	Excellent	Excellent	Excellent	Excellent	Excellent
Optical	Translucent-opaque	Opaque	Transparent-translucent	Transparent	Transparent

As materials which combine ionic and covalent forces in their molecular structure, ionomers can exist in a number of physical states and have a number of physical properties. They may be processed and reprocessed by any thermoplastic processing technique. They are presently more expensive than polyethylene but are available in transparent forms. Ionomers possess a higher moisture vapor permeability than polyethylene. Ionomer applications include safety glasses, shields, toys, containers, packaging films, electrical insulation, and coatings for paper or other substrates. They have found markets in the shoe industry as inner liners and as soles and heels in selected shoe styles (Fig. 13-34). Ionomers are coextruded with polyester films to produce a heat sealable layer while improving package durability. There are increasing applications of homopolymers and copolymers in skin and blister packaging of products.

Table 13-8 gives some properties of ionomers.

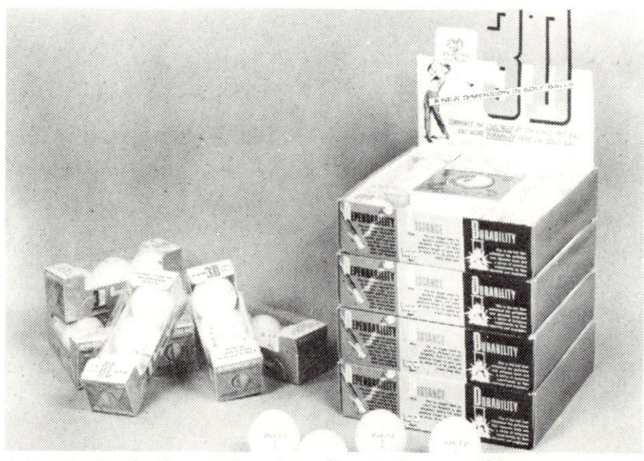

(A) Golf balls with virtually indestructible covers of ionomer resin.

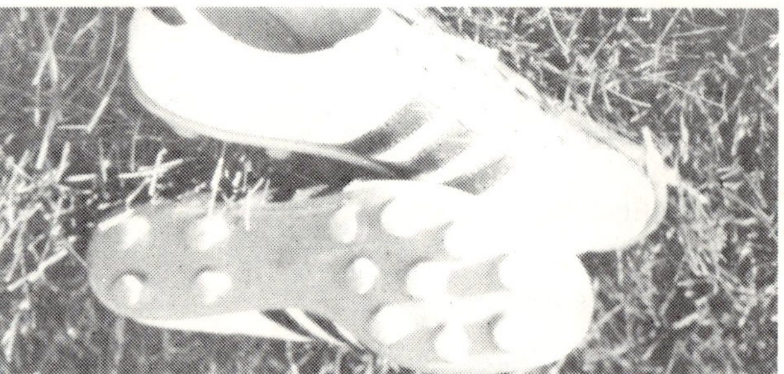

(B) These soccer shoes have ionomer molded soles and heels.

Fig. 13-34. Application of ionomers. (E. I. du Pont de Nemours & Co.)

Phenoxy

A family of resins based on bisphenol-A and epichlorhydrin were introduced in 1962 by Union Carbide. These resins were termed "phenoxy" but

Table 13-8. Ionomer Properties.

Property	Ionomer
Molding qualities	Excellent
Specific gravity (density)	0.93-0.96
Tensile strength (psi)	3500-5000
Izod, impact (ft-lb/in)	6-15
Hardness, Rockwell	D50-D65 (Shore)
Thermal expansion (10^{-5}/°C)	12
Resistance to heat (°F)	160-220
Dielectric strength (volts/mil)	900-1000
Dielectric constant (60 Hz)	2.4-2.5
Dissipation factor (60 Hz)	0.001-0.003
Arc resistance (seconds)	90
Water absorption (24 hrs, %)	0.1-1.4
Burning rate (in/min)	Very slow
Effect of sunlight	Requires stabilizers
Effect of acids	Attacked by oxidizing acids
Effect of alkalies	Very resistant
Effect of solvents	Very resistant
Machining qualities	Fair-good
Optical	Transparent

could be classed as polyhydroxyethers. The plastics structure resembles polycarbonates with similarities in properties (Fig. 13-35). (See Aeromatic Polyethers, Epoxy, and Polysulfones.)

$$\left[-O-C_6H_4-C(CH_3)(CH_3)-C_6H_4-O-CH_2-CH(OH)-CH_2- \right]_n$$

Phenoxy resins are manufactured and sold as thermoplastic epoxide resins. They may be processed on conventional thermoplastic machinery with product service temperatures exceeding 170°F.

Fig. 13-35. This small vent housing grill is made of phenoxy plastics.

Phenoxy resins, because of the reactive hydroxyl groups, may be crosslinked with several agents including diisocyanates, anhydrides, triazines, and melamines.

Homopolymers have good creep resistance, high elongation, low moisture absorption, low gas transmission, high rigidity, tensile strength, and ductility. They find applications as clear or colored protective coatings, molded electronic parts, pipe for gas and crude oil, sports equipment, appliance housings, cosmetic cases, adhesives, and containers for foods or drugs. (See Fig. 13-35.)

Table 13-9 gives some of the properties of phenoxy.

Table 13-9. Phenoxy Properties.

Property	Phenoxy
Molding qualities	Good
Specific gravity (density)	1.18-1.3
Tensile strength (psi)	9000-9500
Izod, impact (ft-lb/in)	2.5
Resistance to heat (°F)	175
Dielectric constant (60 Hz)	4.1
Dissipation factor (60 Hz)	0.001
Water absorption (24 hrs, %)	0.13
Burning rate	Self-exting.
Effect of acids	Resistant
Effect of alkalies	Resistant
Effect of solvents	Soluble in ketones
Machining qualities	Good
Optical	Translucent-opaque

Polyallomers

In 1962 a plastics distinctly different from simple copolymers of polyethylene and polypropylene was produced by Eastman. The process, called *allomerism*, is conducted by alternately polymerizing ethylene and propylene monomers. Allomerism is a variation in chemical composition without a change in crystalline form. Thus the plastics exhibits crystallinity normally associated with the homopolymers of ethylene and propylene. The term *polyallomer* has been used to distinguish this alternately segmented plastics from homopolymers and copolymers of ethylene and propylene.

Although highly crystalline, polyallomers may have a density as low as 0.896. They are available in various formulations with properties which include high stiffness, impact strength, and abrasion resistance. Their flexibility has been used in the production of hinged boxes, looseleaf binders, and various other folders (Fig. 13-36). Polyallomers may be used as food containers or films over a wide service temperature. Other properties are similar to polyethylene but find limited application where other polyolefins are marginal for application.

Fig. 13-36. These notebook folders are made of polyallomer plastics.

Processing may be accomplished on all conventional thermoplastic equipment. Like polyethylene, polyallomers are not cohesively bonded but may be welded.

Table 13-10 lists some of the properties of polyallomers.

Table 13-10. Polyallomer Properties.

Property	Homopolymer
Molding qualities	Excellent
Specific gravity (density)	0.896-0.899
Tensile strength (psi)	3000-3850
Izod, impact (ft-lb/in)	170-250
Hardness, Rockwell	R50-R85
Thermal expansion (10^{-5}/°C)	8.3-10
Resistance to heat (°F)	124-200
Dielectric strength (volts/mil)	800-900
Dielectric constant (60 Hz)	2.3-2.8
Dissipation factor (60 Hz)	0.0005
Water absorption (24 hrs, %)	0.01
Burning rate	Slow
Effect of sunlight	Slight—should be protected
Effect of acids	Very resistant
Effect of alkalies	Very resistant
Effect of solvents	Very resistant
Machining qualities	Good
Optical	Transparent

Polyamides

From research starting in 1928, Wallace Hume Carothers and his colleagues concluded that linear polyesters were not ideally suitable for the commercial production of fibers. He did succeed in producing polyesters of high molecular weight and oriented them by elongation under tension; however, they were still inadequate and could not be successfully spun into fibers. The amino acid units present in the natural fiber silk may have prompted Carothers to investigate synthetic polyamides. Of numerous formulations of amino acids, diamines, and dibasic acids, several indicated promise as possible fibers. By 1938 the first commercially developed polyamide was introduced by the du Pont Company. It was a 6/6 polyamide and was given the tradename Nylon. (The term "nylon" has come to mean any polyamide capable of being processed into filaments, fibers, films, and to some extent molded parts.) This condensation plastics was called Nylon 6,6 (*or* 66) because both the acid and the amine contain six carbon atoms.

$$NH_2(CH_2)_6NH_2 + COOH(CH_2)_4COOH$$
Hexamethylene Diamine   *Adipic Acid*

$$\xrightarrow{\text{heat}} n[NH_2(CH_2)_6NH \cdot CO(CH_2)_4COOH]$$
Nylon Salt

$$\rightarrow [NH(CH_2)_6NH \cdot CO(CH_2)_4CO]_n + n \cdot H_2O$$
Nylon 6,6 Polymer Chain

The repeating —CONH— (amide) link is present in a series of linear, thermoplastic "nylons."

Nylon 6—Polycaprolactam:
$[NH(CH_2)_5CO]_x$

Nylon 6,6—Polyhexamethyleneadipamide:
$[NH(CH_2)_6NHCO(CH_2)_4CO]_x$

Nylon 6,10—Polyhexamethylenesebacamide:
$[NH(CH_2)_6NHCO(CH_2)_8CO]_x$

Nylon 11—Poly(11-aminoundecanoic acid)
$[NH(CH_2)_{10}CO]_x$

Nylon 12—Poly(12-aminododecanoic acid)
$[NH(CH_2)_{11}CO]_x$

There are numerous other types of nylon currently available, including Nylon 8, 9, 12, 46, and copolymers from more sophisticated diamines and acids. In the United States, Nylon 6,6 and 6 are by far the most commonly used. Properties may also be changed by introducing additives. Amino containing polyamide resins will react with a number of materials; consequently, crosslinked, thermosetting reactions are possible. Although developed primarily as a fiber, polyamides find growing applications as molding compounds, extrusions, coatings, adhesives, and casting materials. Acetals and fluorocarbons share some of the same properties and uses as polyamide. Polyamide resins are costly. Consequently, they are selected only when other resins will not meet service requirements. Acetal resins are superior in fatigue endurance, creep resistance, and water resistance. Nylons have met increased competition from these resins.

Molding compounds were offered for sale in 1941 and have grown in applications. They are among the toughest of plastics materials. Nylons are self-lubricating, impervious to most chemicals, and highly impermeable to oxygen. They are not attacked by fungi or bacteria and may be used as food containers.

The largest applications of homopolymer molding compounds (6, 6,6, 6,10, 11, 12) include gears, cams, bearings, valve seats, combs, furniture castors, door catches, or other applications where wearing qualities, quietness of operation, and low coefficients of friction are required (Fig. 13-37).

Because of the crystalline structure polyamide, products made of it usually have a milky-opaque appearance. Transparent films may be obtained from Nylon 6 and 6,6 if they are cooled very rapidly as they emerge from the extruder. Polyamides are clear amorphous materials when melted but on cooling become cloudy because of the crystalline formation. This crystalline structure contributes to the stiffness, strength, and heat resistance. Although the polyamides are probably more difficult to process than other thermoplastic materials, they may be processed in all thermoplastic processing equipment. Moderately high processing temperatures are required, and the melting point is abrupt or sharp. They do not soften or melt over a broad range of temperatures. When sufficient energy has overcome the crystalline and molecular attractions, they suddenly become

13-37. Long-wearing quiet gears made of polyamide require no lubrication. (BASF Corp.)

liquid and may be processed. Because all nylons absorb water, they are usually dried before molding to ensure desirable physical properties of the molded product (Fig. 13-38).

Extruded and blown films are used to package oils, grease, cheese, bacon, and other products where low gas permeability is essential. The high service temperatures of nylon film are used for boil and bake-in-the-bag food products. Although polyamides are hydroscopic (absorb water), they find many applications as electrical insulators primarily because of their toughness and high temperature resistance.

Nylon 11 may be used as a protective coating on metal substrates. Polyamides are used in a powder form by spraying or fluidized bed processes. Typical applications include rollers, shafting, panel slides, runners, pump impellers, and bearings. Water dispersions and organic solvents of polyamide resin permit certain adhesive and coating applications on paper, wood, and fabrics.

Polyamide based adhesives may be of the hot-melt or solution type. Hot-melt types are simply

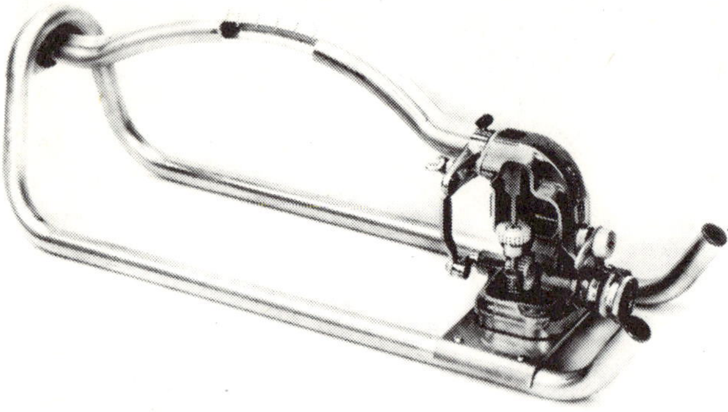

(A) Sprinkler uses two gears molded of polyamide resin in wormgear system.

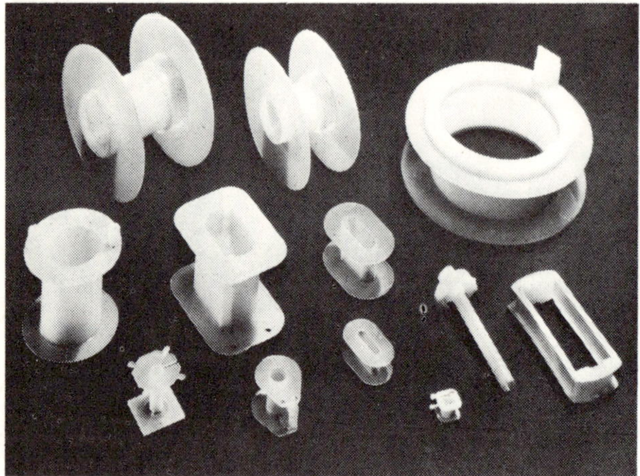

(B) Coil forms provide an excellent example of molding of thin sections.

Fig. 13-38. Molded polyamide resins. (E. I. du Pont de Nemours & Co.)

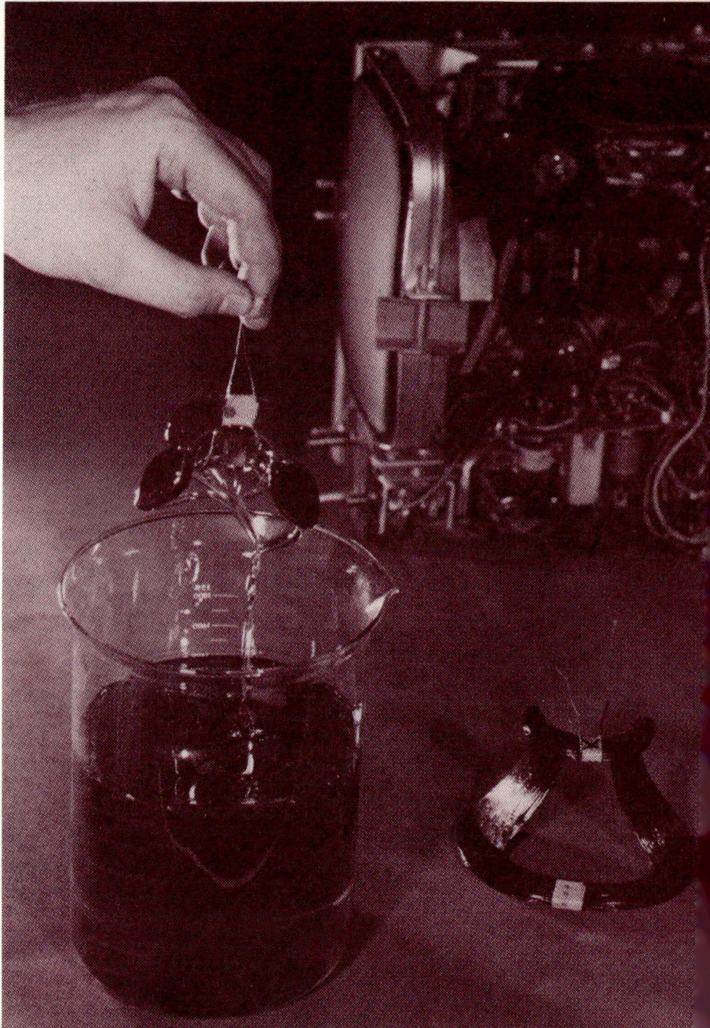

Fig. 13-39. TV yoke and parts are coated with polyamide for insulation and form. (General Mills, Inc.)

heated above the melting point and applied. Amino-polyamide resins may be reacted with epoxy or phenolic resins to produce a thermosetting adhesive. These adhesives find use in bonding of wood, paper laminates, aluminum, and copper to printed circuit boards or as flexible adhesives for bread wrappers, dried soup packets, cigarette packages, and bookbindings. Polyamide-epoxy combinations are used as two-part systems in casting applications such as potting and encapsulation of electrical components (Fig. 13-39). These materials are sometimes used as tools and dies in the metal forming industries.

By combining pigments and other modifying agents, polyamides may be used as printing inks.

The uses of polyamides in the textile and carpet industry are well known and need little discussion. In addition to clothing garments, lightweight tents, shower curtains, and umbrellas are commonly produced from nylon. The monofilaments, multifilaments, and staples are usually made by melt spinning followed by cold drawing to increase tensile strength and elasticity. Monofilaments are used in making fishing lines, surgical sutures, tire cords, rope, sports equipment, and various brushes. Artificial human hair and animal fur are expanding uses of this synthetic fiber.

Polyamides are easily machined (Fig. 13-40). However, drilled or reamed holes are likely to be somewhat undersized because of the resiliency of the material.

Cementing polyamide is difficult because of its excellent solvent resistance. Phenols and formic acid are specific solvents used in cementing polyamides. Epoxy resins are also employed for this purpose.

Table 13-11 gives some of the basic properties of polyamides.

Table 13-11. Polyamide Properties.

Property	Type 6/6 Unfilled	Type 6/10 Unfilled	Type 6/10 Glass Filled
Molding qualities	Excellent	Excellent	Excellent
Specific gravity (density)	1.13-1.15	1.09	1.17-1.52
Tensile strength (psi)	9000-12,000	8500-8600	13,000-35,000
Compressive strength (psi)	6700-12,500	6700-13,000	13,000-24,000
Izod, impact (ft-lb/in)	1.0-2.0	1.2	1.2-6
Hardness, Rockwell	R108-R120	R111	M94, E75
Thermal expansion (10^{-5}/°C)	8	9	1.2-3.2
Resistance to heat	180-300	180-250	300-400
Dielectric strength (volts/mil)	385-470	350-480	400-500
Dielectric constant (60 Hz)	4.0-4.6	3.9	4.0-4.6
Dissipation factor (60 Hz)	0.014-0.040	0.04	0.001-0.025
Arc resistance (seconds)	130-140	100-140	92-148
Water absorption (24 hrs, %)	1.5	0.4	0.2-2
Burning rate (in/min)	Self-exting.	Self-exting.	Self-exting.
Effect of sunlight	Discolors slightly	Discolors slightly	Discolors slightly
Effect of acids	Attacked	Attacked	Attacked
Effect of alkalies	Resistant	None	None
Effect of solvents	Dissolved by phenols & formic acid	Dissolved by phenols	Dissolved by phenols
Machining qualities	Excellent	Fair	Fair
Optical	Translucent-opaque	Translucent-opaque	Translucent-opaque

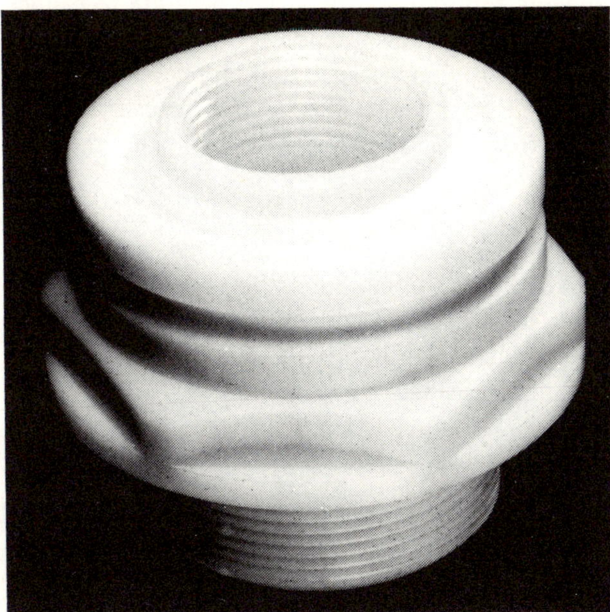

Fig. 13-40. This polyamide article required 18.5 seconds of machining on an automatic lathe. Light metal requires 2½ times as long; free-cutting steel, 13 times as long. (BASF Corp.)

Polycarbonates

An important material in the production of plastics is phenol. It is used in producing phenolic, polyamide, epoxy, polyphenylene oxide, and polycarbonate resins.

Phenol is a compound which has one hydroxyl group attached to an aromatic ring. It is sometimes called *monohydroxy benzene,* C_6H_5OH.

or abbreviated

Phenol or Monohydroxy Benzene

Bisphenol-A (two phenol and acetone), an important ingredient in the production of polycarbonates, may be prepared by combining acetone with phenol as in Fig. 13-41.

Bisphenol-A is sometimes referred to as *diphenylol propane* or *bis-dimethylmethane.*

Another important material used in the production of polycarbonate is phosgene. Phosgene is a

PHENOL ACETONE BISPHENOL-A

Fig. 13-41. Preparation of bisphenol-A.

poisonous gas, which was used in World War I. It is composed of carbon monoxide and chlorine, with the formula $COCl_2$.

As early as 1898 A. Einhorn prepared a polycarbonate material from the reaction of resorcinol and phosgene. Both W. H. Carothers and F. J. Natta performed research on a number of polycarbonates using ester reactions. Carothers discarded the use of linear based polyesters in favor of polyamides.

Research continued after the war in Germany by Farbenfabriken Bayer and in the United States by General Electric. By 1957 both had arrived at the production of polycarbonates produced from bisphenol-A. Volume production in the United States did not begin until 1959.

There are two general methods of preparing polycarbonates: (1) The most common method is the reaction of purified bisphenol-A with phosgene under alkaline conditions. (See Fig. 13-42.) (2) An alternate method involves the reaction of purified bisphenol-A with diphenyl carbonate (metacarbonate) in the presence of catalysts while under a vacuum (Fig. 13-43).

Purity of the bisphenol-A is important if the plastics is to possess high-clarity, long linear chains with no crosslinking substances.

The phosegenation process is preferred because the reaction may be carried out at low temperatures using simple technology and equipment. The process, however, does require the recovery of solvents and inorganic salts. In either method the product is comparatively high in cost compared with other general-purpose thermoplastics.

Polycarbonates may be processed by all conventional thermoplastic processes. It must also be pointed out that their resistance to heat and their rather high melt temperature require higher-temperature processing. The molding temperature is very critical and must be accurately controlled in order to produce usable products. The unique properties of polycarbonate may be attributed to the carbonate groups and the presence of benzene rings in the long, repeating molecular chain. Because of polycarbonate's broad range of properties including high impact strength, transparency, excellent creep resistance, wide temperature limits, high dimensional stability, good electrical characteristics, and self-extinguishing behavior, fabricators provide a diverse group of products. Tough transparent grades are used in lenses, films, windshields, light fixtures, containers, appliance components and tool housings. Temperature resistance is utilized in hot dish handles, coffee pots, popcorn popper lids, hair driers, and appliance housings (Fig. 13-44). These plastics have excellent properties from −275°F to +270°F. Impact and flexural strength is important in pump impellers, safety helmets, beverage dispensers, small appliances, trays, signs, aircraft parts, cameras, and various packaging or film applications. Polycarbonate parts have excellent dimensional stability. Glass-filled grades have improved impact, moisture, and chemical resistance (Fig. 13-45).

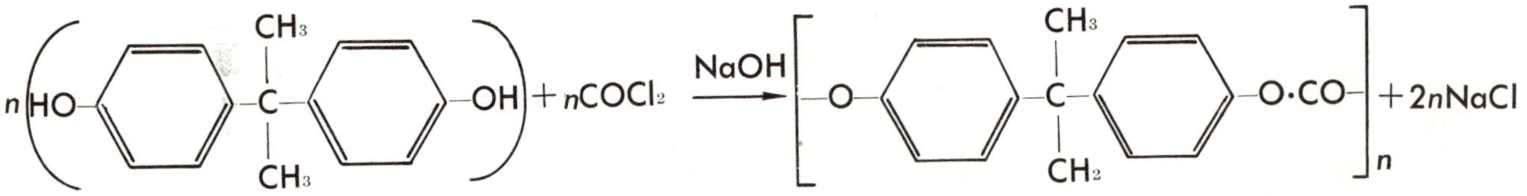

Fig. 13-42. First method of preparing polycarbonates.

BISPHENOL-A DIPHENYL CARBONATE POLYCARBONATE

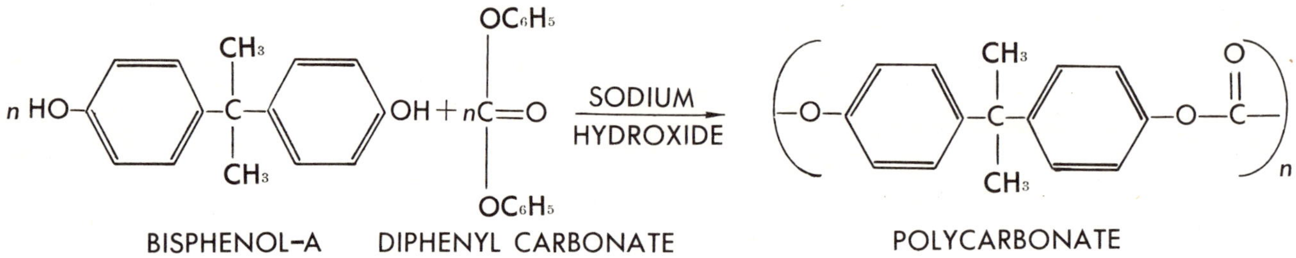

Fig. 13-43. Second method of preparing polycarbonates.

(A) Transparent polycarbonate lids are used on these popcorn poppers.

(B) Power tool housing made of molded polycarbonate.

(C) Lightweight tough appliance housing made of polycarbonate.

Fig. 13-44. Polycarbonate products around the home.

Fig. 13-45. High-impact electrical-grade polycarbonate is used to replace glass insulator and has superior mechanical characteristics and equivalent electrical properties. (H. K. Porter Co., Inc.)

Most aromatic solvents, esters, and ketones will attack polycarbonates. Chlorinated hydrocarbons are used as solvent cements for cohesive bonds.

There are several hundred variations to the polycarbonate structure. The structure may be modified by substituting various *R* side groups or separating the benzene rings by more than one carbon atom. Some possible structural combinations are shown in Fig. 13-46.

POSSIBLE RADICAL SIDE GROUPS

R	*R*₁
—H	—H
—H	—CH_3
—CH_2	—CH_3
—CH_3	—C_2H_5
—C_2H_5	—C_2H_5
—CH_3	—CH_2—CH_2—CH_3
—CH_2—CH_2—CH_3	—CH_2—CH_2—CH_3
—C_6H_5	—C_6H_5

Fig. 13-46. Possible combinations of polycarbonates.

Table 13-12 gives some of the properties of polycarbonates.

Polyimides

Polyimides, laboratory curiosities for many years, were developed by du Pont in 1962. Polyimides are usually obtained from condensation polymerization of an aromatic dianhydride and an aromatic diamine (Fig. 13-47). Although they are linear and thermoplastic, they are difficult to process because of their high melting point. Many polyimides do not melt and must be fabricated by machining or other forming techniques.

Table 13-12. Polycarbonate Properties.

Property	Unfilled	10%-40% Glass-Filled
Molding qualities	Good-excellent	Very good
Specific gravity (density)	1.2	1.24-1.52
Tensile strength (psi)	8000-9500	12,000-25,000
Compressive strength (psi)	10,300-10,800	13,000-21,000
Izod, impact (ft-lb/in)	12-18 (½ × ⅛ in)	1.2-6.5 (¼ × ½ in)
Hardness, Rockwell	M73-78, R115-R125	M88-M95
Thermal expansion (10^{-5}/°C)	6.6	1.7-4
Resistance to heat (°F)	250	275
Dielectric strength (volts/mil)	400	450
Dielectric constant (60 Hz)	2.97-3.17	3.0-3.53
Dissipation factor (60 Hz)	0.0009	0.0009-0.0013
Arc resistance (seconds)	10-120	5-120
Water absorption (24 hrs, %)	0.15-0.18	0.07-0.20
Burning rate (in/min)	Self-exting.	Slow (0.8-1.2)
Effect of sunlight	Slight	Slight
Effect of acids	Attacked slowly	Attacked by oxidizing acids
Effect of alkalies	Attacked	Attacked
Effect of solvents	Soluble in aromatic & chlorinated hydrocarbons bons	Soluble in aromatic & chlorinated hydrocarbons
Machining qualities	Excellent	Fair
Optical	Transparent-opaque	Translucent-opaque

Fig. 13-47. Basic polyimide structure.

Polyimides compete with various fluorocarbons for low friction, good strength, toughness, dielectric strength, and heat resistance. They possess good resistance to radiation but are surpassed in chemical resistance by the fluoroplastics. Polyimide is attacked by strongly alkaline solutions, hydrazine, nitrogen dioxide, or secondary amine compounds.

In spite of their relative high cost and difficulty in processing, polyimides are used in the manufacture of aerospace, electronics, nuclear power, and office and industrial equipment. Various parts include valve seats, gaskets, piston rings, thrust washers, and bushings. Films are made by a casting process (usually from prepolymer form) and are used for laminates, dielectrics, and coatings.

In Fig. 13-48 polyimide is applied (in hot liquid, electrostatic spray form) to electric skillets and other houseware items. After curing and baking at

(A) Cooking items coated with colored polyimide finishes.

(B) Automatic spraying line coating electric skillets.

Fig. 13-48. Examples of polyimide coating (de Beers Labs., Inc.)

550°F for a few minutes, they form a hard and flexible glossy finish similar to porcelain.

Prolonged contact with this resin and reducers may cause serious cracking of the skin. The solvents are no more toxic than other aromatics.

Table 13-13 gives some properties of polyimides.

Table 13-13. Polyimide Properties.

Property	Polyimide (Unfilled)
Specific gravity (density)	1.43
Tensile strength, (psi)	10,000
Compressive strength, (psi)	24,000 +
Izod, impact (ft-lb/in)	0.9
Hardness, Rockwell	E45-E58
Resistance to heat (°F)	570
Dielectric strength (volts/mil)	560
Dielectric constant (60 Hz)	3.4
Arc resistance (seconds)	230
Water absorption (24 hrs, %)	0.32
Burning rate	Nonburning
Effect of acids	Resistant
Effect of alkalies	Attack
Effect of solvents	Resistant
Machining qualities	Excellent
Optical	Opaque

Polyolefins (Polyethylene)

Ethylene gas is a member of an important group of unsaturated, aliphatic hydrocarbons called "olefins" or "alkenes." The word *olefin* means "oil forming." It was originally given to ethylene because oil was formed when treated with chlorine. The term "olefin" now applies to all hydrocarbons with carbon-to-carbon double bonds. Olefins are highly reactive because of this carbon-to-carbon double bond. Some of the principal olefin monomers are shown in Table 13-14.

In the United States ethylene gas is readily produced by cracking higher hydrocarbons of natural gas or petroleum. The importance and relationship of ethylene to other polymers is shown in Fig. 13-49.

Between 1879 and 1900 several chemists experimented with linear polyethylene polymers. In 1900 E. Bamberger and F. Tschirner used the expensive material diazomethane to produce a linear polyethylene they termed "polymethylene."

$$2n\left(\begin{array}{c} CH_2 \\ / \quad \backslash \\ N{=}N \end{array}\right) \rightarrow \left(CH_2{-}CH_2\right)_n + 2n \cdot N_2$$

Diazomethane *Polyethylene*

Table 13-14. The Principal Olefin Monomers.

Chemical Formula	Olefin Name
H H | | C = C | | H H	Ethylene
H H | | C = C | | CH_3 H	Propylene
H H | | C = C | | C_2H_5 H	Butene-1
H H | | C = C | | H_2C H | H—C—CH_3 | CH_3	4-Methylpentene

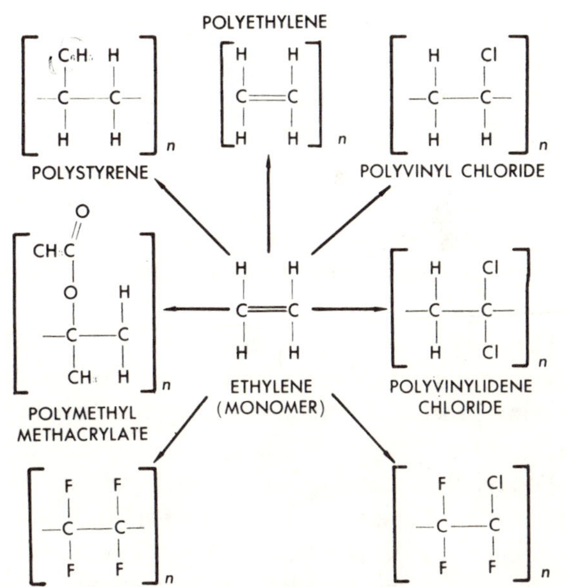

Fig. 13-49. Ethylene monomer is shown indicating the relationship to other monomer resins.

Carothers and his coworkers reported producing polyethylene of low molecular weight in 1930. The actual commercialization of polyethylene came as a result of the research in 1933 by Dr. E. W. Fawcett and Dr. R. O. Gibson of the Imperial Chemical Industries (ICI) in England. Their discovery was a result of investigating the reaction of benzaldehyde and ethylene (obtained from coal) under high pressure and temperature. In September of 1939, on the day Germany invaded Poland, ICI began commercial production of polyethylene. The war efforts consumed all of the polyethylene produced. It was used extensively as insulation on high-frequency radar cables vitally new at the time. By 1943 the United States was producting polyethylene by the high-pressure methods developed by ICI. These early, low-density materials were highly branched with a disorderly arrangement of molecular chains. Low-density materials are softer, more flexible, melt at lower temperatures, and may be more easily processed. By 1954 two new production processes were developed for the production of polyethylene with densities of 0.91 to 0.97.

One process developed by Karl Ziegler and associates in Germany permitted polymerization of ethylene at low pressures and temperatures in the presence of aluminum triethyl and titanium tetrachloride (catalysts). At about the same time the Phillips Petroleum Company developed a polymerization process using low pressures with a chromium trioxide promoted, silica-alumina catalyst. The conversion of ethylene to polyethylene may be accomplished with a catalyst of molybdenum oxide on an alumina support and other promoters. This process was developed by Standard Oil of Indiana; however, only small quantities have been produced in the United States utilizing this process.

The Ziegler process is used more extensively outside the United States, while the Phillips Petroleum process is predominately used by U.S. firms.

Today, it is possible to produce polyethylene with branched chains or linear chains (Fig. 13-50) by either the high pressure (ICI) or low pressure (Ziegler, Phillips, Standard Oil) methods. The differentiation of polymer shape based on pressures used for polymerization is not used today. The American Society for Testing Materials (ASTM) has divided polyethylenes into three groups:

Type 1 (Branched)	0.910-0.925	Low Density
Type 2	0.926-0.940	Medium Density
Type 3 (Linear)	0.941-0.965	High Density

From Fig. 13-51 it can be seen that the physical properties of low-density (branched) and high-

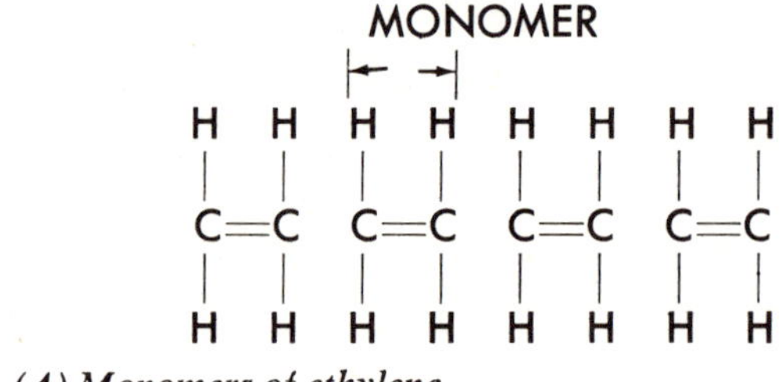

(A) Monomers of ethylene.

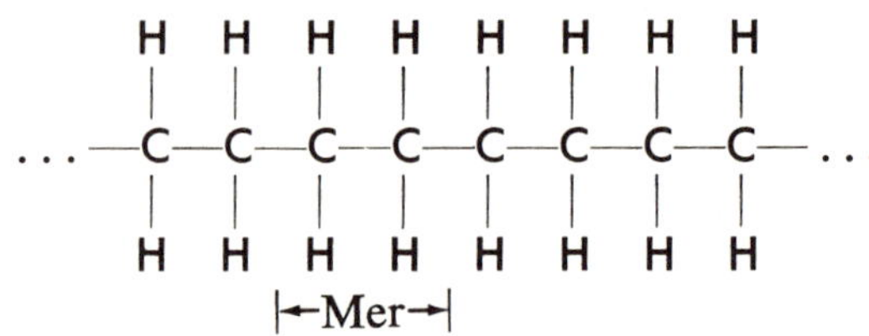

(B) Polymer containing many C_2H_4 mers.

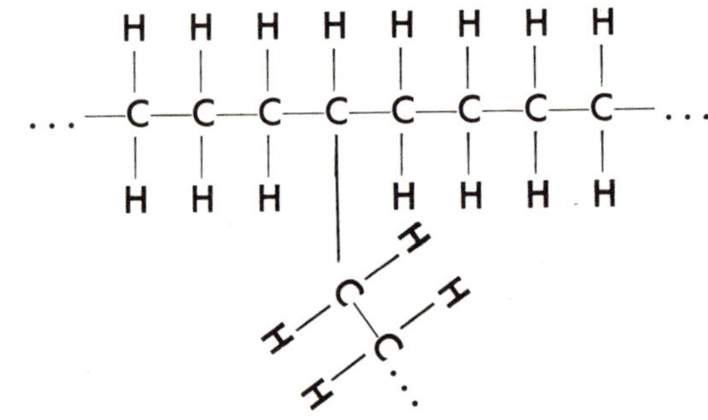

(C) Polymer with branching.

Fig. 13-50. Addition polymerization of ethylene. The original double bond of the ethylene monomer is broken to form two single bonds and thus connect adjacent mers.

density (linear) polyethylenes are different. Low-density polyethylene has a crystallinity of about 60 to 70 percent, while the higher-density polymers may vary from 75 to 95 percent crystallinity. (See Fig. 13-52.)

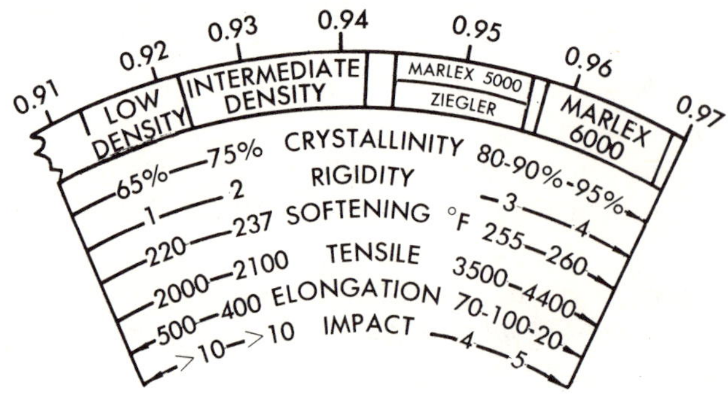

Fig. 13-51. Polyethylene density range. (Courtesy Phillips Petroleum Co.)

(A) Electron micrograph of intercrystalline links bridging radial arms of polyethylene spherulite.

(B) Electron micrograph of small platelet crystals of polyethylene grown on intercrystalline links.

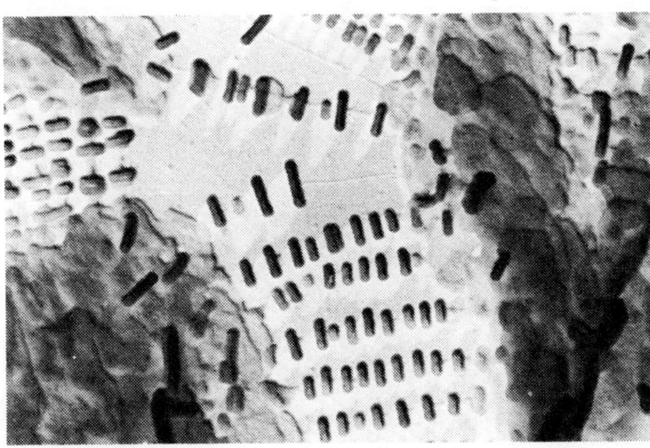

(C) Lamellar crystals of polyethylene were formed by depositing polymer from solution of links.

Fig. 13-52. Closeup of polyethylene. (Bell Telephone Labs.)

Increasing density usually increases stiffness, softening point, tensile strength, crystallinity, and creep resistance. Increased density reduces impact strength, elongation, flexibility, and transparency.

Properties of polyethylene may be controlled and identified by molecular weight and distribution. Molecular weight and molecular weight distribution may have the effects shown in Table 13-15.

Table 13-15. Property Changes Caused by Molecular Weight and Distribution.

Property	As Average Molecular Weight Increases (Melt Index Decreases)	As Molecular Weight Distribution Broadens
Melt viscosity	Increases	
Tensile strength at rupture	Increases	No significant change
Elongation at rupture	Increases	No significant change
Resistance to creep	Increases	Increases
Impact strength	Increases	Increases
Resistance to low temperature brittleness	Increases	Increases
Environmental stress cracking resistance	Increases	Increases
Softening temperature		Increases

In Fig. 13-53 a schematic representation of a narrow molecular weight distribution polymer is compared to a broad molecular weight distribution polymer. The molecular weight distribution is a ratio of the large, medium, and small molecular chains in the resin. If the resin is composed of chains close to the average length, the molecular weight distribution is called "narrow" and the molecular chains can flow past each other much easier than when there are large ones present.

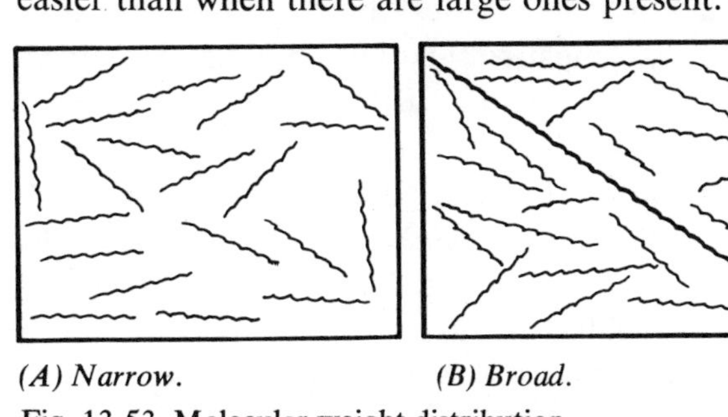

(A) Narrow. *(B) Broad.*

Fig. 13-53. Molecular weight distribution.

A melt index apparatus (Fig. 13-54) is commonly used to measure the melt flow at a specified temperature and pressure. This melt index is largely dependent on molecular weight and distribution.

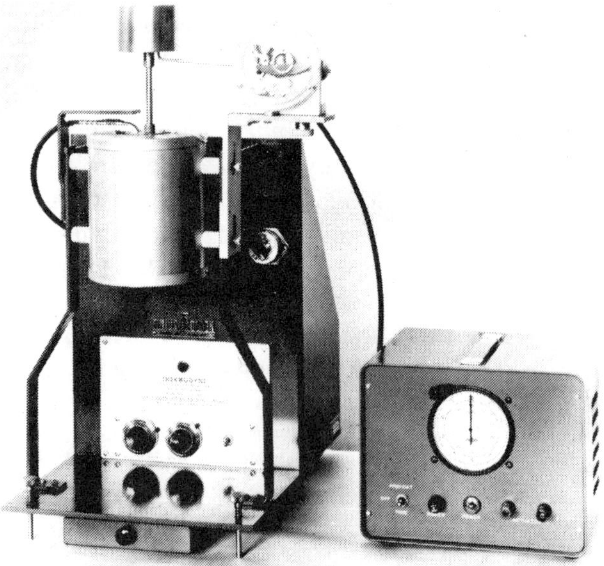

(A) Melt indexer with automatic timer.

(B) Sieglaff-McKelvey rheometer.

Fig. 13-54. These two machines are used to test the flow of plastics materials. (Tinius Olsen Testing Machine Co.)

As melt index decreases, the melt viscosity, tensile strength, elongation, and impact strength increase. For many processing methods it is convenient to have the hot resin flow very easily. This indicates a resin with a high melt index. Usually, polyethylenes with high molecular weight have a low melt index. Thus, by varying density, molecular weight, and molecular distribution, polyethylene may be produced with a wide variety of properties.

The addition of fillers, reinforcements, or monomers may also change properties (Fig. 13-55). Some of the common comonomers are shown in Table 13-16.

Polymethylpentene, developed in 1965, has gained some commercial importance. The materials

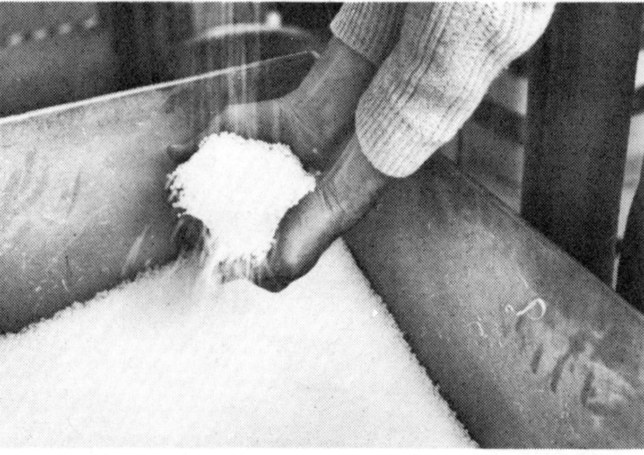

Fig. 13-55. Pure pellets of polyethylene have toughness and resistance to heat and cold. (Western Electric Co.)

Table 13-16. Common Comonomers Used with Olefins.

Formula	Name
H H | | C = C | | C_4H_9 H	1-Hexene
H H | | C = C | | O H | O=C—CH_3	Vinyl acetate
H H | | C = C | | | H | O=C—O—CH_3	Methyl acrylate
H H | | C = C | | | H | O=C—OH	Acrylic acid

are characterized by low density, high transparency, high melting point, and excellent electrical properties. The polymer may be prepared from ethylene or propylene and 4-methylpentene-1. It finds use in applications where clarity and temperature are important to performance of the product.

Most copolymers fall in the range of flexible elastomers. Copolymers of vinyl acetate or methyl acrylate may produce transparent, amorphous materials with rubberlike elasticity. Ethylene-vinyl acetate (Fig. 13-56) has found limited applications replacing plasticized polyvinyl chloride and rubber used in food processing, packaging, or dispensing equipment. Molded parts include toys, gaskets, seals, and disposable medical syringes. Chlorosulphonated polyethylene and other polyolefins are rubber polymers and should not be classed as plastics materials.

$$\left(-\underset{H}{\overset{H}{C}}-\underset{H}{\overset{H}{C}}-\underset{H}{\overset{H}{C}}-\underset{H}{\overset{H}{C}}-\underset{H}{\overset{H}{C}}-\underset{H}{\overset{H}{C}}-\underset{H}{\overset{H}{C}}-\underset{H}{\overset{H}{C}}-\underset{\substack{| \\ O \\ | \\ C=O \\ | \\ CH_3}}{\overset{H}{C}}-\underset{H}{\overset{H}{C}}-\underset{H}{\overset{H}{C}}- \right)_n$$

Fig. 13-56. Ethylene-vinyl acetate (EVA) copolymers offer a wide range of molecular weights.

Polyethylene may be crosslinked from a thermoplastic material to a thermosetting one. It should be obvious that the converting of a thermoplastic material to a thermosetting one, after forming, opens up many new possibilities. This crosslinking may be accomplished by chemical agents (usually peroxides) or by irradiation. There is increasing commercial use of radiation to cause branching of polyethylene products. Radiation crosslinkage is rapid and leaves no objectionable residues. Irradiated parts may be exposed to temperatures exceeding 250°F. (See Fig. 13-57.)

Fig. 13-57. Polyethylene container in center has been exposed to controlled radiation to produce superior heat resistant product. Containers at left and right were not treated and could not hold their shape at 350°F.

Excessive radiation may reverse the effect by breaking the main links in the molecular chain. In order to protect polyethylene from the damaging effects of ultraviolet light radiation, stabilizers or pigments of carbon black must be used to absorb or block the damaging effects of sunlight.

Because of the low price, ease of processing and outstanding broad range of properties, polyethylene has become one of the most used plastics in America. Being one of the lightest thermoplastics, it may be selected where costs are based on cubic weight. Excellent electrical and chemical resistance have led to widespread use of polyethylene as electrical wire coatings and dielectrics or as containers, tanks, pipes, and coatings where chemical agents are present. At room temperature, there is no solvent for polyethylene. It is easily welded, however.

Polyethylene may be easily processed by every method devised for thermoplastics. Probably the largest use is in the manufacture of containers and film consumed by the packaging industry. Blow-molded containers are seen in every supermarket replacing the heavier glass and metal (Fig. 13-58).

(A) Beverage containers. (Uniloy Div., Hoover Ball & Bearing)

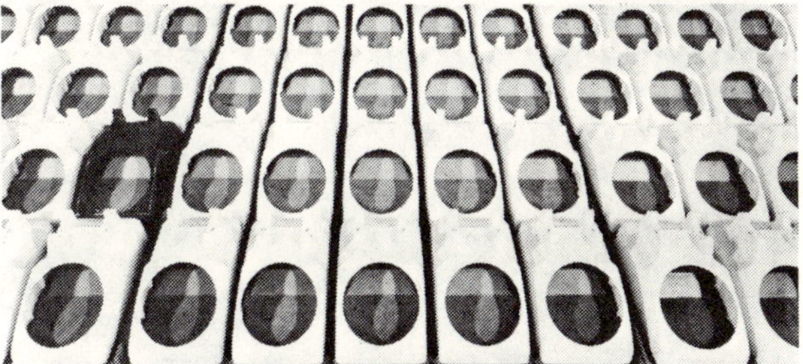

(B) Telephone housings. (Western Electric Co.)

Fig. 13-58. Uses of polyethylene.

Tough plastics bags for packaging foods and films for packaging fresh fruit, frozen foods, bakery products, and other products are but a few of the many applications for polyethylene. Lower-density films are made with good clarity by rapidly chilling the melt as it emerges from the die. As the hot amorphous melt is rapidly chilled, extensive crystallization does not have time to form. Low-density films are used to package new garments including shirts, sweaters, sheets, and blankets. At the laundry and dry cleaner very thin films are used for packaging. Higher-density films are used where greater heat resistance is required as for boil-in-the-bag food packs. Silage covers, reservoir linings, seedbed covers, moisture barriers, and covering for harvested crops are only a few applications in construction, agriculture, or horticulture.

Although polyethylene is a good moisture barrier, it has a high gas permeability. It should not be used under vacuum or for transporting gaseous materials. Although it is permeable to oxygen and carbon dioxide, meats and some produce may require small vent holes to allow ventilation (Fig. 13-59). Oxygen keeps the meat looking red and moisture from condensing on packaged produce.

(A) Fruits.

(B) Meats.

Fig. 13-59. Polyethylene-wrapped grocery products.

Heat sealing and shrink wrapping are easily employed with these films. Electronic or radio-frequency heat sealing is difficult in polyethylene because of the low electrical dissipation (power) factor.

Polyethylene finds applications for the coating of various substrates. It is used to coat paper, cardboard, and fabrics to improve the wet strength and other properties. Coated materials may then be heat sealed. The packaging of milk is a familiar application. Powdered forms are used for dip coating, flame spraying, and fluidized bed coating where a chemicalproof and moistureproof layer is required.

Injection molded toys, small appliance housings, garbage cans, freezer containers, and artificial flowers benefit from polyethylene's toughness, chemical inertness, and low service temperatures (Fig. 13-60).

(A) Ford truck fuel tank.

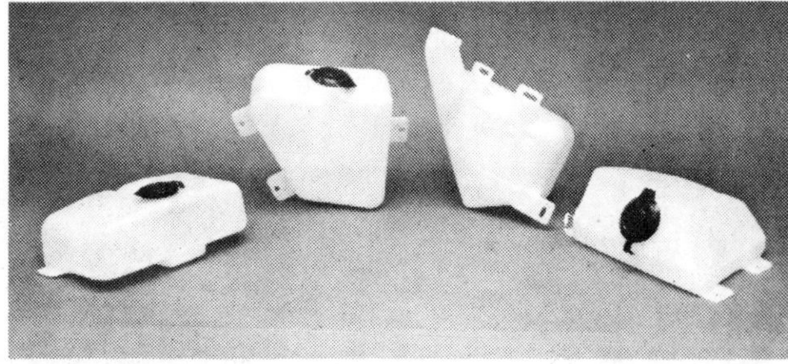

(B) Windshield washer jars.

Fig. 13-60. Blow molded polyethylene products. (Phillips Petroleum Co.)

Extruded pipe and ducting is used in chemical plants and for some domestic cold water service. Corrugated drain pipe is making headway by replacing clay or concrete pipe and tiles; it costs less, is faster to install, and is lighter in weight. Monofilaments find applications for ropes, fishing nets, and as woven materials for lawn chairs. Because of its excellent electrical insulating properties, it is widely used as wire and cable covering. Irradiated or chemically crosslinked films are used as dielectrics in winding electrical coils and for limited packaging applications.

Polyethylene may be foamed by several methods. A foaming agent which decomposes and releases a gas during the molding operation is preferred for commercial manufacture of foamed products. A physical foaming method consists of introducing a gas such as nitrogen in the molten resin under pressure. While in the mold and under atmospheric pressure, the gas filled polyethylene expands. Azodicarbonamide may be utilized for chemical foaming of either low- or high-density resins. Foams may be selected as gasket material, structural foam for furniture components, internal panels in automobiles, and dielectric materials in coaxial cables. Crosslinked foams are suitable for cushioning, packaging, or flotation. Low-density foams find use in wrestling mats, athletic padding, and various flotation equipment. Fig. 13-61 shows some uses of foamed polyethylene.

Table 13-17 gives properties of low-, medium-, and high-density polyethylene.

(A) Ice cream freezer. *(C) Drinking glass.*

(B) Wagon wheel.

Fig. 13-61. Examples of foamed polyethylene. (Phillips Petroleum Co.)

Table 13-17. Polyethylene Properties.

Property	Low Density	Medium Density	High Density
Molding qualities	Excellent	Excellent	Excellent
Specific gravity (density)	0.910-0.925	0.926-0.940	0.941-0.965
Tensile strength (psi)	600-2300	1200-3500	3100-5500
Compressive strength (psi)			2700-3600
Izod, impact (ft-lb/in)	No break	0.5-16	0.5-20
Hardness, Rockwell	D41-D46 (Shore) R10	D50-D60 (Shore) R15	D60-D70 (Shore)
Thermal expansion (10^{-5}/°C)	10-20	14-16	11-13
Resistance to heat (°F)	180-212	220-250	250
Dielectric strength (volts/mil)	450-1000	450-1000	450-500
Dielectric constant (60 Hz)	2.25-2.35	2.25-2.35	2.30-2.35
Dissipation factor (Hz)	0.0005	0.0005	0.0005
Arc resistance (seconds)	135-160	200-235	
Water absorption (24 hr, %)	0.015	0.01	0.01
Burning rate (in/min)	Slow (1.04)	Slow (1-1.04)	Slow (1-1.04)
Effect of sunlight	Crazes—must be stabilized	Crazes—must be stabilized	Crazes—must be stabilized
Effect of acids	Oxidizing acids	Oxidizing acids	Oxidizing acids
Effect of alkalies	Resistant	Resistant	Resistant
Effect of solvents	Resistant (below 60°C)	Resistant (below 60°C)	Resistant (below 60°C)
Machining qualities	Good	Good	Excellent
Optical	Transparent-opaque	Transparent-opaque	Transparent-opaque

Polyolefins (Polypropylene)

Until 1954 most attempts to produce plastics from polyolefins had little commerical success. Only the polyethylene family was commercially important. It was in 1955 that the Italian G. Natta announced the discovery of stereospecific polypropylene. The word "stereospecific" implies that the molecules are arranged in a definite order in space in contrast to branched or random arrangements. He called this regular arranged material *isotactic polypropylene*. Professor Natta received the Nobel prize for chemistry in 1963 for establishing a method for regulating the growing molecular chain during polymerization. Experimenting with Ziegler-type catalysts, Natta replaced the titanium tetrachloride in $Al(C_2H_5)_3 + TiCl_4$ with the stereospecific catalyst titanium trichloride. This breakthrough led to the commercial production of polypropylene.

It is not surprising that polypropylene and polyethylene have many of the same properties because of the similarity in origin and manufacture. Polypropylene has become a strong competitor.

Polypropylene gas, $CH_3 \cdot CH{=}CH_2$, is cheaper than ethylene and is obtained from high-temperature cracking of petroleum hydrocarbons and propane. The basic structural unit of polypropylene is:

$$\left(\begin{array}{cc} CH_3 & H \\ | & | \\ -C- & -C- \\ | & | \\ H & H \end{array} \right)_n$$

In Fig. 13-62 the stereotactic arrangements of polypropylene are shown. In Fig. 13-62A the molecular chains show a high degree of definite order with all the CH_3 groups along one side. Atactic polymers are rubbery, transparent materials of limited commercial value. Atactic and syndiotactic plastics grades are more impact resistant than is the isotactic grade. Both syndiotactic and atactic structures may be present in small quantities in isotactic plastics. Commercially available polypropylene is usually about 90 to 95 percent isotactic.

The general physical properties are similar to to high-density polyethylene. Although polyethylene and polypropylene are very similar, they differ in several important respects. (1) Polypropylene has a density of about 0.90; polyethylene has densities of 0.941-0.965. (2) The service temperature of polypropylene is higher. (3) Polypropylene is harder, more rigid, and has a higher brittle point. (4) Polypropylene is more resistant to environmental stress cracking.

(A) Isotactic.

(B) Atactic.

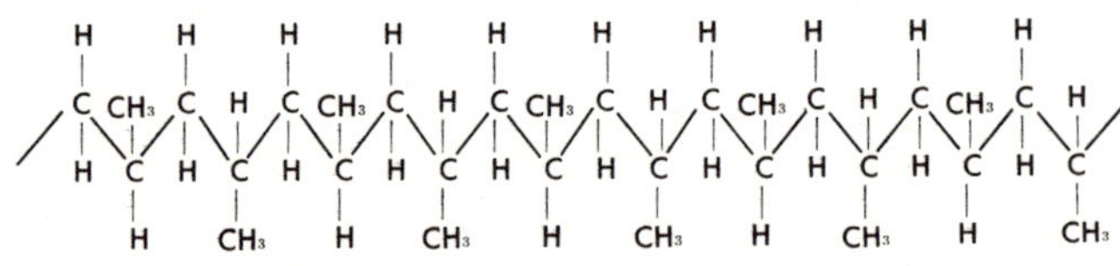

(C) Syndiotactic.

Fig. 13-62. Stereotactic arrangements of polypropylene.

The electrical and chemical properties of polypropylene are very similar to those of polyethylene; polypropylene, however, is more susceptible to oxidation and degrades at elevated temperatures.

Like polyethylene, it may also be manufactured with a variety of properties by the addition of fillers, reinforcements, or blends of special monomers (Fig. 13-63).

It is easily processed in all conventional thermoplastic processing equipment. It cannot be successfully cemented by cohesive means but is readily welded.

Polypropylene is competitive for many applications with polyethylene, especially where it may take advantage of its higher service temperature (Fig. 13-64). Typical applications include sterilizable hospital items, dishes, appliance parts, dishwasher components, containers, items incorporating integral hinges, automotive ducts, and trim. Extruded and cold drawn monofilaments find use as rotproof ropes that will float on water. Some fibers are finding increasing applications for textile

(A) Back and seats of these chairs are molded polypropylene. (Enjoy Chemical Co.)

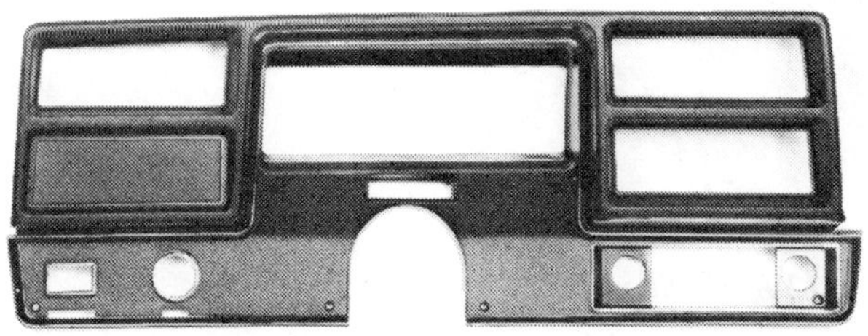

(B) Glass reinforced injection molded polypropylene instrument panels. (AC spark plug Div.)

(C) Pump housing, impeller, magnetic housing, and volute are glass fiber reinforced polypropylene. (Fiberfill Div., Dart Industries)

Fig. 13-63. Polypropylene has a variety of uses.

(A) Coffee pot. *(B) Sterilizable hospital containers.*

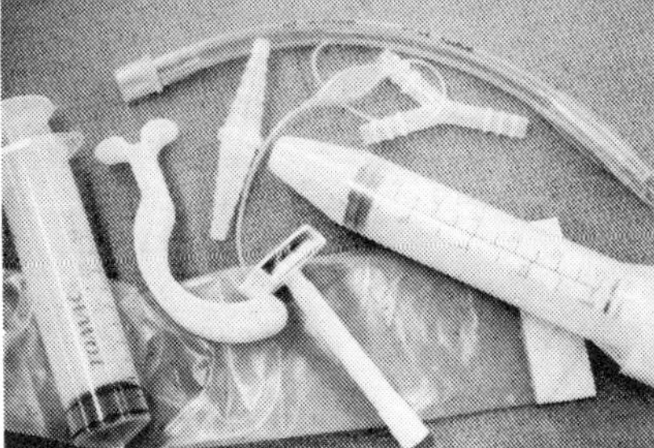

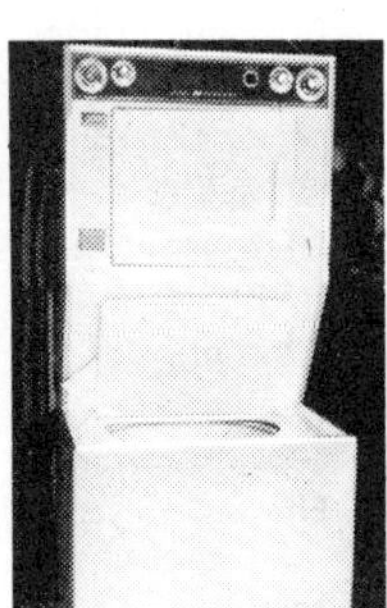

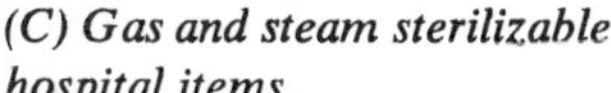

(C) Gas and steam sterilizable hospital items. *(D) Washer-dryer combination.*

Fig. 13-64. High service temperature of polypropylene permits wide applications.

items and for outdoor or automotive carpeting. It may be used as tough packaging film or as electrical insulation on wire and cable.

Because of the excellent abrasion resistance, high service temperature, and potentially lower cost, foamed polypropylene finds growing applications. Cellular polypropylene is foamed in much the same manner as polyethylene.

Table 13-18 gives some of the properties of polypropylene.

Polymethylpentene

This plastics is reported to be an isotactically arranged aliphatic polyolefin of 4-methylpentene-1. Polymethylpentene was developed in the laboratory as early as 1955 but did not gain commercial importance until the Imperial Chemical Industries, Ltd., announced it under the tradename of TPX in 1965.

Ziegler-type catalysts are used to polymerize 4-methylpentene-1 at atmospheric pressures (Fig. 13-65). After polymerization catalyst residues are removed by washing with methyl alcohol. The

Table 13-18. Polypropylene Properties.

Property	Homopolymer (Unmodified)	Glass-Reinforced
Molding qualities	Excellent	Excellent
Specific gravity (density)	0.902-0.906	1.05-1.24
Tensile strength (psi)	4500-5500	6000-9000
Depressive strength (psi)	5500-8000	5500-7000
Izod, impact (ft-lb/in)	0.5-2	1-5
Hardness, Rockwell	R85-R110	R90
Thermal expansion (10^{-5}/°C)	5.8-10.2	2.9-5.2
Resistance to heat (°F)	225-300	300-320
Dielectric strength (volts/mil)	500-660	500-650
Dielectric constant (60 Hz)	2.2-2.6	2.37
Dissipation factor (60 Hz)	0.0005	0.0022
Arc resistance (seconds)	138-185	74
Water absorption (24 hrs, %)	0.01	0.01-0.05
Burning rate (in/min)	Slow	Slow—nonburning
Effect of sunlight	Crazes—must be stabilized	Crazes—must be stabilized
Effect of acids	Oxidizing acids	Slowly by oxidizing acids
Effect of alkalies	Resistant	Resistant
Effect of solvents	Resistant (below 80°C)	Resistant (below 80°C)
Machining qualities	Good	Fair
Optical	Transparent-opaque	Opaque

material is then compounded with stabilizers, pigments, fillers, or other additives into a granular form.

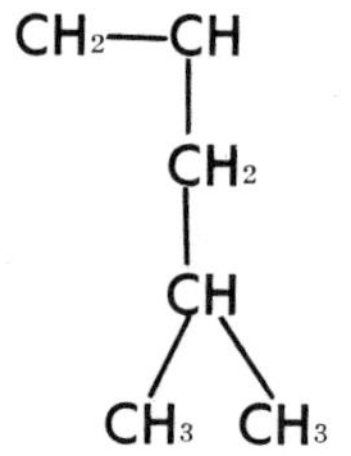

$$\begin{array}{l} CH_2\text{—}CH \\ \quad\quad | \\ \quad\quad CH_2 \\ \quad\quad | \\ \quad\quad CH \\ \quad\; / \quad \backslash \\ CH_3 \quad CH_3 \end{array}$$

Fig. 13-65. Poly (4-methylpentene-1).

In order to avoid confusion and in view of the multitude of saturated hydrocarbons possible, it is necessary to number the carbon atoms of the continuous chain, beginning at the end nearest the branching. This has been done in the formulas illustrated in Fig. 13-66.

$$\begin{array}{cccccc} & CH_3 & & & & \\ & | & & & & \\ CH_3\text{—} & CH\text{—} & CH_2\text{—} & CH_2\text{—} & CH_2\text{—} & CH_3 \\ 1 & 2 & 3 & 4 & 5 & 6 \end{array}$$

(A) 2-methylhexane.

$$\begin{array}{cccc} 1 & 2 & 3 & \\ CH_3\text{—} & CH_2\text{—} & CH\text{—} & CH_3 \\ & & | & \\ & & 4\,CH_2 & \\ & & | & \\ & & 5\,CH_3 & \end{array}$$

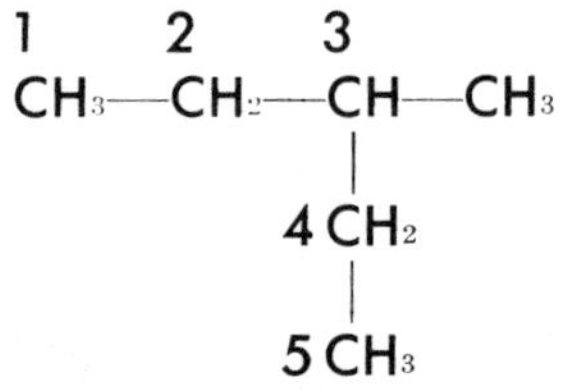

(B) 3-methylpentane.

Fig. 13-66. Continuous-chain formulas with carbon atoms numbered.

Copolymerization with other olefin units including hexene-1, octene-1, decene-1, and octadecene-1 offer varying degrees of enhanced optical and mechanical properties.

Commercial poly-4-methylpentene-1 has a relatively high service temperature which may exceed 320°F. Although the plastics is nearly 50 percent crystalline in structure, it possesses a light transmission value of 90 percent. Spherulite growth may be retarded by rapid cooling of the molded mass. Because of the open packing of the crystalline structure, polymethylpentene has a density of 0.83, which is close to the theoretical minimum for thermoplastics.

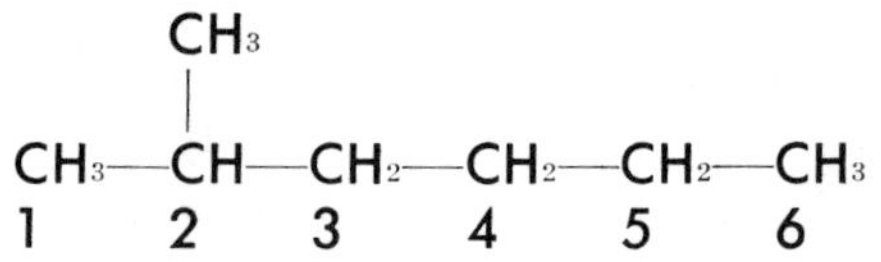

Polymethylpentene may be processed on conventional thermoplastic equipment. However, processing temperatures may exceed 470°F.

In spite of its high cost this plastics has found several uses in chemical plants, transparent optical applications, autoclavable medical equipment, lighting diffusers, encapsulation of electronic components, lenses, and metalized reflectors (Fig. 13-67). A familiar application is in the packaging of bake-in-the-bag and boil-in-the-bag foods. These convenience packages of polymethylpentene are used in the home and in catering services for airlines or manufacturing plants. Packaged foods may be boiled in water or cooked in conventional or microwave ovens. Transparency is useful and helpful in conveying chemicals or foods in dispensing equipment.

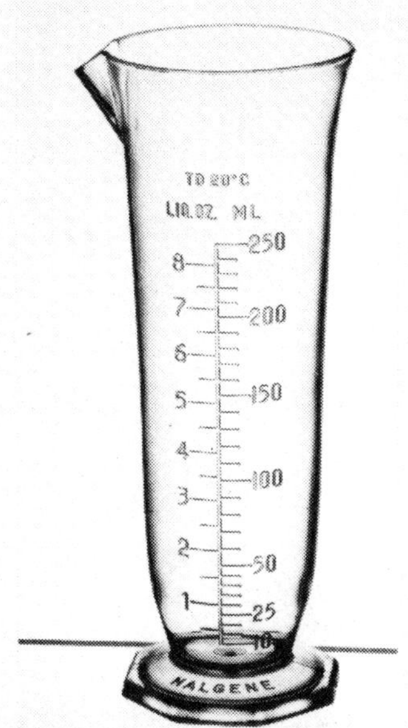

Fig. 13-67. Polymethylpentene is used for various laboratory items where clarity, chemical, and break resistances are important (ICI, Ltd.)

Other side-branched polyolefins are possible. Three possible polymers are shown in Fig. 13-68. The branched side chains increase stiffness and lead to higher melting points. Polyvinyl cyclohexane melts at about 640°F. Table 13-19 gives some of the properties of polymethylpentene.

Polyphenylene Oxides

This family of materials should probably be called "polyphenylene" because a number of plastics have been developed by separating the benzene ring backbone of polyphenylene. By separating these rings the plastics becomes more flexible and may be molded by conventional thermoplastic

```
CH2—CH
     |
     CH
    /  \
  CH3   CH3
```

(A) Poly (3-methylbutene-1).

```
CH2—CH
     |
     CH2
     |
CH3—C—CH3
     |
     CH3
```

(B) Poly (4,4-dimethylpentene-1).

```
CH2—CH
     |
     CH
    /  \
  CH2   CH2
   |     |
  CH2   CH2
    \  /
     CH2
```

(C) Poly (vinylcyclohexane).

Fig. 13-68. Side-branched polyolefin polymers.

Table 13-19. Polymethylpentene Properties.

Property	Polymethylpentene (Unfilled)
Molding qualities	Excellent
Specific gravity (density)	0.83
Tensile strength (psi)	3500-4000
Izod, impact (ft-lb/in)	0.4-1.6
Hardness, Rockwell	1.67-74
Thermal expansion (10^{-5}/°C)	11.7
Resistance to heat (°F)	250-320
Dielectric strength (volts/mil)	700
Dielectric constant (60 Hz)	2.12
Dissipation factor (60 Hz)	0.00007
Water absorption (24 hrs, %)	0.01
Burning rate (in/min)	1.0
Effect of sunlight	Crazes
Effect of acids	Attacked by oxidizing acids
Effect of alkalies	Resistant
Effect of solvents	Attacked by chlorinated aromatics
Machining qualities	Good
Optical	Transparent-opaque

equipment. Polyphenylene with no benzene ring separation is very brittle, insoluble, and infusible.

Polyphenylene

—O— *Poly(phenylene oxide)*

$-CH_2-$ *Poly-p-xylylene*

—S— *Poly(phenylene sulphide)*

Union Carbide introduced a heat resisting plastics in 1964 called *polyphenylene oxide.* It may be prepared by the catalytic oxidation of 2,6-dimethyl phenol as shown in Fig. 13-69.

$$n\,(CH_3)_2C_6H_3-OH + \frac{n}{2}O_2 \longrightarrow \left[(CH_3)_2C_6H_2-O \right]_n + nH_2O$$

Fig. 13-69. Preparation of polyphenylene oxide.

Similar materials have been prepared utilizing ethyl, isopropyl, or other alkyl groups. In 1965 the General Electric Company introduced poly-2,6-dimethyl-1,4-phenylene ether as a polyphenylene oxide material. In 1966 General Electric announced another thermoplastic with the tradename *Noryl.* This material is believed to be produced from *o*-cresol phenol compounds.

OH, CH_3

o-cresol

Because Noryl costs less and has properties similar to polyphenylene oxides, many applications are the same. This modified phenylene oxide material (Noryl) may be processed by conventional thermoplastic equipment with processing temperatures from 375°F to 575°F depending on grade formulations. Modified phenylene oxide parts may be welded, heat sealed, and solvent cemented with chloroform and ethylene dichloride.

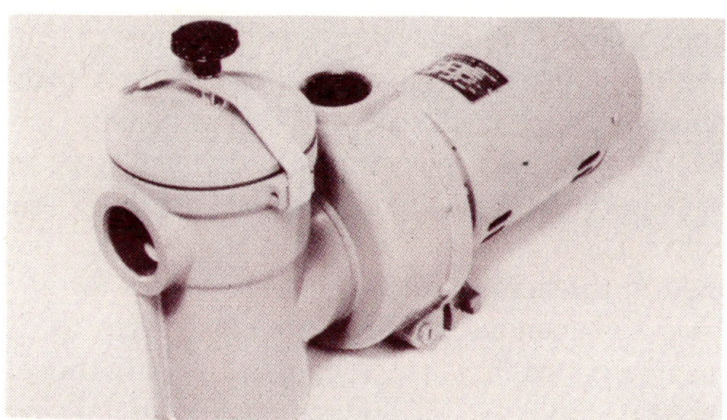

(A) Housing, strainer body, cover, seal plate, and impeller of pump are made of polyphenylene oxide.

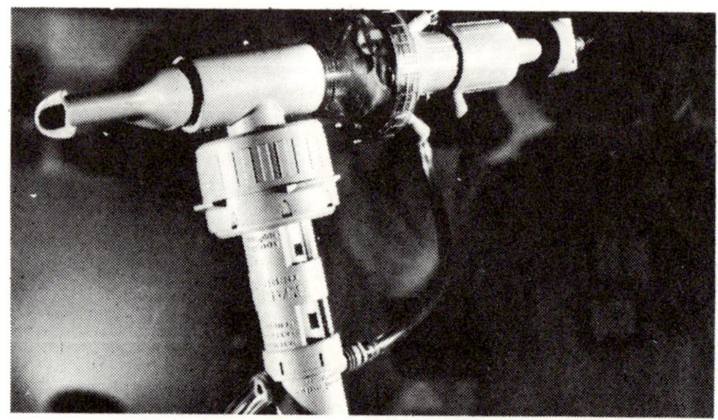

(B) Portable respirator designed for breathing therapy in the home or hospital.

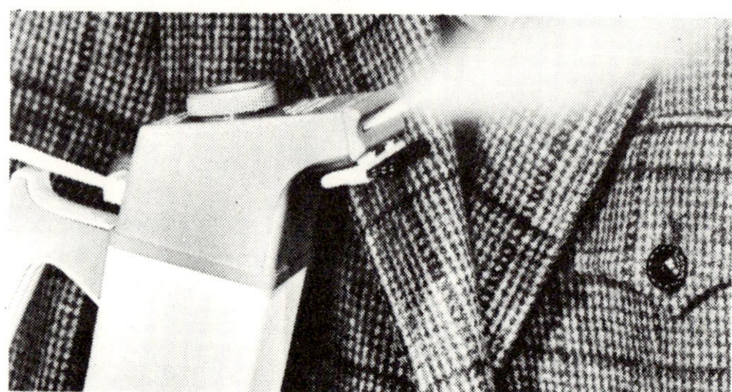

(C) Portable steam presser has a housing molded of polyphenylene oxide.

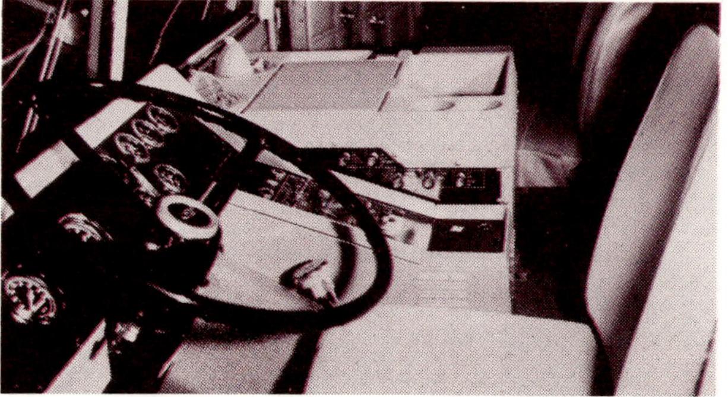

(D) Console of tilt-cab truck is molded of polyphenylene oxide.

(E) Polyphenylene oxide is used for the housing of this desktop computer.

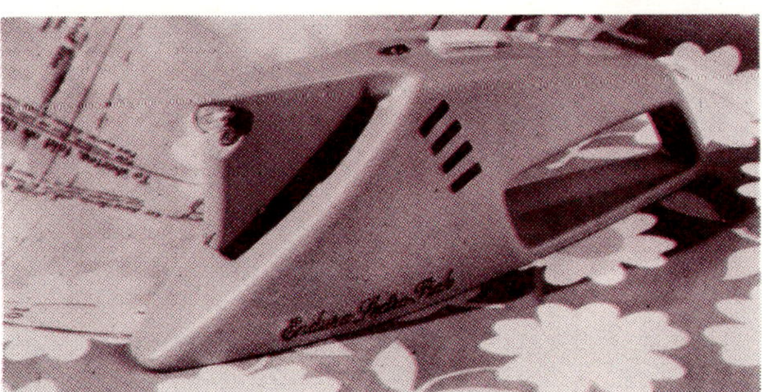

(F) Injection molded housing for automatic electric pinking shears.

Fig. 13-70. Applications of polyphenylene oxide. (General Electric Co.)

In 1965 Union Carbide introduced poly-*p*-xylene under the tradename *Parylene* (Fig. 13-71). Its primary market is for coating and film applications.

Fig. 13-71. Poly-para-xylene structure.

In 1968 the Phillips Petroleum Company announced a material known as *polyphenylene sulphide* with the tradename *Ryton*. The material is available as thermoplastic or thermosetting compounds. Crosslinking is achieved by thermal or chemical means. It has found use as an adhesive, laminating resin, and as coatings for electrical parts.

Table 13-20 gives some of the properties of three polyphenylene oxides.

Polystyrene

Styrene is one of the oldest known vinyl compounds with a long distinguished history; however, industrial exploitation did not start until the late

Table 13-20. Polyphenylene Oxide Properties.

Property	Polyphenylene Oxide (Unfilled)	Noryl SE-1 SE-100	Polyphenylene Sulphides
Molding qualities	Excellent	Excellent	Excellent
Specific gravity (density)	1.06-1.10	1.06-1.10	1.34
Tensile strength (psi)	7800-9600	7800-9600	10,800
Compressive strength (psi)	16,000-16,400	16,000-16,400	
Izod, impact (ft-lb/in)	5.0 (½ × ⅛ in)	5.0 (½ × ⅛ in)	0.3 @ 75°F 1.0 @ 300°F (½ × ¼ in)
Hardness, Rockwell	R115-R119	R115-R119	R124
Thermal expansion (10^{-5}/°C)	5.2	3.3-3.7	5.5
Resistance to heat (°F)	175-220	212-265	400-500
Dielectric strength (volts/mil)	400-550	400-550	595
Dielectric constant (60 Hz)	2.64	2.64-2.65	3.11
Dissipation factor (60 Hz)	0.0004	0.0006-0.0007	
Arc resistance (seconds)	75		
Water absorption (25 hrs, %)	0.066		0.02
Burning rate (in/min)	Self-Exting. Nondrip	Self-Exting. Nondrip	Nonburning
Effect of sunlight	Colors may fade	Colors may fade	
Effect of acids	None		Attacked by oxidizing acids
Effect of alkalies	None		None
Effect of solvents	Soluble in some aromatics	Soluble in some aromatics	Resistant
Machining qualities	Excellent	Excellent	Excellent
Optical	Opaque	Opaque	Opaque

1920's. This simple aromatic compound has been isolated as early as 1839 by the German chemist Edward Simon. Early monomer solutions were obtained from such natural resins as storax and dragon's blood (a resin obtained from the fruit of the Malayan rattan palm). In 1851 French chemist M. Berthelot reported the production of styrene monomers by passing benzene and ethylene through a red-hot tube. This dehydrogenation of ethyl benzene is the basis of commercial methods used today.

By 1925 polystyrene became commercially available in Germany and the United States. For Germany polystyrene had become one of the most important plastics. During World War II Germany had already embarked upon large-scale synthetic rubber production, and styrene was an essential ingredient for the production of styrene-butadiene rubber. When the natural sources of rubber were cut off in 1941 the United States began a crash program for the production of rubber from butadiene and styrene. This synthetic rubber became known as Government Rubber-Styrene (GR-S). Today, there is still a large and growing demand for styrene-butadiene synthetic rubber.

Styrene is chemically known as *vinyl benzene,* with the formula:

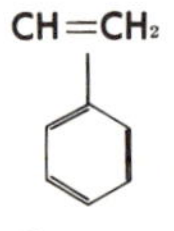

$CH{=}CH_2$

Styrene

This aromatic vinyl compound, in the pure form, will slowly polymerize by addition at room temperature. The monomer is obtained commercially from ethyl benzene (Fig. 13-72).

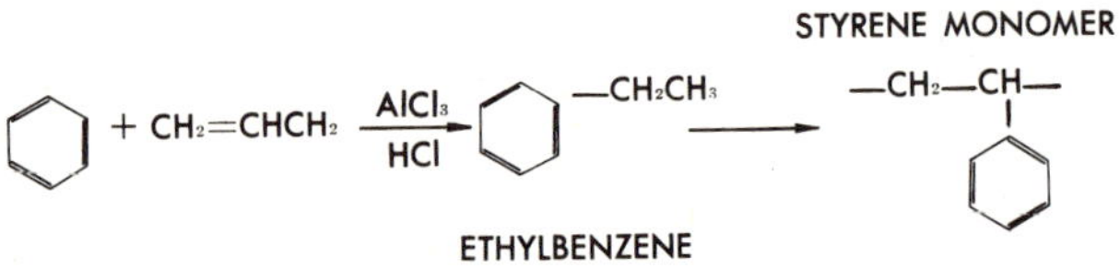

STYRENE MONOMER

$+ CH_2{=}CHCH_2 \xrightarrow[HCl]{AlCl_3}$ —CH_2CH_3 ⟶ —CH_2—CH—

ETHYLBENZENE

Fig. 13-72. Production of the vinyl benzene (styrene) monomer.

Styrene (vinyl benzene) may be polymerized by several methods including bulk, solvent, emulsion, and suspension polymerization. Organic peroxides are used to speed the polymerization process. (See Fig. 13-73.)

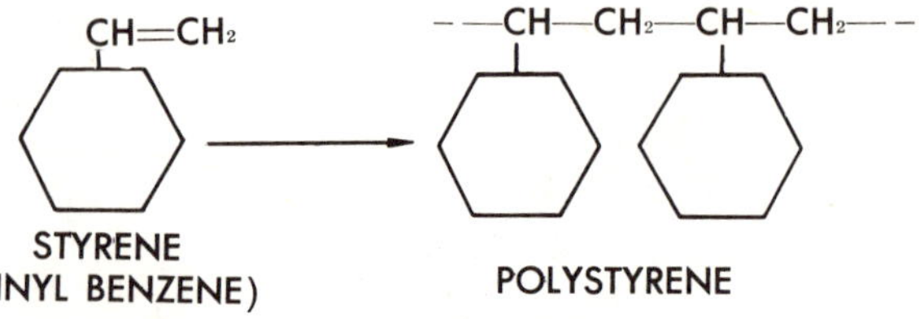

$CH{=}CH_2$ ⟶ - -CH—CH_2—CH—CH_2- - -

STYRENE (VINYL BENZENE) POLYSTYRENE

Fig. 13-73. Polymerization of styrene.

Polystyrene is an atactic, amorphous thermoplastic with the formula shown in Fig. 13-73. It is inexpensive, hard, rigid, transparent, easily molded, and possesses good electrical and moisture resistance (Fig. 13-74). Physical properties vary, depending on the molecular weight distribution, processing, and additives.

(A) Furniture piece with marble top. *(B) Miscellaneous polystyrene parts.*

Fig. 13-74. Objects made of polystyrene.

Polystyrene may be processed by all conventional thermoplastic processes and may be solvent cemented. Some of the more familiar applications may include wall tile, electrical parts, blister packages, lenses, bottle caps, small jars, vacuum formed refrigerator liners, containers of all kinds, and transparent display boxes. Thin films and sheet stock have a metallic ringing sound when struck or dropped. These forms are used in packaging foods and other items including some cigarette packets. Children are familiar with the use of polystyrene in model kits and toys. The housewife may be aware of the many uses of this material in inexpensive dishes, utensils, and glasses. Filaments are extruded and deliberately stretched or drawn to orient molecular chains. This orientation increases tensile strength in the direction of stretching. Filaments may be used for brush bristles.

Expanded or foamed polystyrene is obtained by heating polystyrene containing a gas-producing or blowing agent. The foamed material is accomplished by incorporating a volatile liquid such as methylene chloride, propylene, butylene, or fluorocarbons into the hot melt. As the mixture emerges from the extruder, the blowing agents decompose to release gaseous products, thus yielding the low-density cellular material.

Expanded polystyrene is produced from expandable polystyrene beads containing an entrapped blowing agent. These agents may be pentane, neopentane, or petroleum ether. Upon pre-expanding or final molding, the volatile blowing agent expands, causing individual beads to expand and fuse together. Steam or other heat sources are commonly used to initiate this expanding process. Both expanded forms have a closed cellular structure and may be used as flotation devices. Because of its low thermal conductivity, expanded and foamed polystyrene has found widespread use as thermal insulation (Fig. 13-75). It is used in

Fig. 13-75. Closeup of expanded polystyrene beads. (Sinclear Koppers Co.)

refrigerators, cold storage rooms, freezer display cases, and numerous insulation applications in the building trades. Not only is it an excellent insulator but it has the added advantage of being moisture-proof. There are many packaging applications not only for their thermal insulation value but also for their shock absorption characteristics. By combining these features and their weight factor, manufacturers may package various products in cellular polystyrene and save on shipping and breakage costs.

Foamed and expanded polystyrene sheets may be thermoformed into such familiar packaging items as egg cartons and meat or produce trays. Molded drinking cups, glasses, and "ice" chests are commonly used items. (See Fig. 13-76.)

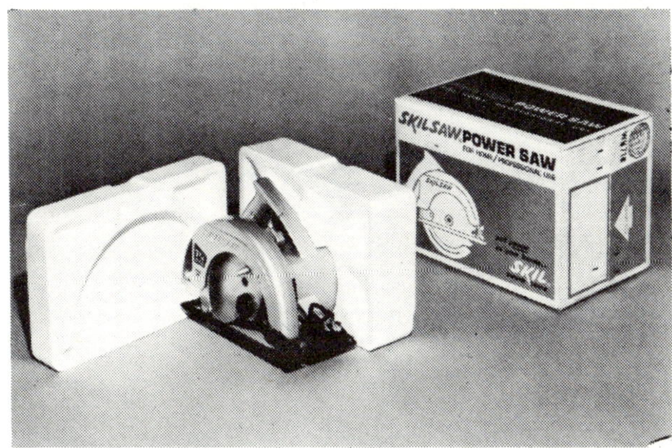

(A) Packing for saw.

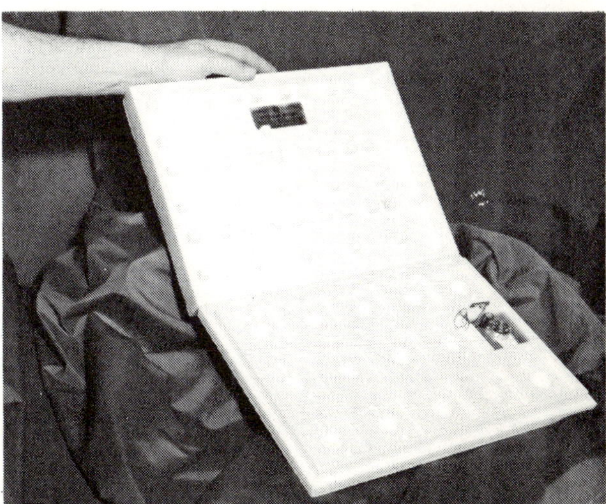

(B) Protection for electronic parts.

Fig. 13-76. Cellular polystyrene as packaging and shipping container. (Sinclair Koppers Co.)

Polystyrene products cannot withstand prolonged heat above 150°F without distortion, nor are they good exterior materials. Special grades and additives may improve this disadvantage. Glass fiber reinforced polystyrenes are used in automotive assemblies, business machines, and appliance housings (see Fig. 13-77).

The properties of polystyrene can be considerably varied by copolymerization and other modi-

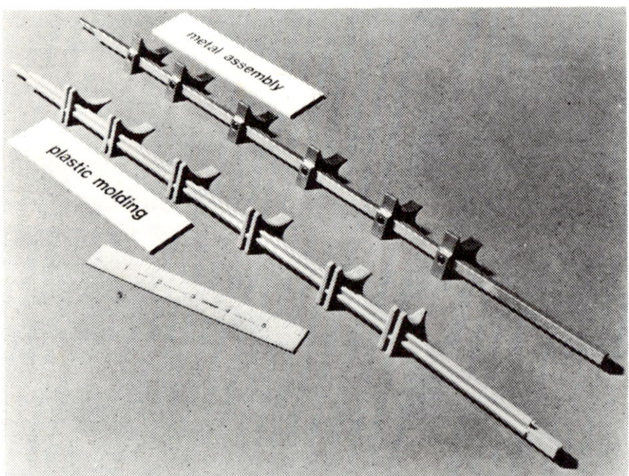

Fig. 13-77. This fiber reinforced polystyrene part replaced a 13-piece metal asembly. (Fiberfill Div., Dart Industries)

fications. Styrene-butadiene rubber has been mentioned. It is used in sporting goods, toys, wire and cable sheathing, shoe soles, and tires. Two of the most important copolymers are styrene acrylonitrile and styrene-acrylonitrile-butadiene (ABS).

Table 13-21 gives some of the properties of polystyrene.

Styrene Acrylonitrile

Acrylonitrile ($CH_2{=}CHCN$) is copolymerized with styrene (C_6H_6) to give products with a higher resistance to various solvents, fats, and other compounds than polystyrene (Fig. 13-78). These products are suitable components for vacuum cleaners and kitchen equipment requiring impact strength and chemical resistance.

```
    H     H     H
    |     |     |
  --C-----C-----C--
    |     |     |
    H   (C6H5)  C≡N
```

Fig. 13-78. SAN mer.

Styrene acrylonitrile (SAN) copolymers may be composed of about 20 to 30 percent acrylonitrile content. By varying proportions of each monomer, a wide range of properties and processability may be obtained. The slight yellow cast of SAN is due to the copolymerization of acrylonitrile with the styrene member.

This copolymer is easily molded and processed by most processing equipment. SAN type materials inherently absorb more moisture than polystyrene. This increased moisture absorption may result in molding defects such as silver streaking. Predrying is highly recommended.

Methyl ethyl ketone, trichloroethylene, and methylene chloride are several of the effective solvents for SAN.

Table 13-21. Polystyrene Properties.

Property	Unfilled Compounds	Impact and Heat Resistant	20-30% Glass-Filled
Molding qualities	Excellent	Excellent	Excellent
Specific gravity (density)	1.04-1.09	1.04-1.10	1.20-1.33
Tensile strength (psi)	5000-12,000	1500-7000	9000-15,000
Compressive strength (psi)	11,500-16,000	4000-9000	13,500-18,000
Izod, impact (ft-lb/in)	0.25-0.40 (1/4 in)	0.5-11	0.4-4.5
Hardness, Rockwell	M65-M80	M20-M80, R50-R100	M70-M95
Thermal expansion (10^{-5}/°C)	6-8	3.4-21	1.8-4.5
Resistance to heat (°F)	150-170	140-175	180-200
Dielectric strength (volts/mil)	500-700	300-600	350-425
Dielectric constant (60 Hz)	2.45-2.65	2.45-4.75	
Dissipation factor (60 Hz)	0.0001-0.0003	0.0004-0.0020	0.004-0.014
Arc resistance (seconds)	60-80	20-10	25-40
Water absorption (24 hrs, %)	0.03-0.10	0.05-0.6	0.05-0.10
Burning rate (in/min)	Slow	Slow	Slow—nonburning
Effect sunlight	Yellows slightly	Yellows slightly	Yellows slightly
Effect of acids	Oxidizing acids	Oxidizing acids	Oxidizing acids
Effect of alkalies	None	None	Resistant
Effect of solvents	Soluble in aromatic & chlorinated hydrocarbons	Soluble in aromatic & chlorinated hydrocarbons	Soluble in aromatic & chlorinated hydrocarbons
Machining qualities	Good	Good	Good
Optical	Transparent	Translucent-opaque	Translucent-opaque

This tough, heat resistant plastics finds applications as telephone parts, containers, decorative panels, food packages and lenses (Fig. 13-79).

Table 13-22 gives some of the properties of SAN.

(A) SAN Vaseline container.

(B) SAN is main component of air precleaner.

Fig. 13-79. Uses of transparent SAN. (Monsanto Co.)

Table 13-22. SAN Properties.

Property	SAN (Unfilled)
Molding qualities	Good
Specific gravity (density)	1.075-1.1
Tensile strength (psi)	9000-12,000
Compressive strength (psi)	14,000-17,000
Izod, impact (ft-lb/in)	0.35-0.50
Hardness, Rockwell	M80-M90
Thermal expansion (10^{-5}/°C)	3.6-3.8
Resistance to heat (°F)	140-205
Dielectric strength (volts/mil)	400-500
Dielectric constant (60 Hz)	2.6-3.4
Dissipation factor (60 Hz)	0.006-0.008
Arc resistance (seconds)	100-150
Water absorption (24 hrs, %)	0.20-0.30
Burning rate (in/min)	Slow—SE
Effect of sunlight	Yellows
Effect of acids	Attacked by oxidizing acids
Effect of alkalies	None
Effect of solvents	Soluble in ketones & esters
Machining qualities	Good
Optical	Transparent

Polysulfones

In 1965 Union Carbide introduced a linear, heat resistant thermoplastic called *polysulfone*. The basic repeating structure consists of benzene rings connected by a sulfone group (SO_2), an isopropylidene group (CH_3CH_3C), and also an ether linkage (O).

One basic polysulfone is prepared by mixing bisphenol-A with chlorobenzene and dimethyl sulphoxide in a caustic soda solution. The resulting condensation polymerization is shown in Fig. 13-80. The light amber color of the plastics is a result of the addition of methyl chloride, which terminates polymerization. The outstanding thermal and oxidation resistance with service temperatures from −150°F to +345°F is the result of the benzene to sulfone linkages. Although polysulfone can be processed by all conventional methods, it must be dried before use and may require processing temperatures exceeding 700°F.

NaO–⟨benzene⟩–C(CH_3)(CH_3)–⟨benzene⟩–ONa$_n$ + Cl–⟨benzene⟩–SO_2–⟨benzene⟩–Cl $\xrightarrow{-2NaCl}$

[–⟨benzene⟩–C(CH_3)(CH_3)–⟨benzene⟩–O–⟨benzene⟩–SO_2–⟨benzene⟩–O–]

Fig. 13-80. Basic polysulfone repeating structure.

(A) Immersible cornpopper has molded polysulfone cover.

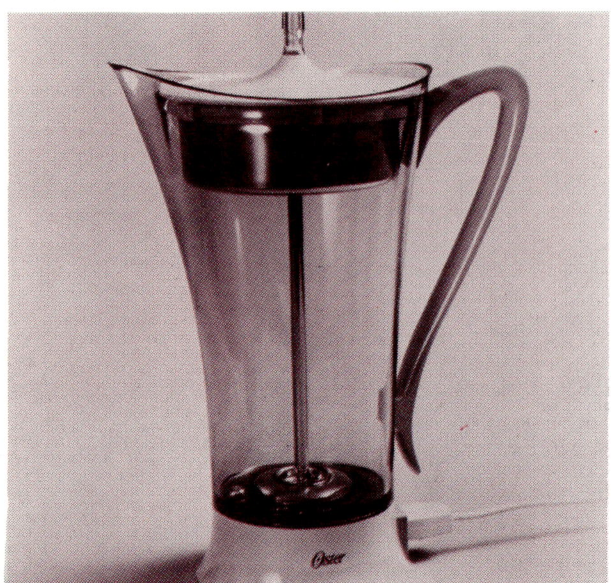

(B) Injection molded polysulfone coffeemaker.

(C) Polysulfone thermos pitcher liner.

Fig. 13-81. Applications of polysulfone. (Union Carbide)

The ether linkage and the carbon of the isopropylidene group imparts toughness and flexibility to the plastics.

Polysulfone can be machined, heat sealed, or solvent cemented by several cements including dimethyl formamide and dimethyl acetamide.

Polysulfones are to some extent competitive with many thermosetting compounds because they may be processed in rapid-cycle thermoplastics equipment.

The excellent mechanical, electrical, and thermal properties are used in such applications as hot-water pipes, alkaline battery cases, distributor caps, face shields for astronauts, electrical circuit breakers, appliance housings, dishwasher impellers, autoclavable hospital equipment, interior components for aerospace craft, shower heads, lenses, and numerous electrical insulating components (Fig. 13-81). If the material is to be used outdoors, it should be painted or electroplated to prevent degradation.

Various other polysulfones have been prepared with a variety of bisphenols with methylene, sulphide, or oxygen linkages. In polyaryl sulfone, the bisphenol groups are linked by ether and sulfone groups. There are no isopropylene (aliphatic) groups present. The term *aryl* refers to a phen*yl* group derived from an *ar*omatic compound. If more than one hydrogen is substituted in the aryl group, a numbering system is normally used. Three possible *di*substituted benzenes are shown in Fig. 13-82. Uses for polyaryl sulfone include electrical connectors, sockets, high-temperature bobbins, and terminal soldering blocks. The basic properties of polysulfone are given in Table 13-23.

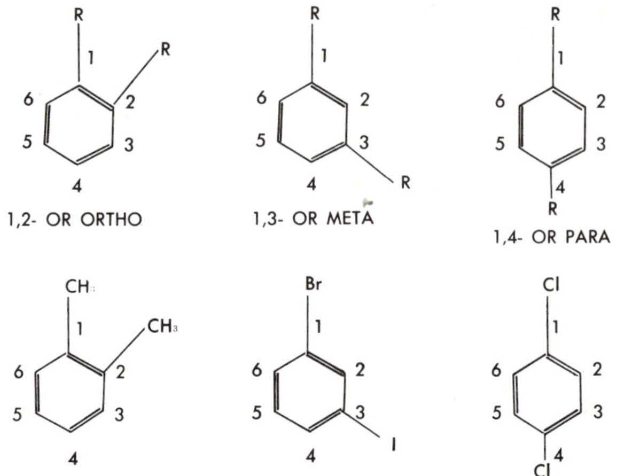

Fig. 13-82. Three possible disubstituted benzenes.

Polyaryl ethers are prepared from aromatic compounds containing no sulphur links. The resulting polymer is more easily processed with service temperatures exceeding 270°F. Applications include business machine parts, recreation helmets, snowmobile parts, pipes, valves, and various appliance components.

Table 13-23 gives some of the properties of polysulfone, polyaryl ether, and polyaryl sulfone.

homopolymers or copolymers may include polyvinyl chloride, polyvinyl acetate, polyvinyl alcohol, polyvinyl butyral, polyvinyl acetal, and polyvinylidene chloride. Fluorinated vinyls are discussed with other fluorine containing polymers.

The history of polyvinyls may have started as early as 1835, when French chemist V. Regnault reported that a white residue could be synthesized from ethylene dichloride in an alcohol solution.

Table 13-23. Polysulfones Properties.

Property	Polysulfone (Unfilled)	Polyaryl Ether (Unfilled)	Polyary Sulfone (Unfilled)
Molding qualities	Excellent	Excellent	Excellent
Specific graivty (density)	1.24	1.14	1.36
Tensile strength (psi)	10,200	7500	13,000
Compressive strength (psi)	13,900	16,000	17,900
Izod, impact (ft-lb/in)	1.3 (¼ in)	8 (½ × ¼ in)	5
Hardness, Rockwell	M69, R120	R117	M110
Thermal expansion (10^{-5}/°C)	5.2-5.6	6.5	4.7
Resistance to heat (°F)	300-345	250-270	500
Dielectric strength (volts/mil)	425	430	350
Dielectric constant (60 Hz)	3.14	3.14	3.94
Dissipation factor (60 Hz)	0.0008	0.006	0.003
Arc resistance (seconds)	75-122	180	67
Water absorption (24 hrs, %)	0.22	0.25	1.8
Burning rate (in/min)	Self-exting.	Slow	Self-exting.
Effect of sunlight	Strength loss Yellows slightly	Slight, yellows	Slight
Effect of acids	None	Resistant	None
Effect of alkalies	None	None	None
Effect solvents	Partly soluble aromatic hydrocarbons	Soluble in ketones, esters, chlorinated aromatics	Soluble in high polar solvents
Machining qualities	Excellent	Excellent	Excellent
Optical	Transparent-opaque	Translucent-opaque	Opaque

Polyvinyls

There is a large and varied group of addition polymers with the formulas $CH_2{=}CH{-}R$ or

$$\begin{array}{c} R \\ | \\ CH_2{=}C \\ | \\ R \end{array}$$

which chemists refer to as *vinyls*. Radical (*R*) side groups may attach to this repeating vinyl group to form several polymers related to each other. Addition polymers are shown in Table 13-24 with the radical side groups attached.

Today, through common usage, the vinyl plastics are those polymers with the vinyl name. Many authorities limit their discussion to include only polyvinyl chloride and polyvinyl acetate. Polyvinyl

This tough white residue was again reported in 1872 by E. Baumann while reacting acetylene and hydrogen bromide in sunlight. For both men, sunlight was the polymerizing "catalyst" which produced the white residue. In 1912 Russian chemist I. Ostromislensky reported the same sunlight polymerization of vinyl chloride and vinyl bromide. Commercial patents were granted in several countries for the manufacture of vinyl chloride by 1930.

In 1933 W. L. Semon of the B. F. Goodrich Company added the plasticizer, tritolyl phosphate, to polyvinyl chloride compounds. The resulting polymer mass could be easily molded and processed without high decomposition or a tendency to adhere to metallic surface on heating.

During the Second World War, Germany, Great Britain, and the United States commercially pro-

Table 13-24. Monofunctional Monomers and Their Polymers.

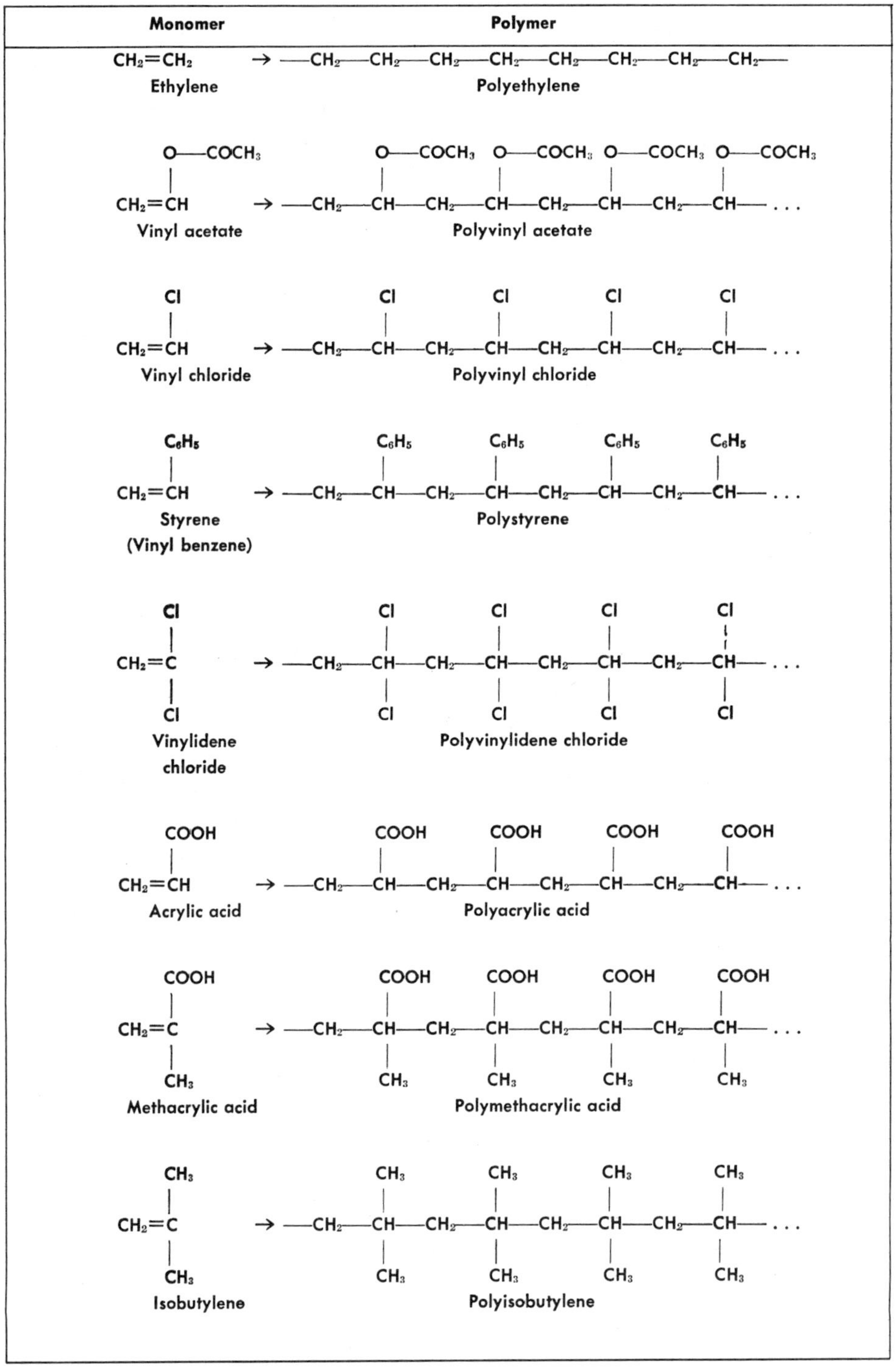

Monomer		Polymer
$CH_2{=}CH_2$ Ethylene	$\rightarrow$	$-CH_2-CH_2-CH_2-CH_2-CH_2-CH_2-CH_2-CH_2-$ Polyethylene
$O-COCH_3$ $\vert$ $CH_2{=}CH$ Vinyl acetate	$\rightarrow$	$-CH_2-\underset{O-COCH_3}{CH}-CH_2-\underset{O-COCH_3}{CH}-CH_2-\underset{O-COCH_3}{CH}-CH_2-\underset{O-COCH_3}{CH}-\ldots$ Polyvinyl acetate
Cl $\vert$ $CH_2{=}CH$ Vinyl chloride	$\rightarrow$	$-CH_2-\underset{Cl}{CH}-CH_2-\underset{Cl}{CH}-CH_2-\underset{Cl}{CH}-CH_2-\underset{Cl}{CH}-\ldots$ Polyvinyl chloride
C_6H_5 $\vert$ $CH_2{=}CH$ Styrene (Vinyl benzene)	$\rightarrow$	$-CH_2-\underset{C_6H_5}{CH}-CH_2-\underset{C_6H_5}{CH}-CH_2-\underset{C_6H_5}{CH}-CH_2-\underset{C_6H_5}{CH}-\ldots$ Polystyrene
Cl $\vert$ $CH_2{=}C$ $\vert$ Cl Vinylidene chloride	$\rightarrow$	$-CH_2-\overset{Cl}{\underset{Cl}{CH}}-CH_2-\overset{Cl}{\underset{Cl}{CH}}-CH_2-\overset{Cl}{\underset{Cl}{CH}}-CH_2-\overset{Cl}{\underset{Cl}{CH}}-\ldots$ Polyvinylidene chloride
$COOH$ $\vert$ $CH_2{=}CH$ Acrylic acid	$\rightarrow$	$-CH_2-\underset{COOH}{CH}-CH_2-\underset{COOH}{CH}-CH_2-\underset{COOH}{CH}-CH_2-\underset{COOH}{CH}-\ldots$ Polyacrylic acid
$COOH$ $\vert$ $CH_2{=}C$ $\vert$ CH_3 Methacrylic acid	$\rightarrow$	$-CH_2-\overset{COOH}{\underset{CH_3}{CH}}-CH_2-\overset{COOH}{\underset{CH_3}{CH}}-CH_2-\overset{COOH}{\underset{CH_3}{CH}}-CH_2-\overset{COOH}{\underset{CH_3}{CH}}-\ldots$ Polymethacrylic acid
CH_3 $\vert$ $CH_2{=}C$ $\vert$ CH_3 Isobutylene	$\rightarrow$	$-CH_2-\overset{CH_3}{\underset{CH_3}{CH}}-CH_2-\overset{CH_3}{\underset{CH_3}{CH}}-CH_2-\overset{CH_3}{\underset{CH_3}{CH}}-CH_2-\overset{CH_3}{\underset{CH_3}{CH}}-\ldots$ Polyisobutylene

duced plasticized polyvinyl chloride (PVC) in aiding their war efforts. It was largely used as a substitute material for rubber.

Today polyvinyl chloride is the leading plastic produced in Europe and ranks second after polyethylene in the United States. The polyvinyl chloride molecule (C_2H_3Cl) is similar to polyethylene, except that one of the four hydrogen atoms of the latter is replaced with the atom chlorine. In Fig. 13-83, this similarity is illustrated.

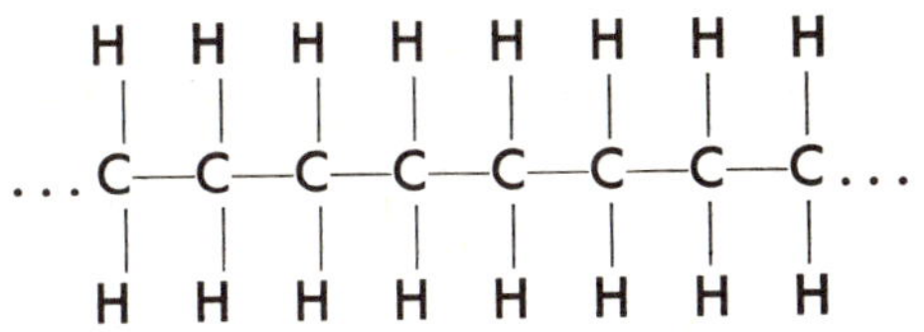

```
      H   H   H   H   H   H   H   H
      |   |   |   |   |   |   |   |
   ...C---C---C---C---C---C---C---C...
      |   |   |   |   |   |   |   |
      H   H   H   H   H   H   H   H
```

(A) Polyethylene.

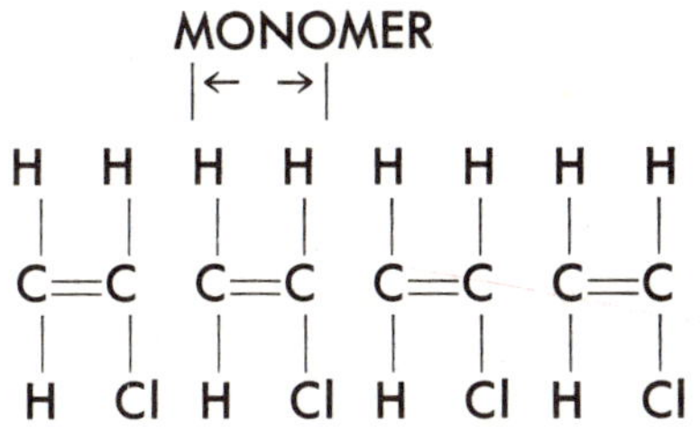

```
          MONOMER
          |<- ->|
  H   H   H   H   H   H   H   H
  |   |   |   |   |   |   |   |
  C===C   C===C   C===C   C===C
  |   |   |   |   |   |   |   |
  H   Cl  H   Cl  H   Cl  H   Cl
```

(B) Vinyl chloride.

```
      H   H   H   H   H   H   H   H
      |   |   |   |   |   |   |   |
   ...C---C---C---C---C---C---C---C...
      |   |   |   |   |   |   |   |
      H   Cl  H   Cl  H   Cl  H   Cl
                  |<-MER->|
```

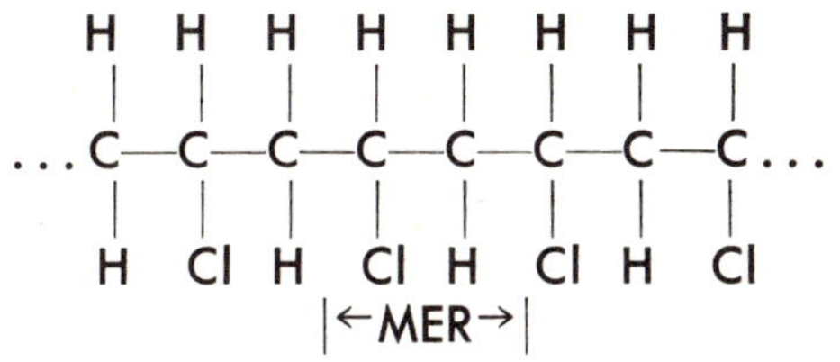

(C) Polyvinyl chloride.

Fig. 13-83. Similarity of polyethylene and polyvinyl chloride.

The basic raw ingredient, depending on availability, is acetylene or ethylene gas. Ethylene is the primary source in the United States. During its manufacture, polymerization may be initiated by peroxides, azo compounds, persulfates, ultraviolet light, or radioactive sources. In order for addition polymerization the double bonds of the monomers must be broken by the application of heat, light, pressure, or a catalyst system.

The versatility of polyvinyl chloride plastics may be broadened by the addition of plasticizers, fillers, reinforcements, lubricants, and stabilizers. They may be formulated into flexible, rigid, elastomeric, or foam compounds.

The largest use of polyvinyl chloride consists of flexible film and sheet applications. These films and sheets are competitive with other films for collapsible containers, drum liners, sacks, and packages. Washable wallpapers and various clothing apparel including handbags, rainwear, coats, and dresses have found growing markets. Sheets are fabricated into chemical tanks and ductwork of all types. They are easily fabricated by welding, heat sealing, or solvent cementing with mixtures of ketones or aromatic hydrocarbons.

Extruded profile shapes of both rigid and flexible polyvinyl chloride find applications as architectural moldings, seals, gaskets, gutters, exterior siding, garden hose, and moldings for movable partitions (Fig. 13-84). Injection molded soles for shoes and other footwear are popular in several countries.

Organosols and plastisols are liquid or paste dispersions or emulsions of polyvinyl chloride used for coating various substrates including metal, wood, plastics, and fabrics. These materials may be applied by dipping, spraying, spreading, or slush and rotational casting. Laminates of polyvinyl film, foam, and fabric are used for upholstery materials in automobiles, buses, aircraft, and furniture. Dip coatings are found on tool handles, sink drainers, and other substrates as a protective layer. Slush and rotational casting of polyvinyls are used to produce hollow articles such as balls, toy dolls, and large containers. Used in the production of floor coverings and tiles are the heavily filled polyvinyls and copolymers. Foams have found limited applications in the textile and carpeting industries. Large quantities of PVC are still used as blow molded containers and as extruded electrical wire coverings (See the various examples shown in Fig. 13-85).

As a rule of thumb, materials in the vinyl family are flame, water, chemical, electrical, and abrasion resistant. They have good weatherability and may be transparent. To aid processing and provide a variety of properties, polyvinyls are commonly plasticized. Both plasticized and unplasticized PVC compounds are available. Unplasticized grades are used in chemical plants and building industries. Plasticized grades are more flexible and soft. With increases in plasticizer, there is more bleeding or migration of the plasticizer chemical into adjacent materials. This is of prime importance in packaging of food products and medical supplies.

All thermoplastic processing techniques are employed with vinyls.

Although polyvinyl chloride (PVC) is the most used and commonly thought-of vinyl, other homopolymer and copolymer polyvinyls are finding increasing applications.

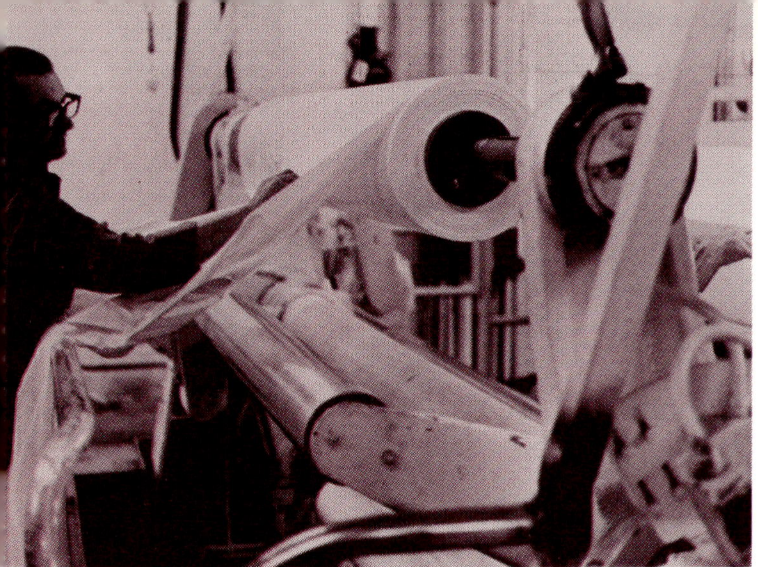

(A) A 200-lb. roll of 0.003-in thick vinyl film on takeoff roller of calendar machine. (Cabot Corp.)

(B) Vinyl-coated glass bottle bounces while identical uncoated bottle, dropped from same height, shatters. (Cabot Corp.)

(C) Polyvinyl coating on wood for weatherproof finish and durability. (Anderson Corp.)

(D) House before application of solid vinyl siding. (Bird & Son, Inc.)

(E) House after application of solid vinyl siding. (Bird & Son, Inc.)

Fig. 13-84. Polyvinyl coating has many applications.

Polyvinyl Acetate

Vinyl acetate ($CH_2{=}CH{-}O{-}COCH_3$) is prepared industrially from liquid or gaseous reactions of acetic acid and acetylene. Homopolymers find only limited applications because of excessive cold flow and low-softening point; however, they are used in paints, adhesives, and various textile finishing operations. Polyvinyl acetates are usually in an emulsion form. "White" liquid adhesives are familiar polyvinyl acetate emulsions. Moisture absorption characteristics are high with selected alcohols and ketones as solvents. Remoistenable adhesives and hot-melt formulations are other familiar applications. Polyvinyl acetates are used as binder emulsions in various paint formulations. Their excellent resistance to degradation by sunlight makes them useful for interior or exterior coatings. Additional applications may include emulsion binders in paper, cardboard, Portland cements, textiles, and chewing-gum bases.

Some of the best-known commercial products are copolymers of polyvinyl chloride and polyvinyl

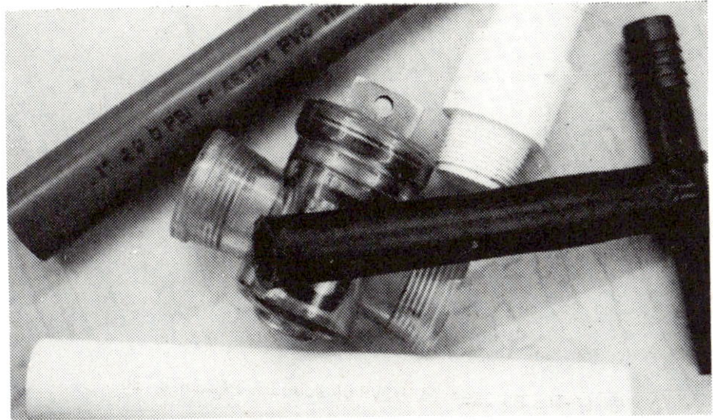

(A) Polyvinyl chloride pipe and plumbing parts.

(B) Portion of record (being held), floor tile, wire cover, embossed film, and molded parts.

(C) Foamed polyvinyl flotation devices.

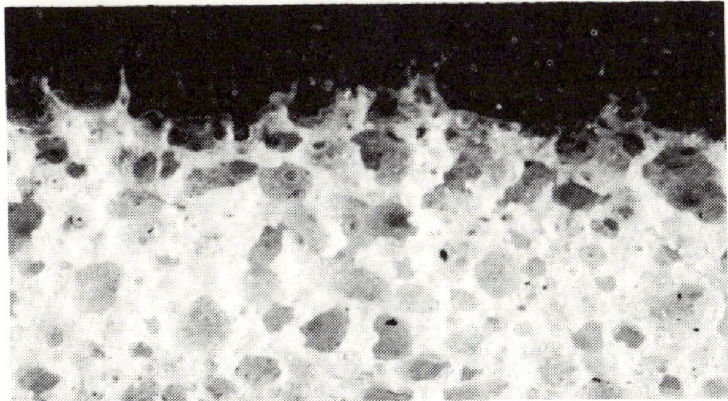

(D) Mechanically frothed or foamed polyvinyl foam on carpet. Magnification 10X. (Firestone Plastics Co.)

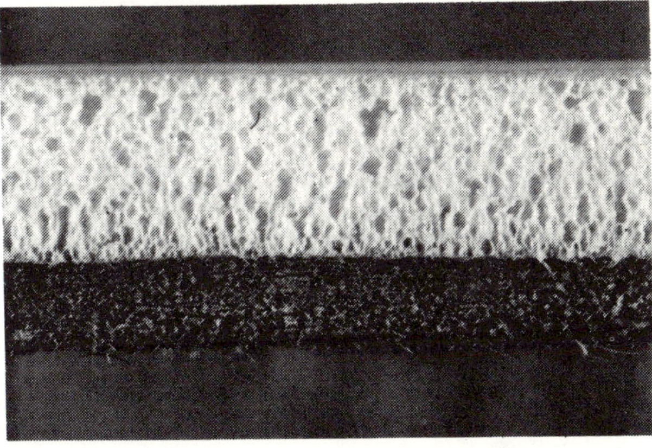

(E) Mechanically frothed or foamed polyvinyl foam on thick flooring construction. Magnification 10X. (Firestone Plastics Co.)

Fig. 13-85. Various uses of polyvinyl.

acetate used as floor coverings and for modern phonograph records (Fig. 13-86). These "vinyl" records have several advantages over polystyrene and the older shellac ones.

$$CH_2{=}CH \cdot OCOCH_3 \longrightarrow -CH_2-\underset{OCOCH_3}{\underset{|}{CH}}-CH_2-\underset{OCOCH_3}{\underset{|}{CH}}-$$

VINYL ACETATE → POLYVINYL ACETATE

$$\left.\begin{array}{l} CH_2{=}CHCl \\ CH_2{=}CHO \cdot COCH_3 \end{array}\right\} \longrightarrow -CH_2-\underset{Cl}{\underset{|}{CH}}-CH_2-\underset{OCOCH_3}{\underset{|}{CH}}-$$

VINYL CHLORIDE + VINYL ACETATE → POLYVINYL CHLORIDE-ACETATE

Fig. 13-86. Production of polyvinyl acetate and polyvinyl chloride-acetate.

Polyvinyl Formal

Polyvinyl formal is generally produced from polyvinyl acetate, formaldehyde, and other additives. The alcohol side groups on the chain are now changed to "formal" side groups. Polyvinyl formal finds its greatest applications as coatings in metal containers and as electrical wire enamels.

Polyvinyl Alcohol

Polyvinyl alcohol is an important derivative produced from the alcoholysis of polyvinyl acetate (Fig. 13-87). Methyl alcohol (methanol) is used in the process. Polyvinyl alcohol (PVA) is alcohol and water soluble. The properties vary depending on the concentration of polyvinyl acetate which remains in the alcohol solution. Polyvinyl alcohol may be used as a binder and adhesive for paper, ceramics, cosmetics, and textiles. It finds applications as water-soluble packages for soap, bleaches, and disinfectants. It is an important mold releasing agent used in the manufacture of reinforced plastics products. There has been only limited use of polyvinyl alcohol for moldings and fibers.

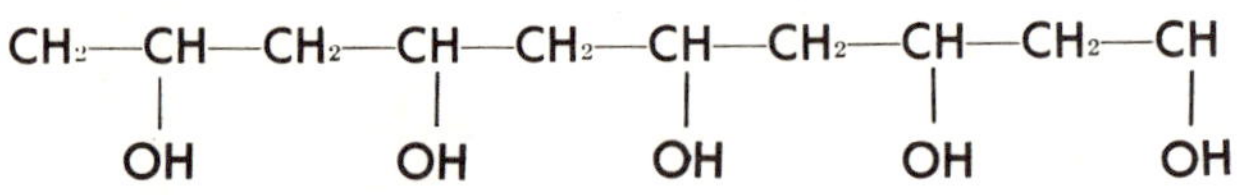

$CH_2—CH—CH_2—CH—CH_2—CH—CH_2—CH—CH_2—CH$

OH OH OH OH OH

Fig. 13-87. Polyvinyl alcohol (OH side groups).

Polyvinyl Acetal

Another important derivative of polyvinyl acetate is polyvinyl acetal. It is produced from the treatment of polyvinyl alcohol (from polyvinyl acetate) with an acetaldehyde (see Fig. 13-88). Polyvinyl acetal materials find only limited use as adhesives, surface coatings, films, moldings, or textile modifiers.

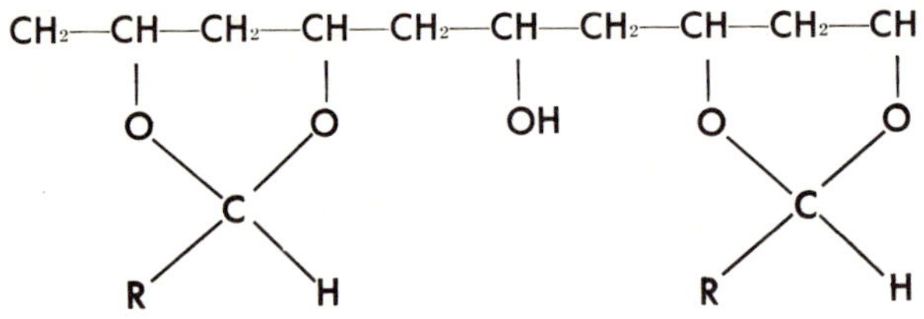

$CH_2—CH—CH_2—CH—CH_2—CH—CH_2—CH—CH_2—CH$

O O OH O O

C C

R H R H

Fig. 13-88. A polyvinyl acetal.

Polyvinyl Butyral

Polyvinyl butyral is produced from polyvinyl alcohol (See Fig. 13-89). This plastics is used as interlayer film in laminated safety glass.

Polyvinylidene Chloride

In 1839 a substance similar to vinyl chloride was discovered, but it contained an additional chlorine atom. This material has become commercially important as vinylidene chloride ($H_2C=CCl_2$). See Fig. 13-90.

Polyvinylidene chloride is costly and difficult to process; consequently, it is normally found as a copolymer. Most of these resins are copolymers of vinylidene chloride with vinyl chloride, acrylonitrile, and acrylate esters. A well-known food wrapping film, Saran, is a copolymer of vinylidene chloride and acrylonitrile. It exhibits exceptional clarity, toughness, and permits little gas or moisture transmission.

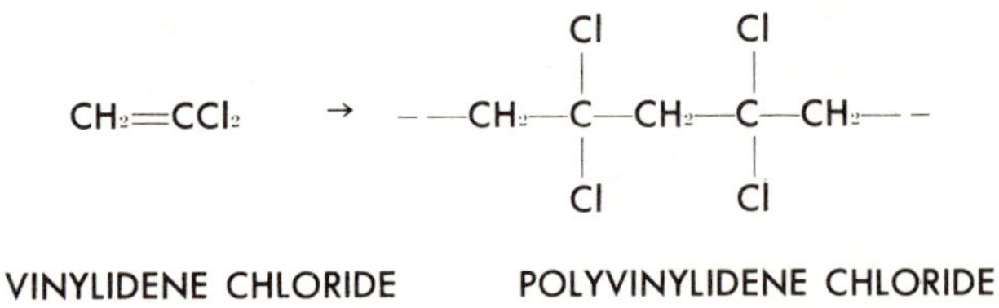

Cl Cl

$CH_2=CCl_2$ → $-—CH_2—C—CH_2—C—CH_2—-$

Cl Cl

VINYLIDENE CHLORIDE POLYVINYLIDENE CHLORIDE

Fig. 13-90. Polymerization of vinylidene chloride.

Although the primary use of polyvinylidene chloride copolymers (see Fig. 13-91) is in the coating and film packaging markets, they have found some success as fibers for carpeting, automobile seat upholstery, draperies, and awning textiles. Their chemical inertness allows them to be used as pipes, pipe fittings, pipe linings, and filters.

There are numerous other polyvinyl polymers which warrant further research and study. Polyvinyl carbazole used for dielectrics, polyvinyl pyr-

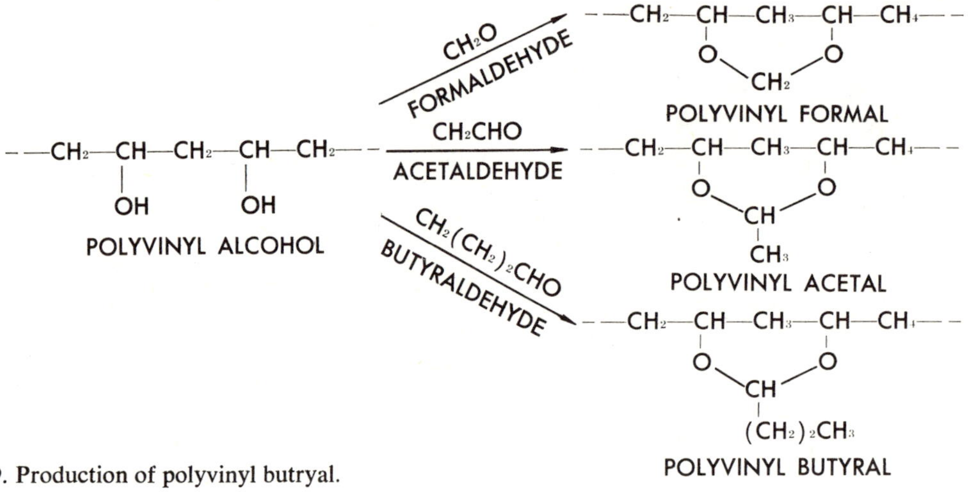

$-—CH_2—CH—CH_3—CH—CH_4—-$

O O

CH_2

POLYVINYL FORMAL

CH_2O

FORMALDEHYDE

$-—CH_2—CH—CH_2—CH—CH_2—-$

OH OH

POLYVINYL ALCOHOL

CH_2CHO

ACETALDEHYDE

$-—CH_2—CH—CH_3—CH—CH_4—-$

O O

CH

CH_3

POLYVINYL ACETAL

$CH_2(CH_2)_2CHO$

BUTYRALDEHYDE

$-—CH_2—CH—CH_3—CH—CH_4—-$

O O

CH

$(CH_2)_2CH_3$

POLYVINYL BUTYRAL

Fig. 13-89. Production of polyvinyl butryal.

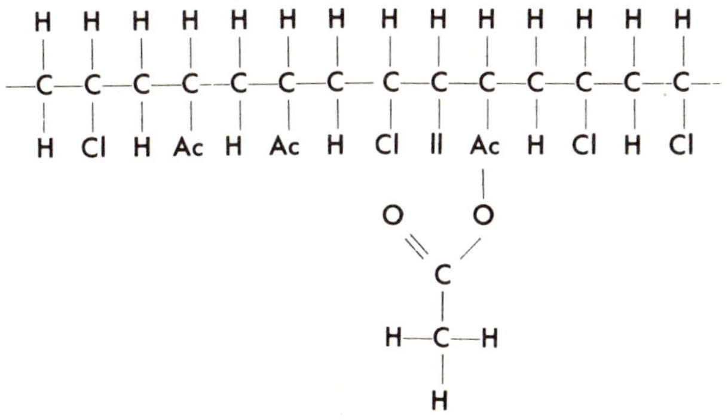

```
 H  H  H  H  H  H  H  H  H  H  H  H  H  H
 |  |  |  |  |  |  |  |  |  |  |  |  |  |
—C—C—C—C—C—C—C—C—C—C—C—C—C—C—
 |  |  |  |  |  |  |  |  |  |  |  |  |  |
 H  Cl H  Ac H  Ac H  Cl ‖  Ac H  Cl H  Cl
                           |
                        O  O
                         \\ /
                          C
                          |
                        H—C—H
                          |
                          H
```

Fig. 13-91. Copolymerization of vinyl chloride and vinyl acetate.

rolidone used as a blood plasma substitute, polyvinyl ethers used as adhesives, polyvinyl ureas, polyvinyl isocyanates, and polyvinyl chloroacetate have been explored for commercial interest.

All chlorinated or chlorine containing polymers may emit toxic chlorine gas upon high temperature degrading. Consequently, adequate venting should be provided to protect the operator during all processing.

Table 13-25 gives some properties of polyvinyl plastics.

Glossary of New Terms

Acetals(poly)—The molecular structure of the polymer is that of a linear acetal (polyformaldehyde) consisting of unbranched polyoxymethylene chains.

Acrylic—A synthetic resin prepared from acrylic acid or from a derivative of acrylic acid.

Acrylonitrile—A monomer with the structure $CH_2{=}CHON$. It is most useful in copolymers. Its copolymer with butadiene is nitrile rubber, and several copolymers with styrene exist that are tougher than polystyrene.

Acrylonitrile-butadiene-styrene—Acrylonitrile and styrene liquids and butadiene gas are polymerized together in a variety of ratios to produce the family of ABS resins.

Allyl—A synthetic resin formed by the polymerization of chemical compounds containing the group $CH_2{=}CH{-}CH_2{-}$. The principal commercially allyl resin is a casting material that yields allyl carbonate polymer.

Amides—Organic compounds containing a $-CONH_2$ group and being derived from organic acids.

Table 13-25. Polyvinyl Properties.

Property	(PVC) Rigid Vinyl Chloride	PVC Acetate (Copolymer)	Vinylidene Chloride Compound
Molding qualities	Good	Good	Excellent
Specific gravity (density)	1.30-1.45	1.16-1.18	1.65-1.72
Tensile strength (psi)	5000-9000	2500-4000	3000-5000
Compressive strength (psi)	8000-13,000		2000-2700
Izod, impact (ft-lb/in)	0.4-20.0		0.3-1.0
Hardness, Rockwell	M110-M120	R35-R40	M50-M65
Thermal expansion (10^{-5}/°C)	5.0-18.5		19.0
Resistance to heat (°F)	150-175	130-140	160-200
Dielectric strength (volts/mil)	400-500	300-400	400-600
Dielectric constant (60 Hz)	3.2-3.6	3.5-4.5	4.5-6.0
Dissipation factor (60 Hz)	0.007-0.020		0.030-0.045
Arc resistance, (seconds)	60-80		
Water absorption (24 hrs, %)	0.07-0.4	3.0+	0.1
Burning rate (in/min)	Self-exting.	Self-exting.	Self-exting.
Effect of sunlight	Needs stabilizer	Needs stabilizer	Slight
Effect of acids	None to slight	None to slight	Resistant
Effect of alkalies	None	None	Resistant
Effect of solvents	Soluble in ketones and esters	Soluble in ketones and esters	None to slight
Machining qualities	Excellent		Good
Optical	Transparent	Transparent	Transparent

Amines—Organic derivatives of ammonia obtained by substituting hydrocarbon radicals for one or more hydrogen atoms.

Azo group—The group —N=N—, generally combined with two aromatic radicals. A whole class of dyestuffs is characterized by the presence of this group.

Cellulosics—A family of plastics with the main constituent being the polymeric carbohydrate cellulose.

Chlorinated polyether—The polymer is obtained from pentaerythritol by preparing a chlorinated oxethane and polymerizing it to a polyether by means of opening the ring structure.

Coumarone—A compound (C_8H_6O) found in coal tar and polymerized with indene to form thermoplastic resins which are used in coatings and printing inks.

Cracking—Thermal or catalytic decomposition of organic compounds to break down the high-boiling constitutents of the compounds into lower-boiling fractions.

Diamines—Compounds containing two amino groups.

Dibasic acid—An acid that has two replaceable hydrogen atoms.

Ester—A compound formed by the replacement of the acidic hydrogen of an organic acid by a hydrocarbon radical; a compound of an organic acid and an alcohol formed with the elimination of water.

Esterification—The process of producing an ester by reaction of an acid with an alcohol with the elimination of water.

Fluoroplastics—A group of plastics materials containing the element fluorene (F).

Homopolymers—A polymer consisting of only one type of monomer.

Ionomer—A polymer with its major component ethylene but containing both covalent and ionic bonds. The polymer exhibits very strong interchain ionic forces.

Methyl methacrylate—A colorless, volatile liquid derived from acetone cyanohydrin, methanol, and dilute sulphuric acid and used in production of acrylic resins.

Monohydric—Containing one hydroxyl (OH^-) group in the molecule.

Phenoxy—A high-molecular-weight thermoplastic polyester resin based on bisphenol-A and epichlorohydrin.

Polyacrylate—A thermoplastic resin made by the polymerization of an acrylic compound.

Polyallomer—Crystalline polymers produced from two or more olefin monomers.

Polyamide—A polymer in which the structural units are linked by amide or thioamide groupings.

Polycarbonate—Polymers derived from the direct reaction between aromatic and aliphatic dihydroxy compounds with phosgene or by the ester exchange reaction with appropriate phosgene derived precursors.

Polyethylene—A thermoplastic material composed by polymers of ethylene (polyolefin).

Polyimide—A group of resins made by reacting pyromellitic dianhydride with aromatic diamines. The polymer is characterized by the fact that it has rings of four carbon atoms tightly bound together.

Polyolefin—A term used to indicate a family of hydrocarbons with carbon-to-carbon bonds.

Polymethyl methacrylate—*See* Methyl methacrylate.

Polymethylpentene—An isotactically arranged aliphatic polyolefin of 4-methyl-pentene-1.

Polyphenylene oxide—Currently made as a polyether of 2,6-dimethyl-phenol by oxidative coupling process by means of air or pure oxygen in the presence of a copper-amine complex catalyst.

Polypropylene—A plastics material made by the polymerization of high-purity propylene gas in the presence of an organometallic catalyst at relatively low pressures and temperatures (polyolefin).

Polystyrene—A thermoplastic material which is produced by the polymerization of styrene (vinyl benzene).

Polysulfone—A thermoplastic consisting of benzene rings connected by a sulfone group (SO_2), an isopropylidene group, and an ether linkage.

Polyvinyls—A broad family of plastics derived from vinyl radical group $CH_2=CH-$.

Radical—A group of atoms of different elements which behaves as a single atom in chemical reactions.

Reagent—A chemical used in chemical analysis for producing a characteristic effect.

Review Questions

13-1. List ten plastics belonging to the family of thermoplastics and give a product application of each.

13-2. List the most important properties of each plastics named for Question 13-1.

13-3. Name five individual polymers which make up the cellulose family.

13-4. Name three resins which may be included in the polyolefin family.

13-5. List several important copolymers and product applications.

13-6. Collect and display products and tradenames of thermoplastic materials.

13-7. What are the sources of cellulose used to product plastics?

13-8. What solvent is used to join cellulose acetate?

13-9. What advantage does cellulose propionate offer to molders?

13-10. What are some of the applications for nonplastics cellulose derivatives?

13-11. What solvents are used to bond polyphenylene oxide?

13-12. Why would it not be advisable to mix polyphenylene oxide and Noryl together for processing?

13-13. What applications are suggested for polysulfone? Why?

13-14. List several reasons why acrylic sheets are widely used for signs.

13-15. What solvents are used to bond acrylic?

13-16. What hazards exist in overheating acetal?

13-17. What effect does moisture have on the processing characteristics of polycarbonate?

13-18. What do the numbers 6,6 and 6,10 mean when referring to polyamides?

13-19. What is the effect of moisture absorption on polyamides?

13-20. What affect does crystallinity of polyamides have on melting, optical, and processing characteristics?

13-21. What is the difference between a fluorocarbon and a fluoroplastic?

13-22. By what processing methods are products made from PTFE?

13-23. Which fluorinated resin is useful over the widest temperature range?

13-24. How does the specific gravity of fluoroplastics compare with those of other thermoplastics?

13-25. What is the largest single market for the vinyl copolymer?

13-26. From what major chemicals is styrene monomer derived?

13-27. What is the hazard that exists in handling and storing expandable beads?

13-28. What other industry contributed to the development of polystyrene?

13-29. What effect does excessive oxidation have on the properties of polyethylene?

13-30. Of what value is melt index rating?

13-31. What additive renders polyethylene suitable for outdoor use?

13-32. What is the difference between isotactic and atactic polypropylene?

13-33. Explain the three classes of polyethylene resins.

14

Thermosetting Plastics

This chapter covers individual groups of thermosetting plastics in alphabetical order.

Alkyds

The history of ester based alkyd resins is somewhat confusing. Traditionally, the term *alkyd* referred to unsaturated polyesters modified with fatty acids or vegetable oils that find applications in paints and other coatings. Today, "alkyd molding compounds" refers to unsaturated polyesters modified with a nonvolatile monomer such as diallyl phthalate and various fillers. These compounds are formed into granular, rope, nodular, putty, and log shapes to facilitate continuous automatic molding (see Fig. 14-1).

(A) Compounding.

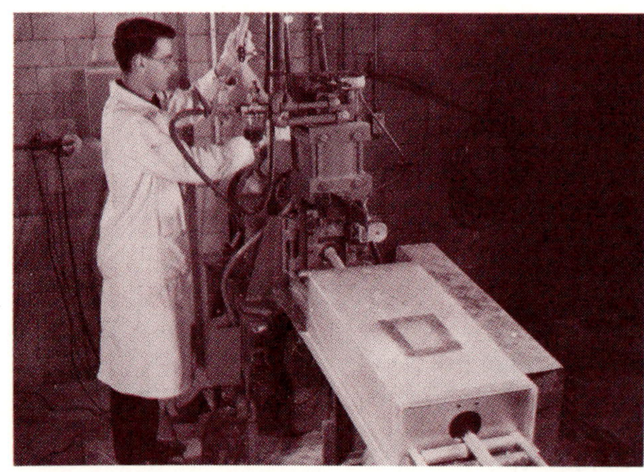

(B) Extruding.

(C) Finished products.

Fig. 14-1. Production of alkyd (rope) molding compounds. (Allied Chemical Corp.)

To obtain resin suitable for molding compounds, it is necessary to modify the crosslinking mechanism so rapid cure will result in the mold. The use of initiators in the resin compound also accelerates the polymerization of the double bonds. These resins should not be confused with saturated polyester molding compounds which are linear and thermoplastic.

R. H. Kienle has been given credit for coining the word *alkyd*. The term is derived from the "al" in *al*cohol and the "cid" from a*cid*. It is usually pronounced *AL-KID*. Alkyd resins may be produced by reacting phthalic acid, ethylene glycol alcohol, and fatty acids of various oils such as linseed oil, soy bean oil, and tung oil (Fig. 14-2). Kienel combined fatty acids with unsaturated esters in 1927 while searching for a better electrical insulating resin for General Electric. The need for materials in both World Wars greatly accelerated the interest and progress of alkyd finishing resins.

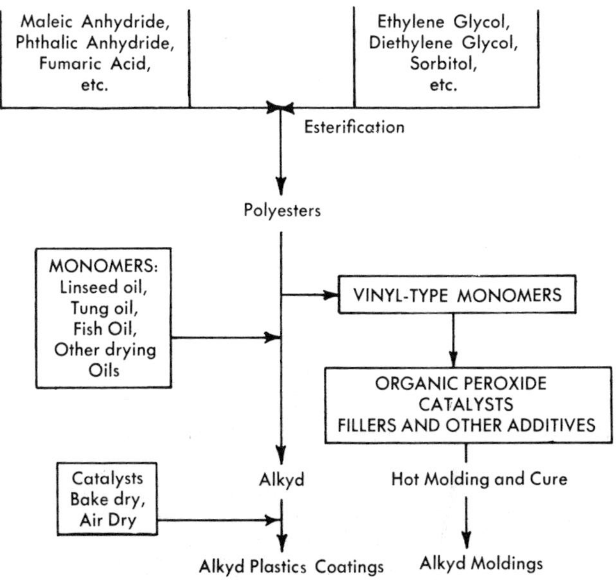

Fig. 14-2 Production of alkyd coatings and moldings.

Alkyd resins may have a structure similar to the one shown in Fig. 14-3. When this resin is applied to a substrate, heat or oxidizing agents are used to initiate crosslinking.

One authority estimates that nearly half of the surface coatings in the United States are of the alkyd polyester type. The extensive use of silicon and acrylic latex coatings in the open market may substantially reduce that estimate.

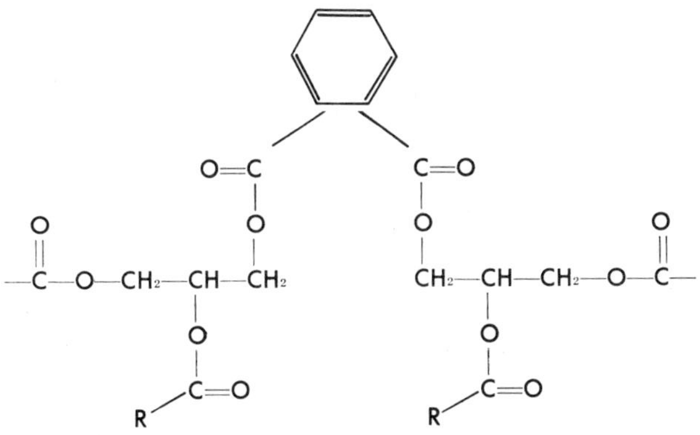

Fig. 14-3. Long chain with unsaturated groups.

Alkyd coatings are of value because of their comparatively low cost, durability, heat resistance, and ability to be modified to meet special coating requirements. To increase durability and abrasion resistance, rosin may be used to modify alkyd resins. Phenolics and epoxy resins may improve hardness and chemical and water resistance. Additional styrene monomers may extend flexibility to the finished coating.

Alkyd resins are used in "oil-based" house paint, baking enamel, farm implement paint, emulsion paint, porch and deck enamel, spar varnish, and chlorinated rubber paint. Modified alkyd resins may produce special coatings such as the *wrinkle* and hammer finish commonly used on various pieces of equipment and machinery.

Alkyd baking finishes are referred to as *heat convertible* because they polymerize or harden when heated. Those resins which harden when exposed to air are referred to as *air convertible* resins.

Miscellaneous uses of alkyd resins include plasticizers for various plastics, vehicles for printing inks, special adhesives for wood, rubber, glass, leather, and textiles, and as binders for abrasives and oils.

New processing technologies and applications for thermosetting alkyd molding compounds are also of current commercial importance.

Alkyd molding compounds are being successfully molded in compression, transfer, and reciprocating screw equipment. By utilizing initiators such as benzoyl peroxide or tertiary butyl hydroperoxide, the unsaturated resins can be made to crosslink at molding temperatures ($<50^\circ$C). Molding cycles may be less than twenty seconds.

Typical applications of alkyd molding compounds include appliance housings, utensil handles, billiard balls, circuit breakers, switches, motor cases, and capacitor and commutator parts (Fig. 14-4). They have been used to replace electrical applications where less expensive phenolic or amino resins are unsuitable.

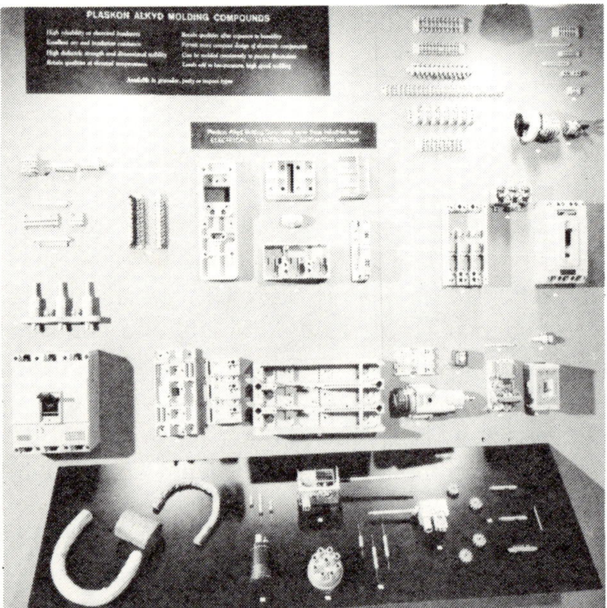

(A) Electrical, electronic and automotive parts. (Allied Chemical Corp.)

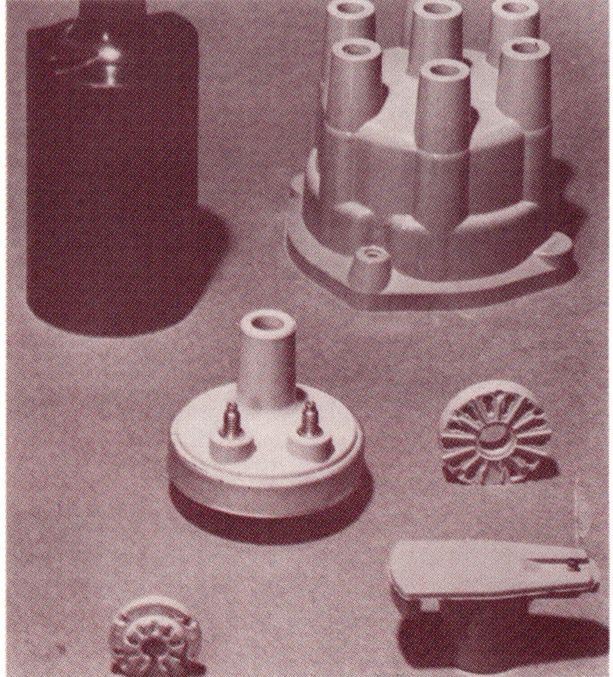

(B) Other automotive applications.

Fig. 14-4. Various alkyd products.

Alkyd putty form compounds are used for encapsulation of electronic and electrical components. The wide range of resin formulations, fillers, and processing techniques provides for a wide range of physical characteristics.

Table 14-1 gives some of the properties of alkyd plastics.

Table 14-1. Alkyd Properties.

Property	Glass Filled	Filled Molding Compound
Molding qualities	Excellent	Excellent
Specific gravity (density)	2.12-2.15	1.65-2.30
Tensile strength (psi)	4000-9500	3000-9000
Compressive strength (psi)	15,000-32,000	12,000-38,000
Izod, impact (ft-lb/in)	0.60-10.0	0.30-0.50
Hardness, Rockwell	60-70 Barcol 98 E scale	55-80 Barcol 95 E scale
Thermal expansion (10^{-5}/°C)	1.5-2.5	2.0-5.0
Resistance to heat (°F)	450	300-450
Dielectric strength (volts/mil)	250-530	350-450
Dielectric constant (60 Hz)	5.7	5.1-7.5
Dissipation factor (60 Hz)	0.010	0.009-0.06
Arc resistance (seconds)	150-210	75-240
Water absorption (24 hrs, %)	0.05-0.25	0.05-0.50
Burning rate (in/min)	Slow to nonburning	Slow to nonburning
Effect of sunlight	None	None
Effect of acids	Fair	None
Effect of alkalies	Fair	Attacked
Effect of solvents	Fair-good	Fair-good
Machining qualities	Poor-fair	Poor-fair
Optical	Opaque	Opaque

Amino Plastics

A large number of polymers have been produced by interaction of amines or amides with aldehydes. The two most significant and commercially successful amino plastics are produced by the condensation of urea and formaldehyde or melamine and formaldehyde.

One authority indicates that urea-formaldehyde polymers may have been produced as early as 1884. By the turn of the century, the German Goldschmidt and other colleagues directed their efforts toward producing a moldable amino plastics.

In the United States urea-formaldehyde resins were commercially produced in 1920, with thiourea-formaldehyde and melamine-formaldehyde resins being produced in the period 1934-1939.

The white crystalline solid, urea (NH_2CONH_2), and aqueous solutions of formaldehyde (formalin) may produce urea resins or modifications of urea-formaldehyde resins by the addition of other reagents. For complete polymerization and cross-linkage of this thermosetting resin, catalysts and/or heat are usually used during the molding operation (Fig. 14-5).

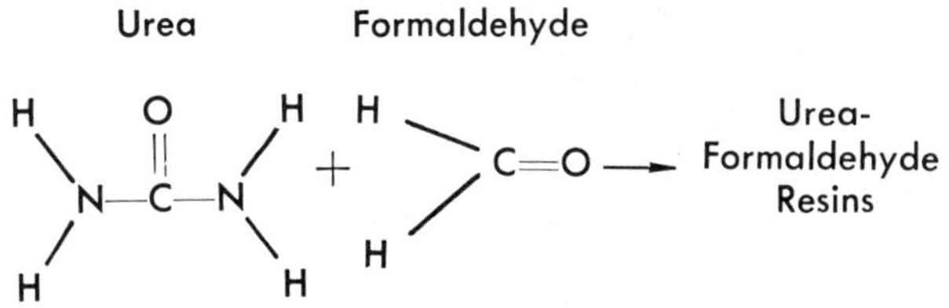

Urea Formaldehyde

Urea-Formaldehyde Resins

Fig. 14-5. Formation of urea-formaldehyde resins.

Urea-formaldehyde polymerization is shown below. The excess water molecule is a result of condensation polymerization. Condensation polymers, by present technology, have a by-product.

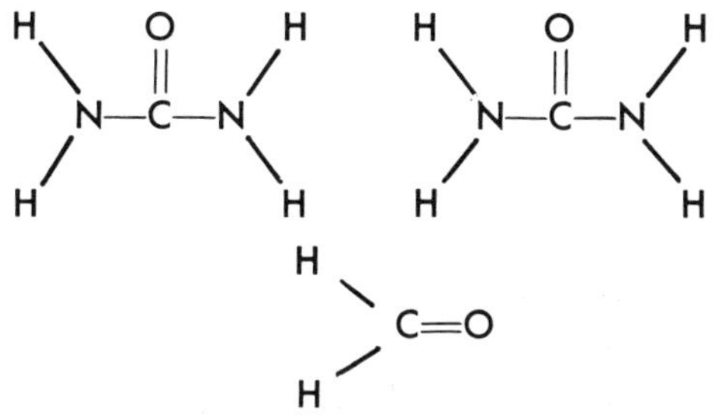

(A) Urea-formaldehyde resin.

$+H_2O$

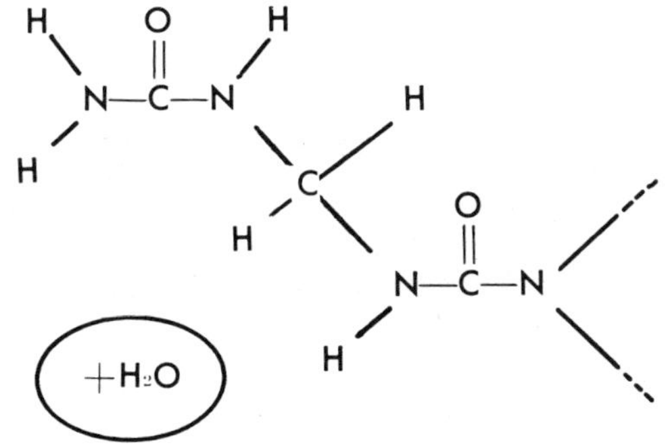

(B) Urea-formaldehyde plastics.

Fig. 14-6. Polymerization of urea and formaldehyde.

Many of the products once produced of urea-formaldehyde plastics are now being made of thermoplastic materials because of the rapid production cycles and higher production rates possible with thermoplastics. Present formulation of amino compounds are being processed in specially designed injection molding equipment, thus increasing productivity and making them competitive with most thermoplastics.

Urea-formaldehyde-base resins are crystal clear and very easily colored. Molding compounds usually contain several ingredients including resin, filler, pigment, catalyst, stabilizer, plasticizer, and lubricant.

In order to accelerate curing and increase production rates molding compounds are produced with latent acid catalysts added to the resin base. This latent catalyst reacts at molding temperatures. Because many urea-formaldehyde molding compounds have prepromoters or catalysts added, they have a limited storage life and should be stored in a cool place. Stabilizers may be added to help control latent catalyst reaction. Lubricants are commonly added to improve molding quality. Plasticizers also improve flow properties and help reduce cure shrinkage. Although the base resin is water-clear, transparent, translucent, or opaque color, pigments may be used. If color is not important, glass fiber, macerated fabric, asbestos, and wood flour may be used to improve physical characteristics and lower cost.

Urea-formaldehyde plastics have very many industrial applications because of outstanding molding qualities and low cost. They are used extensively for various electrical and electronic applications requiring good arc and tracking resistance. They provide good dielectric properties and are unaffected by common organic solvents, greases, oils, weak acids, alkalies, or other hostile chemical environments. Urea compounds do not attract dust from static charges. They will not burn or soften when exposed to an open flame and exhibit very good dimensional stability when filled.

The bulk of urea-formaldehyde molding compounds is used in the manufacture of bottle caps and electrical or thermal insulating applications.

Alpha-cellulose (bleached cellulose), wood flour, or other fillers are commonly used to lower product cost and to improve molding and physical properties. Urea-formaldehyde products do not impart taste or odors to foods or beverages with which they come in contact. They are preferred materials for appliance knobs, dials, handles, pushbuttons, toaster bases, and end plates. Electrical wall plates, switch toggles, receptacles, fixtures, circuit breakers, and switch housings are only a few of the many electrical insulating applications (see Fig. 14-7).

(A) Terminal blocks for electrical connections.

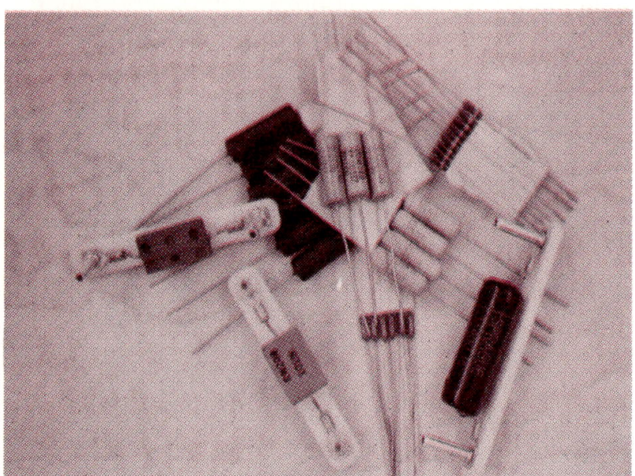

(B) Insulating blocks and packing for various electronic components.

Fig. 14-7. Applications of urea-formaldehyde compounds in electricity and electronics.

At the present time one of the largest applications of urea-formaldehyde resins is in the manufacture of adhesives for furniture glues, plywood, and chipboard. Chipboard is made by combining about ten percent resin binder with wood chips. The resin-bound wooden chips are then pressed into flat sheets. The product has no grain and does not warp because it is free to expand in all directions; however, its water resistance is poor. The adhesive use of this resin for bonding of plywood is suitable only for interior applications (Fig. 14-8).

These resins may be foamed and cured into a plastics state with densities from 0.5 to 3.0 pounds per cubic foot. Such foams are inexpensive and

Fig. 14-8. Applying urea-formaldehyde resin to wood veneer strips during production of plywood. (Monsanto Company)

easily produced either by vigorously whipping a mixture of resin, catalyst, and a foaming detergent or by introducing a chemical agent which generates gas (usually carbon dioxide) as the resin is curing. These foams have found applications as thermal insulating materials in buildings, refrigerators, and as low-density cores for structural sandwich construction (Fig. 14-9). They may be made flame resistant at the expense of foam density.

(A) Block for floral use.

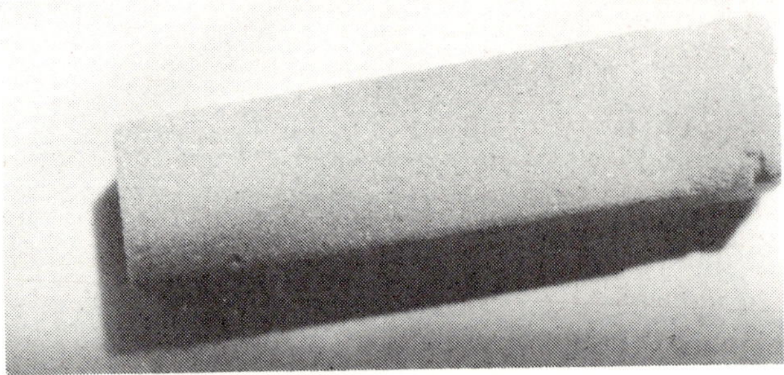

(B) For thermal insulation.

Fig. 14-9. Various uses of open-celled or foamed urea-formaldehyde.

Large amounts of water may be absorbed by these foams because they are open-celled (spongelike) structures. This latter ability to absorb water is utilized by the florist. Stems of cut flowers may be inserted into the water-soaked urea foam, thus prolonging the fresh appearance of cut floral arrangements. Ground foam has been used as an artificial snow in television and other theatrical productions. The open-cellular structure may also be filled with kerosene and used as firelighters for fireplaces.

Urea resins find extensive use in the textile, paper, and coating industries. The existence of crease resistant cellulose shrinkage control and drip-dry fabrics is due largely to these resins.

Urea resins or powders are sometimes used as binders for foundry cores and shell molds.

The surface coating applications of urea-formaldehyde resins are limited to substrates which will withstand curing temperatures of 100°F to 350°F. These resins are combined with a compatible polyester (alkyd) resin to produce enamels of outstanding merit. From five to fifty percent urea resin is added to the alkyd-based resin, resulting in surface coatings which are outstanding in hardness, toughness, gloss, color stability, and outdoor durability.

These surface coatings may be seen on refrigerators, washing machines, stoves, signs, venetian blinds, metal cabinets, and numerous machinery applications. At one time they were widely applied to automobile bodies; however, the curing or baking time required for mass production has made them less popular. The baking time depends on the temperature and the proportion of amino resin to alkyd resin.

Urea resins modified with furfuryl alcohol have been used successfully in the manufacture of coated abrasive papers.

Urea-formaldehyde plastics are easily molded in conventional compression and transfer molding machines. They may also be processed by the reciprocating-screw type injection molding machines. Depending on the grade of resin and the filler used, allowance should be made for shrinkage after removal from the mold. Improved dimensional stability can be accomplished by postconditioning the product in an oven.

Table 14-2 gives some properties of alpha-cellulose filled urea-formaldehyde plastics.

Table 14-2. Urea-Formaldehyde Properties.

Property	Alpha-Cellulose Filled
Molding qualities	Excellent
Specific gravity (density)	1.47-1.52
Tensile strength (psi)	5500-13,000
Compressive strength (psi)	25,000-45,000
Izod, impact (ft-lb/in)	0.25-0.40
Hardness, Rockwell	M110-M120
Thermal expansion (10^{-5}/°C)	2.2-3.6
Resistance to heat (°F)	170
Dielectric strength (volts/mil)	300-400
Dielectric constant (60 Hz)	7.0-9.5
Dissipation factor (60 Hz)	0.035-0.043
Arc resistance (seconds)	80-150
Water absorption (24 hrs, %)	0.4-0.8
Burning rate (in/min)	Self-exting.
Effect of sunlight	Grays
Effect of acids	None to decomposes
Effect of alkalies	Slight to decomposes
Effect of solvents	None to slight
Machining qualities	Fair
Optical	Transparent-opaque

Melamine-Formaldehyde

Until about 1939, melamine-formaldehyde was an expensive laboratory curiosity. Melamine ($C_3H_6N_6$), a white crystalline solid, and formaldehyde result in the formulation of a compound referred to as methylol derivative (Fig. 14-10).

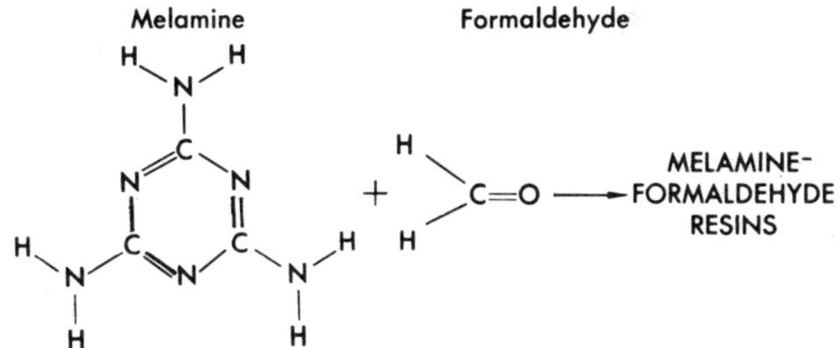

Fig. 14-10. Formation of melamine-formaldehyde resin.

With additional formaldehyde the formulation will react to produce tri-, tetra-, penta-, and hexa-methylol-melamine.

The formation of trimethylol melamine is shown in Fig. 14-11.

NH$_2$ ... NH$_2$—C ... C—NH$_2$ $+3CH_2O \longrightarrow$ NHCH$_2$OH ... HOCH$_2$NHC ... CNHCH$_2$OH $+3H_2O$

Fig. 14-11. Formation of trimethylol melamine.

Commercial melamine resins may be obtained without acid catalysts. Both thermal energy and catalysts, however, are normally utilized to speed polymerization and cure. Polymerization of urea and melamine resins results in the condensation of water. This water will normally evaporate or escape from the molding cavity.

Benzoguanamine ($C_3H_4N_5C_6H_5$) and thiourea ($CS(NH_2)_2$) formaldehyde resins are of minor commercial importance. Development of polymers based on the reaction of formaldehyde and such compounds as aniline, dicyandiamide, ethylene urea, and sulfonamide may provide more complex and varied industrial applications.

In the pure form amino *resins* are colorless and soluble in warm solutions of water and methanol.

Concerning applications for amino resins, urea and melamine plastics are frequently grouped together as one entity. Their similarity in chemical structure, properties, and applications have much in common. Despite the apparent similarity melamine-formaldehyde products are superior to the urea-formaldehyde plastics in several respects.

In general, melamine products are harder, more water resistant, and may be combined with a greater variety of fillers to produce products with better heat, scratch, stain, and chemical resistances. Although melamine products are generally superior to similar urea products they are also significantly more expensive.

Probably the largest single use of melamine-formaldehyde is the manufacture of tableware. For this application molding powders are normally filled with alpha-cellulose. Asbestos or other fillers are sometimes used for handles and utensil housings (Fig. 14-12).

Much resin is used for surface coatings and decorative laminates. Paper-based laminates with such well-known tradenames as Formica and Micarta utilize melamines because of their good water and heat resistance. Photographic prints on fabrics or paper are impregnated with melamine resin and placed on a base or core material and cured in a large press. Kraft paper impregnated with phenolic resin is commonly used for the base material because it is durable, compatible with melamine resin, and costs less than multiple layers of melamine impregnated paper. There is a broad spectrum of applications for these laminates for surfacing wood, metal, plaster, and hardboard. The extensive use of these laminates in the kitchen and for furniture table tops needs little description.

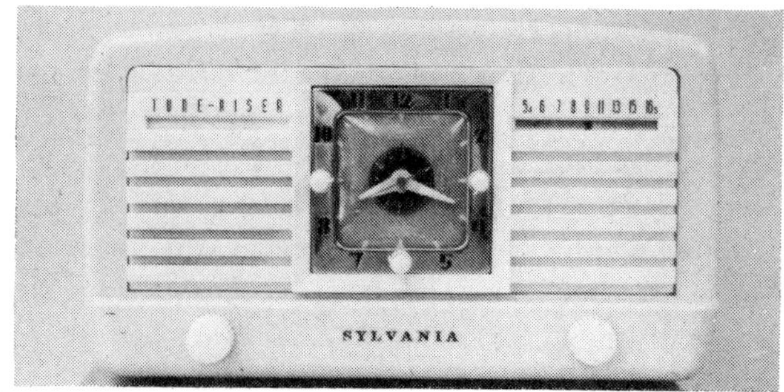

(A) Radio molded in the 1950's.

(B) Kitchen containers.

Fig. 14-12. Applications of melamine plastics.

A three percent melamine resin solution may be added during the preparation of paper pulp to overcome the well-known disadvantage of paper when wet. Paper with this resin binder has a wet strength nearly as great as its dry strength. The crush and folding resistance is also greatly increased without added bitterness.

Water, chemical, alkali, grease, and heat resistant finishes are formulated of amino resins. Heat and/or catalysts are needed to cure the amino resin finish on such familiar products as stoves, washing machines, and other appliances.

Urea or melamine resins may be made compatible with other plastics to produce finishes of outstanding merit. Polyester (alkyd) resins or phenolic resins may be combined to produce finishes with the best features of both resins. These finishes are sometimes used where a hard, tough, mar resistant finish is required.

Melamine resins are frequently used in the manufacture of waterproof exterior plywood, marine plywood, and other adhesive applications requiring a light-colored nonstaining adhesive. Catalysts, heat, or high-frequency energy is commonly used for curing melamine adhesives in plywood and panel assemblies.

Melamine-formaldehyde resins are employed commercially for textile finishing. The well-known drip-dry fabrics, permanent glazing, rotproofing, and shrinkage control owe their existence largely to such resins. Melamine and silicone resins are utilized to produce waterproof fabrics.

Melamine-formaldehyde compounds are easily molded in conventional compression and transfer molding machines. Special reciprocating-screw injection machines are also used.

Table 14-3 gives some of the properties of melamine-formaldehyde plastics.

Table 14-3. Melamine-Formaldehyde Properties.

Property	No Filler	Alpha-Cellulose Filled	Glass Fiber Filled
Molding qualities	Good	Excellent	Good
Specific gravity (density)	1.48	1.47-1.52	1.8-2.0
Tensile strength (psi)		7000-13,000	5000-10,000
Compressive strength (psi)	40,000-45,000	40,000-45,000	20,000-35,000
Izod, impact (ft-lb/in)		0.24-0.35	0.6-18.0
Hardness, Rockwell		M115-M125	M120
Thermal expansion ($10^{-5}/°C$)		4.0	1.5-1.7
Resistance to heat (°F)	210	210	300-400
Dielectric strength (volts/mil)		270-300	170-300
Dielectric constant (60 Hz)		6.2-7.6	9.7-11.1
Dissipation factor (60 Hz)		0.030-0.083	0.14-0.23
Arc resistance (seconds)	100-145	110-140	180
Water absorption (24 hrs, %)	0.3-0.5	0.1-0.6	0.09-0.21
Burning rate (in/min)	Self-exting.	Nonburning	Self-exting.
Effect of sunlight	Color fades	Slight color change	Slight
Effect of acids	None—decomposes	None—decomposes	None—decomposes
Effect of alkalies		Attacked	None to slight
Effect of solvents	None	None	None
Machining qualities		Fair	Good
Optical	Opalescent	Translucent	Opaque

Casein

Casein plastics are sometimes included in the group of natural polymers and referred to as "protein" plastics by many.

Casein is a protein found in a number of sources including human and animal hair, feathers, bones, and other industrial wastes. However, there has been little interest in these sources because of the collection problems and limited use of protein derived plastics. Only skimmed milk is presently of commercial interest in the manufacture of casein plastics.

The history of protein derived plastics can probably be said to date from the work of a German printer, W. Krische, and Adolf Spitteler of Bavaria in about 1895. At about that time there was a demand in Germany for what may be described as a white "blackboard." These boards were thought to possess superior optical properties to those of the normal black variety. By 1897 Spitteler and Krische, in an effort to obtain such a product, developed a casein plastics which could be hardened with formaldehyde. Up until this time experiments with casein plastics proved unsatisfactory because the plastics were soluble in water. Galaith (milkstone), Erinoid, and Ameroid are tradenames of these early protein plastics.

Because casein is not coagulated by heat, it must be precipitated from the milk by the action of rennin enzymes or acids. This powerful coagulate causes the milk to separate into curds and whey. After the whey is removed only the curd which contains the protein is left. This material is washed, dried, and powdered. When kneaded with water the doughlike material may be shaped or molded. A simple drying operation with considerable shrinkage results. Molded products may be made water resistant by soaking them in a formalin solution for varying lengths of time depending on thickness. Casein is thermoplastic while being molded; however, the addition of formaldehyde in a formalin solution creates links which hold the casein molecules together. The long linear chain of casein molecules are known as the *polypeptide chain.* Because there are a number of different peptide groups and possible side reactions, no one formula could represent the interaction between casein and formaldehyde. A very simplified reaction is illustrated in Fig. 14-13.

$$\begin{array}{c} CO \\ | \\ NH \end{array} + CH_2O + \begin{array}{c} CO \\ | \\ NH \end{array} \longrightarrow \begin{array}{c} CO \\ | \\ N \end{array}\!-\!CH_2\!-\!\begin{array}{c} CO \\ | \\ N \end{array}$$

Fig. 14-13. Reaction across the peptide (—CONH—) groups.

It is doubtful that casein will gain in popularity because it is expensive to manufacture and the raw material has value as a food. Casein plastics are seriously affected by humid conditions and cannot be used as electrical insulators. The lengthy hardening process and poor resistance to decomposition by heat make them unpopular for modern processing rates. There are only limited commercial uses of casein plastics in the United States because they offer no property advantages over synthetic polymers and production costs are high.

Casein plastics are used to a limited extent for buttons, buckles, knitting needles, umbrella handles, and other novelty items. They may be reinforced and filled or obtained in transparent colors. Casein has held some of its appeal because it may be colored in beautiful shades such as pearl, onyx, ivory, and imitation horn. Probably the largest proportion of casein is used in the stabilization of rubber latex emulsion, preparation of medical compounds and food products, preparation of paints and adhesives, and the sizing of paper and textiles. Casein finds other applications in insecticides, soaps, pottery, inks, and as modifiers in other plastics. Films and fibers may be produced from these plastics. They are used to produce woollike fibers that are warm, soft, and have properties that compare favorably with natural wool. The films are generally of little use except as convenient forms for the coating of paper and other materials. Casein glue is a well-known wood adhesive (see Fig. 14-14).

Table 14-4 gives some of the properties of casein plastics.

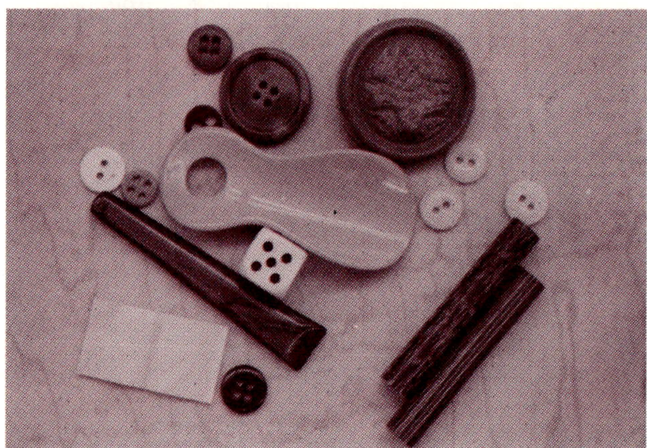

(A) Buttons, shoe horn, dice, tubing, etc.

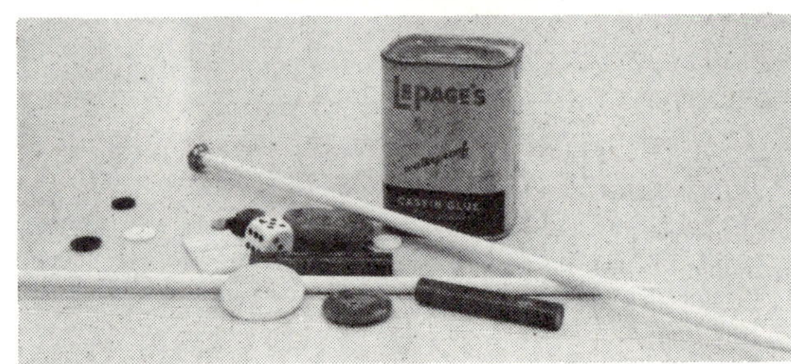

(B) Adhesive, darning needles, etc.

Fig. 14-14. Miscellaneous applications of casein.

Table 14-4. Casein Properties.

Property	Casein Formaldehyde (Unfilled)
Molding qualities	Excellent
Specific gravity (density)	1.33-1.35
Tensile strength (psi)	7000-10,000
Compressive strength (psi)	27,000-50,000
Izod, impact (ft-lb/in)	0.9-1.2
Hardness, Rockwell	M26-M30
Resistance to heat (°F)	275-350
Dielectric strength (volts/mil)	400-700
Dielectric constant (10^6 Hz)	6.1-6.8
Dissipation factor (10^6 Hz)	0.052
Arc resistance (seconds)	Poor
Water absorption (24 hrs, %)	7-14
Burning rate (in/min)	Slow
Effect of sunlight	Yellows
Effect of acids	Decomposes
Effect of alkalies	Decomposes
Effect of solvents	Slight
Machining qualities	Good
Optical	Translucent-opaque

Epoxy

Hundreds of United States patents have been granted for the commercial exploration of epoxide resins. One of the first descriptions of polyepoxides is a German patent by I. G. Farbenindustrie in 1939.

In 1943 the Ciba Company developed an epoxide resin of commercial significance in the United States. By 1948 a variety of commercial coating and adhesive applications were disclosed.

Epoxy resins are thermosetting plastics. There are, however, several thermoplastic epoxy resins used for coatings and adhesives. There are many different epoxy resin structures available today. A large portion of these resins are derived from bisphenol-acetone and epichlorohydrin.

Bisphenol-A (bisphenol-acetate) is made by the condensation of acetone with phenol (see Fig. 14-15).

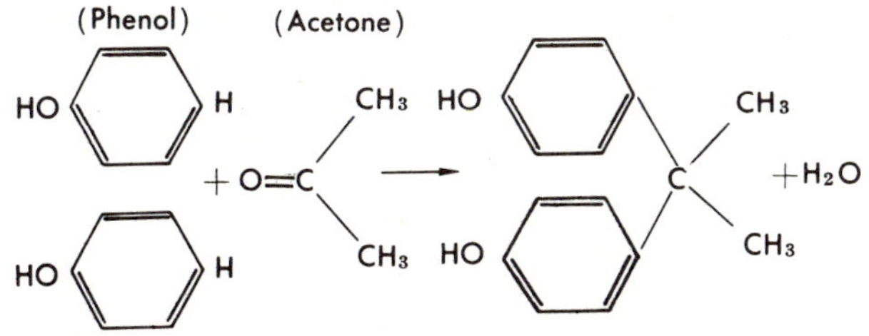

(Phenol) (Acetone)

HO–C₆H₄–H + O=C(CH₃)₂ ⟶ (HO–C₆H₄)₂C(CH₃)₂ + H_2O

Fig. 14-15. Production of bisphenol-A.

Epichlorohydrin based epoxies are widely used because of their availability and lower cost. The epichlorohydrin structure is obtained from the chlorination of propylene:

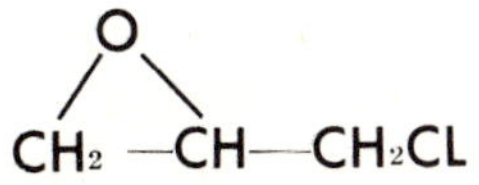

$$\overset{O}{\overbrace{CH_2-CH}}-CH_2CL$$

It will become evident that the epoxy group, for which the plastics family is named, has a triangular structure:

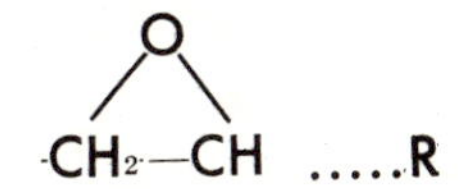

$$\overset{O}{\overbrace{CH_2-CH}}\ .....R$$

Epoxy structures are usually terminated by this epoxide structure; however, numerous other molecular structures may terminate the long molecular chain. A linear epoxy polymer may be formed when bisphenol-A and epichlorohydrin are reacted (Fig. 14-16).

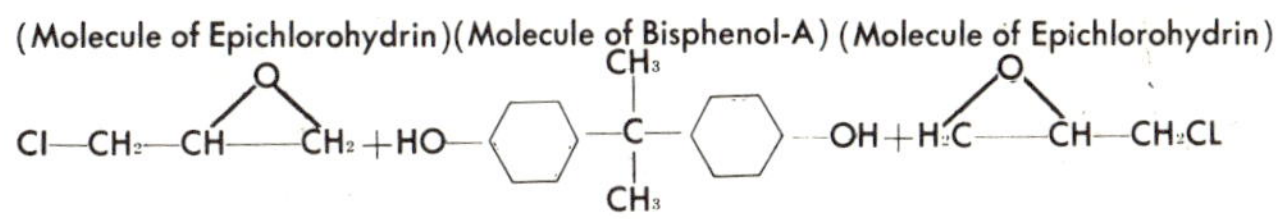

Fig. 14-16. Formation of linear epoxy polymer.

A typical structural formula of bisphenol-A based epoxy resin may be represented as in Fig. 14-17:

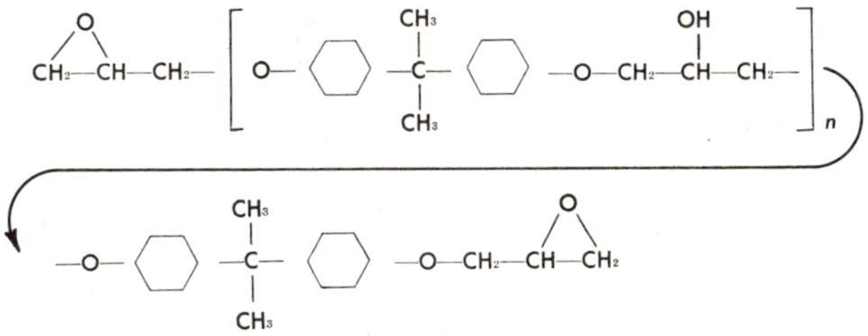

$$\overset{O}{\overbrace{CH_2-CH}}-CH_2-\left[O-C_6H_4-C(CH_3)_2-C_6H_4-O-CH_2-CH(OH)-CH_2-\right]_n-O-C_6H_4-C(CH_3)_2-C_6H_4-O-CH_2-\overset{O}{\overbrace{CH-CH_2}}$$

Fig. 14-17. Bisphenol-A based epoxy resin.

Other intermediate epoxy based resins are possible but are too numerous to mention.

Epoxy resins are usually cured by the addition of catalysts or reactive hardeners. Members of the *aliphatic* and aromatic *amine* family are commonly used "hardening agents." Various *acid anhydrides* are also used to polymerize the epoxide chain.

Epoxy resins will polymerize and crosslink with the addition of thermal energy. The addition of catalysts and heat is often employed to reach a desired degree of polymerization.

Single component epoxy resins may contain latent catalysts which react when sufficient heat is applied. There is a practical shelf life expectancy for all epoxy resins.

Reinforced epoxy *resins* are very strong with good dimensional stability and service temperatures as high as 600°F. Preimpregnated reinforcing materials are used to produce products by hand-layup, vacuum bag, or filament winding processes. Epoxy resins are used to replace the less expensive unsaturated polyester resins for many applications where superior chemical and fatigue resistance is required. A saving of approximately one third of the resin weight is realized when epoxy resins are used in place of polyesters. Consequently, epoxy glass laminates have a high strength-to-weight ratio and are used in applications where the benefits of a high strength-to-weight ratio may be utilized. The superior adhesion to all materials and compatibility make epoxy resins a desirable choice

(Fig. 14-18). Laminated circuit boards, radomes, aircraft parts, and various filament wound pipes, tanks, and containers are only a few of the many uses for this plastics material.

(A) Pressure injected epoxy adhesives penetrate deeply into hairline cracks, permanently bonding and restoring the original load-bearing strength. (Scott J. Saunders Associates)

(B) Boron slat for aircraft with and closeouts and rib caps bonded by epoxy adhesives to the inner skin.

Fig. 14-18. High strength makes epoxy resins invaluable.

Filled epoxy resins are commonly used for special casting purposes. These strong compounds may be used for low cost tooling. Dies, jigs, fixtures, and molds for short production runs are replacing other tooling materials. Faithfully reproduced details are obtained when epoxy compounds are cast against prototype models or patterns (see Fig. 14-19). Various fillers are used in caulking and patching compounds containing epoxy resins. The adhesive quality previously mentioned and low shrinkage during cure make them durable caulking or patching compounds for a multitude of materials.

Fig. 14-19. Epoxy resin casting mold for prototype use or low production of castings or moldings. (General Mills Research Labs.)

Electrical potting is another casting application where the outstanding properties of epoxies are used to protect electronic parts from moisture, heat, and corrosive chemicals (Fig. 14-20). Some of the casting, potting, and encapsulation applications include: electric motor parts, high-voltage transformers, relays, coils, and numerous other components for severe environments.

Fig. 14-20. Selenium rectifier encapsulated in polyamide-epoxy resin blend. (General Mills, Inc.)

Molding compounds of epoxy resins and fibrous reinforcements can be injection, compression, or transfer molded into small electrical hardware, appliance parts, and many modular applications.

By controlling the resin manufacture, the curing agents and the rate of cure, a great versatility is achieved in performance. These resins may be formulated with properties ranging from soft, flexible compounds to hard, chemical resistant products. By incorporating a blowing agent which liberates a gas on heating, low-density epoxy foams may be produced. The outstanding properties offered by the epoxy plastics are adhesion, chemical resistance, toughness, and excellent electrical characteristics.

When the epoxy resins were first introduced in the 1950's, they were recognized as an outstanding *coating* material (Fig. 14-21). Although they are more expensive than other coating materials, their excellent adhesion and chemical inertness make them competitive in the coating field.

(A) Epoxy spray coating (second can from left).

(B) Epoxy coating on metal substrate of modern farm silo storage bins. (Butler Mfg.)

Fig. 14-21. Epoxy coating.

To the consumer today epoxy based finishes for driveways, concrete floors, porches, metal appliances, or wooden furniture are finding additional applications. Epoxy finishes on home appliances are one large application of this durable, abrasion resistant finish. Many products once requiring glass enameled finishes are now using epoxy coatings. Tank cars and other containers may have chemical resistant epoxy lining. Hulls as well as the insides of ships may be coated with epoxy. Longer lasting finishes mean fewer dry dock repairs and reduced surface tension between the ship and water. These factors ultimately reduce maintenance and fuel costs.

The flexibility of many epoxy coatings makes them popular for postforming of coated metal parts. For example, sheets of metal are coated while flat and then formed or bent into shallow pans with no evidence of any damage to the coating.

The outstanding property of epoxy adhesives to bond to dissimilar materials has allowed epoxies to replace many soldering, welding, riveting, and other joining methods. Aircraft and automotive industries use these adhesives where heat or other bonding methods might distort the surface (see Fig. 14-22).

(A) Epoxy adhesive is used to glue or bond vacuum formed automobile body panels together.

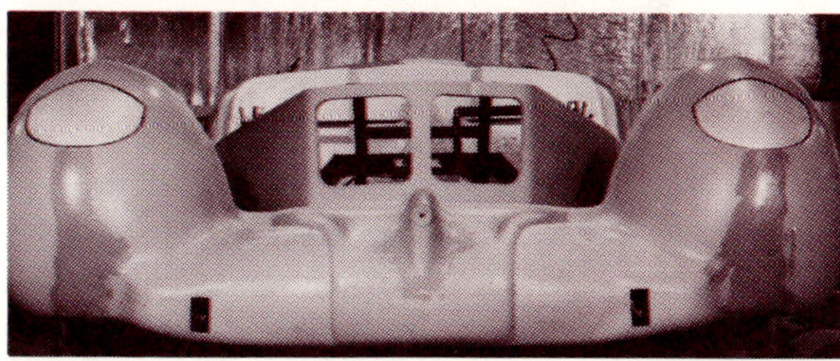

(B) Epoxy is also used to patch or repair panels damaged during fabrication or assembly.

(C) Sanding and finishing the epoxy adhesive and body solder to blend the body contour.

Fig. 14-22. Use of epoxies in automotive industry. (U.S. Gypsum)

Honeycomb or panel structures may utilize the superior adhesive and thermal properties of epoxy fabrication.

Table 14-5 gives some of the properties of various epoxies.

molecular-weight resin that is compounded with fillers and other ingredients. During the molding process, the resin is transformed into a highly crosslinked thermosetting plastic product by heat and pressure.

Table 14-5. Epoxy Properties.

Property	Glass Filled (Molding)	Mineral Filled (Molding)	Microballoon Filled (Molding)
Molding qualities	Excellent	Excellent	Good
Specific gravity (density)	1.6-2.0	1.6-2.0	0.75-1.00
Tensile strength (psi)	10,000-30,000	5000-15,000	2500-4000
Compressive strength (psi)	25,000-40,000	18,000-40,000	10,000-15,000
Izod, impact (ft-lb/in)	10-30	0.3-0.4	0.15-0.25
Hardness, Rockwell	M100-M110	M100-M110	
Thermal expansion (10^{-5}/°C)	1.1-3.5	2.0-5.0	
Resistance to heat (°F)	300-500	300-500	
Dielectric strength (volts/mil)	300-400	300-400	380-420
Dielectric constant (60 Hz)	3.5-5	3.5-5	
Dissipation factor (60 Hz)	0.01	0.01	
Arc resistance, (seconds)	120-180	150-190	120-150
Water absorption (24 hrs, %)	0.05-0.20	0.04	0.10-0.20
Burning rate (in/min)	Self-exting.	Self-exting.	Self-exting.
Effect of sunlight	Slight	Slight	Slight
Effect of acids	Negligible	None	Slight
Effect of alkalies	None	Slight	Slight
Effect of solvents	None	None	Slight
Machining qualities	Good	Fair	Good
Optical	Opaque	Opaque	Opaque

Phenolics (Phenol-Aldehyde)

Phenol-aldehyde or phenolics were one of the first truly synthetic resins produced. Their history dates back to the work of Adolph Baeyer in 1872. Baeyer produced a synthetic, tarlike resinous substance but not for commercial exploitation. It wasn't until 1909 that the Belgian chemist Leo Hendrik Baekeland invented and patented a successful technique for combining the phenol (O_6H_5OH—carbolic acid) and the gaseous formaldehyde (HCCH). Baekeland did much of his work while living in Yonkers, New York.

It was largely the success of the phenol-formaldehyde resins which later stimulated research in urea- and melamine-formaldehyde resins.

The resin formed from the reaction of phenol with formaldehyde (an aldehyde) is commonly known as a *phenolic*.

The reaction of phenol with formaldehyde involves a condensation reaction in which water is formed as a by-product (Fig. 14-23). The initial phenol formaldehyde reactions produce a low-

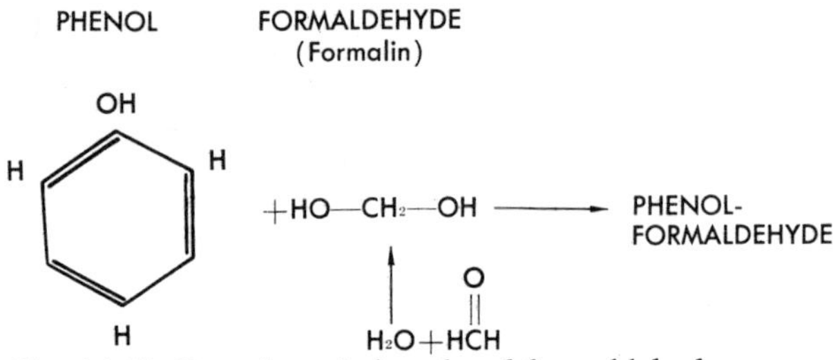

PHENOL FORMALDEHYDE (Formalin)

OH H H $+HO—CH_2—OH$ ⟶ PHENOL-FORMALDEHYDE

H H_2O+HCH (with =O on C)

Fig. 14-23. Reaction of phenol and formaldehyde.

Although the monomer solution of phenol is commercially used, cresols, xylenols, resorcinols, or synthetically produced oil-soluble phenols may be used. Furfural may also replace the formaldehyde.

In one stage resins, a *resol* is produced by reacting a phenol with an excess amount of aldehyde in the presence of a catalyst (not acid). Sodium and ammonium hydroxide are common catalysts. This product is resoluble, low in molecular weight, and will form larger molecules without the addition of hardening agents during the molding cycle.

Two-stage resins are produced when phenol is present in excess with an acid catalyst. The resulting low-molecular-weight and soluble *novolac* resin is the result. This resin will remain a linear thermoplastic resin unless compounds which are capable of forming crosslinkage on heating are added. (Novolacs are resins which will remain thermoplastic and are called *two-stage resins* because it is necessary to add some agent before molding.) See Fig. 14-24.

Resols and novolacs are insoluble, infusible thermosetting plastics when fully polymerzied and cured.

Fig. 14-24. Final curing or heat hardening should be considered a further condensation process. This two-dimensional phenolic plastics is the final product.

Because of the discovery of many new plastics and processes, phenolics do not hold the eminent position they once held (Fig. 14-25). However, their low cost, moldability, and physical properties make them a leader in the thermoset field. These materials are widely used as molding powders, resin binders, coatings, and adhesives.

Molding powders or compounds of novolac resin are rarely used without a filler. The filler is not incorporated simply to reduce cost but is generally necessary to improve physical properties, adaptability for processing, and to reduce shrinkage from the curing process. Curing time, shrinkage, and molding pressures may be reduced by preheating phenol formaldehyde compounds by high-frequency methods. Advances in preheating and equipment technology have kept phenolics competitive with thermoplastics and metals (see Fig. 14-26). Not only may they be used in conventional transfer and compression molding operations but in injection and reciprocating screw principles of molding as well. Molded phenolic parts are usually abrasive and difficult to machine. Although many electrical insulating applications

(A) These billiard balls are molded phenolic.

(B) Molded phenolic parts for humidifier case. (Durez Div., Hooker Chemical Corp.)

(C) Phenolic end panels of pushbutton broiler-oven. (Durez Div., Hooker Chemical Corp.)

Fig. 14-25. Uses of phenolics.

utilize molded phenolics, they possess poor tracking resistance under conditions of high humidity. A few of these applications have been replaced by thermoplastics.

Phenolic resins have a major drawback of being too dark in color for surface layers on decorative laminates and as adhesives where glue joints may show. Cotton fabric, wood, or paper are com-

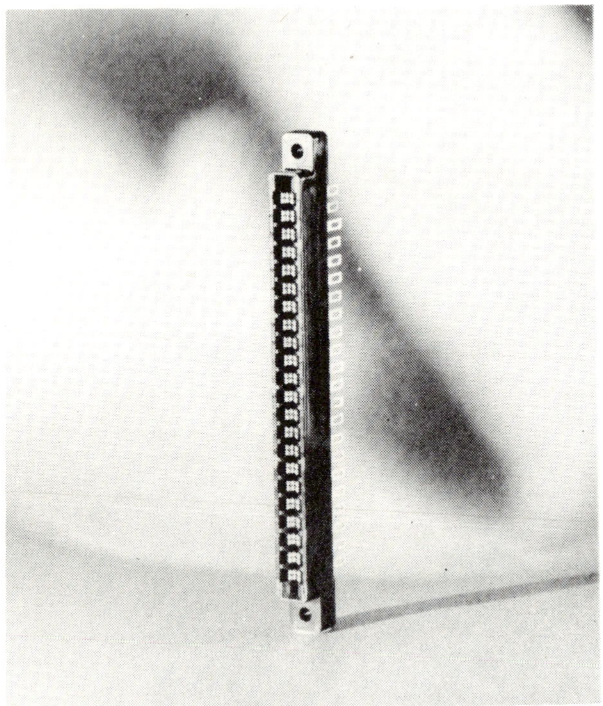

(A) These printed circuit connectors are made of a two-stage phenolic.

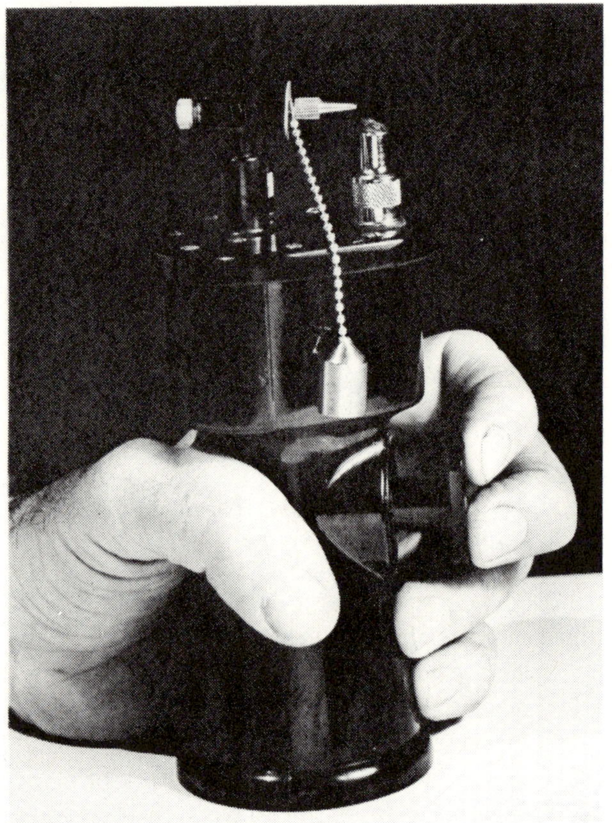

(B) Compact dental torch molded from general-purpose phenolic.

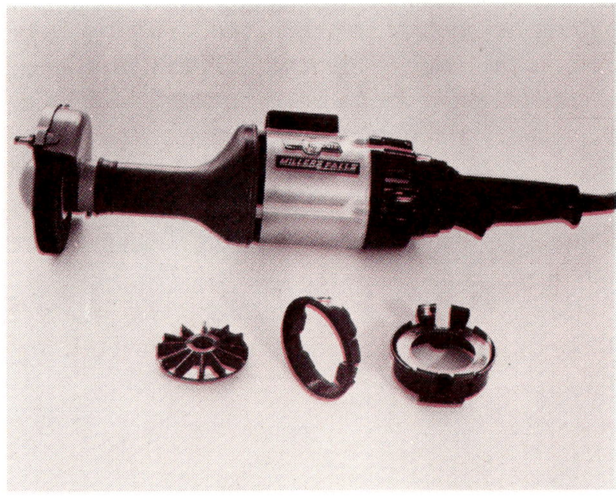

(C) Fan, baffle, and frame insulator of grinder are molded of phenolic.

(D) Phenolic parts of automotive braking systems.

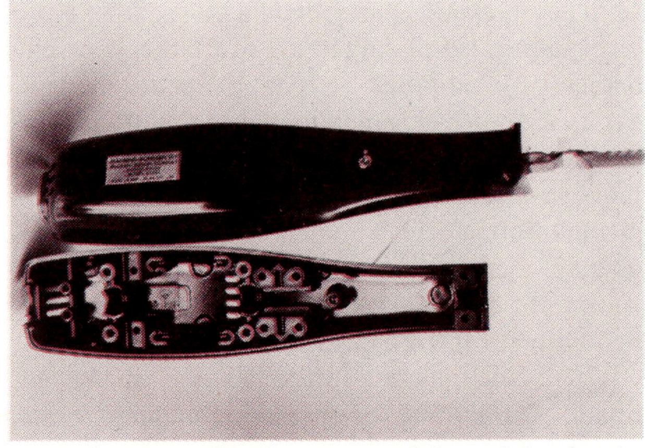

(E) High-impact phenolic compound is used in electric carving knife handle.

Fig. 14-26. Phenolics remain competitive with thermoplastics. (Durez Div., Hooker Chemical Corp.)

monly used laminate materials. Resin impregnated fabric and paper are used in the manufacture of gear wheels, bearings, substrates for electrical circuit boards, and melamine decorative laminates (Fig. 14-27). These laminates are usually produced in large multilayer presses under controlled heat and pressure. Various methods of impregnation are used including dipping, coating, and spreading.

(A) Phenolic impregnated paper and cloth.

(B) Phenolic resin based electrical circuit board.

Fig. 14-27. Phenolic resin impregnated materials.

Phenol formaldehyde resins may be cast and still find applications in various cross sectional shapes, billiard balls, cutlery handles, and novelty items.

Phenolic based resins are available in various physical forms: liquid, powder, flake, or film. The ability of these resins to impregnate and bond with wood and other materials contributes to their success as adhesives. They may be used as additives in nitrile rubber cements to improve adhesion and heat resistance. They are used extensively in the production of plywood and as binder adhesives in wood particle moldings. Wood particle boards are used in many building applications such as heating, subflooring, and core stocks.

Resins are used as binders in the manufacture of abrasive grinding wheels. The desired abrasive grit and resin are simply molded into the required cavity shape and cured. Resin binders are an important ingredient in the application of shell molds and shell cores used in the foundry industry (Fig. 14-28). These molds and cores produce extremely smooth metal castings. As heat-resistant binders, phenolic resins and other ingredients are used in the protection of brake linings and clutch facings.

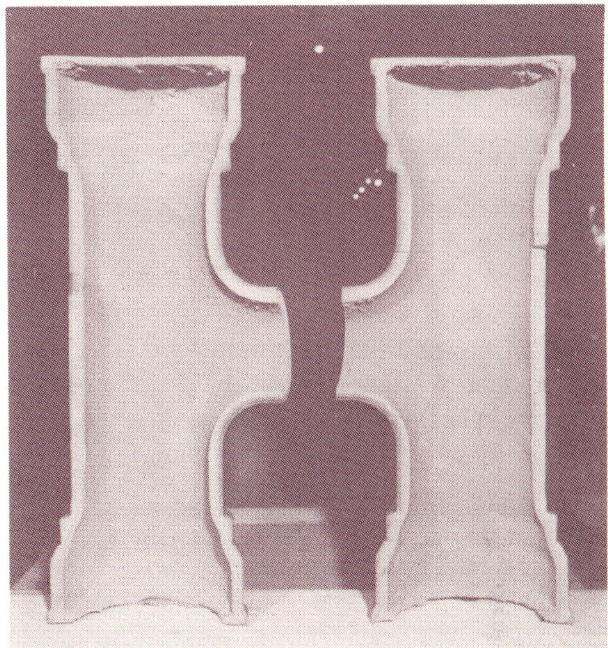

(A) Phenolic resin was used as a binder for this sand foundry core.

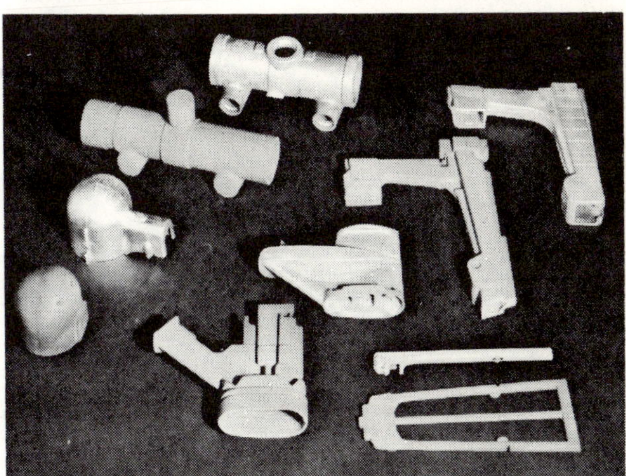

(B) Phenolic resin binder was used for these foundry sand cores. The metallic castings are also shown.

Fig. 14-28. Resin binders in the foundry industry. (Acme Resin Co.)

Because of their high resistance to water, alkali, chemicals, heat, and abrasion, phenolics are sometimes selected for use in finishes. They are useful for coating appliances, machinery, or other applications requiring maximum heat resistance.

A high-strength, heat and fire resistant foam may be produced using phenolic resins. The foam may be produced in the plant or on the site by rapidly mixing a foaming or blowing agent and catalyst with the resin. As the chemical reaction generates heat and begins the polymerization process, the blowing agent vaporizes, causing the resin to expand into a multicellular, semipermeable structure. These foams may be used as fill for honeycomb structures in aircraft, flotation materials, acoustic and thermal insulation, and as packing materials for fragile objects.

Microballoons (small hollow spheres) may be produced of phenolic plastics filled with nitrogen. These spheres vary from 0.0002 to 0.0032 inch in diameter and may be mixed with other resins to produce syntactic foams. They find uses as insulative fillers and as vapor barriers when placed on volatile liquids such as petroleum.

Table 14-6 gives some of the properties of various phenolic materials.

Polyesters (Unsaturated)

The term *polyester resin* encompasses a variety of materials and is often confused with other polyester classifications. By definition a polyester is formed by the reaction of a polybasic acid and a polyhydric alcohol. Modification with acids and/or bases and some unsaturated reactants permit crosslinking to form thermosetting plastics.

Unless otherwise designated, the term "polyester resin" should refer to unsaturated resins based an dibasic acids and dihydric alcohols capable of crosslinking with unsaturated monomers (often styrene). Alkyds and polyurethanes, also of the polyester resin family, are discussed individually.

Sometimes the term "fiber glass" has been used to indicate unsaturated polyester plastics; however, the term should refer to fibrous pieces of glass. Various resins may be used with glass fiber acting as the reinforcing agent. So it is understandable how confusion may occur. The main application for unsaturated polyester resin is in the production of reinforced plastics, and glass fiber is the most favored reinforcement.

Credit for the first preparation of polyester resins (alkyd type) is usually attributed to the

Table 14-6. Phenolic Properties.

Property	Phenol-Formaldehyde (Unfilled)	Phenol-Formaldehyde (Macerated Fabric)	Phenolic Casting (Unfilled) Resin
Molding qualities	Fair	Fair-good	
Specific gravity (density)	1.25-1.30	1.36-1.43	1.236-1.320
Tensile strength (psi)	7000-8000	3000-9000	5000-9000
Compressive strength (psi)	10,000-30,000	15,000-30,000	12,000-15,000
Izod, impact (ft-lb/in)	0.20-0.36	0.75-8	0.24-0.40
Hardness, Rockwell	M124-M128	E79-E82	M93-M120
Thermal expansion (10^{-5}/°C)	2.5-6	1-4	6.8
Resistance to heat (°F)	250	220-250	160
Dielectric strength (volts/mil)	300-400	200-400	250-400
Dielectric constant (60 Hz)	5-6.5	5.2-21	6.5-17.5
Dissipation factor (60 Hz)	0.06-0.10	0.08-0.64	0.10-0.15
Arc resistance (seconds)	Tracks	Tracks	
Water absorption (24 hrs, %)	0.1-0.2	0.40-0.75	0.2-0.4
Burning rate (in/min)	Very slow	Very slow	Very slow
Effect of sunlight	Darkens	Darkens	Color fades
Effect of acids	Decomposed by oxidizing acids	Decomposes by oxidizing acids	None
Effect of alkalies	Decomposes	Attacked	Attacked
Effect of solvents	Resistant	Resistant	Resistant
Machining qualities	Fair to good	Good	Excellent
Optical	Transparent-translucent	Opaque	Transparent-opaque

Swedish chemist Jöns Jacob Berzelius in 1847 and to Gay-Lussac and Pelouze in 1833. Further theories and work were conducted by W. H. Carothers and R. H. Kienle. From that time on through the 1930's most of the work on polyesters was aimed at developing and improving applications for paints and varnishes. Further interest in the resin was stimulated by Carleton Ellis in 1937. He discovered that by adding unsaturated monomers to unsaturated polyesters, crosslinking and polymerization time was greatly reduced. Carleton Ellis has sometimes been called the father of unsaturated polyesters.

Large scale industrial use of unsaturated polyesters developed rapidly as wartime shortages spurred numerous resin applications. Reinforced polyester structures and parts were widely used during World War II.

The word "polyester" is derived from two chemical processing terms: *poly*merization and *ester*fication. In esterfication an organic acid is combined with an alcohol to form an ester and water. A simple esterfication reaction is shown diagrammatically in Fig. 14-29.

R—C—OH + HO—C(=O)—R —ESTERIFICATION→ R—C—O—C(=O)—R + HOH

(ACID) (ESTER) (WATER)

Fig. 14-29. Examples of esterification reaction.

The reverse reaction of esterfication is called *saponification*. In order to obtain a good yield of the ester in a condensation reaction, it is necessary to remove the water to prevent saponification (the reverse reaction) from occurring (see Fig. 14-30).

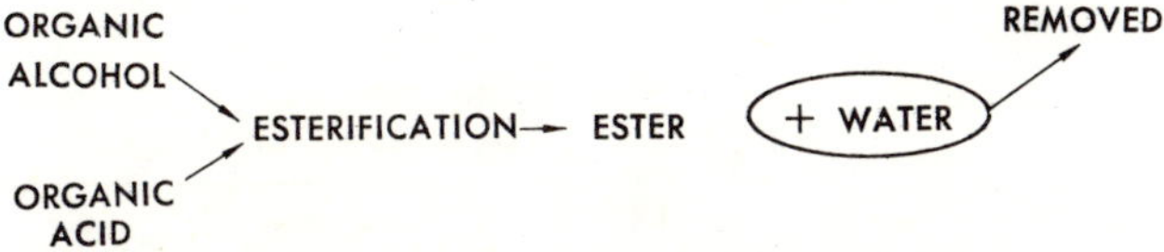

Fig. 14-30. Water should be removed to prevent saponification in an esterification reaction.

Thus, if a polybasic acid such as maleic acid and a polyhydric alcohol such as ethylene glycol are caused to react, and the water is removed as fast as it is formed by heat and condensation, the resulting product will be an unsaturated *polyester*. "Unsaturated" means that the double bonded carbon atoms are reactive or possess unused valance bonds and are avaliable to attach to another atom or molecule. Such a polyester is capable of crosslinkage. There are, however, many other reactive or unsaturated monomers which may be used to modify or tailor the resin to meet a specific purpose or application. Vinyl toluene-chlorostyrene, methyl methacrylate, and diallyl phthalate are commonly used monomers. Unsaturated styrene is an ideal, low-cost monomer most commonly used with polyesters (see Fig. 14-31).

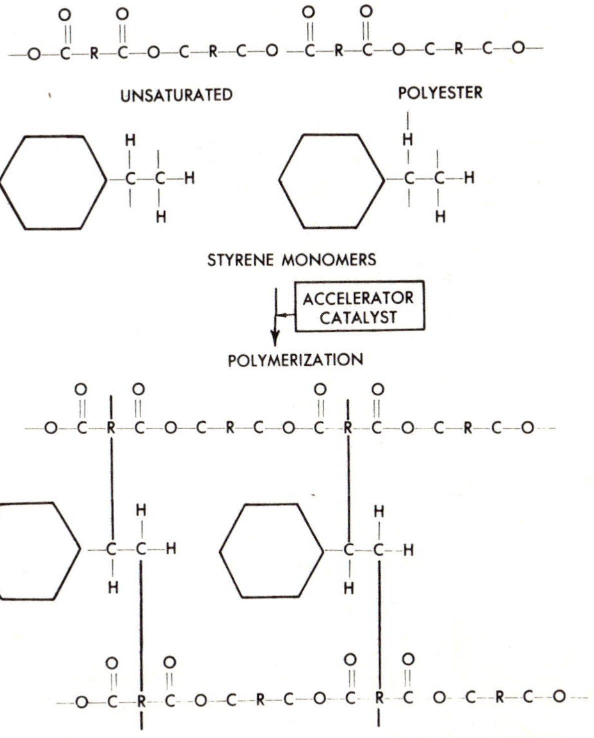

—O—C(=O)—R—C(=O)—O—C—R—C—O—C(=O)—R—C(=O)—O—C—R—C—O—

UNSATURATED POLYESTER

STYRENE MONOMERS

ACCELERATOR CATALYST

POLYMERIZATION

Fig. 14-31. Polymerization reaction with unsaturated polyester and styrene monomers.

The main functions of a monomer are: (1) to act as a solvent carrier for the unsaturated polyester; (2) to lower (thinner) viscosity; (3) to enhance selected properties to meet specific applications; (4) to provide a rapid means of reacting (crosslinking) with the unsaturated linkages in the polyester.

As the molecules randomly collide and occasional bonds are completed, a very slow polymerization (crosslinking) process will occur over a period of days or weeks in simple mixtures of polyesters and monomers.

To accelerate this polymerization reaction at room temperatures, accelerators (promoters) and catalysts (initiators) are added. The accelerators commonly used are cobalt naphthenate, diethyl aniline, and dimethyl aniline. Polyester resins will usually have the accelerator added unless otherwise specified. Resins which contain an accelerator require only the addition of a catalyst to provide rapid polymerization at room temperatures. With the addition of an accelerator the "shelf-life" of the resin is also appreciably shortened. Inhibitors such as hydroquinone or tertiary buty catechol may be added to stabilize or retard premature polymerization while in storage. These additives do not interfere with the final polymerization to any great extent. The speed of cure can be influenced by temperature, light, and the amounts of additives.

Polyester resins may be formulated without accelerators. All resins should be kept in a cool, dark storage area until used. If the accelerator and catalyst are supplied separately, *never* mix them together directly, as a violent explosion may result. Accelerators are ingredients added to the polyester resin to speed up the decomposition of the catalyst and initiate polymerization.

Methyl ethyl ketone peroxide, benzoyl peroxide, and cumene hydroperoxide are three of the most common organic peroxides used to catalyze polyester resins. These catalysts decompose, releasing free radicals, when they come in contact with accelerators in the resin. These free radicals are attracted to the reactive unsaturated molecules and thus initiate the polymerization reaction. By the strictest definition of the word the term "catalyst" is incorrectly used when referring to the polymerization mechanism of polyester resins. A catalyst is a substance which by its mere presence aids a chemical action, without itself being permanently changed. In polyester resins the catalyst decomposes and adds to and becomes a part of the polymer structure. A true catalyst is recoverable at the end of the chemical process. Since these materials are consumed in initiating the polymerization, the term *initiator* might be more accurate.

Exposure to radiation, ultraviolet light, and heat has also been used to initiate the polymerization of the double bonded molecules. If catalysts are used, the resin mix becomes correspondingly more sensitive to heat and light. Consequently, on a hot day or in the sunlight, less catalyst is required for polymerization. On a cold day more catalyst would be needed or the resin and catalyst warmed to produce a rapid cure.

The final curing reaction is called *addition* polymerization because no by-products are present as a result of the reaction. In phenol-formaldehyde reactions the curing reaction is called *condensation polymerization* because there is a water by-product present.

The properties of polyester may be specially formulated for a wide variety of uses by alteration of the chemical structure or by adding selected additives. With higher percentages of unsaturated acid, more crosslinkage is possible and thus a stiffer, harder product. The addition of saturated acids will increase toughness and flexibility. Fillers including thixotropic types, pigments, and lubricants may be added to the resin.

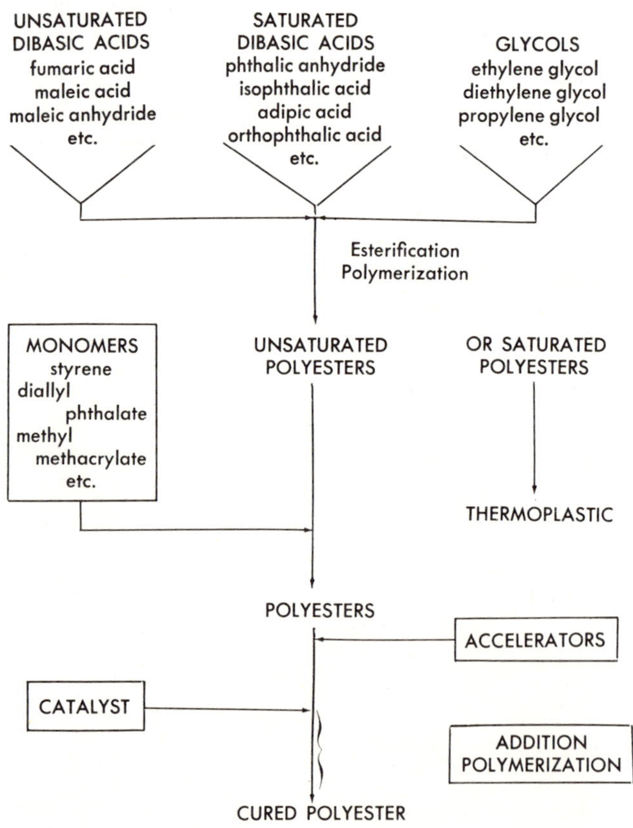

Fig. 14-32. Production scheme of cured polyester.

Saturated polyesters, which are based on the reaction of terephthalic acid ($C_6H_4(COOH)_2$) and ethylene glycol ($(CH_2)_2(OH)_2$), are linear, high-molecular-weight polymers used principally in the field of fibers and film production. It was with linear polyesters that W. H. Carothers did his basic research while attempting to produce polyester textile fibers. After several years of research, Carothers abandoned his attempt to produce polyester fibers and began investigating synthetic polyamides.

Saturated or unreactive polyesters do not undergo any crosslinking. Consequently, these linear polyesters are thermoplastic. Clothing and draperies are common application of these fibers. Industrial applications may include reinforcements for conveyor belting or cords for tires (Fig. 14-33).

Polyester films are used as recording tapes, dielectric insulators, photographic film, and boil-in-the-bag food products (Fig. 14-34).

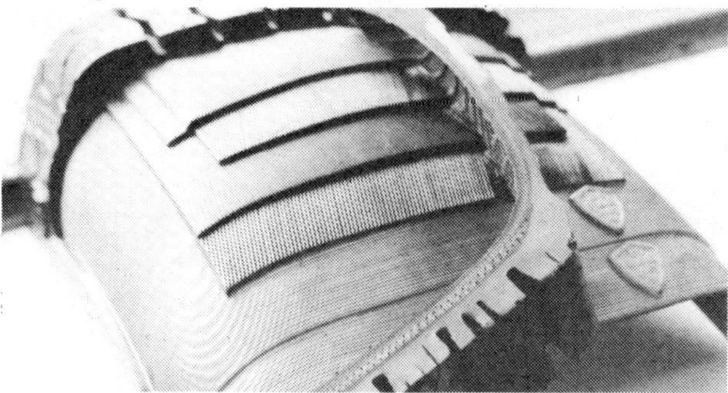

Fig. 14-33. Polyester cords are used in tire construction.

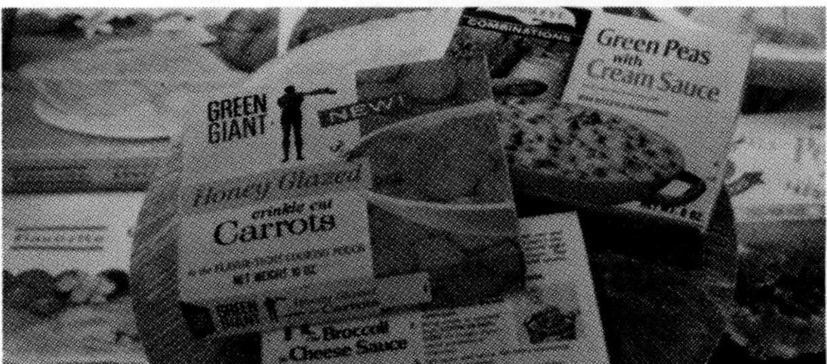

(A) Carrots, peas, and broccoli are packaged in films designed for boil-in-the-bag application.

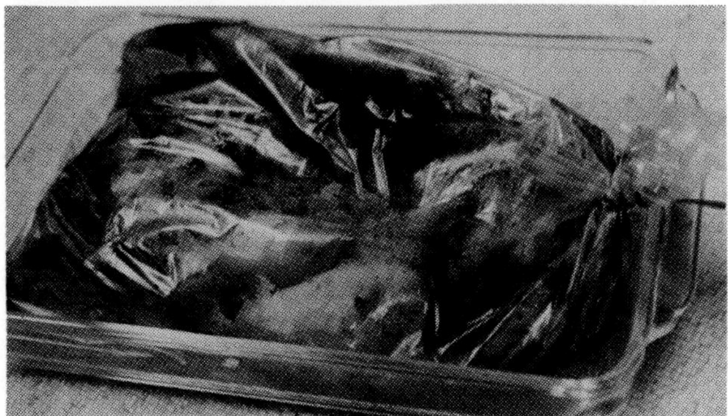

(B) Polyester bake-in-the-bag retains moisture and flavor, with no mess to clean from pans and dishes.

Fig. 14-34. Uses of polyester films.

Because of their thermoplastic nature, compounds based on saturated polyethylene terephthalate may be injection or extrusion molded (see Fig. 14-35).

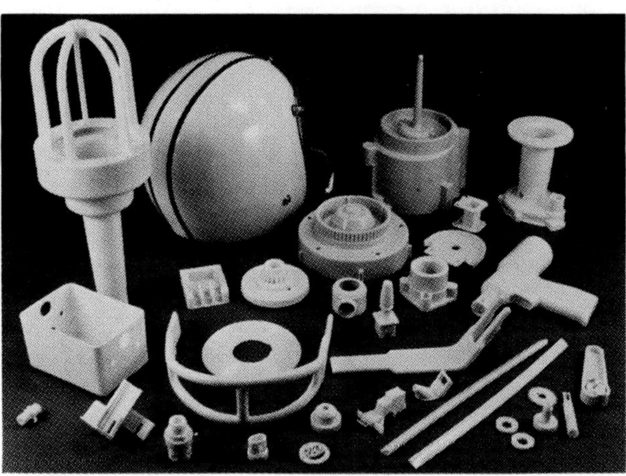

(A) Gears, bearings, machine housings, safety helmets, and other products.

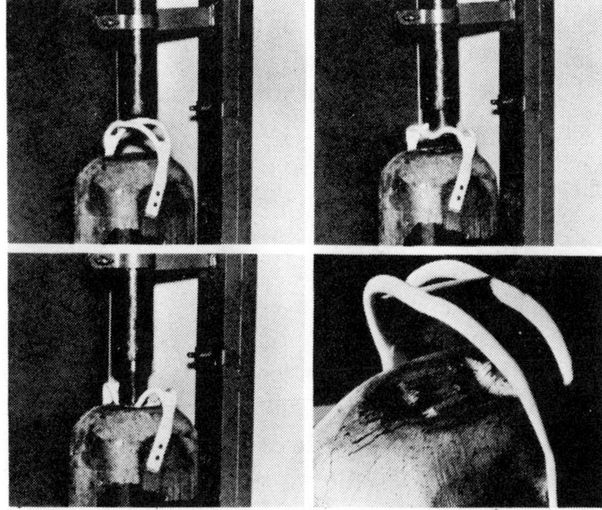

(B) Eight-pound weight, dropped 15 feet, onto football helmet face mask.

Fig. 14-35. Application of polterephthalate. (Eastman Chemical Products, Inc.).

Polyester resins which contain no wax as an ingredient are invariably susceptible to air inhibition. Where the resin is exposed directly to air it is likely to remain undercured, soft, and tacky for some time after setting. This is desirable when multiple layers are to be built up. Resins purchased from the manufacturer without wax are referred to as *air-inhibited* resins. The absence of wax permits better bonds between multiple layers of material in hand layup operations. In some cases a tackfree cure of air exposed surfaces is desired. For one step castings, moldings, or surface coats a non-air-inhibited polyester will produce a tackfree cure. Non-air-inhibited resins contain a wax which floats to the surface during the curing operation, blocking out the air and allowing the surface to cure tack-free. Numerous wax solutions may be successfully used, including household paraffin, carnauba wax, beeswax, steric acid, and others. Adhesion is adversely affected by the use of these waxes. If additional layers are to be added on a non-air-inhibited resin, all wax must be removed from the surface by sanding.

By altering the basic combination of raw materials, fillers, reinforcement, cure time, and processing technique it is possible to obtain unsaturated polyester resins which exhibit a wide range of properties.

Polyester finds its main use in the production of reinforced products. The principle virtue of reinforcement is the high strength-to-weight ratio (see Fig. 14-36). Glass fiber is the most common reinforcing agent. However, asbestos, sisal, various plastics fibers, and exotic whisker filaments are also used. The type of reinforcement selected depends on the end use and the method of fabrication. Reinforced polyesters are among the strongest materials known. They have been successfully used in the manufacture of automobile bodies and in boat hulls over 100 feet in length. Because of their high strength-to-weight ratio they are extensively used in various aircraft and aerospace applications. Other applications utilizing the outstanding properties of reinforced polyesters include radar domes, ducts, storage tanks, sports equipment, trays, furniture, luggage, sinks, and many different kinds of ornaments.

Unreinforced casting grades of polyesters are used for embedding, potting, casting, and sealing. Casting resins filled with wood flour may be cast in silicone molds to reproduce excellent replicas of wood carvings and trim. Special resins may be emulsified with water and further reduce the basic casting costs. These resins are referred to as "water extended resins" because they may contain up to 70 percent water. The castings do undergo some shrinkage due to water loss.

(A) There are 640 panels of fiber glass reinforced plastics in the roof of this pavilion. (Hooker Chemical Corp.)

(B) Polyester glass reinforced welding helmet.

Fig. 14-36. Polyester fiber glass reinforced products have high strength-to-weight ratio.

Fabrication methods for reinforced polyesters include hand layup, sprayup, matched molding, premix molding, pressure-bag molding, vacuum-bag molding, casting, and continuous laminating. Numerous other molding modifications are sometimes used. Compression molding equipment is sometimes utilized when doughlike premixes containing all ingredients are used.

Table 14-7 gives some of the properties of various polyesters.

Polyurethane

The term "polyurethane" refers to the reaction of polyisocyanates (–NCO–) and polyhydroxyl (–OH–) groups. A simple reaction of isocyanate and an alcohol is shown below. The reaction product is urethane, not a polyurethane.

$$\underset{\text{(poly) isocyanate}}{R \cdot NCO} + HOR_1 \xrightarrow{\text{(poly) hydroxyl}} R \cdot NH \cdot COOR_1 \text{ (poly) urethane}$$

Table 14-7. Polyester Properties.

Property	Thermoplastic Polyester	Thermosetting Polyester (Cast)	Thermosetting Polyester (Glass Cloth)
Molding qualities	Good	Excellent	Excellent
Specific gravity (density)	1.37-1.38	1.10-1.46	1.50-2.10
Tensile strength (psi)	10,400	6000-13,000	30,000-50,000
Compressive strength (psi)	18,600	13,000-36,500	25,000-50,000
Izod, impact (ft-lb/in)	0.8	0.2-0.4	5.0-30.0
Hardness, Rockwell	R120, M94	M70-M115	M80-M120
Thermal expansion ($10^{-5}/°C$)	6	5.5-10	1.5-3
Resistance to heat (°F)	175-250	250	300-350
Dielectric strength (volts/mil)	350-400	380-500	350-500
Dielectric constant (60 Hz)	3.65	3.0-4.36	4.1-5.5
Dissipation factor (60 Hz)	0.0055	0.003-0.028	0.01-0.04
Arc resistance (seconds)	40-120	125	60-120
Water absorption (24 hrs, %)	0.02	0.15-0.60	0.05-0.50
Burning rate (in/min)	Slow burning	Burns—self-exting.	Burns—self-exting.
Effect of sunlight	Discolors slightly	Yellows slightly	Slight
Effect of alkalies	Attacked by oxidizing acids	Attacked by oxidizing acids	Attacked by oxidizing acids
	Attacked	Attacked	Attacked
Effect of solvents	Attacked by halogen hydrocarbons	Attacked by some	Attacked by some
Machining qualities	Excellent	Good	Good
Optical	Transparent-opaque	Transparent-opaque	Translucent-opaque

German chemists Wurtz, in 1848, and Hentschel, in 1884, produced the first isocyanates which later lead to the development of polyurethanes. It was Dr. Otto Bayer and his coworkers who actually instigated the commercial development of polyurethanes in 1937. Since that time polyurethanes have developed into a variety of commercially available forms including coatings, elastomers, adhesives, molding compounds, foams, and fibers.

It is because the isocyanates and di-isocyanates are so highly reactive and react with compounds containing reactive hydrogen atoms that various polyurethane polymers may be produced. The recurring link of the polyurethane chain is NHCOO or NHCO.
$$\overset{\displaystyle O}{\underset{}{\|}}$$

HO(CH_2)$_4$OH + [ring with CH_3; OCN NCO] —— ··O(CH_2)$_4$O·CONH [ring with CH_3] NHCO·O(CH_2)$_4$O . . .

Dihydric Alcohol — Toluene Di-isocyanate — Linear Polyurethane

More complex polyurethanes have been developed based on toluene di-isocyanates (TDI) and polyester, diamine, castor oil, or polether chains. Other isocyanates used are diphenylmethane di-isocyanate (MDI) and polymethylene polyphenyl isocyanate (PAPI).

The first polyurethanes were produced in Germany and were to be competitive with other polymers produced at that time. These plastics were linear aliphatic polyurethanes used to make fibers. The linear polyurethanes are thermoplastic and may be processed by all conventional thermoplastic processing techniques including injection and extrusion. Because of their relative cost when compared with other plastics having comparable properties, they find only limited use as fibers or filaments.

Polyurethane coatings are noted for their high abrasion resistance, unusual toughness, hardness, good flexibility, chemical resistance, and weatherability (Fig. 14-37). The ASTM has designated polyurethane coatings into five distinct types as shown in Table 14-8.

These resins are used as clear or pigmented finishes for home, industrial, or marine use. They are used to improve the chemical and ozone resistance of rubber and other polymers. These coatings and finishes may be simple solutions of linear polyurethanes or complex systems of polyisocyanate and such OH groups as polyesters, polyethers, or castor oil.

Table 14-8. ASTM Designations for Polyurethane Coatings.

ASTM Type	Components	Pot Life	Cure	Clear or Pigmented Uses
(I) Oil-modified	One	Unlimited	Air	Interior or exterior wood and marine. Industrial enamels
(II) Prepolymer	One	Extended	Moisture	Interior or exterior. Wood, rubber and leather coatings
(III) Blocked	One	Unlimited	Heat	Wire coatings and baked finishes
(IV) Prepolymer + Catalyst	Two	Limited	Amine/ Catalyst	Industrial finishes and leather, rubber products
(V) Polyisocyanate + Polyol	Two	Limited	Air NCO/OH reaction	Industrial finishes and leather, rubber products

(A) Foams, insulation, sponges, belts, and gaskets.

(B) This man is showing how a painting might typically be packed inside fiber glass reinforced plastics and urethane shipping crate. (Poly-Con Industries, Inc.)

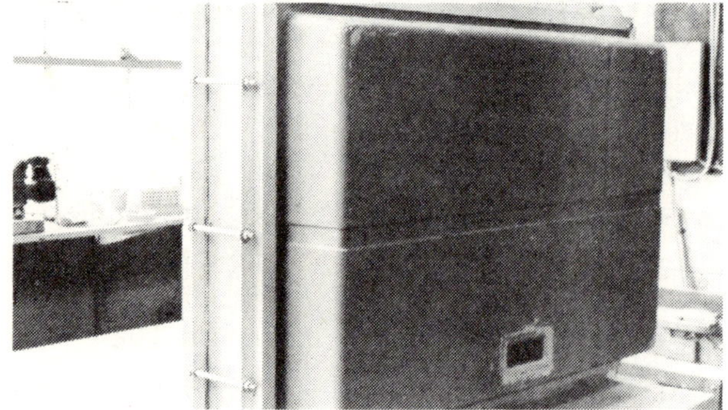

(C) The lightweight self-locking container is shockproof as well as fire retardant and moistureproof. (Poly-Con Industries, Inc.)

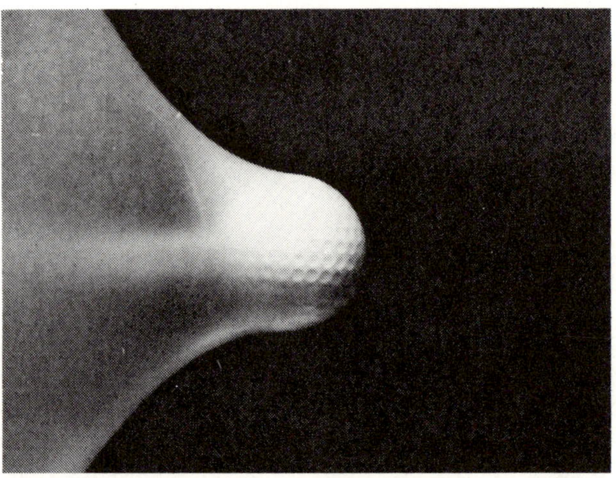

(D) Puncture resistance and tensile strength of polyurethane is demonstrated as though 0.003-in film stops a hard-driven golf ball. (B.F. Goodrich Chemical Co.)

Fig. 14-37. Various applications of polyurethane.

Numerous polyurethane elastomers (rubbers) may be prepared from di-isocyanates, linear polyesters or polyether resin and curing agents (Fig. 14-38). If formulated into a linear thermoplastic urethane, they may be processed by conventional thermoplastic processing equipment and find applications as shock absorbers, bumpers, gears, cable cover, hose jacketing, elastic thread (Spandex), and diaphragms. The more common applications of crosslinked thermosetting elastomers include industrial tires, shoe heels, gaskets, seals, O-rings, pump impellers, and tread stocks for passenger and truck tires. Polyurethane elastomers are especially noted for extreme resistance to abrasion, ozone aging, and hydrocarbon fluids; however, polyurethane elastomers cost more than conventional rubbers. These elastomers are tough, elastic, and demonstrate a wide range of flexibility at temperature extremes.

$$nHOR—(OR—)_x—OH + nOCNR_1NCO \rightarrow (—OR—(OR)_xOCONHR_1NHCO—)_n$$

Polyester *Di-isocyanate* *Polyurethane*

$$nHOR(—OCOR_2CO \cdot OR—)_xOH + nOCNR_1NCO \rightarrow$$

Polyether *Di-isocyanate*

$$—OR(—OCOR_2CO \cdot OR)_xOCONHR_1NHCO—_n$$

Polyurethane

Fig. 14-38. Production of polyurethane elastomers.

Polyurethane foams are widely used and well-known materials. They are available in flexible, semirigid, and rigid forms with a wide variety of densities. Various-density flexible foams used as cushioning for furniture, automobile seating, and mattresses are produced by reacting toluene di-isocyanate (TDI) with polyester and water in the presence of catalysts. At higher densities they are cast or molded into drawer fronts, doors, moldings, and complete pieces of furniture. Flexible foams are open-celled structures. Consequently, they may be used as artificial sponges. Flexible foams used by the garment and textile industry for backing and insulation are produced by reacting TDI with polyester and catalysts. (See the carpet backing in Fig. 14-39.) Semirigid foams find use as energy absorbing materials in automobile crash pads, arm rests, and sun visors.

Fig. 14-39. Polyurethane foam backing is shown on this carpet.

The three largest uses of rigid polyurethane foam are in the manufacture of furniture, automotive and construction moldings, and various thermal insulation applications. Replicas of wood carvings, decorative parts, or moldings are produced from high-density, self-skinning foams. The excellent insulating value of these foams makes them an ideal choice for insulation of refrigerators, refrigerated trucks, and railroad cars. They may be foamed in place for various architectural applications, or placed on vertical surfaces by spraying the reaction mixture through a spray nozzle. They have found use as flotation devices, packing, and structural reinforcements.

Rigid polyurethane is a closed cellular material produced by the reaction of TDI (prepolymer form) with polyethers and reactive blowing agents such as monofluorotrichloromethane (fluorocarbon). Dipehenylmethane di-isocyanate (MDI) and polymethylene polyphenyl isocyanate (PAPI) are also used in the formulation of some rigid foams (Fig. 14-40). MDI foams have better dimensional stability while PAPI foams have high temperature resistance.

(A) Polyurethane closed-cell insulation protects 10-ton cargo container in test.

(B) TV cabinet made of fire retardant polyurethane foam.

Fig. 14-40. Uses of polyurethane foams. (Hooker Chemical Corp.)

Polyurethane based caulks and sealants are inexpensive polyisocyanate materials used for encapsulation and construction manufacturing applications. Various polyisocyanates are also important adhesives. They produce strong bonds between flexible fabrics, rubbers, foams or other materials.

Many blowing or foaming agents are explosive and toxic. When mixing or processing polyurethane foams, one should make certain that proper ventilation is provided.

Table 14-9 gives some properties of urethane plastics.

Table 14-9. Polyurethane Properties.

Property	Cast Urethane	Urethane Elastomer
Molding qualities	Good	Good-excellent
Specific gravity (density)	1.10-1.50	1.11-1.25
Tensile strength (psi)	175-10,000	4500-8400
Compressive strength (psi)	2000	2000
Izod, impact (ft-lb/in)	5 to flexible	Does not break
Hardness, Rockwell	10A-90D (Shore)	30A-70D (Shore) M28, R60
Thermal expansion (10^{-5}/°C)	10-20	10-20
Resistance to heat (°F)	190-250	190
Dielectric strength (volts/mil)	400-500	330-900
Dielectric constant (60 Hz)	4-7.5	5.4-7.6
Dissipation factor (60 Hz)	0.015-0.017	0.015-0.048
Arc resistance (seconds)	0.1-0.6	0.22
Water absorption (24 hrs, %)	0.02-1.5	0.7-0.9
Burning rate (in/min)	Slow to self-exting.	Slow to self-exting.
Effect of sunlight	None to yellows	None to yellows
Effect of acids	Attacked	Dissolves
Effect of alkalies	Slight—attacked	Dissolves
Effect of solvents	None to slight	Resistant
Machining qualities	Excellent	Fair—excellent
Optical	Transparent—opaque	Transparent-opaque

Silicones

In the branch of science known as organic chemistry, carbon is studied because of its capacity to form molecular structures with many other elements. Carbon compounds are considered reactive elements because they are capable of entering more molecular combinations than other elements. Life on earth is based on the element carbon.

The second most abundant element on earth is silicon. It has the same number of available bonding sites as carbon. Some scientists have speculated that life on other planets may be based on the element silicon. This possibility seems difficult to believe because silicon is an inorganic solid with a metallic appearance. Most of the earth's crust is composed of SiO_2 (silicon dioxide) in the form of sand, quartz, and flint.

The tetravalent capacity of silicon interested chemists as early as 1863. Friedrich Wohler, C. M. Crafts, Charles Friedel, F. S. Kipping, W. H. Carothers, and many other contributors lead to the development of silicone polymers.

By 1943 the newly formed Dow Corning Corporation was producing the first commercially produced silicone polymers in the United States. Today there are thousands of uses for silicone polymers. The word "silicone" should be applied to polymers containing silicon-oxygen-silicon bonding; however, it is used to denote any polymer containing silicon atoms.

In many carbon-hydrogen compounds, silicon may replace the element carbon. Methane (CH_4) may be changed to silane or silico methane (SiH_4). Many structures similar to the aliphatic series of saturated hydrocarbons may be formed.

The following general types of bonds may be of value for future understanding of the formation of silicone polymers:

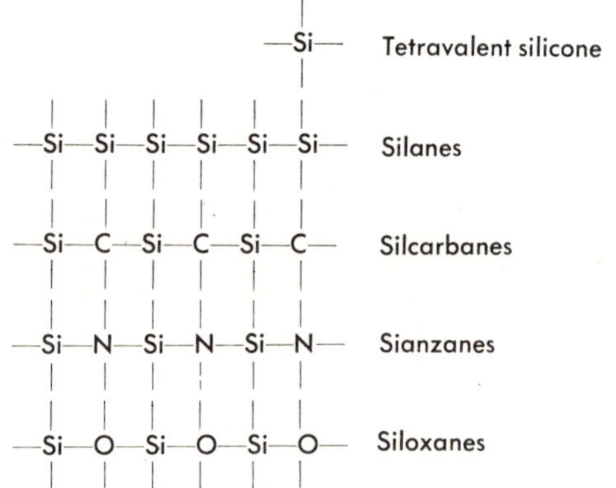

Compounds with only silicon or hydrogen atoms present are called "silanes." When the silicon is separated by carbon atoms, the structure is called "silcarbanes" (sil-*carb*-anes).

A *polysiloxane* is produced when more than one oxygen atom separates the silicon chain.

—Si—O—O—O—Si—O—O—O—

A polymerized silicone molecular chain could be based on the structure shown in Fig. 14-41 and be modified by radical (R) or organic groups.

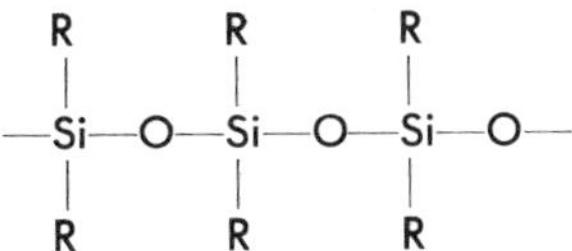

Fig. 14-41. Example of polymerized silicone molecular chain.

Many silicone polymers are based on chains, rings, or networks of alternating silicon and oxygen atoms. Common ones contain methyl, phenyl, or vinyl bonds on the siloxane chain (Fig. 14-42).

CH_3 CH_3 CH_3
—Si—O—Si—O—Si—O—
CH_3 CH_3 CH_3

(A) Based on methyl (CH_3) radical.

C_6H_5 C_6H_5 C_6H_5
—Si—O—Si—O—Si—O—
C_6H_5 C_6H_5 C_6H_5

(B) Based on phenyl (C_6H_5) radical.

Fig. 14-42. Two siloxane polymers.

A variety of polymer compounds are formed by varying the organic radical groups of the "silicon" chain. Numerous copolymers are also available.

Because of the amount of energy required to produce a pound of silicone plastics, the price per pound is high. Depending on the type and grade of resin, silicone plastics may cost from two to five dollars per pound. With improved methods of manufacture and greater volume of use, this price may drop. Silicone plastics may still be the most economical material considering longer product life, higher service temperatures, and flexibility at temperature extremes.

Silicones are produced in five commercially available categories: fluids, compounds, lubricants, resins, and elastomers (rubber).

Probably the best known silicone plastics is associated with oils and ingredients for polishes. The layman may associate lens-cleaning tissues or water repellent fabrics as being treated with a thin "film" coating of silicone.

Silicone *fluids* are added to various liquids to prevent foaming (antifoaming), prevent transmission of vibrations (damping), and improve electrical and thermal limits of various liquids. These fluid silicones are used as additives in paints, oils, inks, mold releasing mixtures, finishes for glass, fabrics, and coatings for paper.

Silicone *compounds* are usually granular or fibrous filled materials. Because of their outstanding electrical and thermal properties, mineral- and glass-filled silicone compounds are used for encapsulation of electronic components. Fig. 14-43 illustrates such uses of silicone compounds.

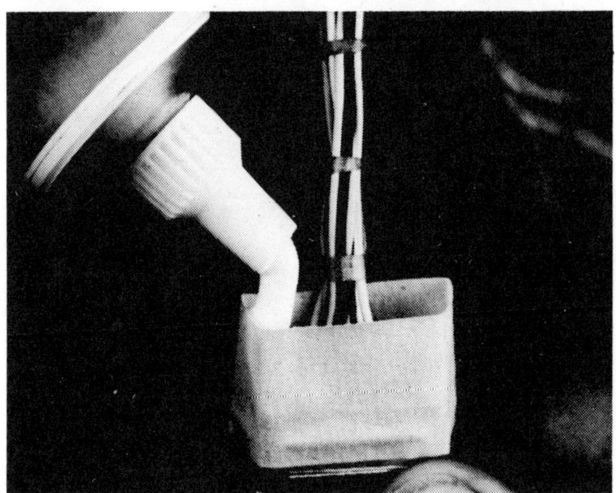

(A) Potting of small electrical component with silicone compound.

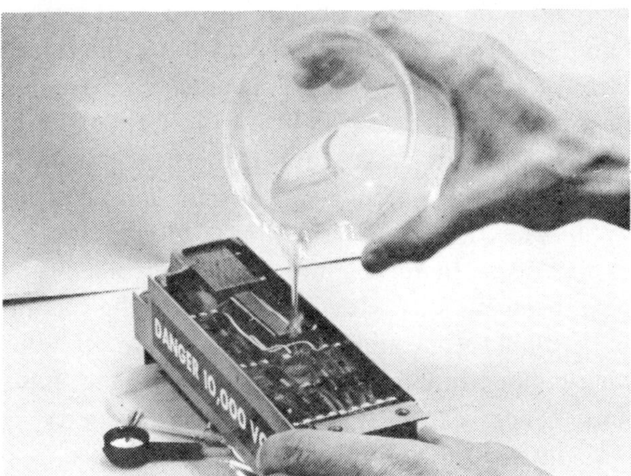

(B) Potting of electronic components with silicone casting resin.

Fig. 14-43. Use of silicone compounds. (Dow Corning Corp.)

As adhesives and sealants, silicones are limited primarily by their high cost. Their high service temperature and elastic properties, however, make them useful for sealing, gasketing, caulking, encapsulating and repairing all types of materials (Fig. 14-44).

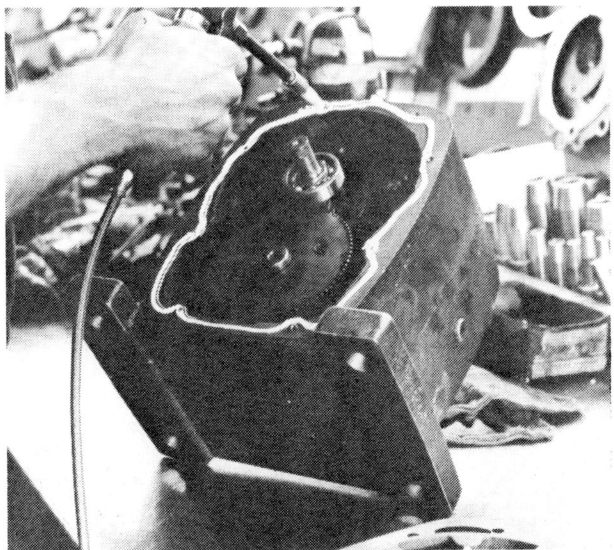

Fig. 14-44. Silicone sealants are used for this gear box. (Dow Corning Corp.)

The chemical inertness of foamed silicone is useful in breast and facial implants in plastic surgery. Its main applications include electrical and thermal insulation of electrical wires and electrical components.

As *lubricants,* silicones have been prized for their nondeteriorating qualities at extreme service temperatures. Lithium soaps and other fillers are often added to provide compounds with service temperatures from $-100°F$ to $+400°F$. Silicones are used for lubricating rubber, plastics, ball bearings, and as valve-sealing and vacuum pump grease.

Silicone *resins* are used for a broad class of applications. In the picture below, silicone resins have been used as releasing agents for the baking of bread. These flexible, tough coatings are used in high-temperature paints for manifolds and mufflers. Their excellent waterproofing property is used in treating masonry and concrete walls.

Their excellent waterproofing, thermal, and electrical properties make silicone resins valuable material for electrical insulation in motors and generators.

Glass cloth reinforced laminates find convenient uses for structural parts, ducts, radomes, and electronic panel boards. These silicone laminates are characterized by their excellent dielectric and thermal properties and strength-to-weight ratio.

Diatomaceous earth, glass fiber, or asbestos may be used as fillers in preparing a premix or putty for the molding of small parts.

Some of the best known silicones are in the form of *elastomers.* Few industrial rubbers or elastomers can withstand prolonged exposure to ozone (O_3) or hot mineral oils. Silicone "rubbers" are stable at elevated temperatures and remain flexible when exposed to ozone or oils.

Silicone elastomers find use as artificial organs, O-rings, gaskets, diaphragms and as flexible molds for casting of plastics and low-melting-point metals (Fig. 14-45).

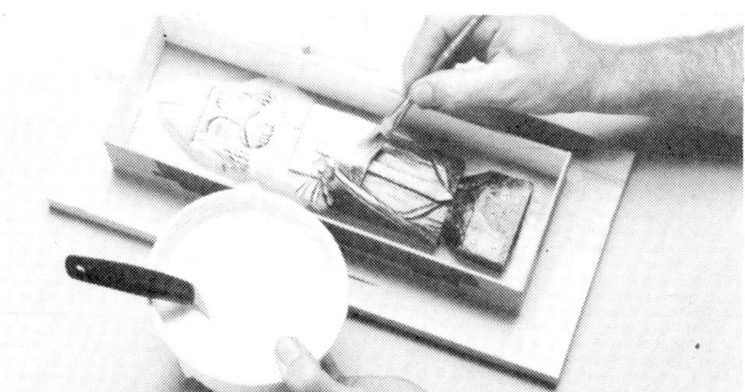

(A) After thoroughly mixing the silicone mixture, brush a thin layer on the prepared part, being careful to avoid air bubbles.

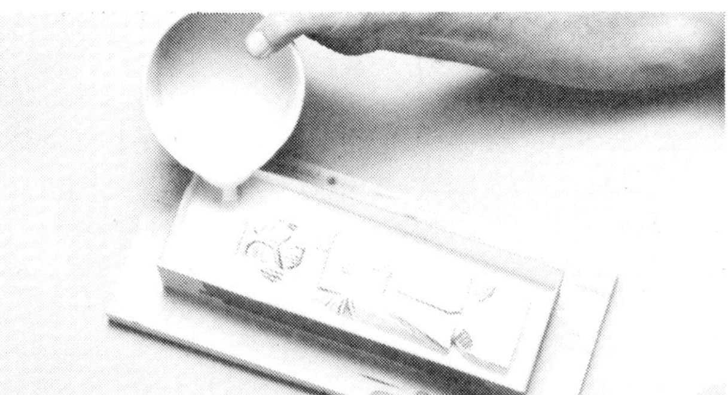

(B) Pour the remainder of the mixture into mold until at least ½ inch is covering model. Cure for 24 hours, then mold may be removed.

Fig. 14-45. Use of silicone elastomer as mold. (Dow Corning Corp.)

Room-temperature vulcanizing (RTV) elastomers are used to duplicate intricate molded parts, seal joints, and adhere parts (Fig. 14-46).

(A) The decorative details of this frame were captured in a mold made of silicone RTV material.

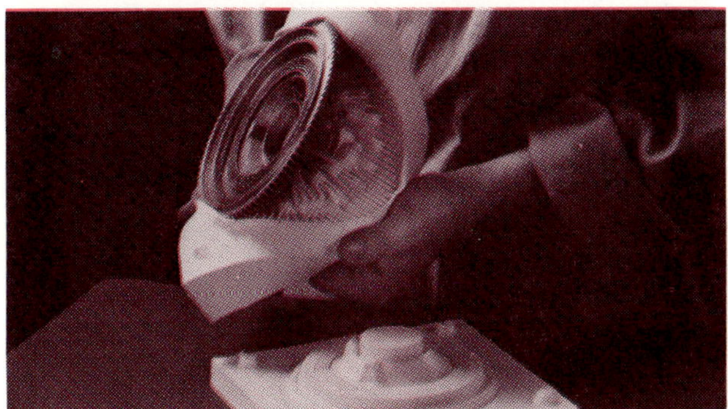

(B) A metal pattern is being removed from this two-part silicone mold.

(C) Silicone RTV reproduces fine details and permits severe undercuts.

Fig. 14-46. Use of silicone RTV in molding. (Dow Corning Corp.)

"Silly Putty" and "Crazy Clay" are two nationally known novelty silicone products (Fig. 14-47). This "bouncing putty" is a silicone elastomer used for damping noise and as sealing and filling compounds. A hard type of bouncing putty will rebound to 80 percent of the height from which it is dropped. Other "super" rebounding novelty items are produced from this compound.

Silicone molding compounds may be processed in the same way as other thermosetting organic

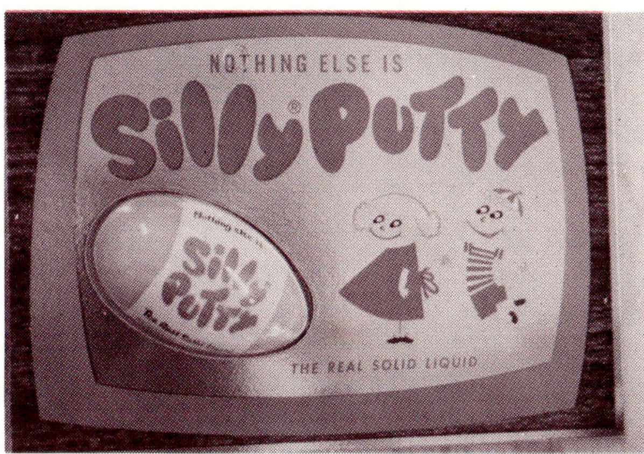

Fig. 14-47. Silicone compound used as a novelty toy.

plastics. Much of the silicone is utilized in resin form as castings, coatings, adhesives, or laminating compounds.

Research and development is being carried out utilizing various elements with covalent bonding capacity. Boron, aluminum, titanium, tin, lead, nitrogen, phosphorus, arsenic, sulphur, and selenium may be considered for inorganic or semi-inorganic plastics. The formulas in Fig. 14-48

$$\left(-\underset{C_4H_9}{N}-\overset{C_6H_5}{B}-\underset{C_4H_9}{N}-\right)_n$$

(A) Boron (monomer).

$$C_2H_5-\overset{C_2H_5}{\underset{C_2H_5}{Si}}-O-\underset{\underset{C_2H_5-\underset{C_6H_5}{Si}-C_2H_5}{|}}{\underset{O}{Al}}-O-\overset{C_2H_5}{\underset{C_2H_5}{Si}}-C_2H_5$$

(B) Aluminum.

$$\left(-\overset{C_2H_5}{\underset{C_6H_5}{Sn}}-O-\right)_n$$

(C) Tin (monomer).

$$\left(-\overset{O}{\underset{O}{\overset{\|}{\underset{\|}{S}}}}-\right)_n$$

(D) Sulphur (monomer).

$$\left(-\overset{C_6H_5}{\underset{C_6H_5}{Pb}}-O-\right)_n$$

(E) Lead (monomer).

$$-O-\overset{R}{\underset{R}{Si}}-O-\overset{OC_4H_9}{\underset{OC_4H_9}{Ti}}-O-$$

(F) Titanium.

$$\left(-N=\overset{Cl}{\underset{Cl}{P}}-\right)_n$$

(G) Phosphorous (monomer).

$$\left(-\overset{O}{\underset{O}{\overset{\|}{\underset{\|}{Se}}}}-\right)_n$$

(H) Selenium (monomer).

Fig. 14-48. Possible chemical structures of inorganic or semi-inorganic plastics.

indicate a few of the many possible chemical structures of inorganic or semi-inorganic plastics.

Table 14-10 gives some of the properties of silicone plastics.

Table 14-10. Silicone Properties.

Property	Cast Resin (Including RTV)	Molding Compounds (Mineral Filler)	Molding Compounds (Glass Filler)
Molding qualities	Excellent	Excellent	Good
Specific gravity (density)	0.99-1.50	1.7-2	1.68-2
Tensile strength (psi)	350-1000	4000-6000	4000-6500
Compressive strength (psi)	100	13,000-18,000	10,000-15,000
Izod, impact (ft-lb/in)		0.26-0.36	3-15
Hardness, Rockwell	15-65 (Shore A)	M71-M95	M84
Thermal expansion (10^{-5}/°C)	8-30	2-4	0.24-0.30
Resistance to heat (°F)	500	600	600
Dielectric strength (volts/mil)	550	200-400	200-400
Dielectric constant (60 Hz)	2.75-4.20	3.5-3.6	3.3-5.2
Dissipation factor (60 Hz)	0.001-0.025	0.004-0.005	0.004-0.030
Arc resistance (seconds)	115-130	250-420	150-250
Water absorption (24 hrs, %)	0.12 (7 days)	0.08-0.13	0.1-0.2
Burning rate (in/min)	Self-exting.	None to slow	None to slow
Effect of sunlight	None	None to slight	None to slight
Effect of acids	Slight to severe	Slight	Slight
Effect of alkalies	Moderate to severe	Slight to marked	Slight to marked
Effect of solvents	Transparent-opaque	Attacked by some	Attacked by some
Optical	Swells in some	Opaque	Opaque
Machining qualities		Fair	Fair

Glossary of New Terms

Alkyd—Polyester resins made with some fatty acid as a modifier.

Amino—Indicates the presence of an NH_2 or NH group. Materials with these groups.

Casein—A protein material precipitated from skimmed milk by the action of either rennin or dliute acid. Rennet casein is made into plastics.

Condensation—A chemical reaction in which two or more molecules combine with the separation of water or some other simple substance. If a polymer is formed the condensation process is called *polycondensation*.

Epoxy—Material based on ethylene oxide, its derivatives or homologs. Epoxy resins form straight-chain thermoplastics and thermosetting resins.

Formalin—A commercial 40 percent solution of formaldehyde in water.

Isocyanate resins—Resins synthesized from isocyanates and alcohols. Most applications are based on their combination with polyols. *See* Polyesters, Polyethers, and Polyurethanes.

Novolac—A phenolic-aldehyde resin which, unless a source of methylene groups is added, remains permanently thermoplastic.

Phenolic—A synthetic resin produced by the condensation of an aromatic alcohol with an aldehyde, particularly of phenol with formaldehyde.

Polyester—A resin formed by the reaction between a dibasic acid and a dihydroxy alcohol, both organic. Modification with multifunctional acids and/or bases and some unsaturated reactants permit crosslinking to thermosetting resins. Polyesters modified with fatty acids are called *alkyds*.

Polyurethane—A family of resins produced by reacting di-isocyanate with organic compounds containing two or more active hydrogens to form polymers having free isocyanate groups. These groups, under the influence of heat or certain catalysts, will react with each other, or with water, glycols, etc. to form a thermosetting material.

Potting—An embedding process for parts that are assembled in a container or can into which the insulating material is poured.

Rennin—An enzyme of the gastric juice which causes the coagulation of milk.

Saturated compounds—Organic compounds which do not contain double or triple bonds and thus cannot add elements or compounds.

Silicone—One of the family of polymeric materials in which the recurring chemical group contains silicone and oxygen atoms as links in the main chain.

Review Questions

14-1. List eight plastics belonging to the family of thermosetting plastics and give a product application of each.

14-2. List the most important properties of each plastics named in Answer 14-1.

14-3. Name three thermosetting plastics used extensively in the wood and plywood industries.

14-4. Collect and display thermosetting plastics products with plastics tradenames.

14-5. What is the basic difference between thermoplastic and thermosetting materials?

14-6. Where are silicone liquids used?

14-7. What catalysts are used to cure silicone rubbers?

14-8. What is meant by the term *elastomer?*

14-9. Where are silicone foams used?

14-10. What does *RTV* mean?

14-11. What chemical is reacted with bisphenol-A to produce an epoxy?

14-12. Which processes utilize solid forms of epoxy resins?

14-13. What safety precautions should be taken when working with chemicals with which you are not familiar?

14-14. For what purposes is styrene monomer added to polyester resin?

14-15. What is the difference between a catalyst and a promoter?

14-16. List the dangers that must be recognized in working with catalysts and promoters.

14-17. Which gases are used to produce urea resin?

14-18. Which urea or phenolic offers the wider choice of color?

14-19. Of urea, phenolic, or melamine, which has the highest heat resistance?

14-20. How does urea resin impart crush resistance to fabrics?

14-21. Of melamine or phenolic plastics, which exhibits better water resistance?

14-22. What are the two major sources for phenol?

14-23. By what other name is phenol known?

14-24. What is formalin?

14-25. What is a resole?

14-26. What is the difference between a casting resin and a molding compound?

14-27. What is the purpose of using phenol furfural as a binder for molding compounds?

14-28. What is meant by condensation reaction?

15

Molding Processes

Previous chapters have discussed the history, chemistry, properties, and product applications of various plastics. It is the objective of this chapter to discuss how the various plastics are formed from resins, powders, granules, etc. into end products or components of end products. Decorating, finishing, and fabrication of parts are highly specialized fields.* Many of the processes are analogous to metalworking techniques, paper making, glass blowing, and other areas; many, however, apply only to the processing of plastics.

According to the Society of the Plastics Industry, Inc., it is the processor who converts plastics into products and the fabricator and finisher who further fashion and decorate the plastics products.

The low-cost mass production of quality plastics products is dependent on the machinists and toolmakers. New, specially formulated plastics and processing techniques have helped change plastics processing from a craft to a true technology.

The basic processes of the plastics industry may be summarized by these eight major headings: (1) molding, (2) casting, (3) thermoforming, (4) expanding, (5) coating, (6) decorating, (7) machining and finishing, and (8) the assembly or fabrication.

Molding

One basic assumption of molding all plastics is that they may be made fluid sometime during the molding operation and will solidify to the plastics state on cooling. In order for an operation to be classed as molding, force is required. Molding processes include: (1) injection molding, (2) compression molding, (3) transfer molding, (4) extrusion molding, (5) blow molding, (6) calendering, (7) laminating and reinforcing, (8) cold molding, (9) sintering, and (10) liquid resin molding.

*See also Chaps. 20-22.

Injection Molding

This process was used as early as the Civil War. In 1872 the Hyatt brothers were awarded a patent covering injection molding in the United States. Today injection molding is used to mold many thermosetting compounds and all thermoplastics except TFE fluorocarbons (Fig. 15-1).

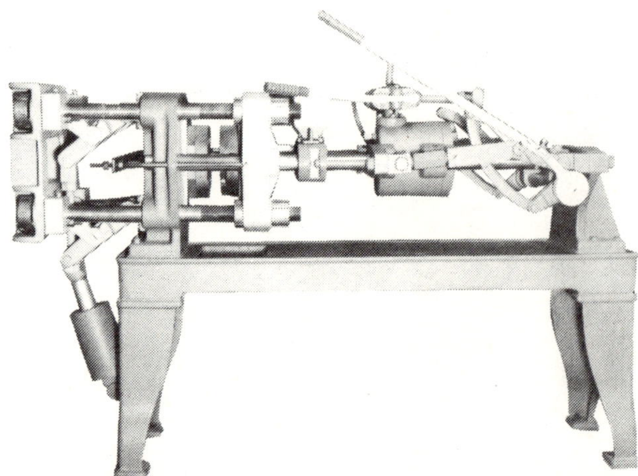

Fig. 15-1. This manual, air-operated machine was the first injection molding machine utilized for plastics. (Foster Grant Co., Inc.)

During injection molding, granular plastics is heated and forced through a heated cylinder. The hot mass is then injected into the closed mold cavity. After a cooling cycle, the plastics part is removed automatically or by the operator. The process is something like that of an injection glue gun in which hot glue is forced into a crack or joint to solidify. (See Fig. 15-2.) Diecasting of metals is a very similar technique from which most of the early injection molding mechanisms were patterned. Injection molding is similar to transfer molding except there is a large bulk of molten plastics in the heating chamber.

There are two basic injection techniques used with various modifications: (1) plunger type and (2) reciprocating screw type. Both may have preplasticizer sections or they may be used in combination together.

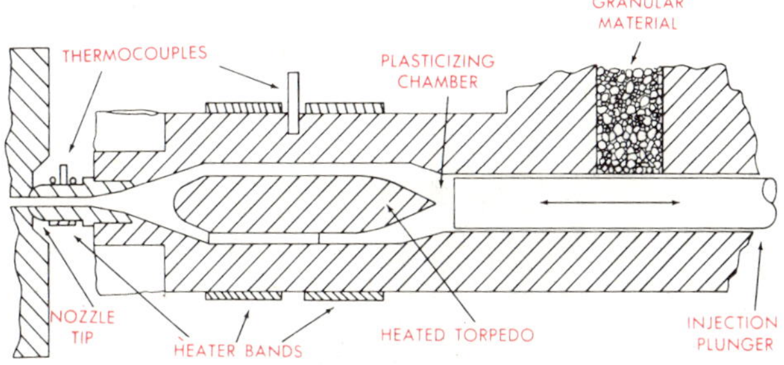

(A) Simple schematic.

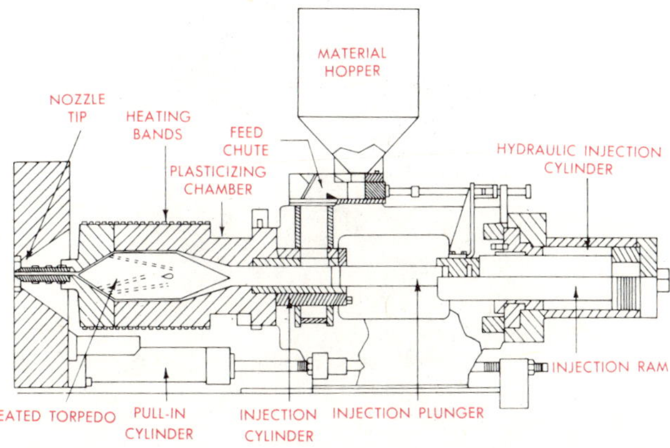

(B) Complete with ram and hydraulic cylinder. (Hydraulic Press Mfg. Co.)

Fig. 15-2. Injection molding machine (plunger type).

The major difference between the two methods of injection molding is in the way the molten mass is forced through the heating cylinders and injection chamber. In plunger machines the material is forced around the torpedo and then into the mold cavity. In reciprocating screw machines the granular material is made molten more rapidly and will blend colors or other materials more rapidly due to the action of the screw. (See Fig. 15-3.)

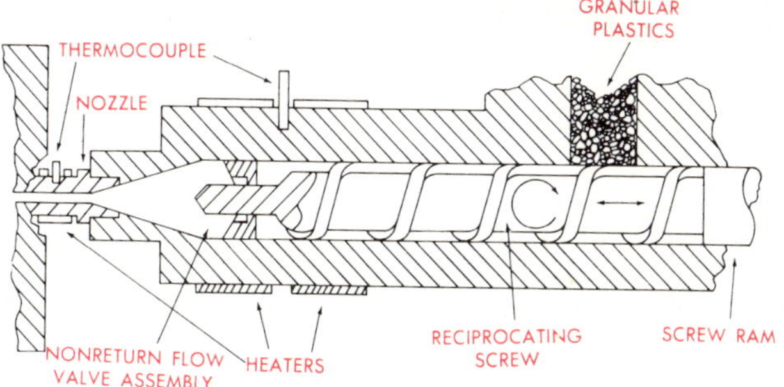

(A) Simple schematic. (Eastman Chemical Products, Inc.)

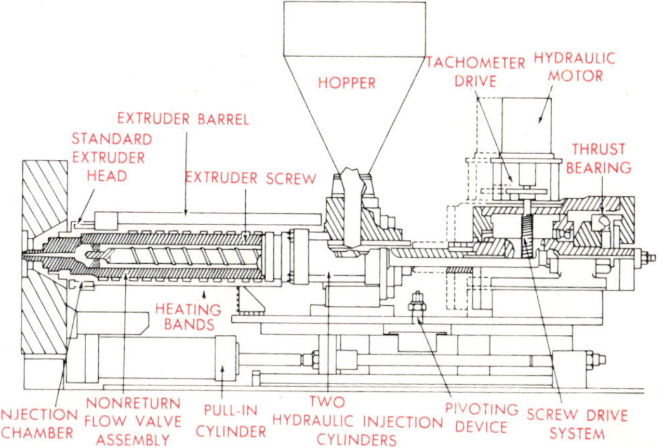

(B) Schematic of machine. (Hydraulic Press Mfg. Co.)

Fig. 15-3. Injection molding machine (reciprocating screw type).

Some modern injection molding machines are shown in Fig. 15-4.

If the time taken to melt the plastics granules is reduced, cycle (molding) times may be reduced. This increase in melting capacity usually involves a separate heating chamber where the material is preplasticized (melted) and then transferred to the main cylinder. See Fig. 15-5.

A rotary shear-cone preplasticizer (Fig. 15-5E) may be used with the plunger action of an injection molding machine. The rotary shear machine takes up less floor space and does not use an extruder screw. The plastics material is metered into the plunger chamber which houses a revolving cone. As the plunger forces the materials past the opening between the spinning cone and the housing, the molding compound shears, plasticizes, and homogenizes. This hot plasticized material passes into the

(A) Horizontal injection molding press. (Pennwalt Stokes)

(B) Ram (or plunger) injection molding machine. (Hull Corp.)

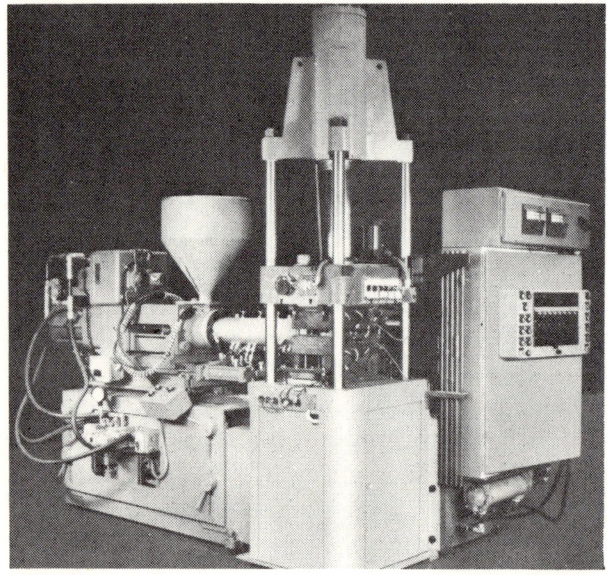

(C) This machine uses reciprocating screw to plasticize material and inject it through a sprue brushing. (Hull Corp.)

(D) Machine shown in (C) showing sprue bushing at the parting line. (Hull Corp.)

(E) Fully automatic injection press. (Pennwalt Stokes)

Fig. 15-4. Modern injection molding machines.

injection plunger chamber. It is then forced by the plunger action into mold cavities (not shown). Faster molding cycles and more efficient viscosity control are an advantage in this technique. Rotary

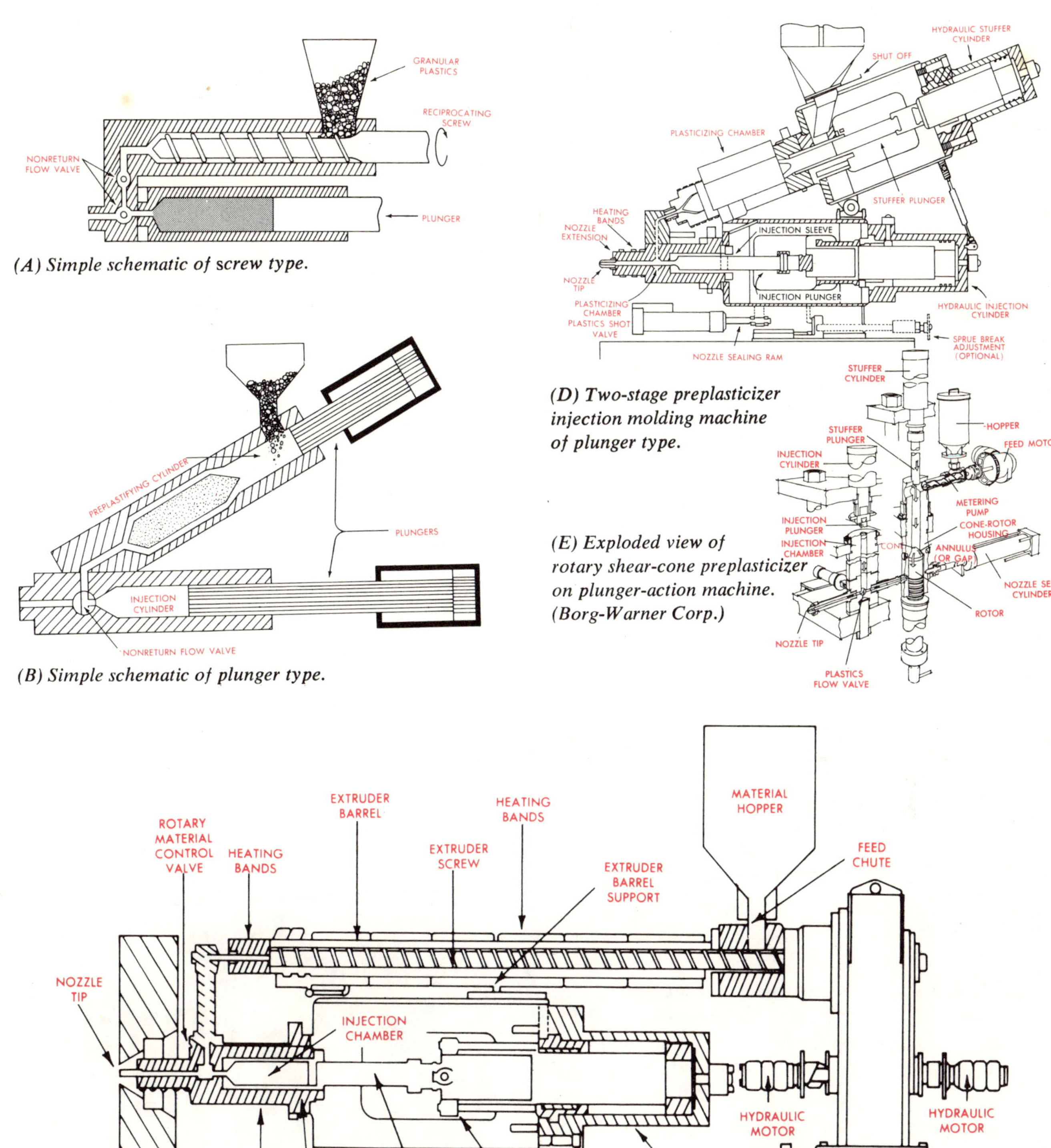

(A) Simple schematic of screw type.

(B) Simple schematic of plunger type.

(D) Two-stage preplasticizer injection molding machine of plunger type.

(E) Exploded view of rotary shear-cone preplasticizer on plunger-action machine. (Borg-Warner Corp.)

(C) Complete schematic of machine.

Fig. 15-5. Injection molding machines with two-stage preplasticizer.

shear machines take up less floor space than extruder type preplasticizer units.

Injection molding is popular because metallic inserts may be used, production rates are high, surface finish can be controlled to produce any desired texture, dimensional accuracy is good, and gates, runners, or rejected parts may be ground and reused. Any scrap generated using thermosetting resins is lost because it cannot be ground up and reused.

Mold design is another important consideration and an important factor in determining the output of injection molding. Because designing of injection molds is a complex subject with numerous modifications, only a broad discussion of mold design and terms will be used.* A typical two-piece, two-cavity injection mold design is shown in Fig. 15-6.

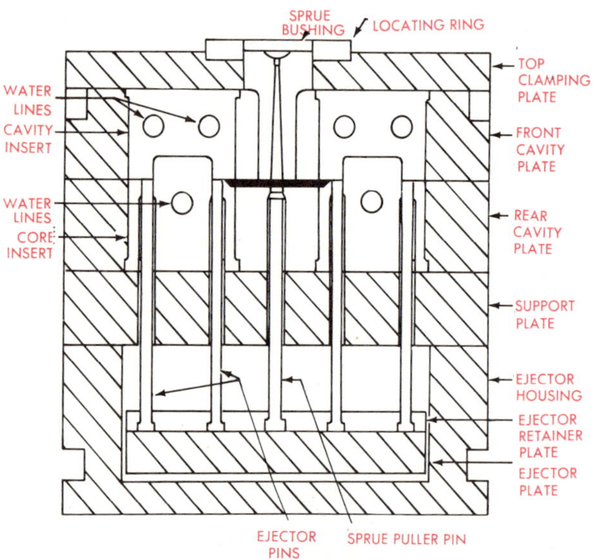

Fig. 15-6. Two-plate injection mold. (Dow Chemical Co.)

Because injection molding machines may use pressures five to ten times greater than those required on compression presses, a hydraulic, mechanical (toggle), or a combination of mechanical-hydraulic clamping units is used to hold the mold halves closed during injection. These pressures may range from 10,000 to 30,000 psi, depending on materials being used.

As the hot, molten material is forced through the nozzle into the mold, it flows through several channels or passageways before it enters the mold cavity. The terms *sprue, runner,* and *gate* are used to designate these channels (Fig. 15-7A). The plastics in them is sometimes removed with the plastics product (Fig. 15-7B, Fig. 15-8). These parts are

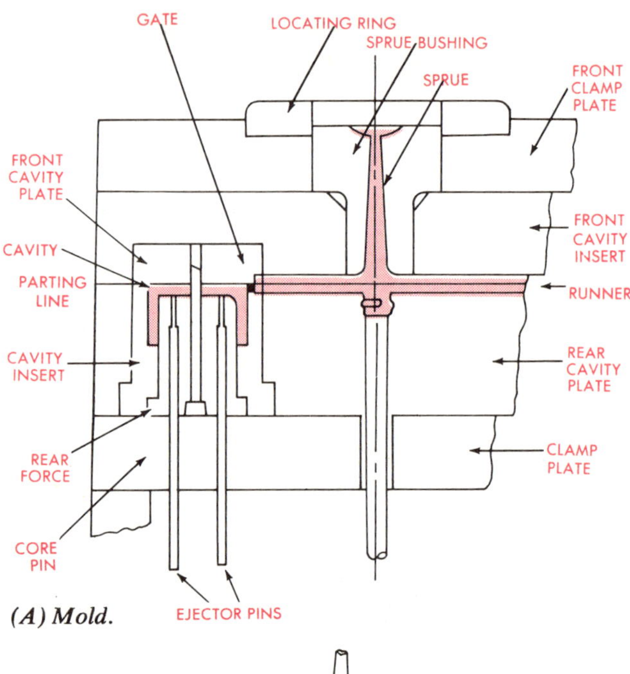

(A) Mold.

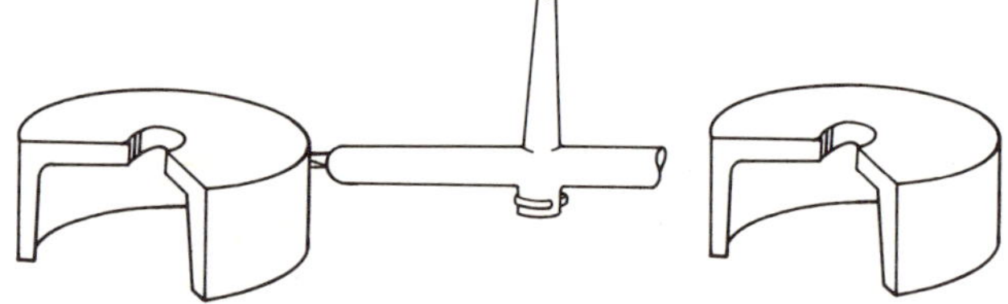

(B) Molded part, runner, gate, and sprue. *(C) Typical part.*

Fig. 15-7. Construction of injection mold showing molded part.

Fig. 15-8. Operator removes injection molded part. Note sprue sticking out from part. (Shell Oil Co.)

*See Chaps. 10 and 11.

then ground into granular form and reprocessed in the molding cycle.

In some molds there is only one cavity while others have many. The *gate* is the point of entry into the mold cavity. In multicavity molds there is a gate entering each mold cavity. The gate may be of various shapes and sizes. As a rule gates should be small in order to leave as small a blemish as possible and yet allow a smooth flow of molten material into the cavity. (See Fig. 15-9.) A small gate will help the finished article break away from the sprue and runners (Fig. 15-10).

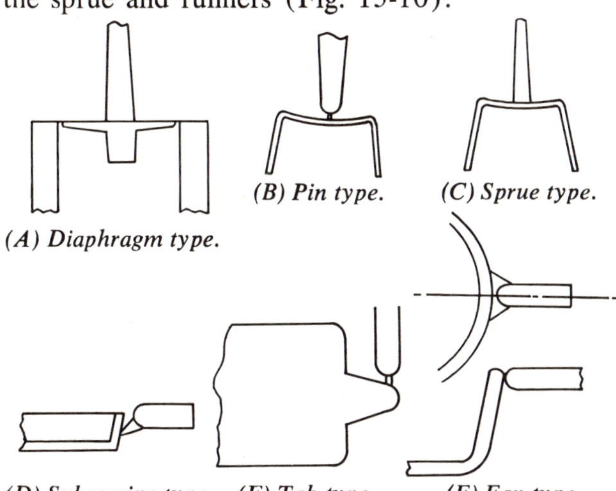

(A) Diaphragm type. *(B) Pin type.* *(C) Sprue type.*

(D) Submarine type. *(E) Tab type.* *(F) Fan type.*

Fig. 15-9. A few of the many possible gating systems.

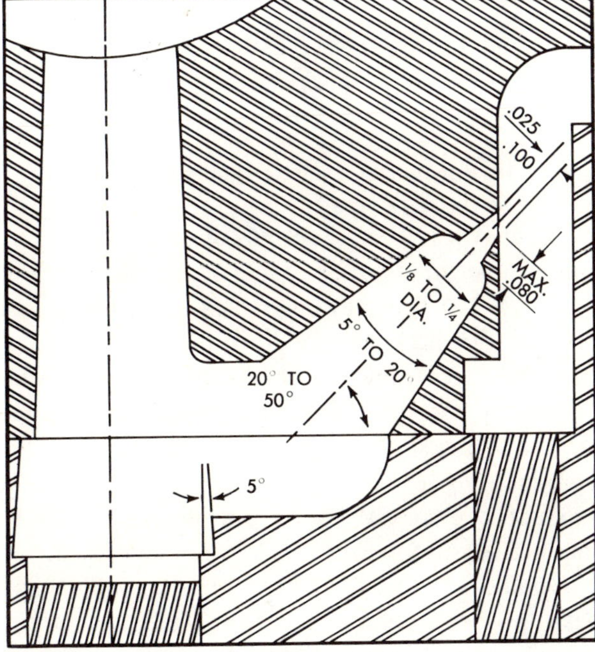

Fig. 15-10. Typical tunnel gate which may be considered a pinpoint gate variation design. (Mobay Chemical Co.)

Runners are narrow channels that convey the molten plastics from the sprue to each cavity. In multicavity molds the runner system should be designed so all materials will travel the same distance from the sprue to each cavity. See Fig. 15-11.

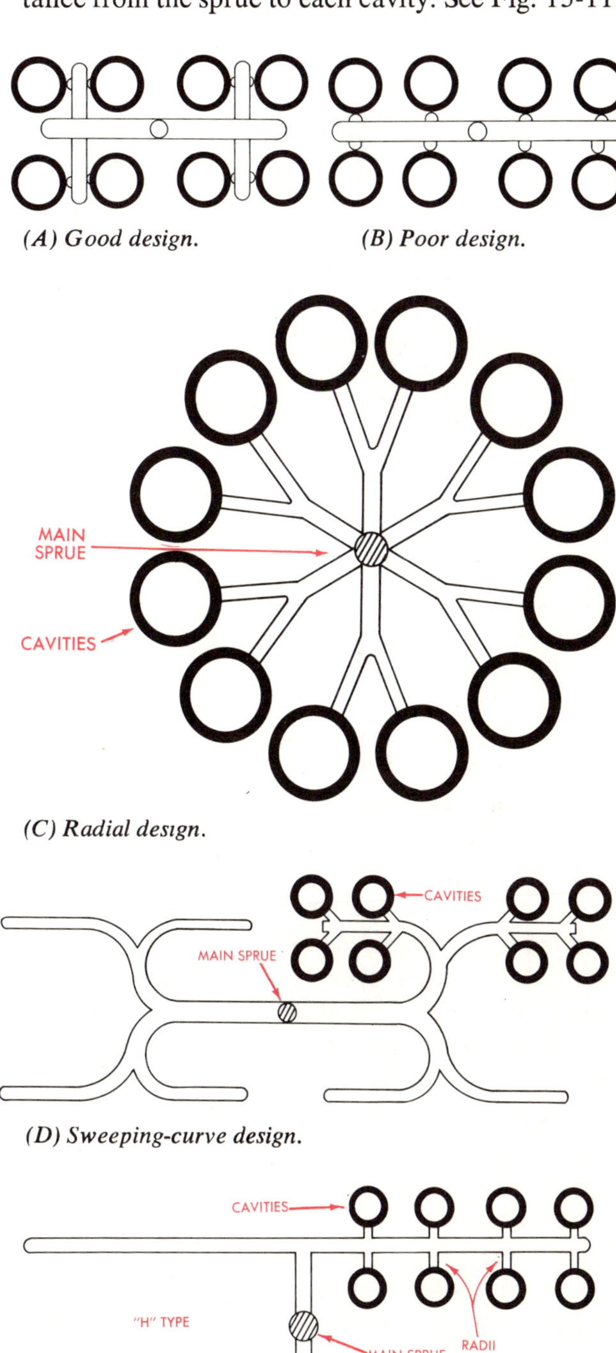

(A) Good design. *(B) Poor design.*

(C) Radial design.

(D) Sweeping-curve design.

(E) "H" design.

Fig. 15-11. Typical runner designs. (E. I. Du Pont de Nemours & Co., Inc.)

The heavy tapered channel connecting the nozzle with the runners is called the *sprue*. In a single-cavity mold the sprue channel feeds the material either directly or through a gate into the mold cavity (see Fig. 15-12).

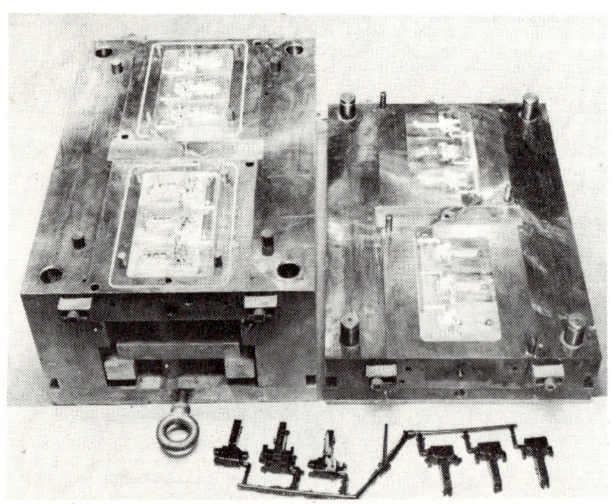

Fig. 15-12. Injection mold with molded items shown below. Note flash (or fins) on runner and parts. (Hull Corp.)

Normally the sprues, gates, and runners are cooled and removed with each cycle, and reground for molding. The reprocessing of sprues and runners is costly and restricts the weight of molded articles per cycle of the injection molding machine. In *hot runner molding* the sprues and runners are kept hot by means of heating elements built into the mold. As the mold opens, the hardened part with gate pulls free from the still-molten hot runner system. On the next cycle the mold closes and the hot material left in the sprue and runner is forced into the mold cavity. See Fig. 15-13.

A similar system called *insulated runner molding* is used in molding polyethylene or other materials with low thermal transfer (Fig. 15-14). In this design large runners are used. As the molten material is forced through these runners the material begins to solidify, forming a plastics lining. The lining serves as insulation for the inner core of molten material. This inner core remains hot and continues to flow through the tunnellike runner to the mold cavity. This system is also called *runnerless molding*.

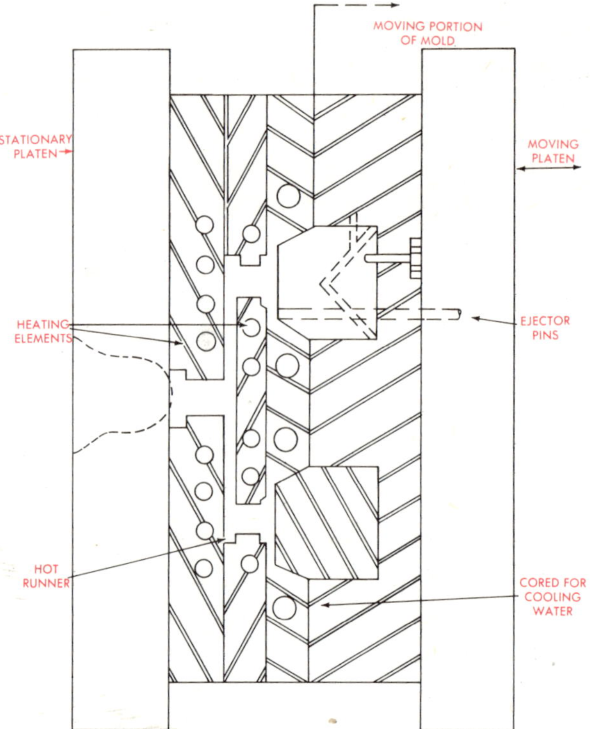

Fig. 15-13. Typical hot runner mold design.

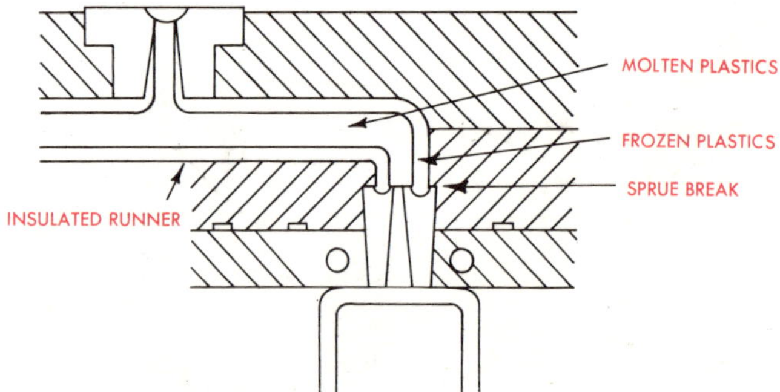

Fig. 15-14. Principle of insulated runner molding.

Other types of mold designs or modifications include valve gated molds, single or multiplate molds (Fig. 15-15), unscrewing molds (for internal or external threads), cam and pin molds (for undercuts and cores), and multicolor or material molds (where a second color or material is injected around the first molding, leaving portions of the first molding exposed). Examples of multicolor products are special knobs, buttons, and letter or number keys. (See Fig. 15-16.)

Good draft, suitable gating, constant wall thickness, proper cooling, sufficient ejection, proper steels, and ample mold support are all important factors to consider in mold design.

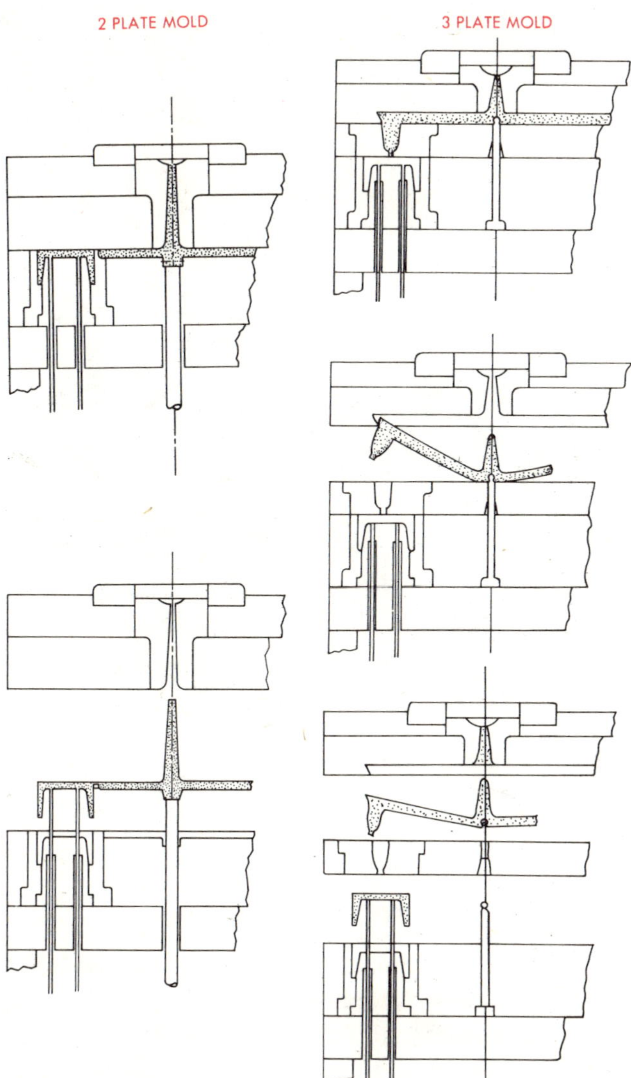

Fig. 15-15. Multiplate mold design for injection molding. (E. I. du Pont de Nemours & Co., Inc.)

(A) Pocket calculator keys. *(B) Pocket calculator with keys.*

Fig. 15-16. Examples of multicolored injection molded products.

Temperature controllers, coolant conditioners, dehumidifying dryers, scrap regrinders, and other peripheral equipment are important and necessary to the injection molding process.

In Table 15-1 a few selected molding difficulties, their causes, and possible remedies are shown.

Table 15-1. Problems in Injection Molding.

Difficulty	Cause	Possible Remedy
Black specks, spots, or streaks	Flaking off of burned plastics on cylinder walls	Purge heating cylinder
	Air trapped in mold causing burning	Vent mold properly
	Frictional burning of cold granules against cylinder walls	Use lubricated plastics
Bubbles	Moisture on granules	Dry granules before molding
Flashing	Material too hot	Reduce temperature
	Pressure too high	Lower pressure
	Poor parting line	Reface the parting line
	Insufficient clamp pressure	Increase clamp pressure
Poor finish	Mold too cold	Raise mold temperature
	Injection pressure too low	Raise injection pressure
	Water on mold face	Clean mold
	Excess mold lubricant	Clean mold
	Poor surface on mold	Polish mold
Short moldings	Cold material	Increase temperature
	Cold mold	Increase mold temperature
	Insufficient pressure	Increase pressure
	Small gates	Enlarge gates
	Entrapped air	Increase vent size
	Improper balance of plastics flow in multiple cavity molds	Correct runner system
Sink marks	Insufficient plastics in mold	Increase injection speed, check gate size
	Plastics too hot	Reduce cylinder temperature
	Injection pressure too low	Increase pressure
Warping	Part ejected too hot	Reduce plastics temperature
	Plastics too cold	Increase cylinder temperature
	Too much feed	Reduce feed
	Unbalanced gates	Change location or reduce gates
Surface marks	Cold material	Increase plastics temperature
	Cold mold	Increase mold temperature
	Slow injection	Increase injection speed
	Unbalanced flow in gates and runners	Rebalance gates or runners

Injection molding is not practical for short production runs because of high machine, tooling, and mold costs.

Typical applications include toys, bathroom, and kitchen wall tile, cases, housings for radio cabinets and appliances, refrigerator parts, handles, battery cases, electrical parts, gears, impellers, tail lights, instrument panels, steering wheels, pump parts, fasteners, grills bearings, and containers of all types (Fig. 15-17).

(A) Uppers are polyurethane and vinyl, heel traction bar is acetal polymer, and swivel unit is injection molded polyamide. (E. I. du Pont de Nemours & Co., Inc.)

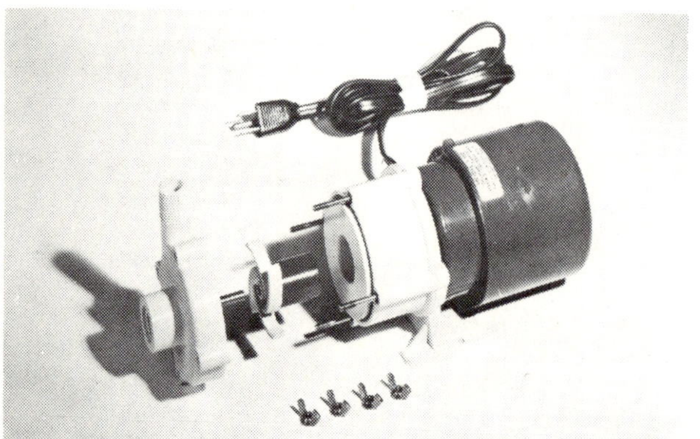

(B) Injection molded pump of glass fiber reinforced polypropylene for pump housing, magnet housing, impeller, and volute. (Fiberfill Div., Dart Industries)

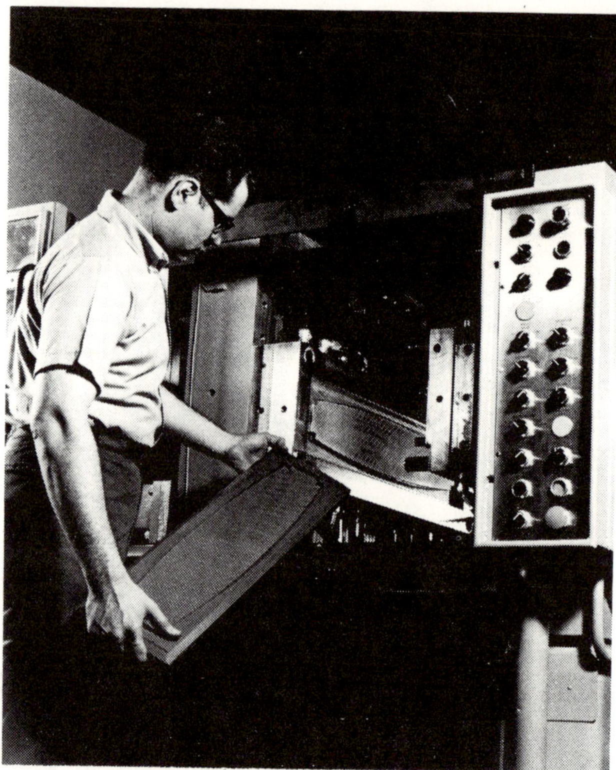

(C) Injection molding of large plastics part. (Cincinnati Milling Machine Co.)

Fig. 15-17. Injecting molding produces a large variety of products.

Compression Molding

Compression molding may be considered one of the oldest known molding processes. In this process the plastics material is placed in a mold cavity and formed by heat and pressure. Thermosetting compounds are normally used for compression molding but thermoplastics may be used. The process is remotely like making waffles. Heat and pressure force the materials into all areas of the mold. The heat hardens the material, and the plastics part is removed from the mold cavity. (See Fig. 15-18.)

In order to reduce pressure requirements and production (cure) time in the mold the plastics material is usually preheated by infrared, induction, or other methods before it is placed in the mold cavity. A screw type extruder is sometimes used to reduce cycle time and increase mold productivity. The screw extruder is often used to produce preformed slugs which may be manually or automatically loaded in the molding cavity. The screw-compression process will greatly reduce cycle time, thus eliminating the greatest disadvantage of com-

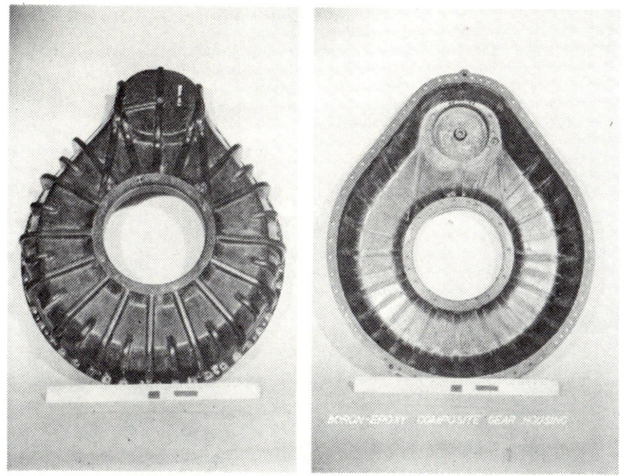

Fig. 15-18. Reinforced plastics gear housing is lighter and stronger than metal housing it replaces. Many similar parts are compression molded.

pression molding of previous years. Under optimum conditions good compression molded parts with heavy wall thickness may be produced with up to 400 percent greater product output per mold cavity.

During preforming and the actual molding process, heat and various catalysts initiate crosslinking of molecules. During the crosslinking reaction, gases, water, or other materials may be liberated. If these materials are trapped in the mold cavity, the plastics part may be damaged, of poor quality, or marked by surface blisters. In order to prevent the entrapment of gases or other unwanted materials produced during the operation, molds are vented or specially designed to allow the escape of these foreign substances.

In the molding operation the thermosetting plastics compound becomes crosslinked and infusible; consequently, the product may be removed from the molding cavity while it is still hot, without fear of distortion. Thermoplastic materials, because they do not crosslink to any extent, must be cooled before removal.

There are three different types of compression mold designs. The molds are usually produced of hardened steel to withstand the great pressure and abrasive action of the hot plastics compound as it liquifies and flows into all parts of the mold cavity.

(1) The least complex and most economical from the standpoint of original mold cost is the *flash* type mold (Fig. 15-19). In this type of construction, excess material is forced out of the molding cavity forming a flash. The flash must be removed from the molded part and is wasted.

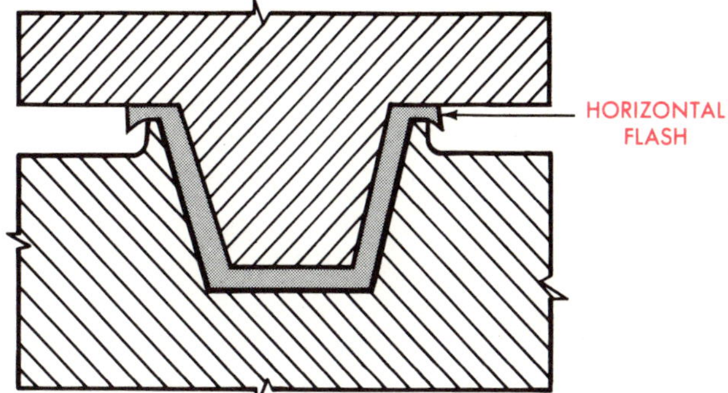

Fig. 15-19. Simplified diagram of flash type mold.

(2) In *positive* type mold designs no provision is made for placing excess material in the cavity (Fig. 15-20). The preforms must be carefully measured if each part is to be the same density and

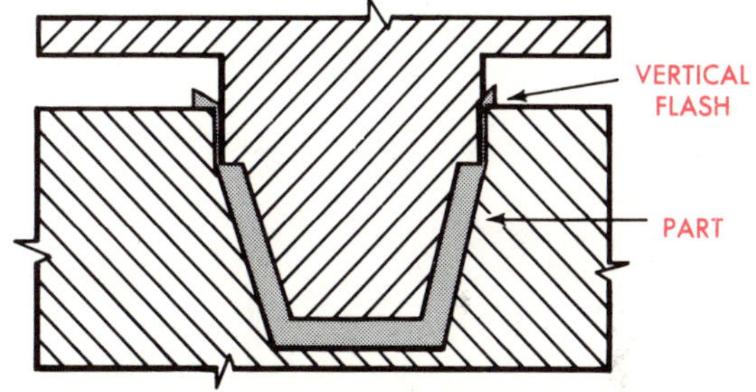

(A) Simplified diagram.

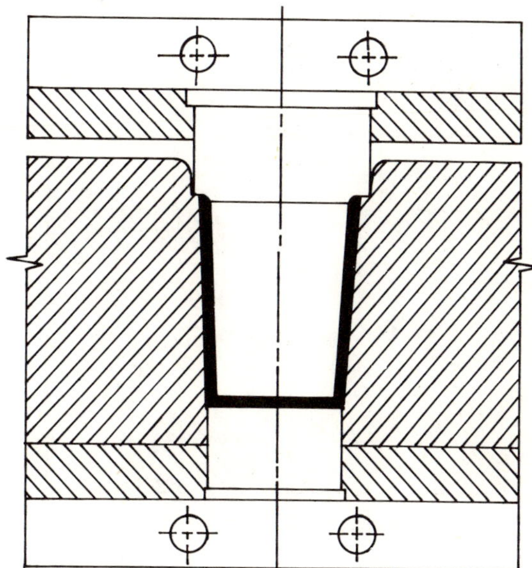

(B) Positive mold with land.

Fig. 15-20. Positive type mold design.

thickness. If too much material is loaded in the cavity, the mold may not fully close. In positive molds little or no flash occurs. The design is used for molding laminated, heavily filled, or high-bulk materials. If the materials are carefully measured, very dense, uniform parts may be formed. With fully positive molds gases liberated during the chemical curing of thermosets may be trapped in the mold cavity. During compression the mold may be opened briefly to allow gases to escape. This operation is sometimes called *breathing*.

(3) *Semipositive* molds have horizontal and vertical flash waste (Fig. 15-21). This design is expensive to manufacture and maintain but is the most practical where many parts or long runs are required. The design allows for some inaccuracy of charge by allowing flash, yet it gives a dense, uniform molded part. As the mold charge is compressed in the cavity, excess material is allowed to escape. As the mold body continues to close very little material is allowed to flash. When the mold fully closes, the telescoping male is stopped by the *land*.

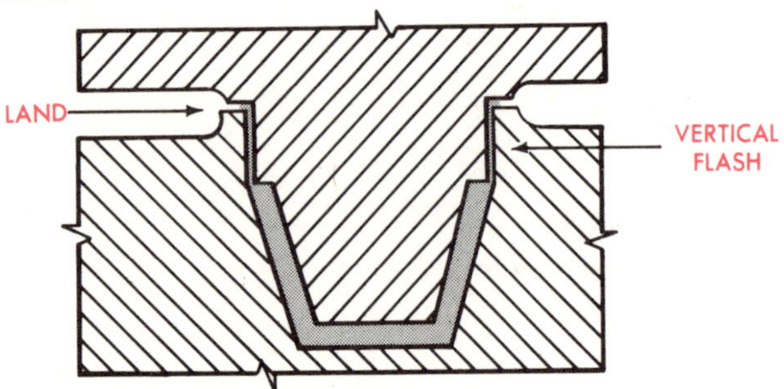

(A) Simplified diagram.

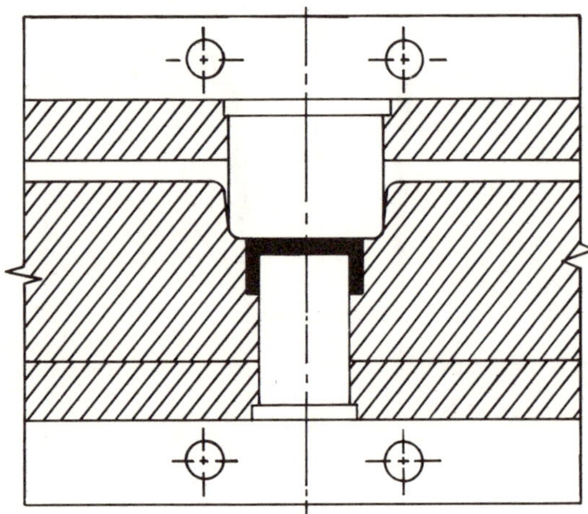

(B) Semipositive mold with land.

Fig. 15-21. Semipositive mold design.

In all three mold designs heat for molding is normally provided by electrical heating elements or steam cores. In the compression molding of thermoplastics cold-water cooling cores are also included in the mold design. In order to apply sufficient force during the molding operation hydraulic or pneumatic presses are used. Fig. 15-22 shows such a setup.

(A) This photo shows mold for making three ashtrays, which are about to be ejected. (Dake Corp.)

(B) A 200-ton-capacity automatic compression molding press with integral powder preheater. (Stokes-Pennwalt Corp.)

(C) *Large compression molding press being tested. (Hull Corp.)*

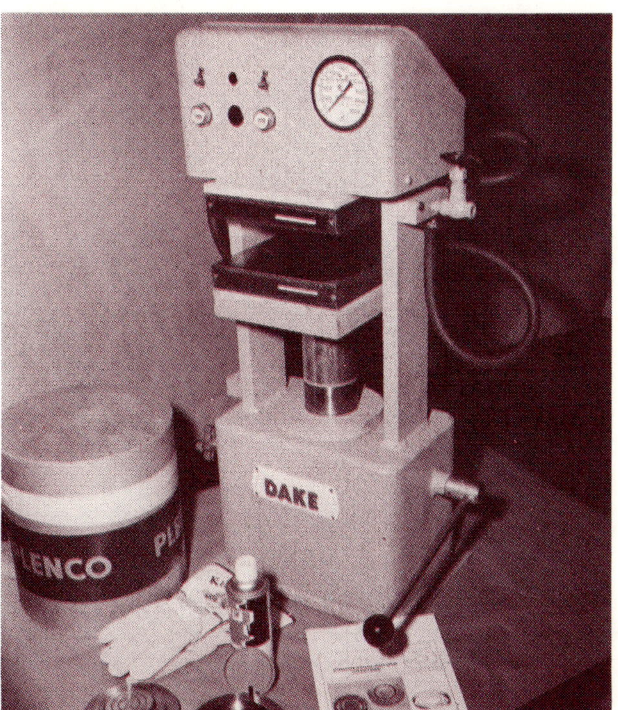

(D) *Small laboratory press and mold. (Dake Corp.)*

(E) *Laboratory press used for manufacturing ABS test samples. (Wilson Instrument Div. of ACCO)*

Fig. 15-22. Compression molding presses.

Long runs of parts of moderate complexity are produced by compression molding because mold maintenance costs and initial costs are low. There is little waste of material, finishing costs are low, and large, bulky parts are practical. The greatest limitation is that extremely intricate parts with inserts, undercuts, side draws, and small holes are not practical while trying to maintain close tolerances. See Fig. 15-23.

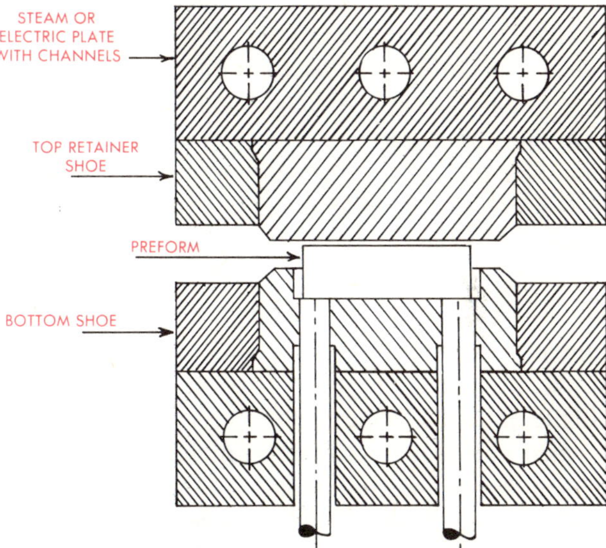

(A) *Preform about to be molded.*

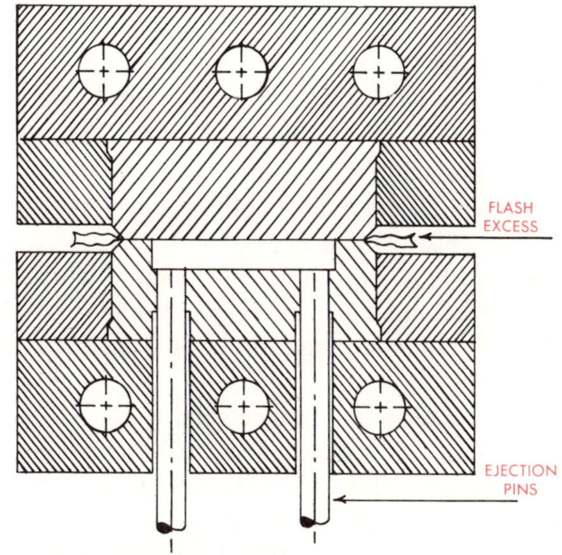

(B) *Closed mold showing flash.*

Fig. 15-23. Principle of compression molding.

The flash or parting line is usually located at the point of greatest diameter and is removed by hand or machine operations.

Compression molded parts include dinnerware, buttons, buckles, knobs, handles, appliance housings, drawers, parts bins, radio cases, large containers, and numerous electrical parts.

Transfer Molding

Transfer molding has been known and practiced since World War II. The process is sometimes called *plunger molding, duplex molding, auxiliary-plunger transfer molding, step molding, injection transfer molding,* or *impact molding.* It is actually a variation of compression molding but differs from it in that the material is loaded in a chamber outside the mold cavity. Heat and pressure are applied to the molding compound in this exterior chamber (may be plunger or screw type) forcing the plasticized (fluid) mass into the mold cavity. Transfer molding techniques have the advantage over compression molding in that the molten mass is fluid when it enters the mold cavity; consequently, fragile, intricate shapes with inserts or pins may be formed with accuracy. Transfer molding techniques are much like injection molding except thermosetting compounds are normally utilized.

According to the American Society of Tool and Manufacturing Engineers, there are two basic types of transfer molds: (1) pot or sprue type and (2) plunger type.

Plunger type molds (Fig. 15-24) differ from the sprue type (Fig. 15-25) in that the plunger or force is pushed to the parting line of the mold cavity. In plunger type molds only runners and gates are left as waste on the molded part.

A third type may also be included in which the molding compound is preplasticized by extruder action and then a plunger forces the melt into the mold.

When a series of runners and gates are used trapezoidal, half-round and full-round runner grooves are machined into the mold. With trapezoidal and half-round runners and gates only one half of the mold or die plate is grooved (Fig. 15-26). Trapezoidal and half-round grooves are easy to machine but generally require more molding pressure. Round runners are recommended for transfer molding if extruder type plasticizers are used. Pinpoint or submarine gating as shown in

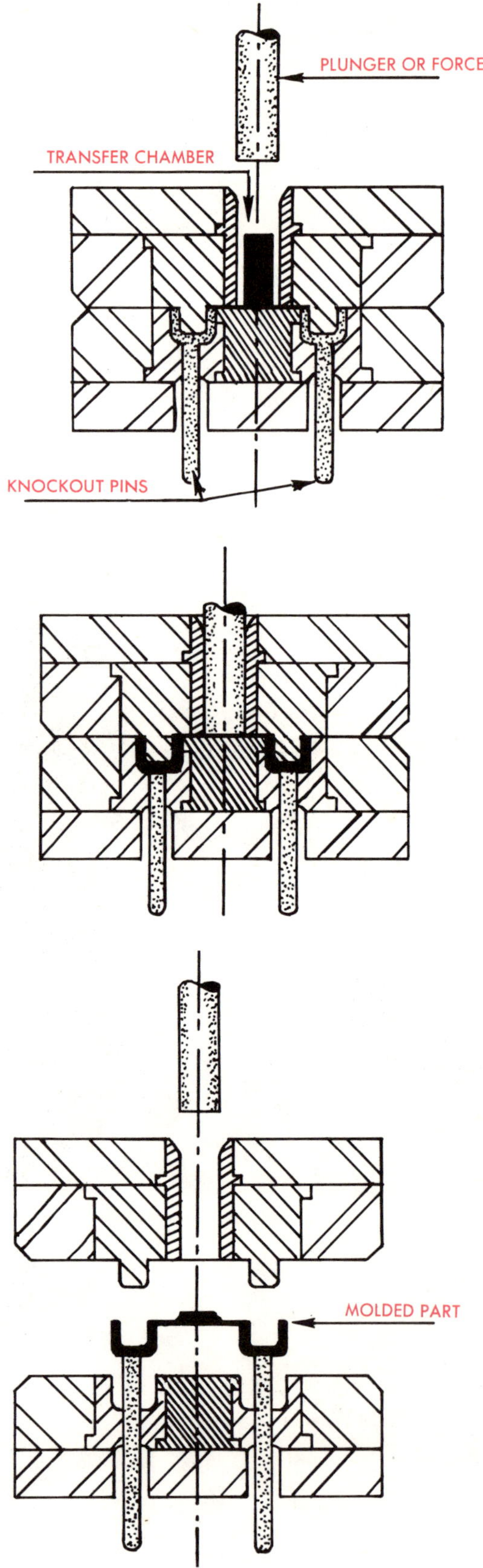

Fig. 15-24. Plunger type transfer mold.

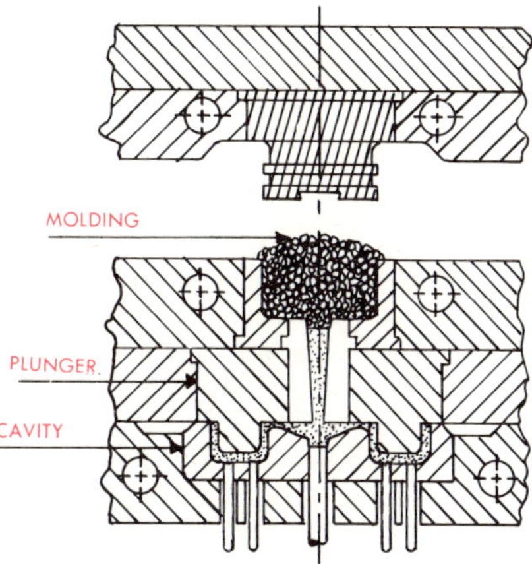

(A) Open position.

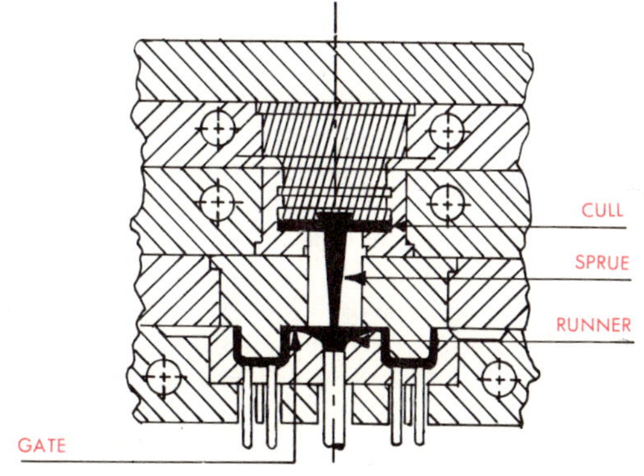

(B) Closed position.

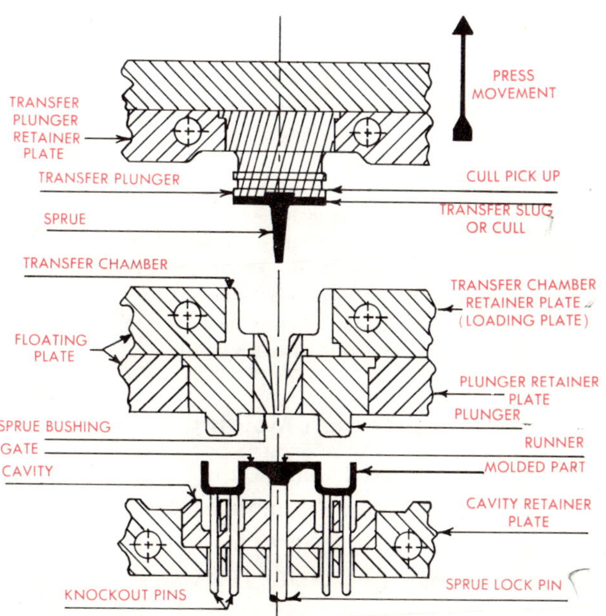

(C) Release position.

Fig. 15-25. Sprue type integral transfer mold. (American Technical Society)

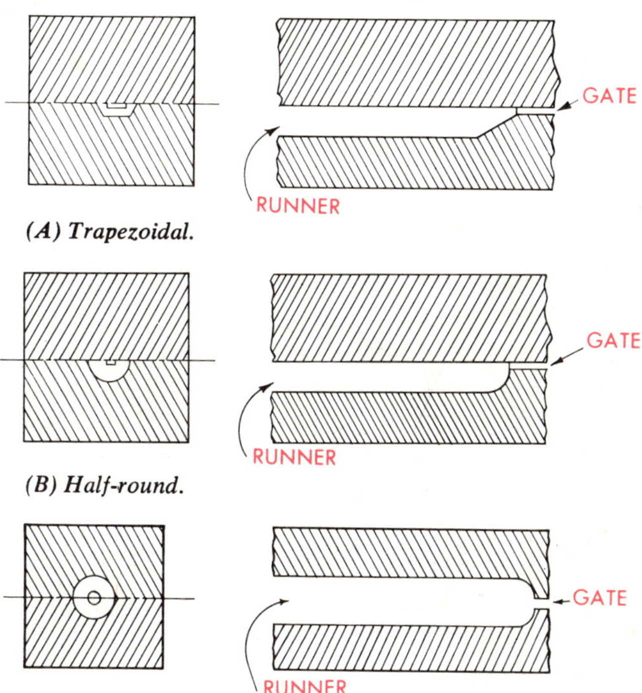

(A) Trapezoidal.

(B) Half-round.

(C) Full-round.

Fig. 15-26. Three basic runner systems used for transfer and injection molding.

Fig. 15-27 may be used. However, this system also requires greater molding pressures.

The cost of elaborate mold designs and high waste from culls, sprues, and flash are two major

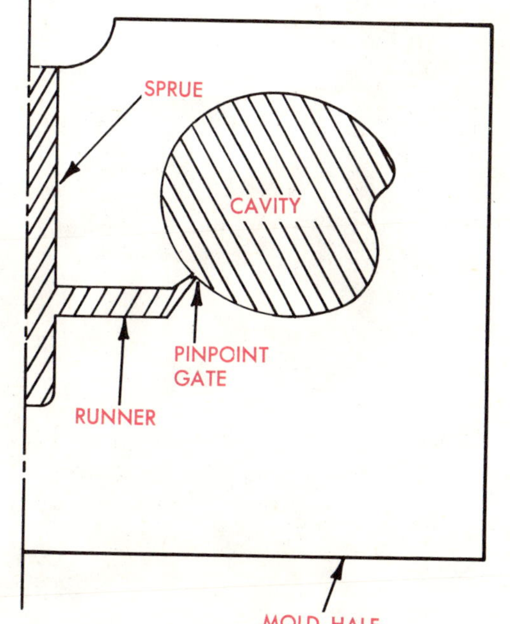

Fig. 15-27. Pinpoint or submarine gate system is used to eliminate gate breakoff problems and reduce or eliminate finishing problems caused by "large" gate marks.

limitations of transfer molding. See Fig. 15-28. Although most parts are limited in size, there are numerous applications including distributor caps, camera parts, switch parts, buttons, coil forms, terminal block insulators, and intricate shapes such as cups and caps for cosmetic containers.

(A) A 50-ton screw transfer molding press. (Stokes-Pennwalt Corp.)

(B) This part was transfer molded in the press and mold shown. (FMC Corp.)

Fig. 15-28. Transfer molding.

Table 15-2 lists some common problems and remedies in transfer molding.

Table 15-2. Problems in Compression and Transfer Molding.

Defect	Possible Remedy
Cracks around inserts	Increase wall thickness around inserts
	Use smaller inserts
	Use more flexible material
Blistering	Decrease cycle and/or mold temperature
	Vent mold—breathe mold
	Increase cure—increase pressure
Short and porous moldings	Increase pressure
	Preheat material
	Increase charge weight of material
	Increase temperature and/or cure time
	Vent mold—breathe mold
Burned marks	Reduce preheating & molding temperature
Mold sticking	Raise mold temperature
	Preheat to eliminate moisture
	Clean mold—polish mold
	Increase cure
	Check knockout pin adjustments
Orange peel surface	Use a stiffer grade of molding material
	Preheat material
	Close mold slowly before applying high pressure
	Use finer ground materials
	Use lower mold temperatures
Flow marks	Use stiffer material
	Close mold slowly before applying high pressure
	Breathe mold
	Increase mold temperature
Warping	Cool on jig or modify design
	Heat mold more uniformly
	Use stiffer material
	Increase cure
	Lower temperature
	Anneal in oven
Thick flash	Reduce mold charge
	Reduce mold temperature
	Increase high pressure
	Close slowly—eliminate breathe
	Increase temperature
	Use softer grade material
	Increase clamping pressure

Extrusion Molding

The process of extrusion molding is a continuous one, forming primarily thermoplastic materials into three main groups: (1) profile shapes, (2) films and sheets, and (3) coatings around wire and cables. Extruders are also utilized in the blow, compression, and injection molding processes.

Extrusion is similar to the process used for producing wire and other metallic profile shapes. Except for the heating of materials the process is something like a sausage-stuffing machine or a hand-operated cake decorator.

The fundamental concept is that dry, powder, or granular plastics is heated and forced through an orifice in a die. The heart of the process is the extruder which plasticizes (melts and mixes) the material and forces it through the die. Although screw type extruders are the most common, ram or plunger types are used for special applications including extrusion of thermosetting plastics and TFE fluorocarbon shapes.

Gutta percha, rubber, and shellac were extruded as early as 1845 by ram type machines. Screw type machines began to appear in Germany and the United States in the early 1930's. Today the extruder is probably the most widely used piece of plastics process equipment. It is employed for plasticizing units in other processing techniques. Injection molding is the most widely used molding process. Extrusion molding is popular because extrusion dies are relatively simple, inexpensive, and large amounts of material may be forced through these dies in a continuous form.

A screw type extruder is shown in Figs. 15-29 and 15-30. Primarily the speed and shape of the close fitting screw indicate the output, milling rate, and die pressure of the extruder. The polymer is compacted, heated, deaerated, compressed, and plasticized by the action of the screw as it passes various zones in the extruder barrel. Screws are characterized by their length to diameter *(L/D)* ratio. A 20:1 screw could be two inches in diameter and forty inches long. Screws of 16:1 and 40:1 are also used. Various screw designs are shown in Fig. 15-31.

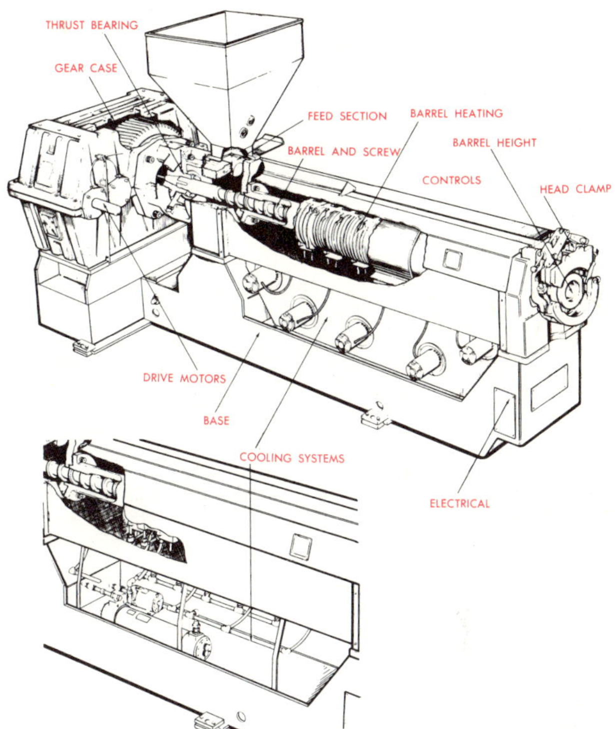

Fig. 15-30. Illustration of extruder with parts labeled. (Davis-Standard/Goulding)

In each design the channel depth of the screw decreases at the metering section. This continuous reduction forces out air and compacts the material (Fig. 15-31A). A breaker plate acts as a screen,

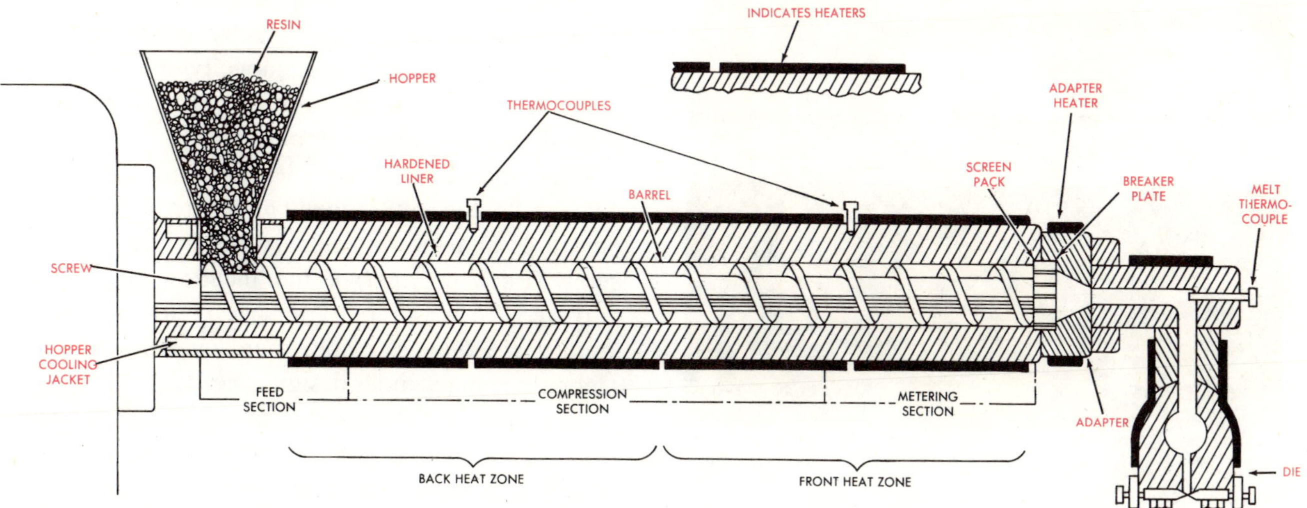

Fig. 15-29. Cross section of a typical screw type extruder with die turned down. (USI)

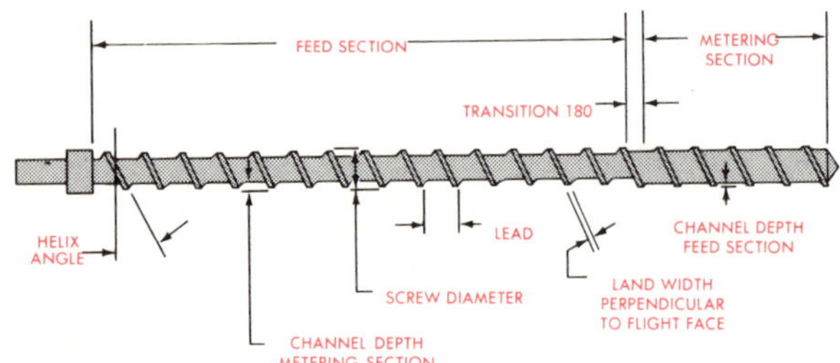

(A) Metering type screw. (Processing of Thermoplastic Materials)

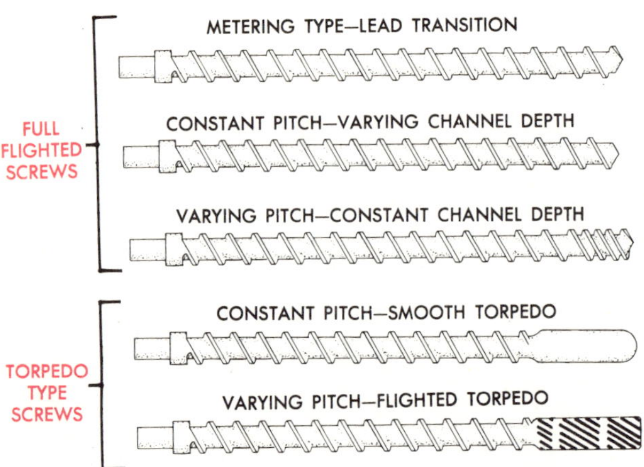

(B) Common extruder screws. (Processing of Thermoplastic Materials)

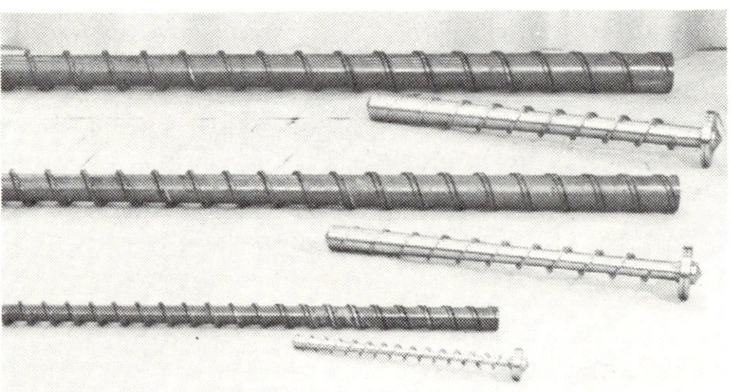

(C) Side view of extruder screws. (Waldron-Hartig Div., Midland-Ross Corp.)

(D) End view of extruder screws. (Waldron-Hartig Div., Midland-Ross Corp.)

Fig. 15-31. Common screw-types.

filtering out pieces of foreign material and creating back pressure by restricting the flow of molten plastics (Fig. 15-32). Heaters are used around the barrel to help melt the plastics and start the process.

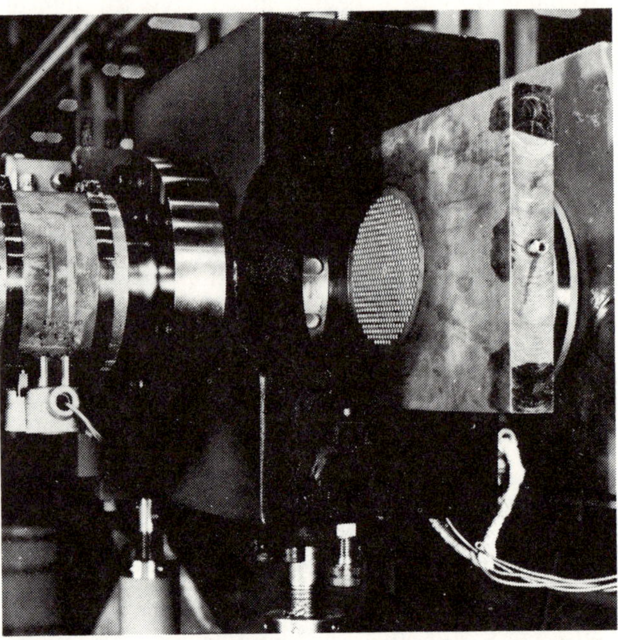

Fig. 15-32. A breaker plate acts as a screen, creating back pressure of material in the extruder barrel. (Waldron-Hartig Div., Midland-Ross Corp.)

Once the extruder is mixing, blending, and forcing the material through the die, frictional heat is generated by the screw and this may be sufficient to plasticize the material. External heaters are used to maintain a fixed temperature in various zones once the process is started. Processing of some materials requires special screw designs. Some utilize twin, parallel screws for greater capacity and better mixing. (Fig. 15-33). Extruder capacity of low-density polyethylene may vary from less than 2 to more than 10,000 pounds per hour.

Polyvinyl chloride has a tendency to decompose or degrade due to prolonged exposure to high temperatures and mechanical actions of the extruder screw. Each polymer or modification of polymer may require a different or special screw design. With new advances in polymer development, new technology in extruder design is needed.

Extruders are also used to compound and blend basic plastics with plasticizers, fillers, colorants, and other ingredients. Granular molding materials are produced by extruding these materials through dies.

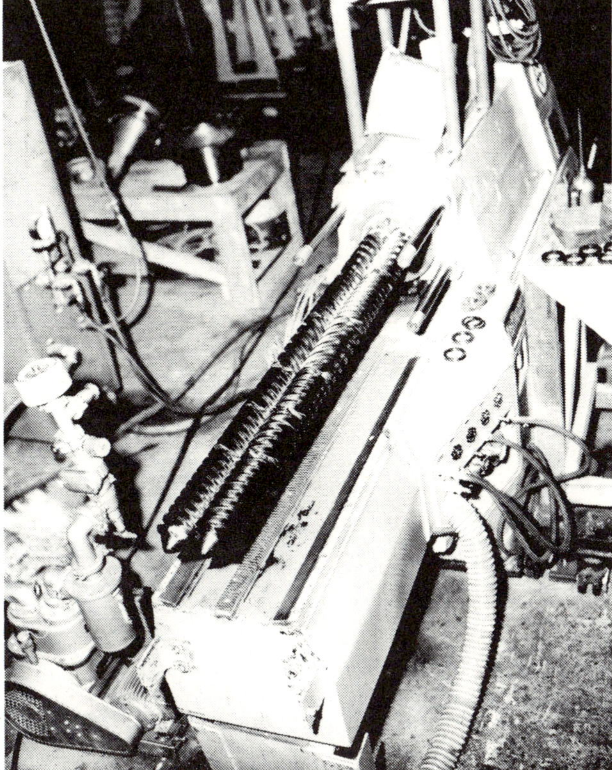

Fig. 15-33. Twin parallel extruder screws. (Fellows Gear Shaper Co.)

(A) Die and cutter inside pelletizing head.

(B) Hot pelletizing from extruder (center).

Fig. 15-34. Extrusion of material through dies. (Fellows Gear Shaper Co.)

The spaghettilike extrusion is then chilled and chopped into granular molding form. Granular forms are used extensively in extruders and injection molding machines.

The product is dependent on the die to actually form the molten plastics as it emerges from the extruder. (See Fig. 15-34.) Dies may be made of mild steel or of chromium-molybdenum steel for long runs. Stainless alloys are used for corrosive materials. Die design and construction is a very broad and complex topic.

Allowances must be made in the orifice design to produce the exact cross-sectional dimensions after the extrudate has cooled. In complex cross sections where thin sections or sharp edges are formed, cooling occurs more rapidly at such portions. These areas shrink first, causing them to be smaller than the rest of the section. To compensate for this problem the orifice in the die is made larger at these points. See Fig. 15-35.

Extruder products may be divided into six basic areas of application: (1) rod and profiles, (2) pipe, (3) film and sheet, (4) monofilaments, (5) extrusion coating, and (6) wire and cable covering.

Extrusion of rod shapes from a circular orifice opening in a die is probably the simplest of the ex-

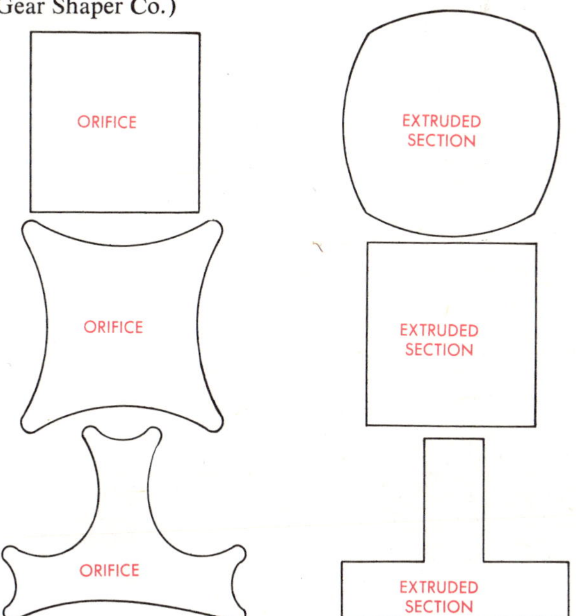

Fig. 15-35. Relationships between die orifices and extruded sections.

trusion processes. Many extrusions are produced horizontally through dies and cooled by air jets, water troughs, or cooling sleeves. Numerous profile shapes are produced for a variety of applications. Vinyl siding and tracks or channels for sliding doors and windows are familiar examples.

Extrudates are sometimes postformed into different shapes by the use of sizing plates, shoes, or rollers. A flat-tape shape may be postformed into corrugated form or round rods postformed into oval or other new shapes while the extrudate is still hot (see Fig. 15-36).

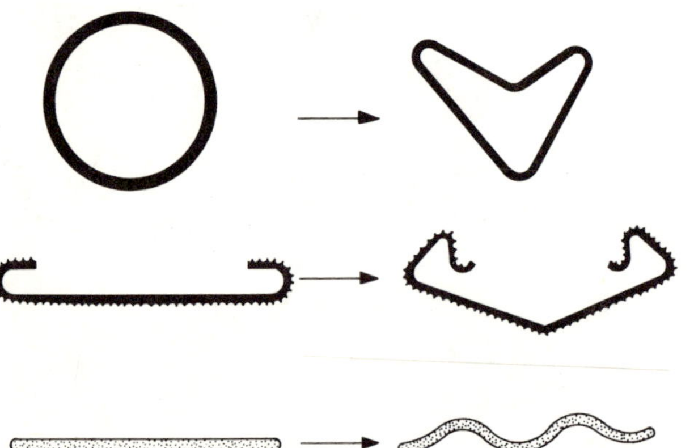

Fig. 15-36. The extrudates on the left are postformed by various shaped rollers to form the product.

Pipe or tubular forms are shaped by the exterior dimensions of the orifice opening and by the mandrel (sometimes called a *pin*) for inside dimensions. See Fig. 15-37. In order to prevent the tube from collapsing before cooling the tube is pinched shut on the end and air is forced into the extrudate through the die. Although this air pressure expands the pipe slightly, the thickness of the pipe wall is controlled by the mandrel and die size.

There are two basic methods of producing film: (1) slot or cast extrusion and (2) blow (lay-flat or tubing) extrusion. Both sheets and film forms are produced by extruding molten thermoplastic materials through dies with a long horizontal slot.

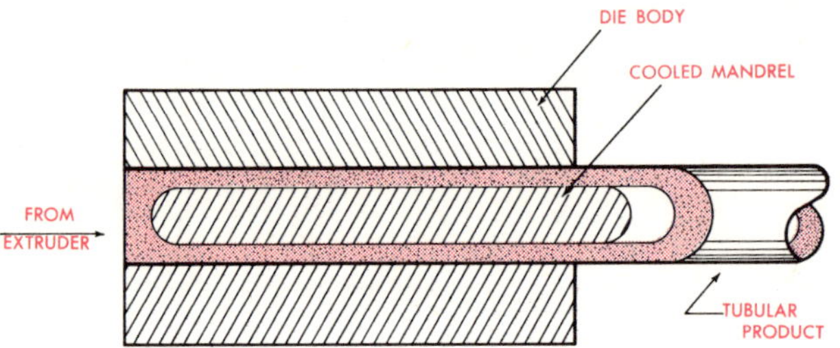

Fig. 15-37. Hot molten material is extruded around cold mandrel or pin in pipe forming operation.

In both "T" and "coathanger" type die construction the molten material is fed (predominately) to the center of the die before being formed by the die lands and adjustable jaw (Fig. 15-38). The width may be controlled by the external deckle bars or the actual die width.

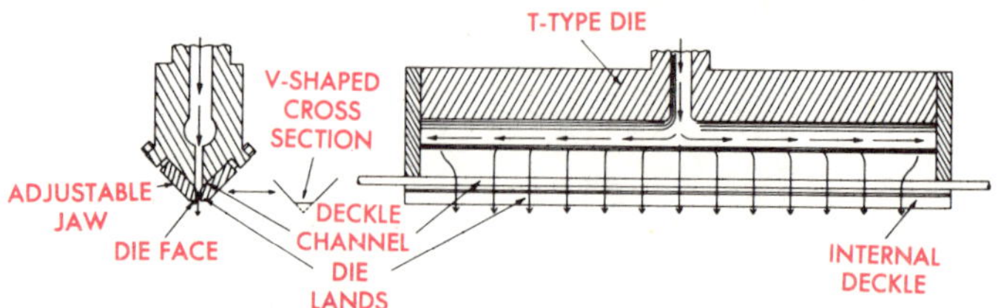

(A) Cross section of T type die.

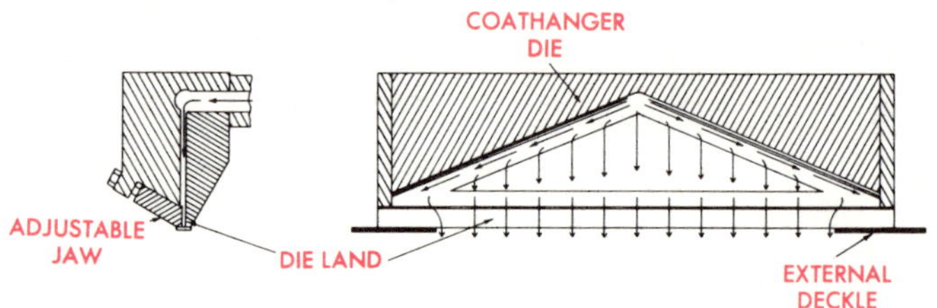

(B) Cross section of coathanger type die.

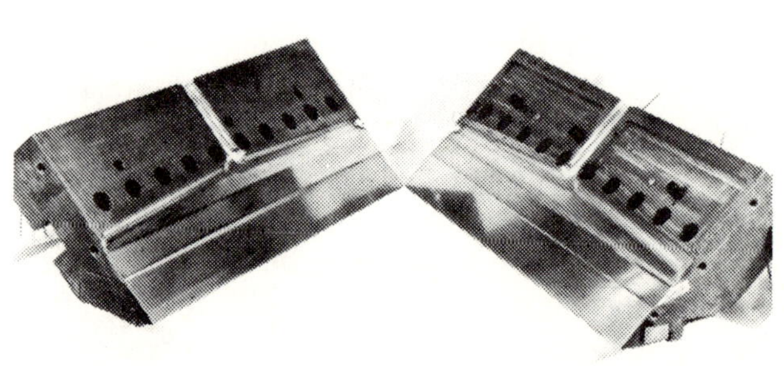

(C) Sheet dies of T type construction. (Waldron-Hartig Div., Midland-Ross Corp.)

Fig. 15-38. Two types of extrusion dies.

In Fig. 15-39A an adjustable choke or restrictor bar is used in the extrusion of sheeting. Sheeting type dies are usually more heavily constructed and possess longer die lands than do the film extruder type dies.

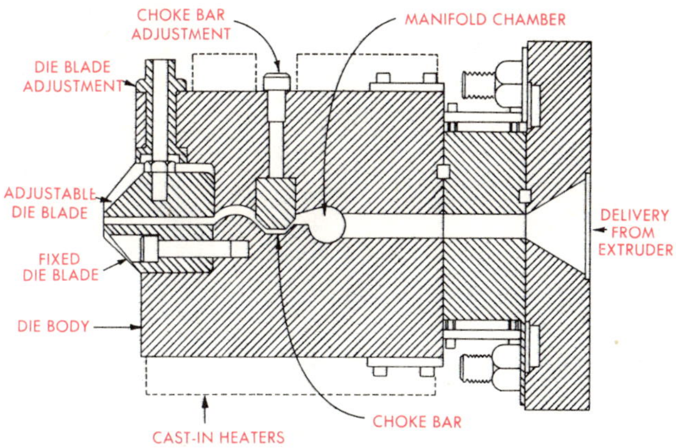

(A) Sheet extrusion die. (Phillips Petroleum Co.)

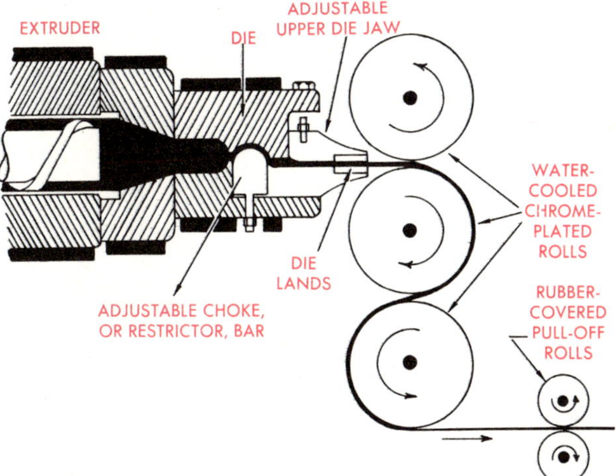

(B) Sheeting die with takeoff unit. (USI)

(C) Sheet is limited in width by the opening of the die orifice. (Waldron-Hartig Div., Midland-Ross Corp.)

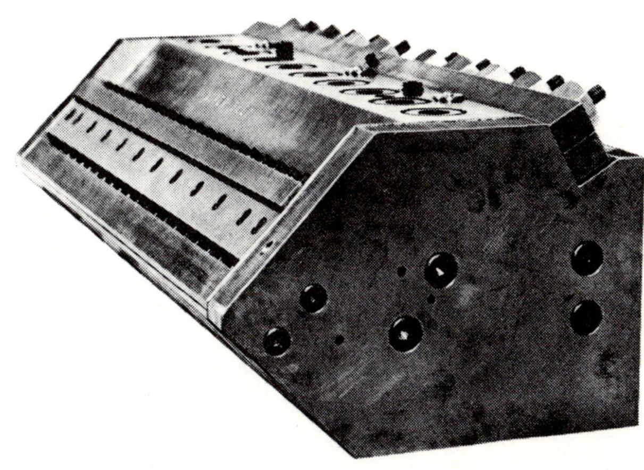

(D) Extrusion sheet die with adjustable jaw. (Waldron-Hartig Div., Midland-Ross Corp.)

Fig. 15-39. Dies with adjustable choke bars.

In Fig. 15-40A the sheet extruded into a quench tank while in Fig. 15-40B the sheet is being drawn over chill rollers (sometimes called *film casting*). Both chill roller and water tank sheet extrusions are used commercially. Water temperature, vibration, and currents must be carefully controlled when using the water tank method, if clear, defect-free sheeting is to be produced. Heavy-gauge films are currently being made by this high-speed method.

With chill roller methods there is a substantial improvement in optical properties of the sheet or film. This method also lends itself to the production of stiffer film extrudate. Both slot type methods provide close thickness tolerances of the film and orientate the molecular structure in one direction.

Table 15-3 gives some common sheet-extrusion problems and their remedies.

Several resins may be extruded at one time into a multimanifold die to produce a sheet or film composed of different resins (see Fig. 15-41).

Blown or tubular extruded film is distinctly different from slot extruded films. The majority of large or inexpensive films are produced by the blow film extrusion method (see Fig. 15-42).

The film is produced by forcing the molten material through a die and around a mandrel before it emerges through the orifice opening in the

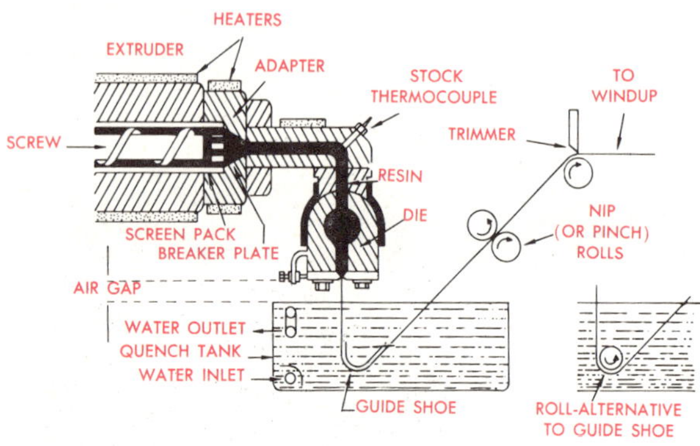

(A) Schematic cross section of the front part of a flat-film extruder, the quench tank, and the takeoff equipment. (USI)

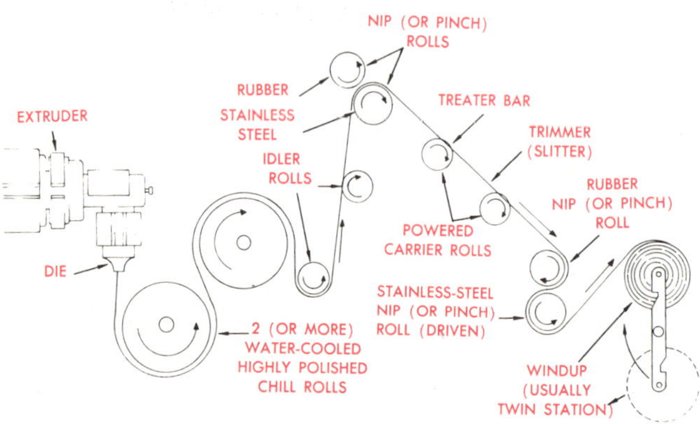

(B) Schematic drawing of chill-roll film extrusion equipment.

(C) Sheet extrusion line with extruder and die at left-center. (Chemplex Co.)

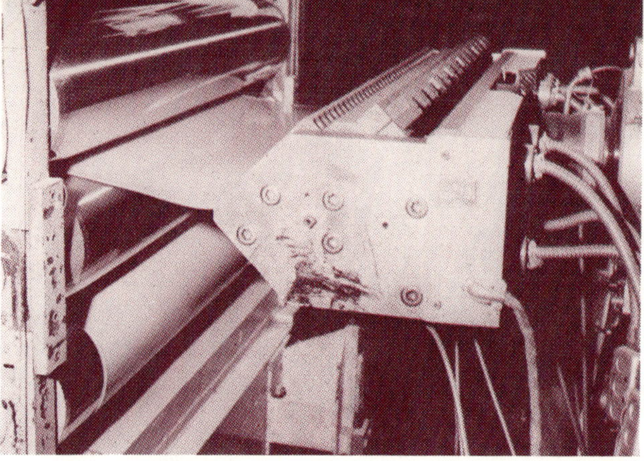

(D) Sheet extrusion from die into nip of the chill rollers. (Fellows Gear Shaper Co.)

(E) Sheet extrusion, looking toward extruder. Finished sheet will be sheared into desired lengths. (Fellows Gear Shaper Co.)

Fig. 15-40. Film extrusion processes.

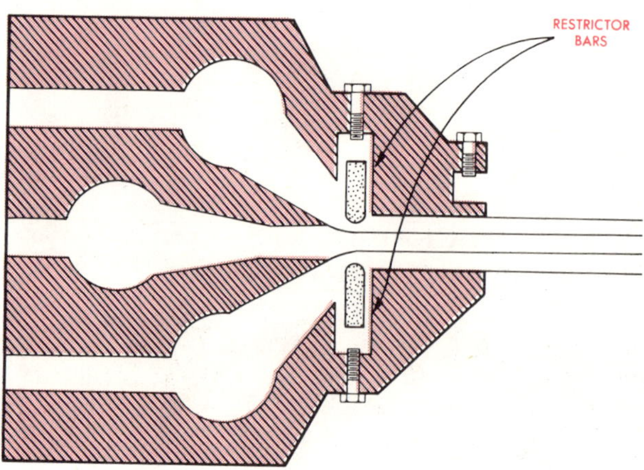

Fig. 15-41. Typical multimanifold die coextruding sheet or film of three different resins.

(A) Extruder and blown film being taken off. (Chemplex Co.)

(B) Closeup of extruded blown film.

(C) View looking down at extruder die and gauge bars. (Chemplex Co.)

(D) Note how large this blow extruded film gets after blow. (BASF Corp.)

Fig. 15-42. Blown film extrusion.

Table 15-3. Troubleshooting Sheet or Film Extrusion Equipment.

Defect	Possible Remedy
Continuous lines in direction of extrusion	Repair or clean out die
	Die contamination or scored polish rollers
	Reduce die temperatures
	Use properly dried materials
Continuous lines across sheet	Jerky operation—adjust tension on sheet
	Reduce polishing roll temperatures or increase roll temperatures
	Check back-pressure gauge for surging
Discoloration	Use proper die and screw design
	Minimize material contamination
	Temperature too high—too much regrind
	Repair and clean out die
Dimensional variation across sheet	Adjust bead at polishing rolls
	Balance die heats
	Reduce polishing roll temperatures
	Check temperature controllers
	Repair or clean out die
Voids in sheet	Balance extruder line conditions
	Use proper screw design
	Minimize material contamination
	Reduce stock temperature
Dull strip	Die set too narrow at this point
	Minimize material contamination
	Increase die temperature
	Repair or clean out die
Pits, craters	Balance extruder line conditions
	Minimize material contamination
	Use properly dried materials
	Control stock feed
	Reduce stock temperature

form of a tube (Fig. 15-43). At this point it is similar to producing pipes or tubes. In blow extrusion this tube (bubble) is then expanded by blowing air through the center of the mandrel until the desired film thickness is reached (something like blowing up a balloon). The tube is usually cooled by air from a cooling ring placed around the die. The *frost line* is the zone where the temperature of the plastics tube has fallen below the softening point of the plastics. In polyethylene or polypropylene film extrusion, the frost zone is evident and actually appears "frosty" because of

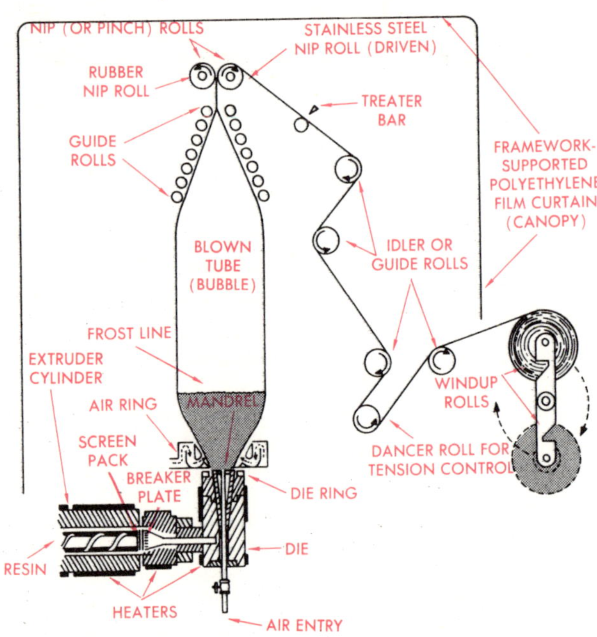

(A) Basic apparatus.

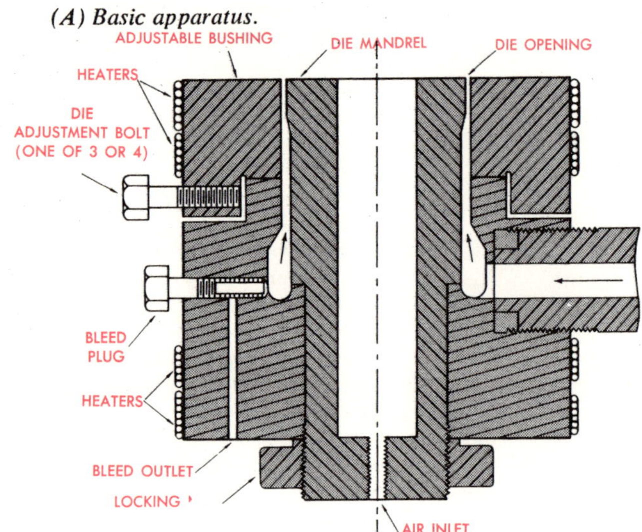

(B) Side-fed manifold-type blown film die. (Phillips Petroleum Co.)

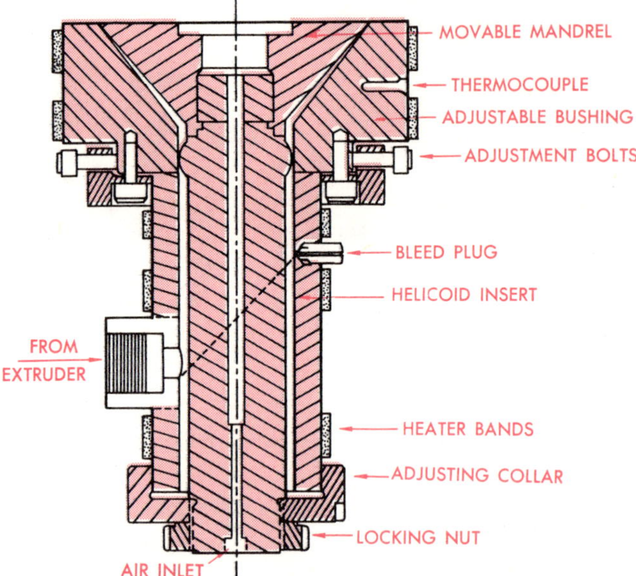

(C) Adjustable die-opening type blown film die. (Phillips Petroleum Co.)

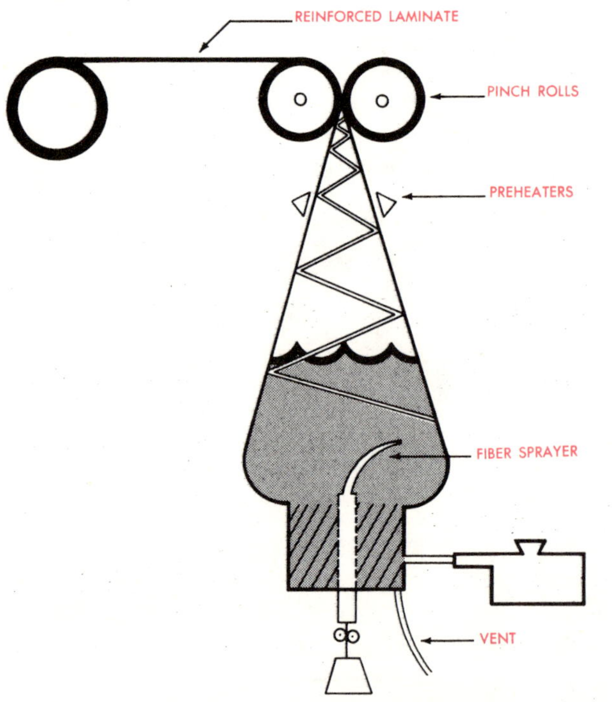

(D) Filament reinforced sheet by blown film process.

Fig. 15-43. Schematic drawings of the blown film extrusion procedure. (U.S. Industrial Chemicals)

the change taking place as the plastics cools from an amorphous to a crystalline state. With some plastics there will be no visible frost line.

The size and thickness of the finished film is controlled by the extrusion speed, takeoff speed, the die (orifice) opening, temperature of the material, and by the air pressure inside the bubble or tube. Blow extruded film is sold as seamless tubing, flat film, and film folded in various ways. Film producers may slit the tubing on one edge during windup. If the tube is blown to a diameter of 8 feet, the flat film will have a width (slit and opened) of over 25 feet. Slot dies of this size are not practical. For the packaging of some foods and garments tubular films are desirable and low cost. Only one heat seal is required in the production of bags from blown tubing.

Blown films are semioriented (less orientation of molecules in a single direction) because they are stretched as the tube is expanded by air pressure. Such bidirectional stretching results in a more balanced molecular orientation in all directions. Improved physical properties are generally considered an asset in blow film production. However, clarity, surface defects, and film thickness are more difficult to regulate than with slot extrusion methods.

Monofilaments are produced much like profile shapes except that a multiorifice die is used. These dies contain a large number of small openings from which the molten material emerges. Multiorifice dies are used in the production of granular pellets, monofilament, and multifilament strands (see Fig. 15-44).

Fig. 15-44. Heavy filaments emerge from the compounding extruder into bath. (Chemplex Co.)

Monofilaments range from 0.005 to about 0.060 inch in diameter and may be handled individually by takeoff equipment. Multifilaments are usually very small or fine and are not easily handled individually; consequently, the molten plastics is forced through a multiorifice die called a *spinneret* (Fig. 15-45). This process produces a multifilament yarn or thread. Many textiles, carpets and ropes are made from these yarns and threads.

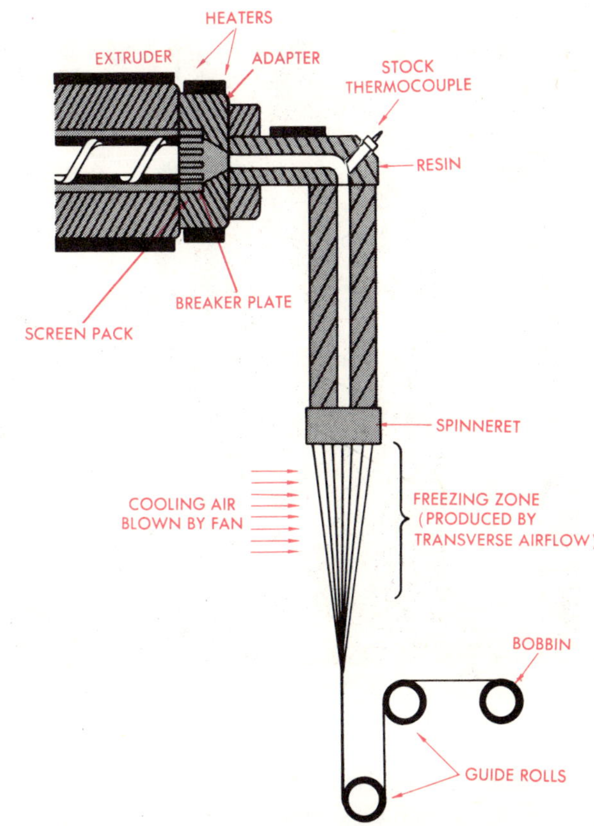

Fig. 15-45. Multifilament extrusion through a spinneret.

Filaments and films may be produced by cold extrusion. In this process solutions of plastics are extruded and caused to solidify by evaporation of solvents or by chemical coagulation in a hardening bath (Fig. 15-46). Rayon filaments are commonly produced by forcing solutions of plastics through spinnerets and into a chemical hardening bath. Films may be cold extruded (sometimes called *casting*) in solution form on polished conveyors (Fig. 15-47). The solvents are removed by evaporation and collected for reuse. Polyvinyls, polycarbonates, acetates, and polysulfones are cold extruded plastics.

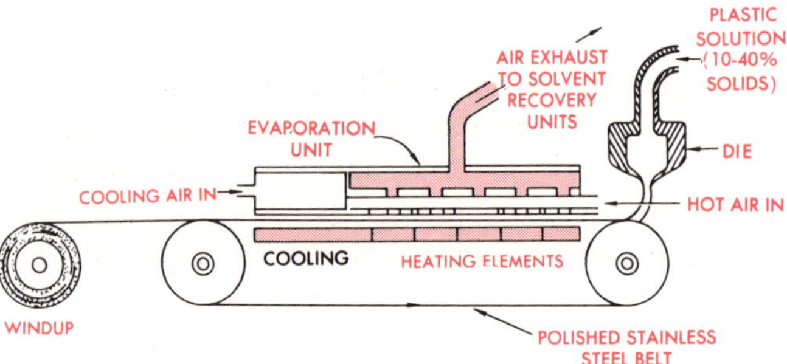

Fig. 15-46. Cold extrusion of film showing evaporation of solvents.

Fig. 15-47. Example of cast film extrusion (hot method on chill roller). (Chemplex Co.)

Table 15-4 gives some common defects and remedies in blow extruding film.

Paper, fabric, cardboard, plastics, and metal foils are common substrates for extrusion coating operations shown in Fig. 15-48. In extrusion coating a thin film of molten plastics is extruded on to the substrate and pressed between rollers without the use of adhesives. On special applications adhesives may be used to ensure proper bonding. Some substrates are preheated and primed with adhesion promoters to ensure proper adhesion. Slot type extruder dies are used for this process.

When two or more plastics coatings or substrates are used, the process is similar to other laminating processes. These laminates may be composed of several different layers of plastics or layers of other materials (see Fig. 15-49B).

Table 15-4. Troubleshooting Blown Film.

Defect	Possible Remedy
Black specks in film	Clean die and extruder
	Change screen pack
	Check resin for contamination
Die lines in film	Lower die pressure
	Increase melt temperature
	Polish all rough edges in film path
	Check nip rolls—make smooth
Bubble bounces	Increase screw rpm and nip roll speed
	Enclose tower or stop drafts
	Adjust cooling ring to obtain constant air velocity around ring
Poor optical and physical properties	Raise melt temperature
	Increase blow-up ratio
	Increase frost line height
	Clean die lips, extruder, and rollers
Failures at fold	Decrease nip roll pressure
Failure at weld lines	If possible, bleed die at weld line
	Heat die spiders—insulate air lines there
	Increase melt temperature
	Check for contamination
Film won't run continuously	Clean die and extruder
	Lower melt temperature
	Increase film thickness

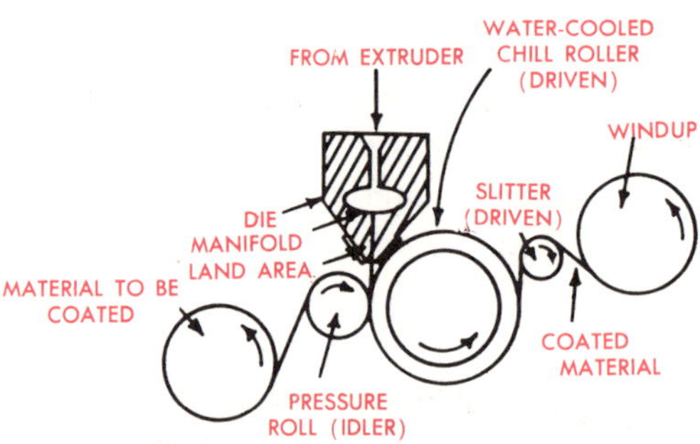

(A) Diagram of operation.

(B) The man on the ladder is watching the extrusion coating of a paper substrate. (Chemplex Co.)

Fig. 15-48. Extrusion coating of substrates.

A three-ply laminate film is used to wrap a variety of bread products (Fig. 15-49A). This film laminate is composed of an inner core (ply) of polypropylene with an outer layer (ply) of polyethylene. A vinyl plastics film laminate was developed specifically for packaging meat products by the Dow Chemical Company and the Oscar Mayer Company. Three different plies are utilized in this laminate: Saran 18 (polyvinylidene chloride) for the outer layer, polyvinyl chloride 88 for the core, and Saran 22 is used for the inner sealing layer (see Fig. 15-50).

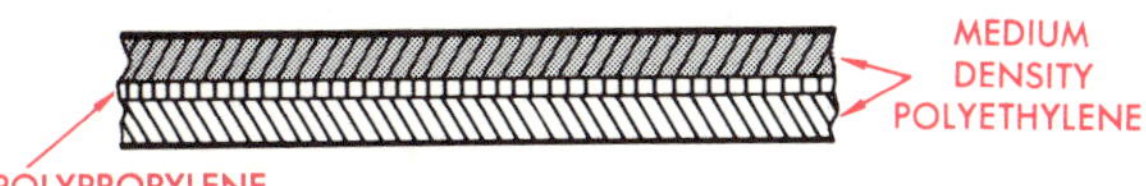

(A) Three-ply film (bread bag use).

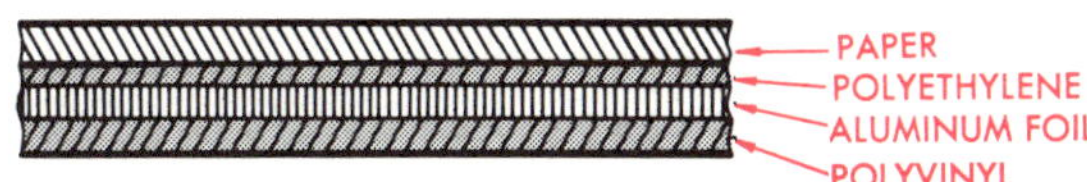

(B) Four-ply film (food pouch use).

(C) Two-ply film (boil-in-bag use).

Fig. 15-49. Plastics laminates.

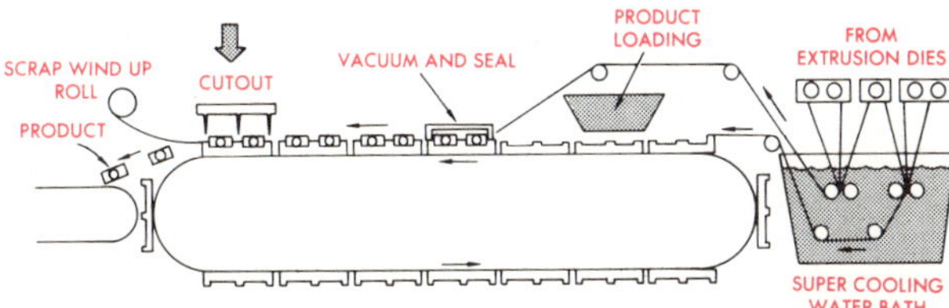

(A) Extrusion and packing.

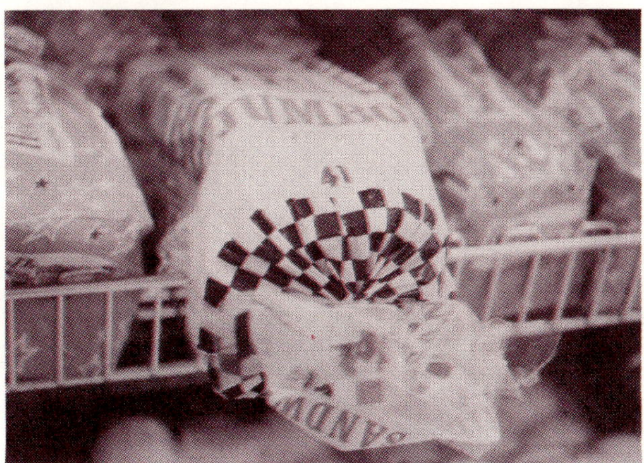
(B) Three-ply laminate film is used on this product.

(C) Saranpac process is used to wrap these products.

(D) Acetate-polyethylene film ply for packaging nuts.

(E) Foil-paper polyethylene packages for dry meats.

Fig. 15-50. Extrusion and product packing technique along with some products.

In the "Saranpac" process all three film layers are extruded and pressed together in a cooling tank. The laminated film is then formed to contain the meat product and vacuum sealed.

Because there are numerous applications of extruded film laminates only a selected few examples are shown in Table 15-5.

Table 15-5. Selected Extruded Film Laminates and Applications.

Laminate Material	Application
Paper-polyethylene-vinyl	Sealable pouches for dried milk, soups, etc.
Acetate-polyethylene	Tough, heat-sealable packing for nuts
Foil-paper-polyethylene	Moisture barrier pouch for soup mixes, dry milk, etc.
Polycarbonate-polyethylene	Tough, puncture resistant skin packages
Paper-polyethylene-foil-polyethylene	Strong heat seals for dehydrated soups
Paper-polyethylene-foil-vinyl	Heat-sealable pouches for instant coffee
Polyester-polyethylene	Tough, moistureproof boil-in-bag pouches for foods
Cellulose-polyethylene-foil-polyethylene	Gas and moisture barrier for pouches of ketchup, mustard, jam, etc.
Acetate-foil-vinyl	Opaque, heat-sealing pouches for pharmaceuticals
Paper-acetate	Glossy, scratch resistant material for record covers, paperback books

Extrusion coating of wire and cable is shown in Fig. 15-51. During this process molten plastic is forced around the wire or cable as it passes through the die. The die actually controls and forms the coating on the wire. Wires and cables are usually heated before coating to remove moisture and ensure adhesion. As the coated wire emerges from the crosshead die it is cooled in passing through a water bath.

Two or more wires may be coated at one time. Television and appliance cords are familiar examples. Wooden strips, cotton rope, and various plastics filaments may also be coated by this process.

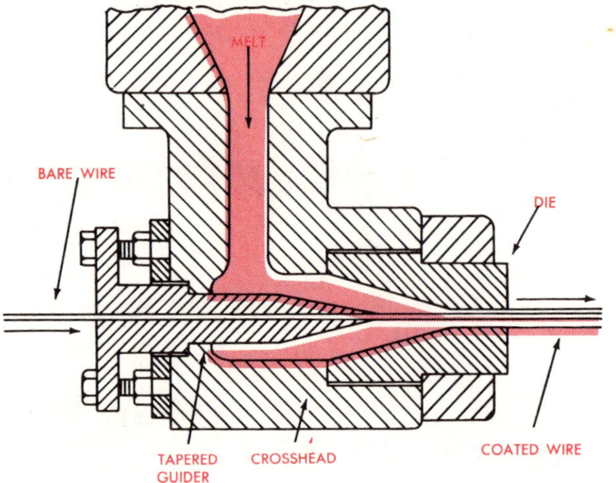

(A) A crosshead holds the wire-coating die and the tapered guider as the soft plastics flows around the moving wire.

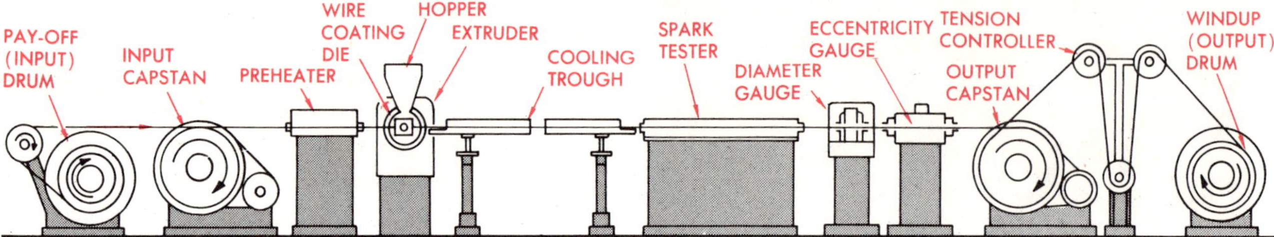

(B) A general layout of the components in a wire-coating extrusion plant. (U.S. Industrial Chemicals Co.)

Fig. 15-51. Extrusion coating of wire and cable.

Blow Molding

Blow molding is a technique adopted and modified from the glass industry for producing one-piece containers and other articles. Although the process has been used for centuries for blowing bottles of glass, blow molding of thermoplastics did not develop until the late 1950's. In 1880 blow molding was accomplished by heating and clamping two sheets of celluloid in a mold. Air was then forced in to form a blow molded baby rattle—possibly the first blow molded thermoplastic article produced in the United States.

The basic principle of blow molding is simple (see Fig. 15-52). A hollow tube (parison) or balloon of molten thermoplastic is placed in a female mold and forced (blown) with air pressure against the walls of the mold. After a cooling cycle the mold opens and ejects the finished product. Although the process is used to produce various containers, objects such as toys, packaging units, automobile parts, and appliance housings may also be blow molded.

Construction of the blow mold is inexpensive, utilizing aluminum, beryllium-copper, or steel as basic materials. Aluminum is a favorite and is one of the lowest-cost blow molded materials. It is most common because it is lightweight and transfers heat

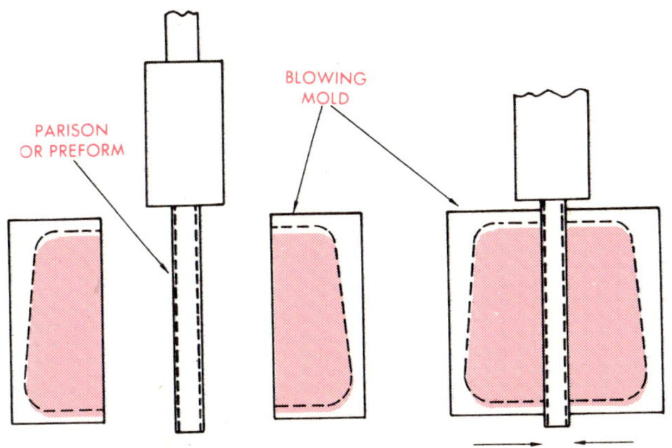

(A) Molded hollow tube (parison) placed between two halves of mold.

(B) Mold closes around parison.

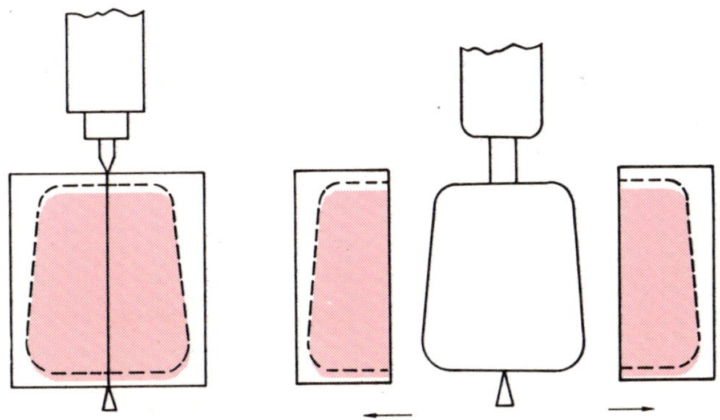

(C) Parison, still molten, is pinched off and inflated by air blast which forces walls against inside contours of cold mold.

(D) When the piece has cooled enough to become solid, the mold is opened and the product ejected.

Fig. 15-52. Blow molding sequence. (USI)

rapidly. Beryllium copper is harder, more expensive, and more wear resistant. Steel is used at pinch-off points. If ferrous molds are used, they are usually plated to prevent rusting or pitting.

There are two basic blow molding methods: (1) injection blowing and (2) extrusion blowing. The major difference in the two methods is in the production of the hot, hollow tube or parison.

Injection blow molding is used because it can produce more accurately the material thickness in various areas of the part. The major advantage is that any shape with varying wall thickness can be made exactly the same each time. There is no bottom weld or scrap to reprocess. The major disadvantage is that two different molds are required. One to mold the preform (parison, see Fig. 15-53A) and the other for the actual blowing operation (Fig. 15-53B). During the actual molding operation the injection molded preform is quickly placed in the blowing mold while it is still hot. Air is then forced into the preform, forcing it to expand against the walls of the mold. The injection blow process has been called *transfer blow* because the injected preform must be transferred to the blow molding cycle. At one time this transfer was made by hand (see Fig. 15-54).

In extrusion blowing, a hot tubular parison is continuously extruded (except when using accum-

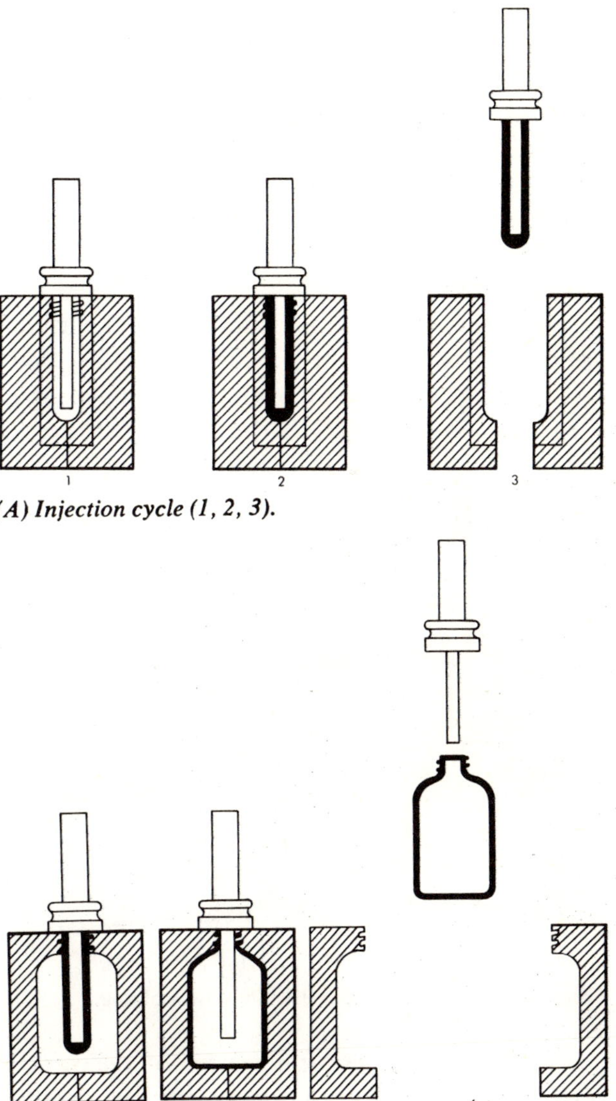

(A) Injection cycle (1, 2, 3).

(B) Blowing cycle (4, 5, 6).

Fig. 15-53. Schematic of the injection blow molding process.

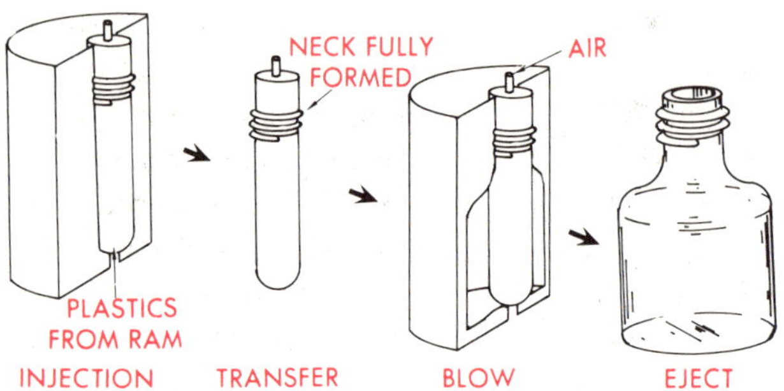

Fig. 15-54. Injection blow process. (Monsanto Co.)

ulator or ram systems). The mold halves then close, sealing off the open end of the parison (Fig. 15-55). Air is injected and the hot parison expands against the walls of the mold where it cools and is then ejected. Extrusion blow molding can mold articles holding up to one hundred gallons of water. Preforms of this size are too costly. Blow extrusion also offers strain-free articles at a high production rate. Controlling wall thickness is the largest disadvantage. Scrap reprocessing is also required. By controlling (sometimes called *programming*) the wall thickness of the extruded parison the disadvantage of thinning in large portions of the mold is reduced. If, for example, a part required an extremely large body but needed strength at corners of this extreme, a parison could be produced with those areas much thicker (see Fig. 15-56).

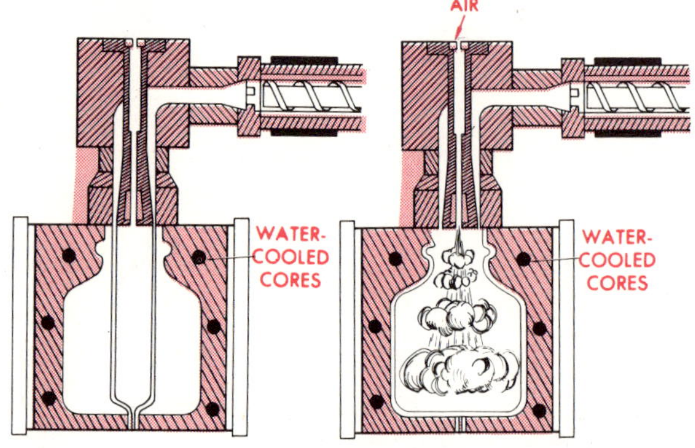

(A) Closing of mold halves. *(B) Injection of air.*

Fig. 15-55. Schematic of extrusion blow molding.

In Fig. 15-57 the arrangement of the various extruder and die parts is shown. By this method one or more continuous parisons are extruded. In Fig. 15-58 hot plastics is fed into an accumulator and then forced through the die. By this method a

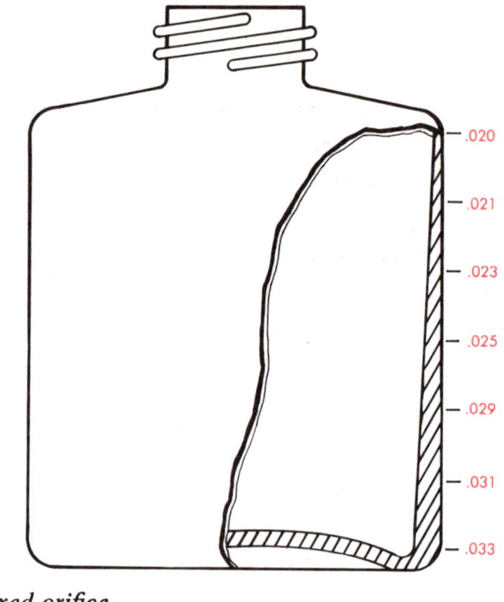

(A) Fixed orifice.

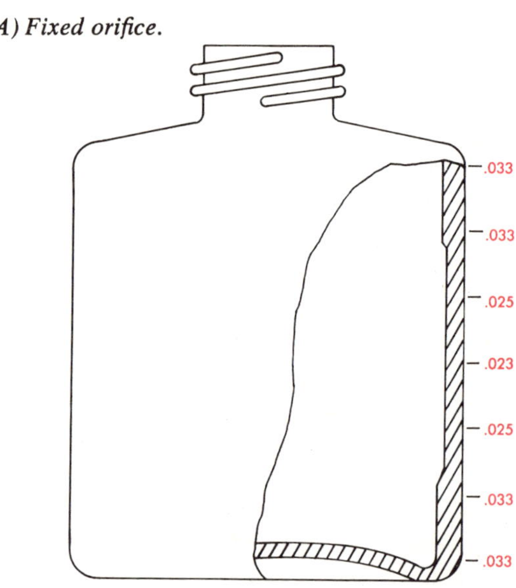

(B) Programmed orifice.

Fig. 15-56. Parison programming with variable die orifice.

controlled length of parison is produced when the ram or plunger is activated. The extruder then fills the accumulator and the cycle begins again.

The wall thickness of the tube or parison may be controlled (programmed) to suit the container configuration by using a die with a variable orifice as shown in Fig. 15-59.

Various blow molding processes or modifications of forming the blown product have been developed by manufacturers (Fig. 15-60). Each process may have a particular advantage for molding of special products. One manufacturer fills and forms the container in one operation. The product is forced in the parison in place of compressed air.

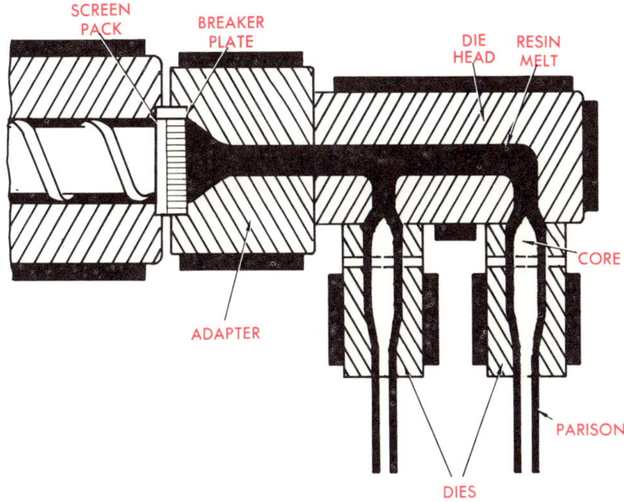

Fig. 15-57. The various machine parts between the extruder face and the die faces, as found in most extrusion blow molding presses.

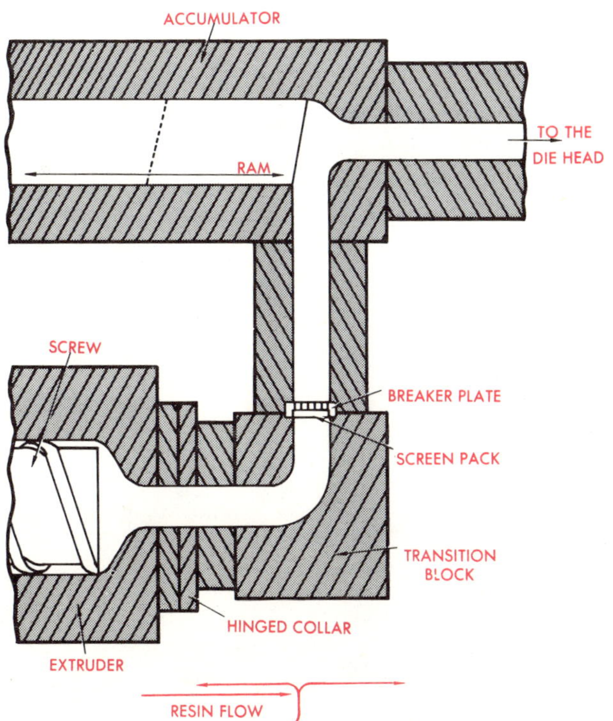

Fig. 15-58. The extruder collar, transition block, screen pack, and breaker plate setup in the blow molding press with accumulator.

Two additional blow molding variations should be mentioned: (1) cold parison and (2) sheet blowing.

Cold parison is a process in which the parison is extruded by normal methods (either injection or

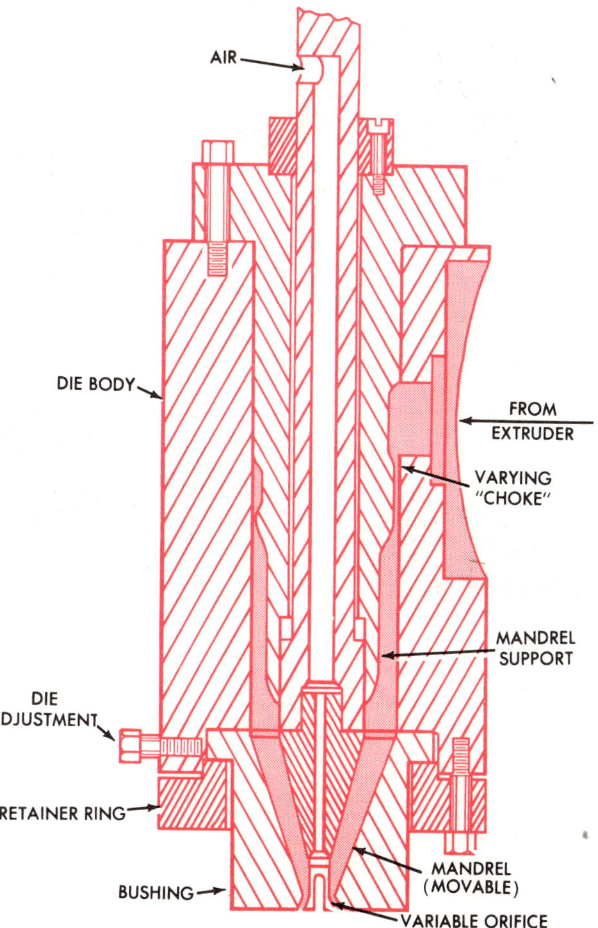

Fig. 15-59. Programming die for blow molding. (Phillips Petroleum Co.)

extrusion) but is cooled and stored until needed. The parison is later heated and blown to shape. The major advantage is that the parison can be shipped to other locations or stored in case of a breakdown or material shortage.

Sheet blowing is based on the blow forming of hot extruded sheets as they are pinched between mold halves. The edges are fused together by the pinching action of the mold. Two different colored sheets may be extruded and formed into a product of two separate colors (Fig. 15-61). It should be apparent that the pinch-welded seams are the largest disadvantage. Two extruders are generally required and there is much scrap to be reprocessed.

Many blow molded products are given a flame treatment to enhance antistatic properties and to make the surface more receptive to ink or other decorative media. (See Chap. 20.) Fig. 15-62

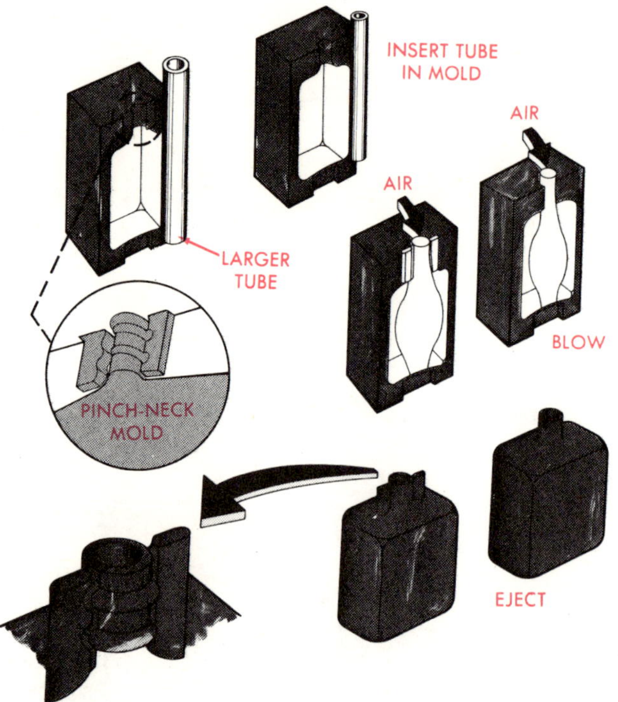

(A) Pinch-neck and "regular" processes.

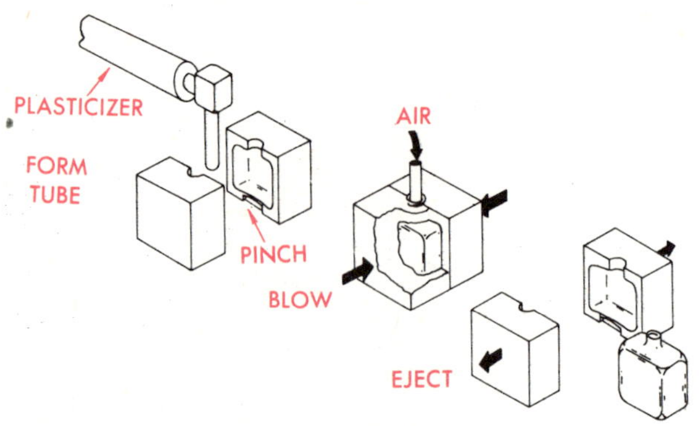

(B) Basic pinch-tube process.

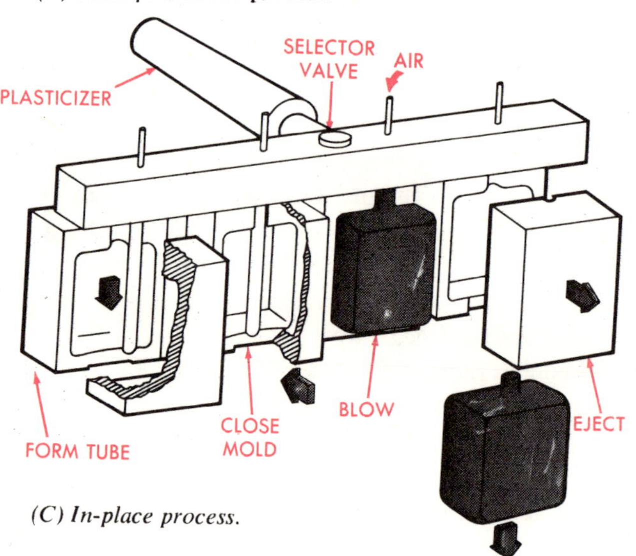

(C) In-place process.

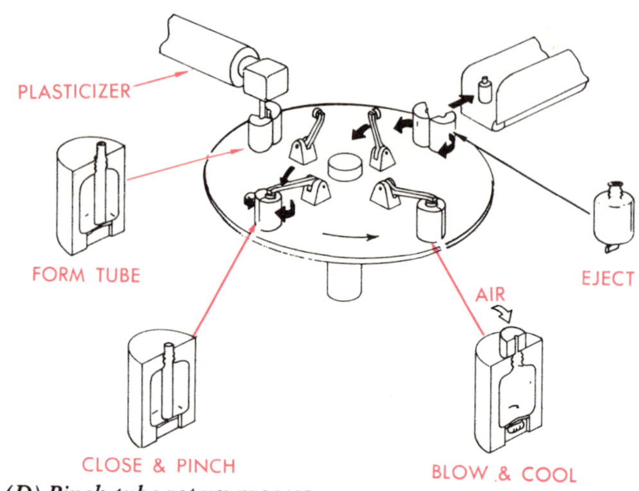

(D) Pinch-tube rotary process.

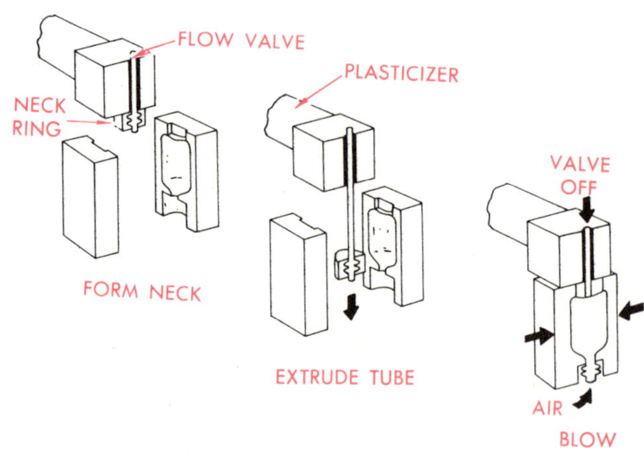

(E) Neck-ring process.

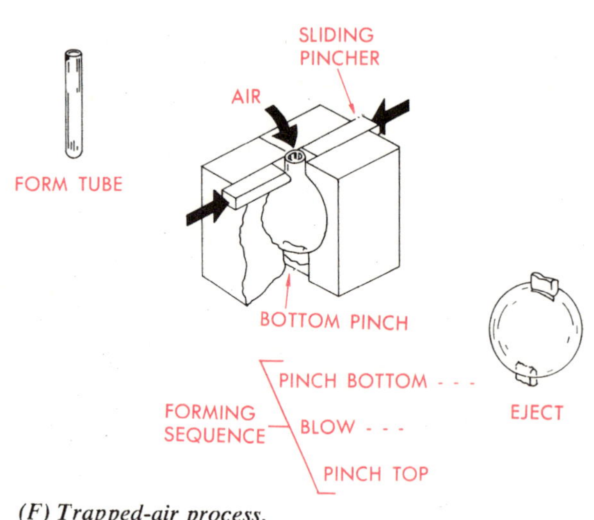

(F) Trapped-air process.

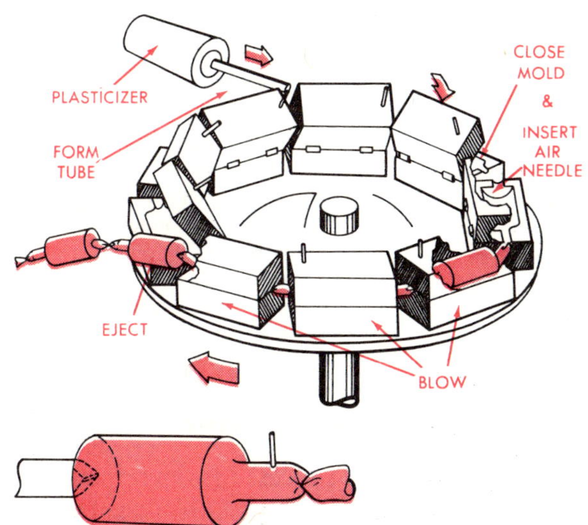

(G) Continuous-tube process (I).

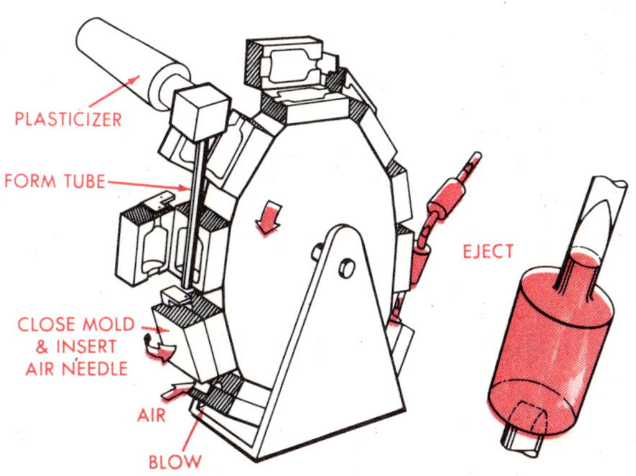

(H) Continuous-tube process (II).

Fig. 15-60. Various blow molding processes. (Monsanto Company)

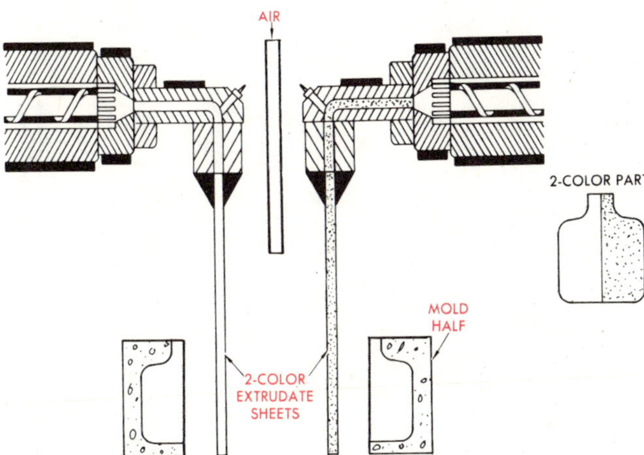

Fig. 15-61. Two-color extrusion is pinched between mold halves to form sheet-blown two-color part.

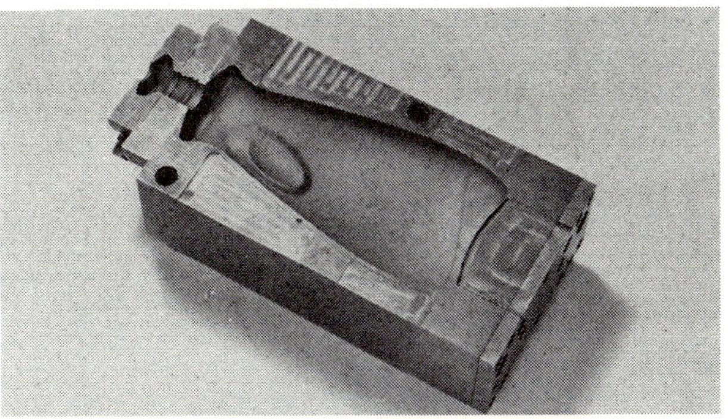

(A) Half the mold for a container with a handle and threaded top. (Unilog Div., Hoover Ball & Bearing Co.)

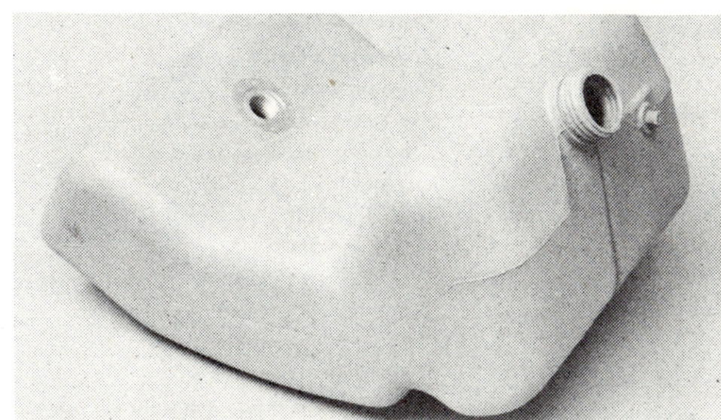

(B) Blow molded snowmobile fuel tank. (Phillips Petroleum Co.)

(C) Blow molded fuel tank for yard tractor. (Phillips Petroleum Co.)

(D) Blow molded high-density polyethylene containers. (Unilog Div., Hoover Ball and Bearing Co.)

Fig. 15-62. Blow mold and blow molded products.

shows some blow molded products and a mold, and Table 15-6 gives some common problems and remedies in blow molding.

Table 15-6. Troubleshooting Blow Molding.

Defect	Possible Remedy
Excess parison stretch	Reduce stock temperature Increase extrusion rate Reduce die tip heat
Die lines	Die surface poorly finished or dirty Blowing air orifice too small—more air Extrusion rate too slow—parison cooling
Uneven parison thickness	Center mandrel and die Check heater bands for uneven heating Increase extrusion rate Reduce melt temperature Program parison
Parison curls up	Excessive temperature difference between mandrel and die body Increase heating period Uneven wall thickness or die temperature
Bubbles (fisheyes) in parison	Check resin for moisture Reduce extruder temperatures for better melt control Tighten die tip bolts Reduce feed section temperature Check resin for contamination
Streaks in parison	Check die for damage Check melt for contamination Increase back pressure on extruder Clean and repair die
Poor surface	Extrusion temperature too low Die temperature too low Poor tool finish or dirty tools Blowing air pressure too low Mold temperature too low Blowing speed too slow
Parison blowout	Reduce melt temperature Reduce air pressure or orifice size Align parison and check for contamination Check for hot spots in mold and parison
Poor weld at pinch-off	Parison temperature too high Mold temperature too high Mold closing speed too fast Pinch-off land too short or improperly designed
Container breaks on weld lines	Increase melt temperature Decrease melt temperature Check pinch-off areas Check mold temperature and decrease cycle time
Container sticks in mold	Check mold design—eliminate undercuts Reduce mold temperature and melt temperature Increase cycle time
Part weight too heavy	Parison temperature too low Melt index of resin too low Annular opening too large
Warpage of container	Check mold cooling Check for proper resin distribution Lower melt temperature Reduce cycle time for cooling
Flashing around container	Lower melt temperature Check blowing pressure and air start time Check for molds closing on parison Check air start time and pressure

Calendering

Calendering is a process in which thermoplastic materials are squeezed to thickness by two or more heated rollers (Fig. 15-63). Films and sheet forms may be produced by this method with a glossy matt or embossed finish. Expensive garment materials are common examples. Floor tiles and solid plastics linoleum may also be produced by this method. As early as 1836 two-roller calenders were being used to process sheet rubber. Today, predominately vinyls and a variety of polyurethanes, polyethylene-polypropylene copolymers, polyethylene-vinyl acetate copolymers, polyethylene-ethyl acrylate copolymers, modified polystyrenes, and other synthetic rubbers are being processed by this method into film and sheet stock.

The calendering process consists of blending a hot mix of resin, stabilizers, plasticizers, pigments, etc. in a continuous or Banbury type mixer. This

Fig. 15-63. Calendering of thermoplastic material. (Monsanto Co.)

hot mix is then fed through a two-roll mill to give a heavy sheet stock. Because the rollers of the calender equipment are very costly and easily damaged by metallic contaminants, a metal detector is often used to scan the sheet before entering the calender. Once the heavy sheet enters the calendering equipment a series of heated, revolving rollers squeeze the material into the desired thickness. As the material passes through the calendering rollers, the hot sheet becomes progressively thinner. A special pair of precision, high-pressure finish rollers are used as gauging and embossing rolls. Finally, the hot sheet is cooled on the chilled roll and taken off in sheet or film forms. Calender rollers are usually arranged in an inverted *L* or a *Z* arrangement. (See Fig. 15-64.)

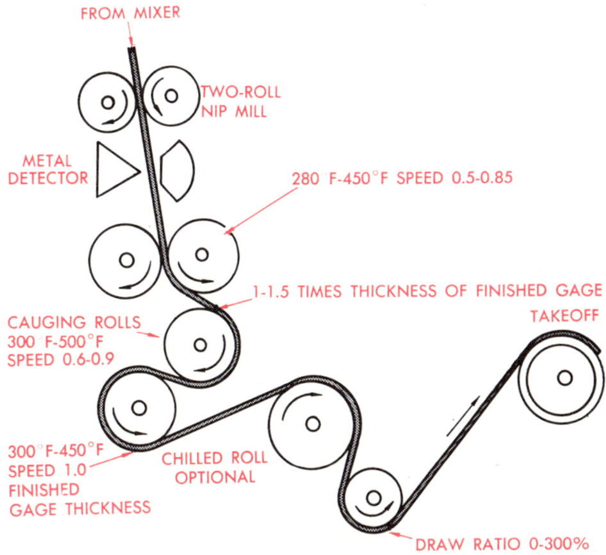

Fig. 15-64. Calendering of thermoplastic sheet while passing through a series of rollers (a simplified Z-type arrangement of rolls).

Calendering equipment together with accessory control equipment is very expensive. This high cost has discouraged the installation of calendering equipment for a number of years. Replacement costs for a calender line may exceed 400 million dollars. Calendering has some advantages over extrusion and other methods when producing colored or embossed films or sheets. Calendering is one of the best methods for producing PVC sheets and film. When changing colors, a calender requires a minimum of cleaning, while an extruder must be purged and cleaned thoroughly.

The term "calendering" is also used (incorrectly) in discussing the application of a plastics film to one or both sides of fabric or paper. These operations should be classed as either a coating or laminating operation. As coating machines calenders may press paper, fabric, or other substrates together as they pass through the rollers.

Table 15-7 gives some common calendering problems and remedies.

Table 15-7. Troubleshooting Calendering.

Defect	Possible Remedy
Blistering of film or sheet	Reduce melt temperature Reduce program speed of rolls Check for resin contamination Reduce temperature of chill rolls
Thick section in center, thin edges	Use crowned rolls Increase nip opening Check bearing load of rolls
Cold marks or crow's feet	Increase stock temperature Decrease program feed
Pin holes	Check for contamination in resin Blend plasticizers more thoroughly in resin
Dull blemishes	Check for lubricant or resin contamination Check roll surfaces Increase melt temperature Increase roll temperature
Rough finish	Raise roll temperature Rolls too cold Excessive nip between rolls
Roll bank on sheet	Incorrect stock temperature—increase/decrease Provide constant takeoff speeds Reduce roll temperatures and nip clearance

Laminating and Reinforcing

By definition a laminated plastics is "a plastics material consisting of superimposed layers of a synthetic resin impregnated or coated filler which has been bonded together, usually by means of heat and pressure, to form a single piece."

Lamination may be divided into two distinct processes: (1) high pressure (more than 1000 psi) and (2) low pressure (less than 1000 psi). There has been a tendency to use the terms "structural laminates" and "reinforced plastics laminates" in referring to low-pressure laminates. The actual process of reinforcing is accomplished through many processing operations including laminating, compression, transfer, injection, casting, calendering, rotational molding, and other selected plastics processes.

The result of the laminating process should not be confused with "reinforced plastics" in which the reinforcement is not a laminated form.*

Traditionally, laminated plastics were products produced by high molding pressures exceeding 1000 pounds per square inch. Today, however, laminated products may also include some materials used with processing pressures of less than 1000 psi. Various reinforcements of plastics may be used in both high- and low-pressure lamination processes.

Simply stated, lamination is the process of combining two or more composite layers into one piece. A laminate may consist of several layers of resin impregnated paper, cloth, asbestos, wood, or glass fibers. These composite layers may consist of polyethylene film on paper, cloth, wood, or metal foils. In other laminates the plastics is used as an adhesive, holding various substrates together as in plywood, some metal laminates, and honeycomb construction panels.

Cams, pulleys, gears, fan blades, decorative tops, printed circuit boards, and nameplate stock are familiar laminated products.

High-Pressure Lamination

In high-pressure lamination thermosetting resins are widely used to impregnate the base materials. Urea, melamine, phenolic, polyester, epoxy, and other resins are used (Table 15-8).

*See Reinforced and Laminated Plastics, p. 77.

For some products molds are used to form the impregnated stock into various shapes including rods, tubes, cups, plates, and cones. In this case, matched metal molds must be used in fusing and compressing the mass into a laminate structure.

High-pressure industrial laminates (Fig. 15-65) possess an extremely wide range of properties and may be considered for special nondecorative applications. One of the earliest phenolic laminates as we know it today was produced in 1905 by J. P. Wright, the founder of the Continental Fiber Company. This was five years before Baekeland had patented his ideas of using phenolic resin

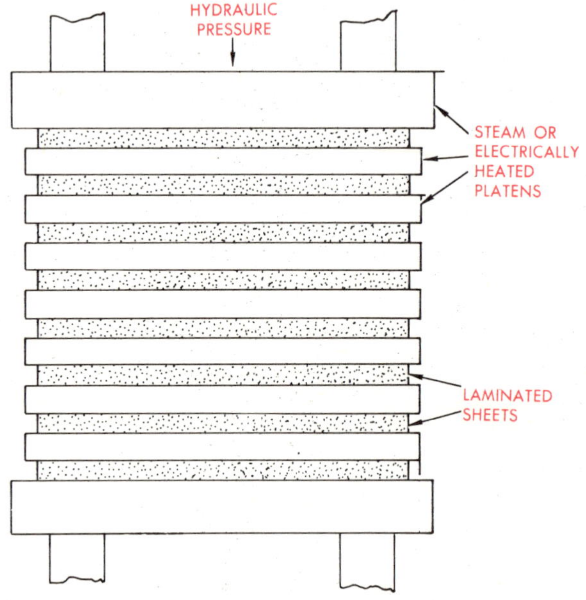

Fig. 15-65. Multiple high-pressure laminating in a hydraulic press.

Table 15-8. Selected Resin/Plastics and Reinforcements Used in Lamination.

Resin/Plastics	Paper	Cotton Fabric	Asbestos Fabric	Fibrous Glass Fabric/Mat	Metallic Foils	Composites, Honeycomb, etc.
Acrylic	LP	—	—	LP		—
Polyamide	LP	—	—	LP	LP	—
Polyethylene	LP	—	—	LP	LP	—
Polypropylene	LP	—	—	LP	LP	—
Polystyrene	—	—	—	LP		
Polyvinyl chloride	LP	—	—	—	LP	LP
Polyester	LP-HP	LP	LP	LP	—	LP
Phenolic	LP-HP	LP-HP	HP	HP-LP	HP	LP-HP
Epoxy	LP	LP-HP	—	LP-HP	LP	LP
Melamine	HP	HP	—		HP	
Silicone	—	—	—	HP	—	LP

LP = Low-pressure (laminate).
HP = High-pressure (laminate).
— = Only limited amounts manufactured in this category.

impregnated sheets as laminates. Various high-pressure industrial laminates were made from a layup of paper, cloth asbestos, synthetic fiber, or fibrous glass using this basic patent. Today there are over fifty standard industrial grades of laminates for electrical, chemical, and mechanical applications. Initially high-pressure industrial laminates took the place of "mica" as a quality electrical insulation material. In about 1913 the the Formica Corporation emerged, producing high-pressure laminates to replace (FOR) mica (MICA) in many electrical and mechanical applications. By 1930 the National Electrical Manufacturers Association (NEMA) realized the potential of decorative laminates, and in 1947 a separate NEMA section was established in close association with numerous government agencies and associations which were also interested in establishing product standards for decorative laminates (see Fig. 15-66).

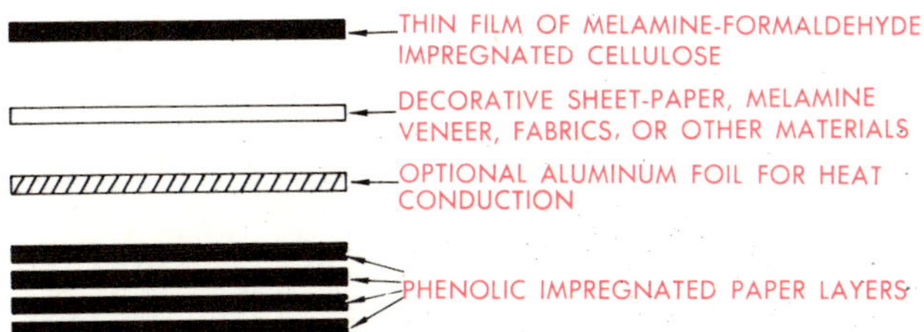

Fig. 15-66. Composition of decorative high-pressure laminates, showing various layers.

The major disadvantage of high-pressure lamination is that production rates are slow when compared with those of high-speed injection molding of plastics parts. In the production of matched mold laminates and many flat laminates, production rates may be competitive with other molding rates. The actual impregnation of the reinforcing layer is accomplished by various methods including premix, dipping, coating, or spreading (see Fig. 15-67).

After a drying period the impregnated laminating stock is cut to desired sizes and placed in multisandwich form between metal plates of the press. These plates may be glossy, matt, or embossed. Metal foils are sometimes used between surface layers to produce a decorative finish. With the well-known decorative laminates a printed pattern layer and a protective overlay sheet is

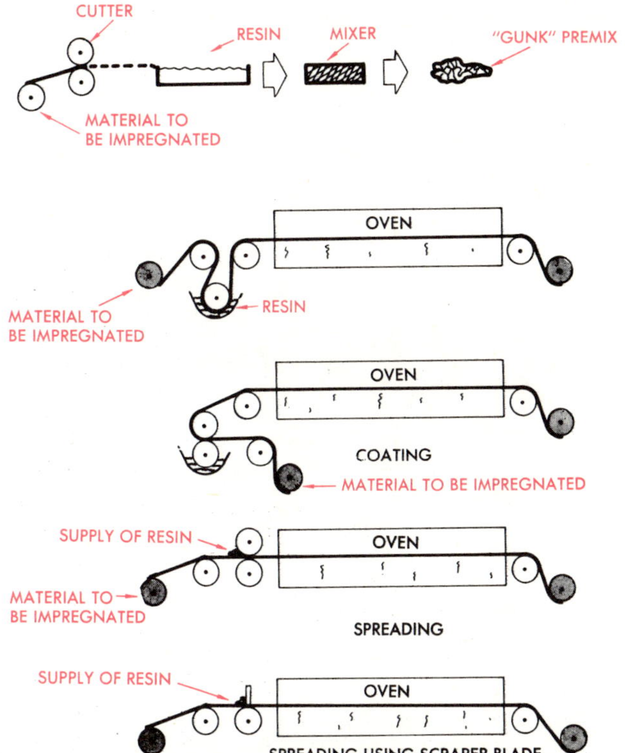

Fig. 15-67. Various impregnation methods.

superimposed on the base material. The prepared stock is then heated and compressed under high pressure and temperature. The combined heat and pressure cause the resin to flow and the layers to compact into one polymerized mass. Polymerization may also be accomplished by chemical or radiation sources. When the thermosetting resins have cured or the thermoplastic resins have cooled, the laminate is removed from the laminating press.

A few of the many industrial high-pressure laminates are shown in Fig. 15-68. These high-pressure laminates were made by impregnating a reinforcing material with a thermosetting resin, semicuring, and bonding layers in a press with high temperature and pressure (see Fig. 15-69). Various reinforcing materials are used, including papers, cotton, asbestos, and glass fabrics.

Low-Pressure Laminates

It has previously been pointed out that low-pressure laminates are sometimes referred to as "reinforced plastics." A reinforced plastics or resin may be processed by several processing techniques. "Low-pressure laminates" refers only to the process

Fig. 15-68. Miscellaneous examples of reinforced plastics parts. (Spaulding)

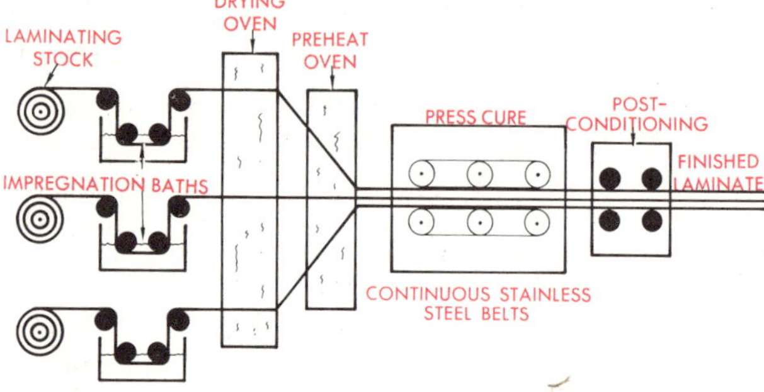

Fig. 15-69. Continuous high-pressure lamination.

of forming multiple layers of reinforcement and plastics or resin.

In low-pressure lamination inexpensive molds made of wood, concrete and plaster, or other plastics may be used to shape the product. Low-pressure lamination may be used for fabrication of boat hulls, automobile bodies, luggage, airplanes, storage tanks, and other large structures. Production of high-pressure laminates of this size is not practical or economical. In low-pressure lamination only enough pressure is applied to keep the resin impregnated layers together until the resin cross-links or cools (thermoplastics).

The use of thermoplastic materials in low-pressure lamination has some limitation but interest and product application are growing in demand. Hot melts, from plasticizing equipment, are used in producing composite layered laminates.* Acrylic monomers have been used where color or clarity is important in the laminated product. Selected thermoplastic powders and films may be heated by various energy sources and used in fusing and compressing the mass in a laminate structure. Many of these thermoplastic laminates may then be postformed, hot or cold, on dies or molds with varying cross-sectional thickness.**

In the past only thermosetting plastics were commercially reinforced in large quantities. Today, low-pressure laminates (primarily thermosetting) are produced by modifications of about ten molding processes (see Fig. 15-70): (1) matched molding, (2) hand layup, (3) vacuum bag molding, (4) pressure bag molding, (5) spray-up molding, (6) filament winding, (7) centrifugal laminating, (8) pultrusion, (9) continuous laminating, and (10) cold laminating. In each example, molds must be carefully prepared in order to ensure proper release of the finished product. Film, wax, and silicone releasing agents are commonly used on mold surfaces.

Matched male and female molds are used to shape the reinforced resin (Fig. 15-70A). "Gunk" molding compounds and saturated reinforcements may be used. With fibrous glass preforms, a measured amount of catalyzed resin is added and the mold closed. After heat and pressure causes the resin to polymerize, the article is removed. The resulting product may not be considered a laminate in the strictest sense. A similar technique using "gunk" molding compounds is very common in the compression molding industry.***

A variation to this process is *macerated* laminate molding. Macerated parts are produced by chopping the reinforcing materials into ⅛ to ½ inch squares and then molding in the matched molds. Products produced from matched mold

---

*See Extrusion Coating, p. 309.

**See Cold Molding, p. 276.

***See Compression Molding, p. 240.

laminating are strong and may have excellent surface finish on both inside and out; however, mold and equipment costs are high.

Thermosetting resins are also employed during *hand layup molding* (Figs. 15-70B and 15-71). Because there is no pressure (only atmospheric) used in applying the saturated reinforcing material to the mold, this process is not truely a molding process by definition. *Contact molding* is a more descriptive term. After the mold is properly prepared with a releasing agent, a layer of catalyzed resin is applied and allowed to polymerize to the gel state (tacky).

(A) Large hydraulic molding press. The matched mold is designed to produce a reinforced chair shape. (Cincinnati Milacron)

(B) Hand layup and contact molding process is utilized in the application of fiber glass to a structure of titanium and honeycomb. (Bell Helicopter Co.)

(C) Low-pressure molded reinforced fairing along with helicopter nose section, firewall, fuselage, and deck panels. (Bell Helicopter Co.)

(D) Men placing fluted-core construction reinforcements in a female mold. (McMillan Radiation Labs., Inc.)

(E) Contact mold for hand layup of honeycomb reinforcements for radome. (McMillan Radiation Labs., Inc.)

Fig. 15-70. Molding processes of low-pressure laminates.

Specially formulated gel-coat resins are commonly used in industry to improve flexibility, blister resistance, stain resistance, and weatherability of the product. Neopentyl glycol, trimethyl-pentane-diol gylcol, and propylene glycol based gel coats represent a major advantage in product surface treatment of reinforced polyesters.

This gel coat of resin provides a protective surface layer through which fibrous reinforcements are not to penetrate. (A prime cause of deterioration of fibrous reinforced plastics is penetration of water as fibers protrude at the surface.) A layer of reinforcement is then applied and additional catalyzed resin is poured, brushed, or sprayed over the reinforcement. This sequence is repeated until

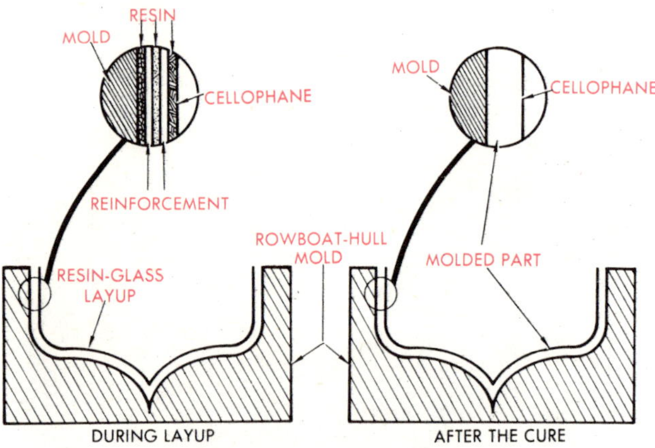

(A) Reinforcing material in mat or fabric form applied to a mold and then saturated with a selected resin.

(B) Hand layup of honeycomb reinforcement, with inner skin and vacuum bag and oven cure to follow. (McMillan Radiation Labs., Inc.)

Fig. 15-71. Hand layup of reinforced thermoset.

the desired thickness is obtained. In each layer the mixture is worked to the mold shape manually with rollers. The reinforced laminate is then allowed to harden or cure. External heating is sometimes used to speed polymerization.

During vacuum bag molding a plastic film (polyvinyl alcohol, polyethylene, polyester, etc.) is placed over the layup and about 12 psi of vacuum is drawn between the film and the mold (see Fig. 15-72). Vacuum is usually measured in inches or centimeters of mercury drawn in a graduated tube. The ratio of pounds pressure and inches or centimeters is expressed in this formula:

$$\frac{14.7}{12} = \frac{29.9}{x}$$

or $x = 24.4$ inches of mercury

$24.4 \times 2.54 = 61.976$ centimeters of mercury

where

$x =$ unknown in inches of mercury,
$14.7 =$ known psi pressure,
$29.9 =$ inches of mercury corresponding to 14.7 psi.

The plastics film forces the reinforcing material against the mold surface producing a high-density air-bubble-free product. Tooling for vacuum bag molding is expensive when large pieces are made, and production is slow compared to the high-speed production rates of injection molding.

Pressure bag molding is also expensive and slow in comparison to high-speed injection molding; however, large, dense products with good finishes both inside and out are possible. Pressure bag molding is a process where a rubber bag is used to force the laminating compound against the contour of the mold. Approximately five pounds of pressure is applied to the bag during heating and curing cycle (see Fig. 15-73).

Sprayup methods are commonly used to spray catalyst, resin, and chopped roving simultaneously onto complex mold shapes (Fig. 15-74). The mixture is hand-rolled to compact the mixture in the mold, removing air bubbles and ensuring complete wetting of the reinforcement. Heat may be applied to speed cure and production rates. This method is low cost, provides for an extremely high degree of complex shapes, and production rates are relatively rapid in comparison to hand

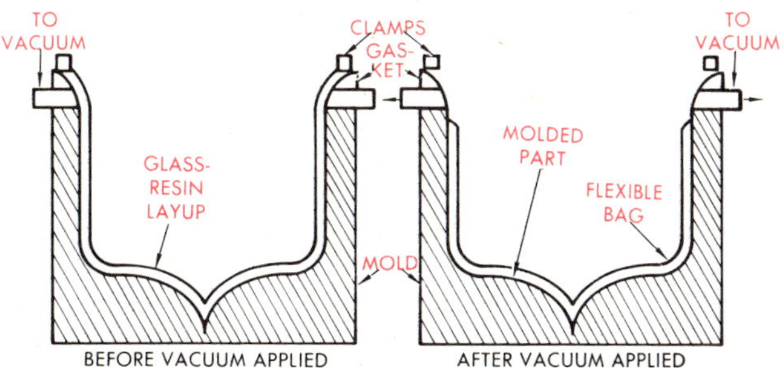

(A) Principle involves application of pressure to a layup, resulting in improved strength and surface on unfinished side of molded product.

(B) Cargo carrier is vacuum bag molded from three layers of prepreg and one layer of dry glass cloth. (FMC Corp.)

(C) Cargo carrier in (B) in use on aircraft. (FMC Corp.)

Fig. 15-72. Vacuum bag molding.

laying methods. Care must be taken to ensure that uniform layers of materials are applied or mechanical properties may not be consistent throughout the product.

Filament winding may produce strong parts by winding fibrous reinforcements on a mold. Many

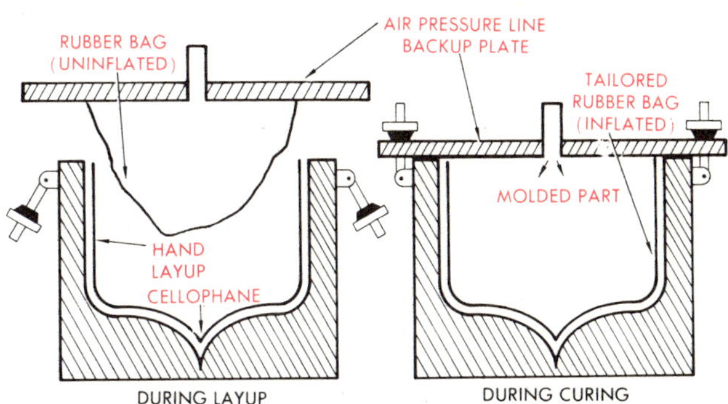

(A) Pressure and heat are applied to a hand layup during the curing stage.

(B) An unfinished, untrimmed aircraft part is show after removal from a pressure bag mold. Illumination from the opposite side shows honeycomb orientation. (McMillan Radiation Labs., Inc.)

Fig. 15-73. Pressure bag molding method.

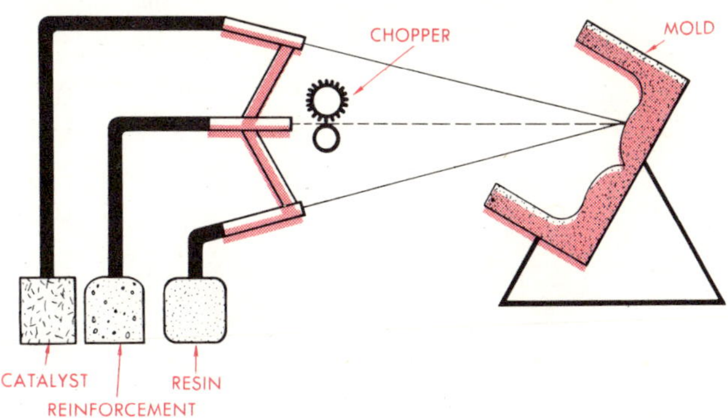

Fig. 15-74. Sprayup method covers complex shapes as readily as simple configurations, an advantage over hand layup.

cylindrical laminated forms are produced by this method. Various yarns or filament reinforcing materials are used. These strands of reinforcement are saturated or coated with catalyzed resin and mechanically wound onto a collapsible mandrel (Fig. 15-75). Specially designed winding machines may lay down these strands in a predetermined pattern to give maxmium strength in the desired direction (Fig. 15-76). The collapsible mandrel must have the desired shape of the finished product. Soluble or low temperature melting mandrels may also be used for special complex shapes or sizes.

The advantage of filament winding is that it allows the designer to place the reinforcement in the areas where the greatest stress is to be applied. Containers made by this process usually have a higher strength-to-weight ratio than other materials and may be produced at a lower cost in virtually any size. Fig. 15-77 illustrates the importance of winding patterns on pressure vessels.

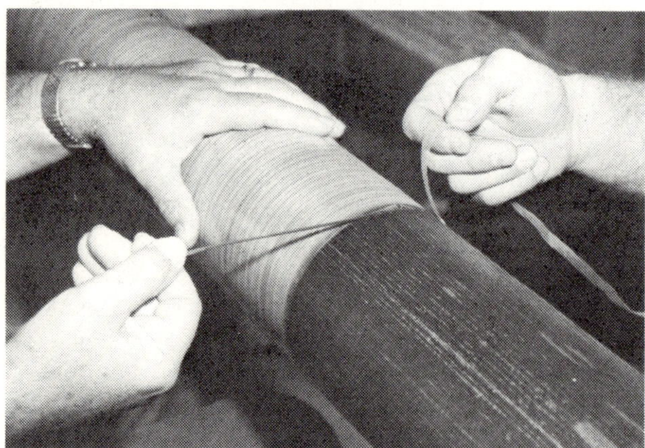

Fig. 15-75. Manual winding of 1/8-in wide boron-epoxy prepeg tape is used in the production of this helicopter tail-rotor driveshaft. (Wittaker Corp.)

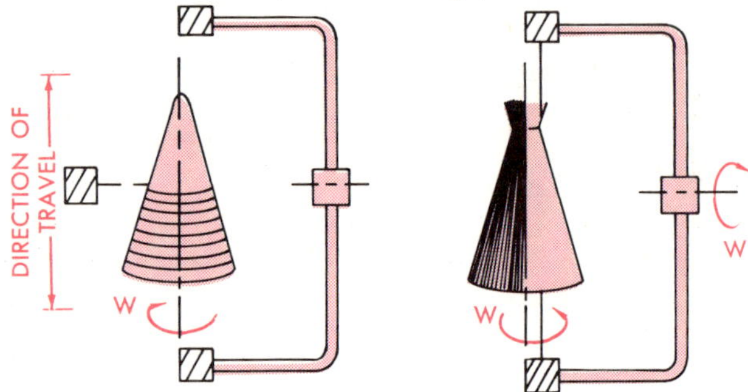

(A) Strickland B winding techniques involve pulling fibers lengthwise over a mandrel and overwrapping with circumferential winding. (CIBA Corp.)

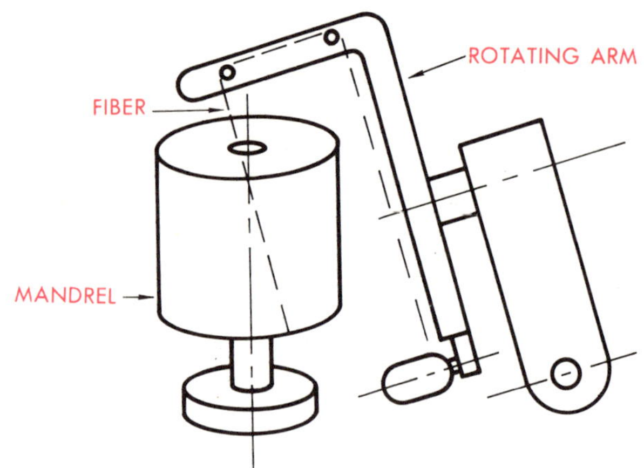

(B) Planetary machine technique has reinforcing wrapped end-to-end or pole-to-pole. (CIBA Corp.)

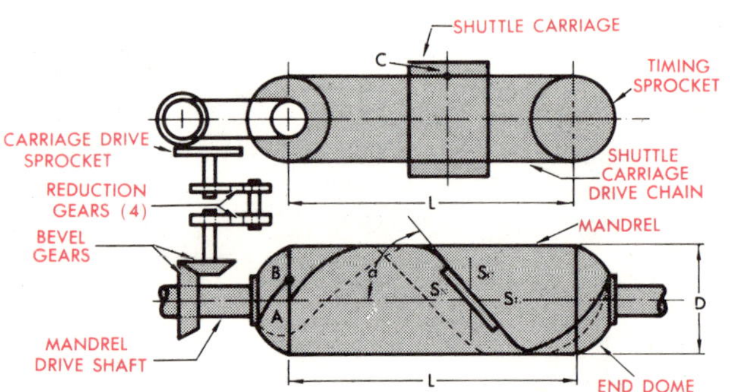

(C) This method involves a helical pattern along a cylindrical mandrel. (CIBA Corp.)

(D) This machine is used in the filament winding of large solid-propellant rocket motor cases. (SCI, Inc.)

Fig. 15-76. Filament winding techniques.

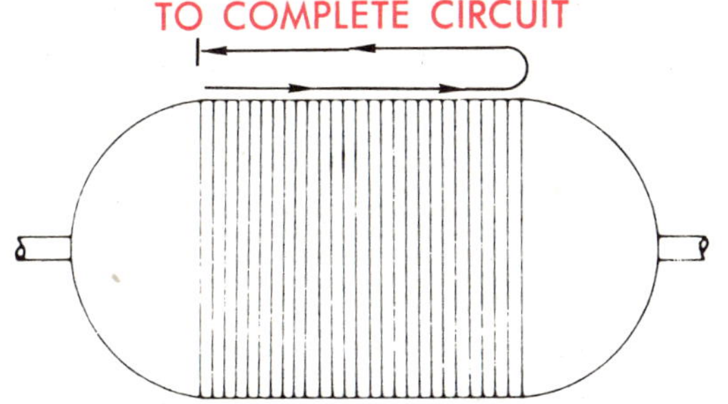

(A) Circular loop windings provide optimum girth or loop strength in a filament wound structure.

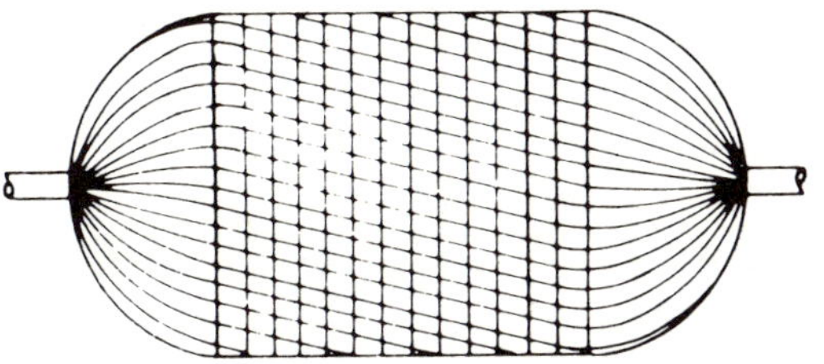

(B) Single circuit helical windings combined with circular loop windings provide high axial tensile strength.

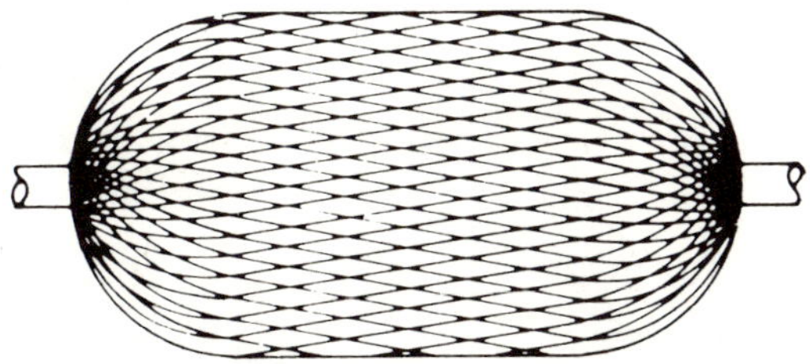

(C) Multiple circuit helical windings permit optimum use of the glass filament's strain characteristics without the addition of loop windings.

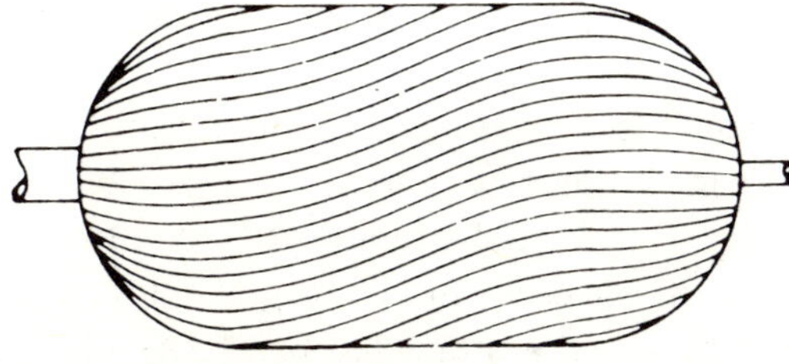

(D) Dual helical windings are used when openings at ends of the structure are of a different diameter.

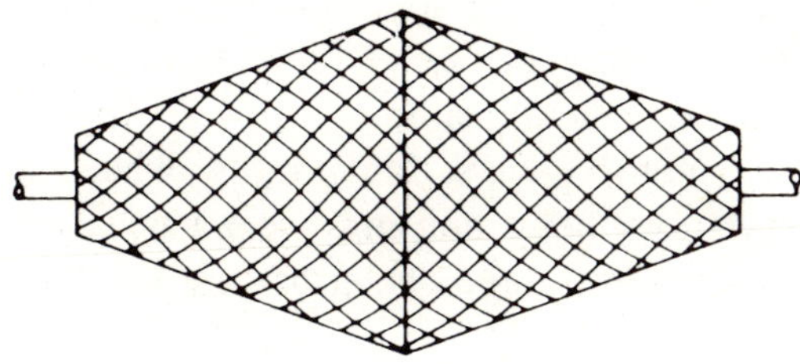

(E) Variable helical winding pattern is used to produce odd-shaped filament wound structures.

(F) Planar windings provide optimum longitudinal strength (with respect to winding axis).

Fig. 15-77. The advantage of various types of filament winding. (CIBA Corp.)

In *centrifugal* laminating, the resin and reinforced materials are formed against the mold surface as it rotates (Fig. 15-78). During this rotation a resin mix is distributed uniformly through the reinforcement by centrifugal force. Heat is then applied to help polymerize the resin. Tanks and tubing may be produced in this manner of lamination.

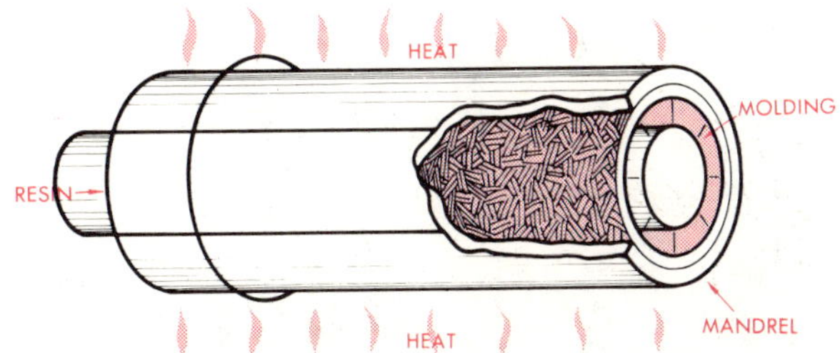

Fig. 15-78. In the centrifugal casting method the part is formed against the inner surface of a hollow mandrel when the chopped reinforcement and resin inside the mandrel are distributed uniformly while the mandrel rotates in an oven.

In pultrusion resin soaked matting or rovings, along with other fillers, are pulled through a long heated die. The heated die (between 250° and 300°F) forms the product shape and polymerizes the resin as it is drawn through the die. Radio-frequency heating may also be used to speed production rates.

Production rates vary from a fraction of a foot to over ten feet per minute. Various resins are used, including vinyl esters, polyesters, and epoxies. Although fibrous glass is the most used reinforcement, asbestos, graphite, carbon, boron, polyester, and polyamide fibers may be used. Reinforcements can be positioned in the pultrusion product where strength requirements are the highest. The process appears to be similar to extrusion. In the extrusion process the homogeneous material is *pushed* through the die opening. In pultrusion the resin soaked reinforcements are *pulled* through a heated die where the resin is cured (Fig. 15-79). Siding, gutters, I-beams, fishing rods, and other applications for unidirectional strength are provided in this process (Fig. 15-79B).

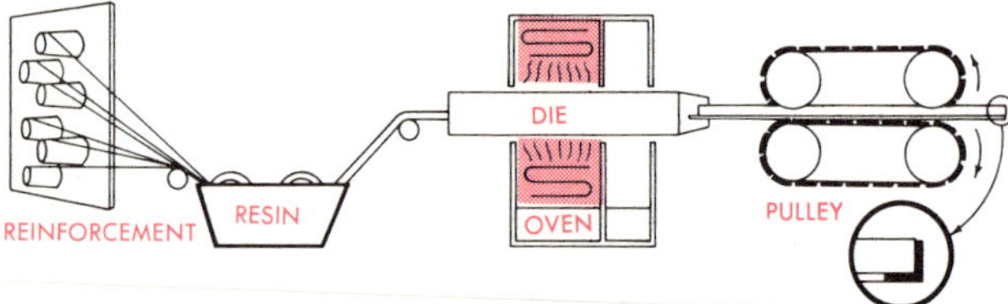

(A) Basic scheme of continuous pultrusion. (Koppers Co.)

(B) Pultruded structural supporting members of fiber glass reinforced polyester support operating floor of a chemical mixing plant. (Morrison Molded Fiber Glass Co.)

Fig. 15-79. Use of pultrusion methods.

In *continuous* (low-pressure) *laminating* (Fig. 15-80) fabrics or other reinforcements are saturated with resin and passed between two plastics film layers (cellophane, ethylene, vinyl, etc.). The laminated composite is controlled by the number of layers and by the squeeze rollers through which the laminate passes. The laminate is then drawn through a heating zone to speed polymerization and onto a takeoff roll.

Corrugated awnings, skylights, and structural panels are some familiar examples of continuous laminating.

A similar continuous laminating process is used in high-speed matched-die molding. As shown in Fig. 15-80B the laminate blank is fed to molds, which pass through a pressing station, curing, and demold in a continuous cycle. This molding technique eliminates waiting full cure cycle times at the press. The production of the laminate molding stock can be classed as a low-pressure laminating process, while the actual forming or molding operation may be classed as either high or low pressure, depending on the desired properties of the molding technique. These laminates, sometimes called sheet molding compounds (SMC), normally require molding pressures in the area of 1000 psi. These sheet molding compounds are commonly processed in compression molding presses and dies.*

Multilayered or sandwich construction of composite laminates is common. Extruders are commonly utilized to deposit a thin layer of plastics material on both sides of a continuous extruded or roller-fed core. In Fig. 15-81 three extruders are used to produce a laminate, possibly made from more than one polymerized material and with either a solid-center core or one produced of low-density materials.

As a result of this laminated material, a new processing technique was made possible. When thermoplastic laminates are reinforced, many can be postformed on metal stamping presses. Some of these thermoplastic laminates may be draped or thermoformed.**

Fibrous glass reinforced thermoplastics may be cold formed much as metal (Fig. 15-82). This specially formulated reinforced compound is available in sheet form. During the forming operation the sheet is preheated to about 400°F and formed on conventional metal stamping presses. It is possible with this method to produce parts with intricate designs and varying wall thickness. Production rates may exceed 360 parts per hour on conventional metal stamping presses. Applications include motor covers, fan guards, wheel covers, battery trays, lamp housings, seat backs, and numerous interior and exterior automotive trim panels.

---

*See Compression Molding, p. 240.
**See Thermoforming Processes, Chap. 17.

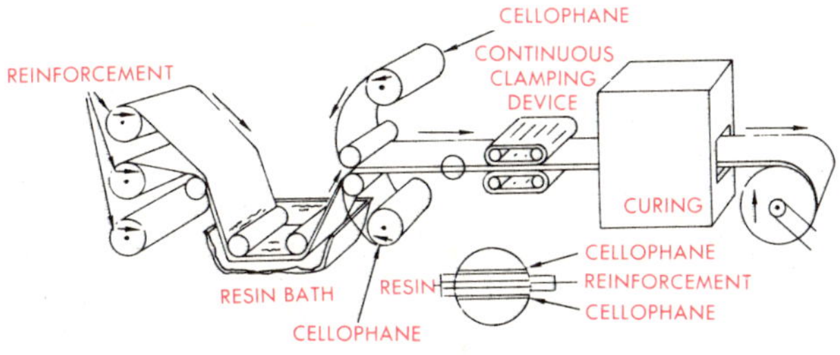

(A) Basic scheme.

Reinforced molding compounds are utilized in numerous molding processes and should not be confused with laminates.*

Short, milled, or chopped glass fibers are most commonly used to reinforce these molding compounds. (See Table 15-9.) However, other plastics fibers as well as exotic metallic and crystalline whiskers are employed.

*See other molding processes in this chapter.

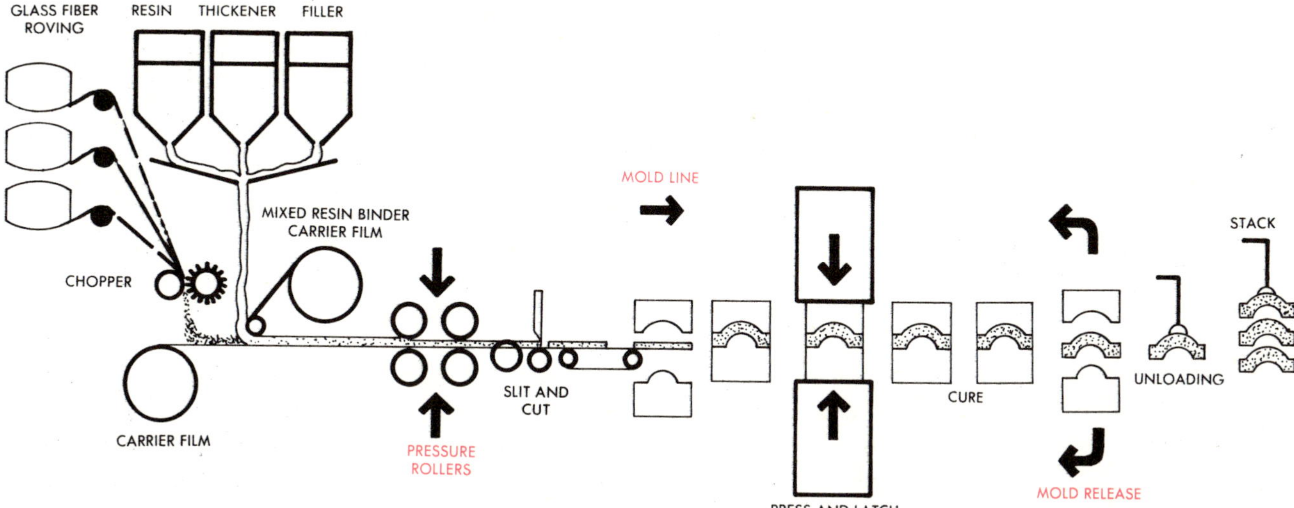

(B) High-speed molding concept with molding system that eliminates full cure cycle at press. (Owens/Corning Fiberglass)

Fig. 15-80. Methods of continuous laminating.

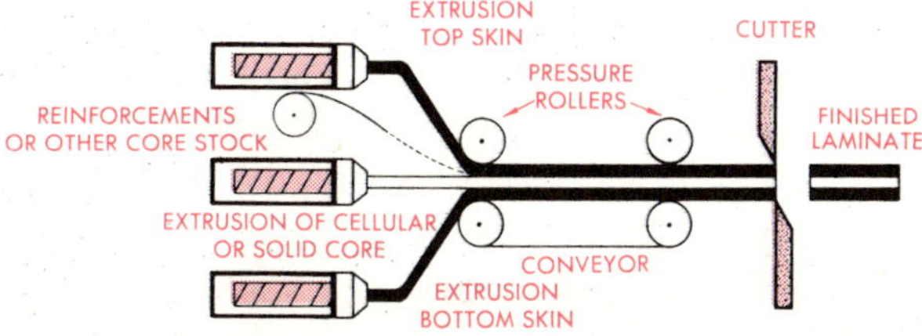

Fig. 15-81. Continuous low-pressure laminate production utilizing extruders and reinforcements or other core stock.

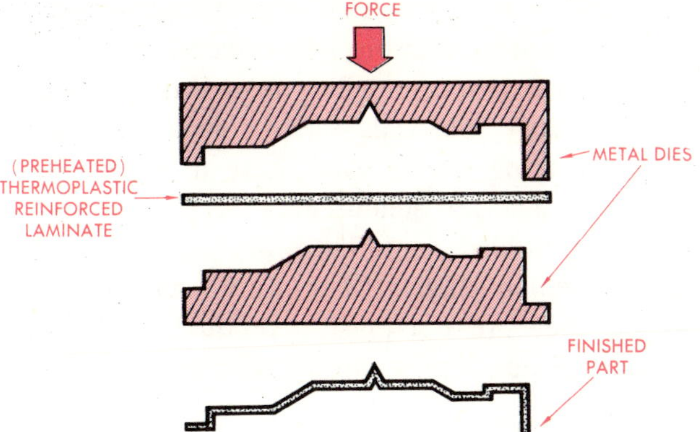

(A) Laminate is preheated and formed on cold conventional metal-stamping dies and equipment.

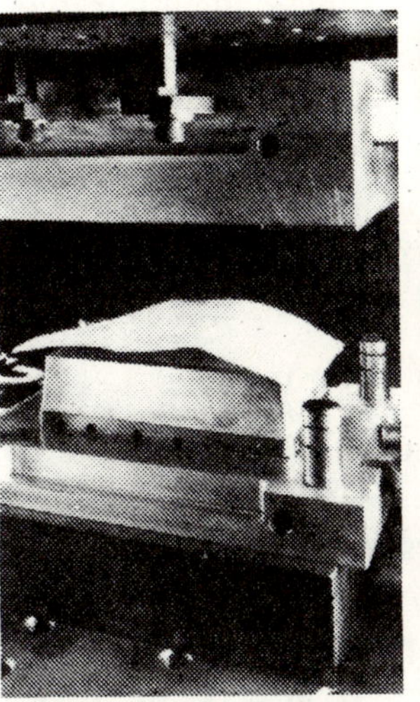

(B) Preheated sheet is formed between cooled matched metal dies mounted in a conventional metal stamping press. (G.R.T.L. Company)

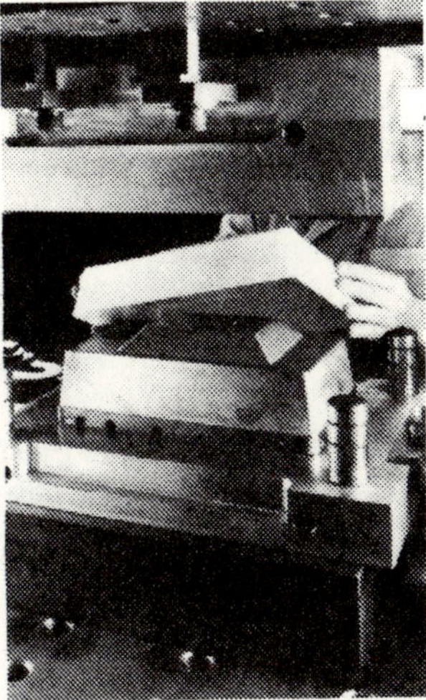

(C) Blank which is smaller than the part, flows out to the periphery of the dies without forming trim or flash. (G.R.T.L. Company)

Fig. 15-82. Cold formable laminates.

Table 15-9. Typical Properties of Fibrous Glass Reinforced Plastics.

Plastics	Specific Gravity	Tensile Strength (1000 psi)	Compressive Strength (1000 psi)	Thermal Expansion (10^{-5} in/in/°C)	Deflection Temperature at 264 psi (°C)
Acetal	1.54-1.69	9-18	12-12.5	1.9-3.5	154-232
Epoxy	1.8-2.0	1.4-30	30-38	1.1-3.5	121-232
Phenylene oxide	1.21-1.36	14-17	18-30	1.1-2.2	132-143
Polycarbonate	1.34-1.58	13-21	17-18	1.4-2.0	140-145
Polyester (thermoplastic)	1.48-1.63	10-17	18-19.5	1.4	200-230
Polyester (thermosetting)	1.35-2.3	25-30	15-30	1.5-2.5	204-260
Phenolic	1.75-1.95	5-10	17-26	0.8-2.05	149-316
Melamine	1.8-2.0	5-10	20-35	1.5-1.7	> 204
Silicone	1.87	4-6	12-20		< 482
Polyethylene	1.09-1.28	7-11	5-6	1.7-2.7	116-127
Polypropylene	1.04-1.22	6-9	6.5-7	1.6-2.4	132-149
Polystyrene	1.20-1.34	10-15	13-19	1.7-2.2	99-104
Polysulfone	1.31-1.47	11-17	19-21	1.7	171-177

Because the thermoplastic materials may be processed in several different ways and because the usefulness of the many thermoplastic applications is extended, many new innovative uses are evolving.

Reinforced molding compounds may be injection, matched-die, transfer, compression, and extrusion molded to produce products of intricate shapes with a broad range of physical properties. All of the molding processes may utilize reinforcement compounds. There is some difficulty in blow molding small, thin-walled articles. Thick-walled gasoline tanks and other products, however, may find reinforcements acceptable with this process. Injection molding is the most common method of processing reinforced thermoplastics.

Cold Molding

Cold molding of plastics materials began in the United States in about 1908. The process was adapted from the ceramics industry, which was well established at the time. Ancient man utilized this basic principle by using the pressure of his hands to form and shape pottery from clay. In Biblical times duplicate, ornamental clay parts were compressed in wooden male and female molds. The formed part was then removed from the mold and baked in kilns (ovens).

The process of cold molding is similar to compression molding except the molds are not heated. Materials are compounded and molding is done in conventional presses. From 2000 to 12,000 pounds per square inch are used to rapidly press the compound into a solid cake (something like a preform). After being formed to shape in the unheated molds the material is hardened into an infusible mass by baking in an oven (Fig. 15-83).

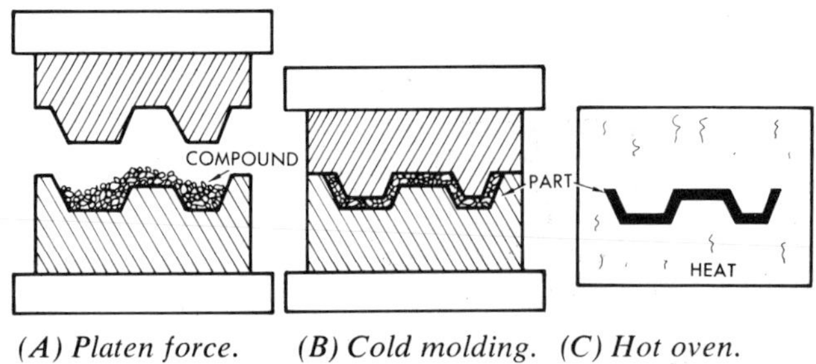

(A) Platen force. *(B) Cold molding.* *(C) Hot oven.*
Fig. 15-83. Principle of cold molding.

Cold molding materials may be divided into two distinct groups according to the ingredients used: (1) nonrefractory (organic) and (2) refractory (inorganic).

Compounds for organic or nonrefractory cold molded parts may include a formulation of various plastics (mostly phenolics), raw linseed oil, asbestos, various bitumens, sulphur, or other additives. Cold molded products—both organic and inorganic types—generally lack luster and are generally produced in dark colors. To help improve the surface finish of organic molded parts, molds may be heated briefly during the rapid compression cycle. The actual cure of the compound is accomplished in ovens. Electrical insulator parts, utensil handles, battery boxes, and valve wheels are typical applications.

Inorganic or refractory cold molding compounds may contain formulations of cement, asbestos fibers, clay, and lime or silica in a water blend. Once refractory materials are molded, they are placed in baking ovens which have a wet steam atmosphere. The hot steam cures the cement and hardens the molded piece. These products are used where extremely high temperatures may exist; however, they are not plastics products and are included because they are examples of cold molding.

Cold molding has several advantages because low-cost materials may be used, cooling and equipment costs are low, electrical properties are good, and because parts are cured in a baking oven and not in the mold, production rates are high.

Sintering

Sintering is the process of compressing powdered plastics in a mold at temperatures just below its melting point for about one-half hour (Fig. 15-84). The powdered particles are fused (sintered) together, but the mass as a whole does not melt. Bonding is accomplished by the exchange of atoms between the individual particles. Following the fusion process the material may be postformed under heat and pressure to form the desired dimensional size. The three most important variables governing the sintering process are temperature, time, and the composition of selected plastics.

The process is adapted from the sintering operations of powder metallurgy. Sintering can process polytetrafluoroethylene, polyamides, and other specially filled plastics. It is the primary method by which polytetrafluoroethylene is processed. Although dense parts of excellent electrical and mechanical properties may be produced, the cost of tools and production is high and thin walls or variations in cross-sectional thickness are difficult to form.

Typical applications may include bushings, bearings, hubs, and various electrical insulating parts.

Liquid Resin Molding

Liquid resin molding is a low-pressure method of producing products. The method is sometimes called *liquid transfer molding* or *liquid injection*

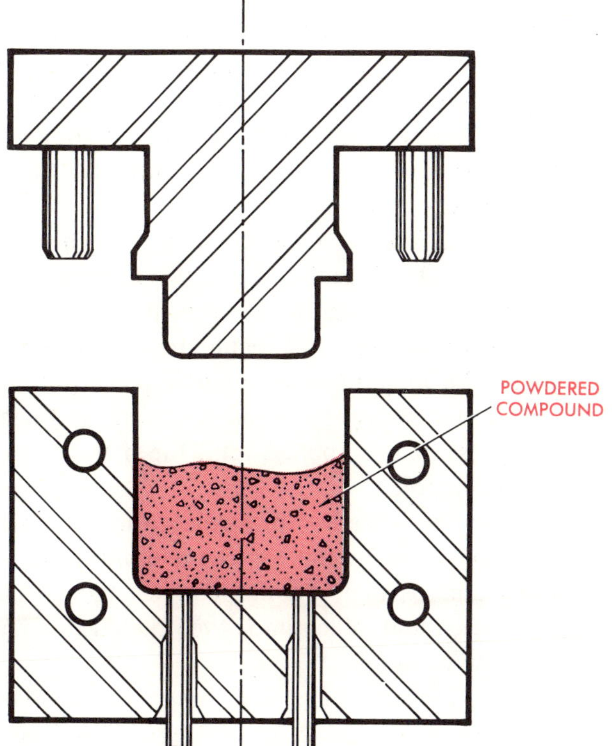

(A) Before compression.

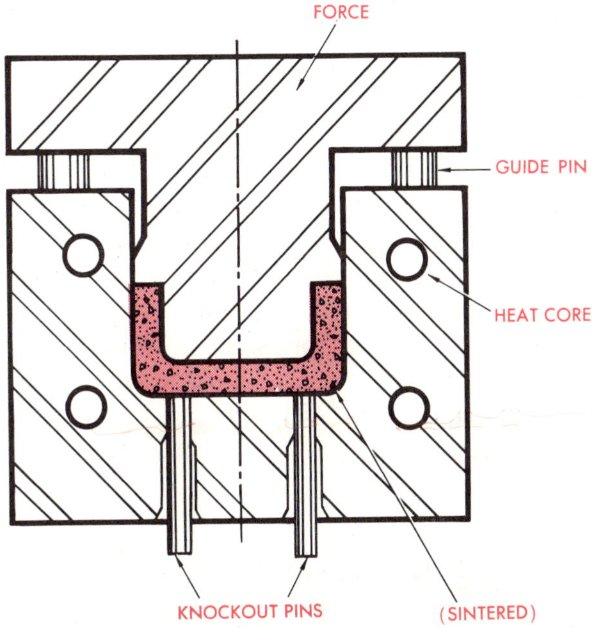

(B) Compression and heating.

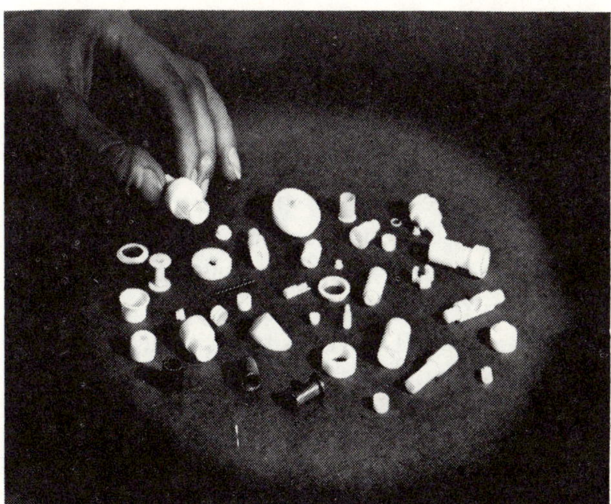

(C) PTFE parts have been sintered to basic shape before machining. (Chemplast, Inc.)

Fig. 15-84. Sintering plastics parts.

molding. It is a very specialized process used primarily in the electronics industry for potting and encapsulating components. The major difference is that liquid resins, not powders or pellets, are used in the technique. The resins are then forced (little pressure needed) into the molding cavity where they are rapidly cured. Epoxies, silicones, polyesters, and polyurethanes lend themselves to both liquid-resin molding-design concepts and applications.

Various delicate electronic components, including coils, transformers, resistors, diodes and other nonelectrical parts, may be produced by liquid molding. This molding technique does not distort windings or shift delicate devices in the molding cycle (see Fig. 15-85).

The main advantages of liquid resin molding include: (1) elimination of plasticizing stage necessary with dry compounds, (2) permitting of encapsulation of delicate or fragile parts, (3) no hand mixing, (4) elimination of preheating and preforming, (5) lower pressures are needed. (6) minimum material waste, (7) rapid cure of resins at low temperatures, (8) improved reliability and dimensional stability, and (9) reduction of material handling.

Glossary of New Terms

Air-assist forming—A method of thermoforming in which airflow or air pressure is employed to partially preform the sheet immediately prior to the final pull-down into the mold using vacuum.

Annealing—A process of holding a material at a temperature near, but below, its melting point, the objective being to permit stress relaxation without distortion of shape.

Automatic cycle—A machine operation that will perform repetitive cycles.

Back pressure—The viscosity resistance of a material to continued flow when a mold is closing. In extrusion, the resistance to the forward flow of molten material.

Baffle—A device used to restrict or divert the passage of fluid or gases through a pipeline or channel.

Banbury—An apparatus for compounding materials composed of a pair of contrarotating rotors which masticate the materials to form a homogeneous blend.

Barrel—The cylindrical housing in which the extruder screw rotates.

Barrel vent—An opening through the barrel wall to permit the removal of air and volatile matter from the material being processed.

Breaker plate—A perforated metal plate located between the end of the screw and die head.

Blow molding—A method of fabrication in which a parison is forced into the shape of the mold cavity by internal air pressure.

Blowup ratio—In blow molding, the ratio of the mold cavity diameter to the parison diameter. In blow film, the ratio of the final tube diameter to the original die diameter.

Breathing—The opening and closing of a mold to allow gases to escape early in the molding cycle. *Also called* degassing.

Calender—To prepare sheets of material by pressure between two or more contrarotating rolls.

Cast—To form an object by pouring a fluid monomer-polymer solution into an open mold where it finishes polymerizing or is melted and cooled into a plastics form. The word "casting" should not be used for the word "molding."

Coating—Placing a permanent layer of material on a substrate. It is not meant to be removed.

Cold molding—A procedure in which a composition is shaped at room temperature and cured by subsequent baking.

Compression mold—A mold which is open when the material is introduced and which shapes the material by heat and by the pressure of closing.

Compression molding—A technique of molding in which the molding compound is placed in the open mold cavity, the mold is closed, and heat and pressure are applied until the material has cured or cooled.

Drape forming—Method of forming thermoplastic sheet in which the sheet is clamped into a movable frame, heated, and draped over high points of a male mold. Vacuum is then pulled to complete the forming.

Drawing—The process of stretching a thermoplastic sheet or rod in order to reduce its cross-sectional area.

(A) Mold half for liquid molded ignition coil.

(B) First liquid molded ignition coil to meet all high voltage tests.

(C) Liquid molded electrical coil completely encapsulated.

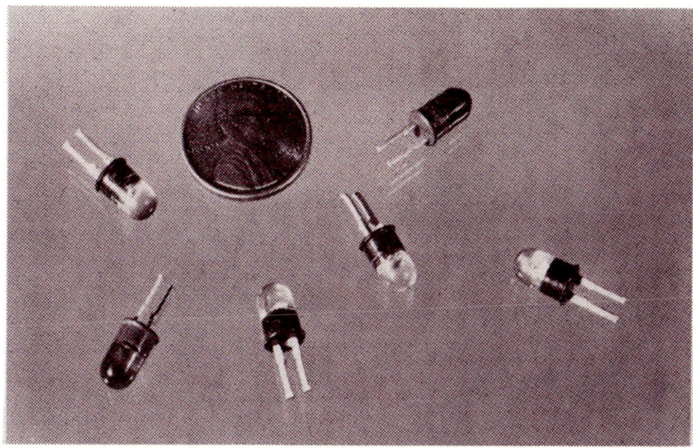

(D) Liquid molding translucent resins for both electrical and optical properties permits production of inexpensive light emitting diodes.

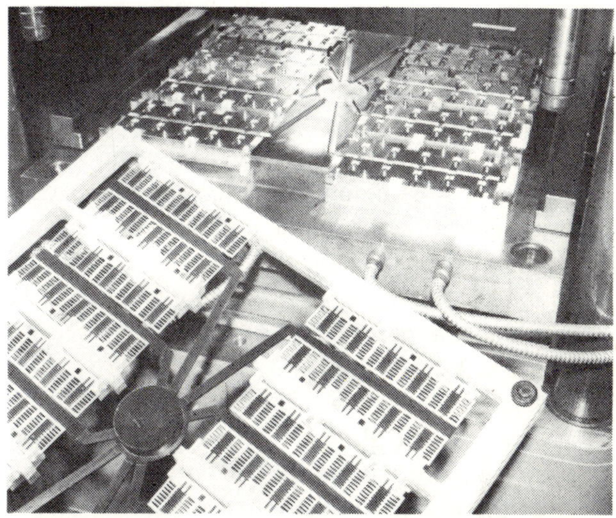

(E) Liquid resin molding is used in molding of intricate parts. High-speed fully automatic liquid resin machines may be preferred.

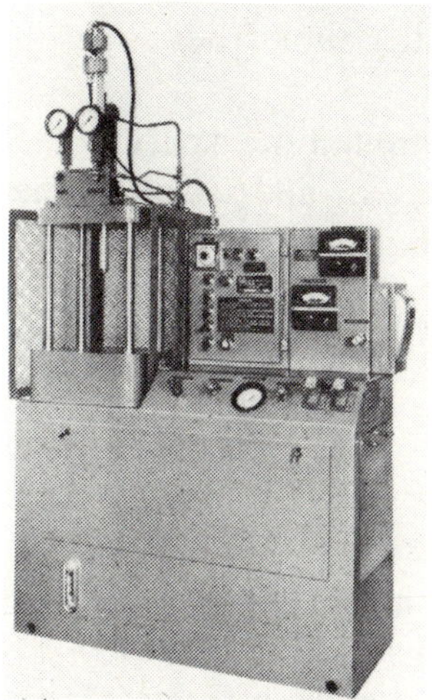

(F) Machine for liquid resin molding.

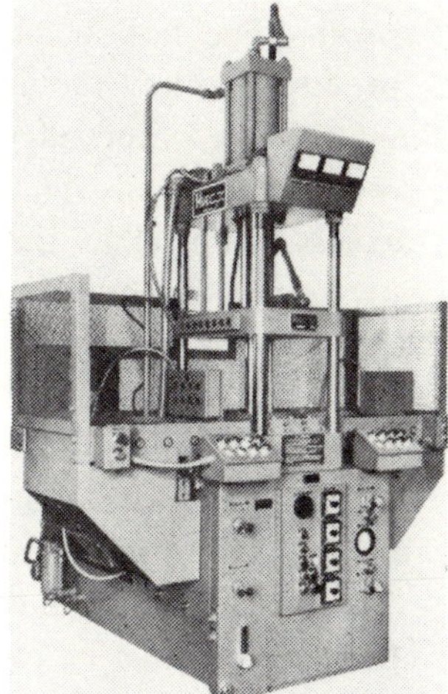

(G) Another machine for liquid resin molding.

Fig. 15-85. Liquid resin molds and moldings. (Hull Corp.)

Dwell—A pause in the application of pressure to a mold, made just before the mold is completely closed, to allow the escape of gas from the molding material.

Extrudate—The product or material delivered by an extruder.

Extrusion—The compacting of a plastics material and the forcing of it through an orifice in more or less continuous fashion.

Female mold—The indented half of a mold designed to receive the male half.

Flash—Extra plastics attached to a molding along the parting line. It must be removed.

Frost line—In extrusion, a ring-shaped zone located at the point where the film reaches its final diameter. It is characterized by a frosty appearance because of cooling in polyethylene plastics.

High-pressure laminates—Laminates formed and cured at pressures higher than 1000 psi.

Hopper dryer—A combination feeding and drying device for extrusion and injection molding of thermoplastics.

Injection blow molding—A blow molding process in which the parison to be blown is formed by injection molding.

Injection molding—A molding procedure whereby a heat-softened plastics material is forced from a cylinder into a relatively cool cavity which gives the article the desired shape.

Low-pressure laminates—Laminates formed and cured with pressures less than 1000 psi. In hand layup, no pressure is used.

Parison—The hollow plastics tube from which a product is blow molded.

Pinch-off—A raised edge around the cavity in the mold which seals off the part and separates the excess material as the mold closes around the parison in the mold.

Plattens—The mounting plates of a press to which the entire mold assembly is bolted.

Purging—Cleaning one color or type of material from the cylinder of a molding machine.

Sintering—Forming articles from fusible powders. The process of holding the pressed powder at a temperature just below its melting point.

Spinneret—A type of extrusion die with many tiny holes, through which a plastics melt is forced to make fine fibers and filaments.

Thermoforming—Processes of forming a thermoplastic sheet by heating and pulling it down onto a mold surface.

Transfer molding—A method of molding plastics by softening by heat and pressure in a transfer chamber, then forcing it at high pressure through the sprues, runners, and gates into a closed mold for final curing.

Review Questions

15-1. What is meant by *cycle time* in injection molding?

15-2. Why is the process of injection molding used to produce many plastics articles?

15-3. What are the channels called that carry material from the nozzle to the cavity?

15-4. Give five typical products produced by injection molding.

15-5. Define the terms *gate, sprue,* and *runner* as used in injection molding.

15-6. What is the major difference between plunger type injection molding and screw type injection molding?

15-7. What kinds of materials are used in compression molding?

15-8. How is production time reduced in the compression molding process?

15-9. How is cycle time reduced in the compression molding process?

15-10. What are the three types of compression mold designs?

15-11. What does the term *breathing* mean in compression molding?

15-12. What is the function of a *land* in a compression mold?

15-13. Name five typical compression molded parts.

15-14. How does transfer molding differ from compression molding?

15-15. What advantage does transfer molding have over compression molding?

15-16. What types of materials are used in transfer molding?

15-17. What is the major difference between transfer molding and injection molding?

15-18. What are the primary disadvantages of transfer molding?

15-19. Give five typical products produced by transfer molding.
15-20. What types of materials are used in extrusion molding?
15-21. What causes the plastics to move through extruder barrels?
15-22. What is the function of the breaker plate in the extrusion process?
15-23. What name is given to the material which is extruded?
15-24. What are the two film producing methods in extrusion molding?
15-25. What is a coathanger die?
15-26. What is the zone called that appears frosty in extrusion-blown film?
15-27. What is the basic principle behind cold extrusion?
15-28. Blow molding is a refinement of what ancient art?
15-29. What are the two basic blow molding methods?
15-30. Why are two different molds required in the injection blow molding process?
15-31. How does the heated-tube method of blow molding differ from extrusion blow molding?
15-32. How is product wall thickness controlled in extrusion blowing?
15-33. What is calendering, and what limitations does it have as a process?
15-34. How are patterns placed on calendered films?
15-35. What materials and products are produced on calendering equipment?
15-36. What is the meaning of the term *lamination? Reinforcing?*
15-37. Why are products laminated? Name a common consumer product which is laminated.
15-38. What is the difference between low-pressure and high-pressure lamination of plastics?
15-39. What is the major disadvantage of high-pressure lamination?
15-40. Why are gel-coat resins used on laminated products?
15-41. What are some of the advantages of sprayup molding?
15-42. What is the major difference between cold molding and other molding processes?
15-43. What materials are cold molded? Name a product.
15-44. What other molding process is similar to cold molding?
15-45. What is sintering? Name a product application of sintering.
15-46. What plastics are used in liquid resin molding?
15-47. For which industry is liquid resin molding primarily used?

16

Casting Processes

Casting includes a number of processes where monomers, modified monomers, powders, or solvent solutions are poured into a mold and caused to become a solid plastics mass. No pressure is applied in casting operations (only the weight of the mass). The cast material is solidified in the mold. The plastics state may be accomplished by evaporation, chemical action, cooling, or by the addition of external heat. For a product to be classed as a casting, no pressure is required and the product is removed from the mold or form. The process is similar to casting of metals.

The early castings of phenolic by Baekeland were dark colored products with restricted limitations. Today, both thermosetting and thermoplastic resins are cast into a variety of useful products. The major advantage of casting is that large, inexpensive molds can be used when only a small number of articles are required. Parts may also be stress-free.

Casting processes or techniques may be placed into six distinct groups including: (1) simple casting, (2) film casting, (3) hot-melt casting, (4) slush casting, (5) rotational casting (sometimes called *rotational molding*), and (6) dip casting.

Simple Casting

Simple casting is the process in which liquid resins or molten plastics are poured into molds and allowed to polymerize or cool. The plastics part is then removed from the mold for further processing if necessary. Simple castings find applications as rods, tubes, cylinders, sheets, and other shapes. Many are easily fabricated or machined into finished products. Potting, encapsulation, and embedment are also classed as simple casting. Foams are cast but are discussed under the foaming processes. Some cast and reinforced toolings and fixtures are included as simple castings.

In simple casting operations wood, metal, plaster, selected plastics, selected elastomers, and glass may be used as molds. Common casting resins include acrylics, polyesters, silicones, epoxies, phenolics, ethyl cellulose, cellulose acetate butyrate, and polyurethanes. Probably the most familiar is polyester resin used in crafts and hobby work. Silicones are used (cast over patterns) extensively for making molds into which plastics or other materials are cast. Water extended polyesters and polyurethanes are used to cast furniture and cabinet parts. (See Fig. 16-1.)

Examples of simple casting include jewelry, billiard balls, cast sheets for windows, furniture parts, watch crystals, sunglass lenses, handles for tools, desk sets, knobs, table tops, sinks, and fancy buttons. Fig. 16-2 shows the basic principle of simple casting.

A number of thermoplastic and thermosetting plastics may be cast into sheet form. Acrylic sheets are commonly produced by pouring a catalyzed monomer or partially polymerized resin between two parallel plates of glass (Fig. 16-3). The glass is usually sealed with a gasket material to prevent

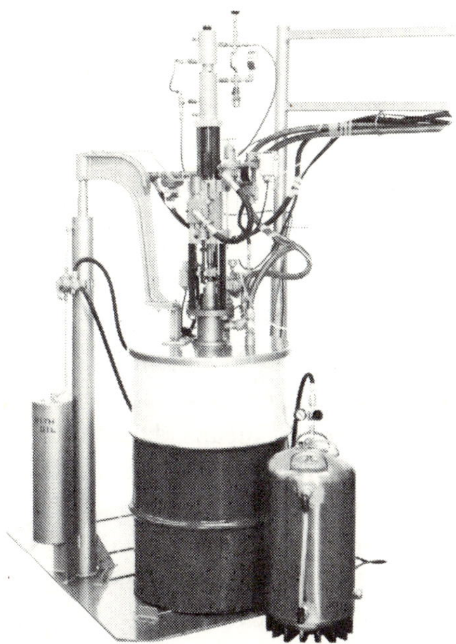

(A) This machine is designed to blend components for use in producing polyester products by simple casting methods. (Pyles Industries, Inc.)

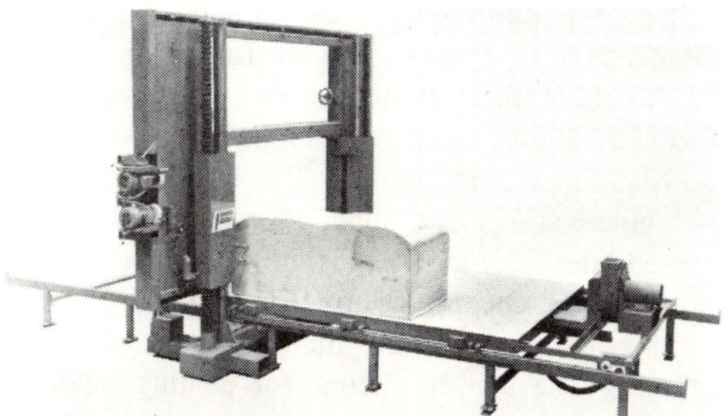

(B) This large block of cellular plastics has been cast in an open mold and is about to be cut into slabs on this machine. (McNeil Femco Corp.)

(C) These pictures and mirror frames were made of water extended polyester resin by simple casting techniques.

Fig. 16-1. Simple casting and its products.

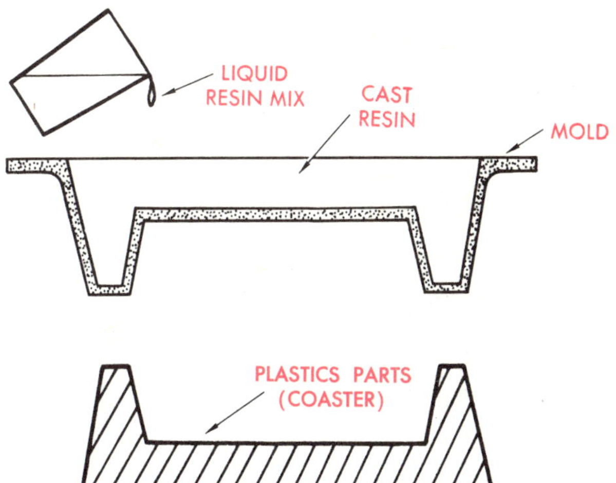

Fig. 16-2. Solid casting in an open one-piece plastics mold.

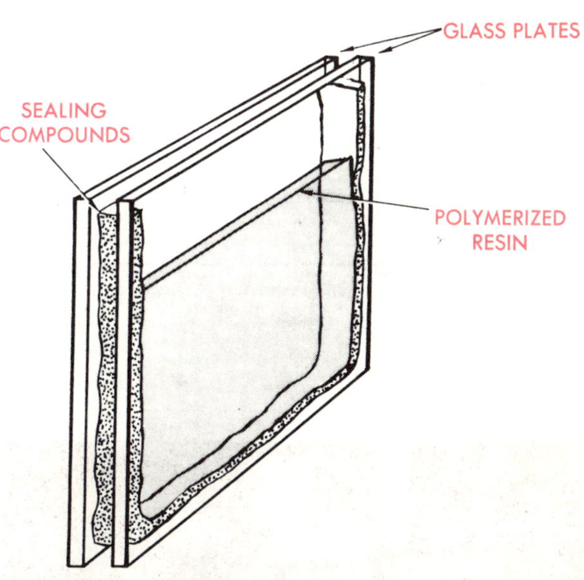

Fig. 16-3. Casting plastics sheets.

leakage and help control the thickness of the cast sheet.

Film Casting

The casting of film involves dissolving plastics granules or powder, plasticizers, colorants, or other additives in a suitable solvent. The solvent solution of plastics is poured onto a stainless steel belt and the solvents evaporated by the addition of heat. The film deposit left on the moving belt is stripped or removed and wound on a takeup roller (see Fig. 16-4). This film may be "cast" (coating or laminate) directly on fabric, paper, or other substrates.

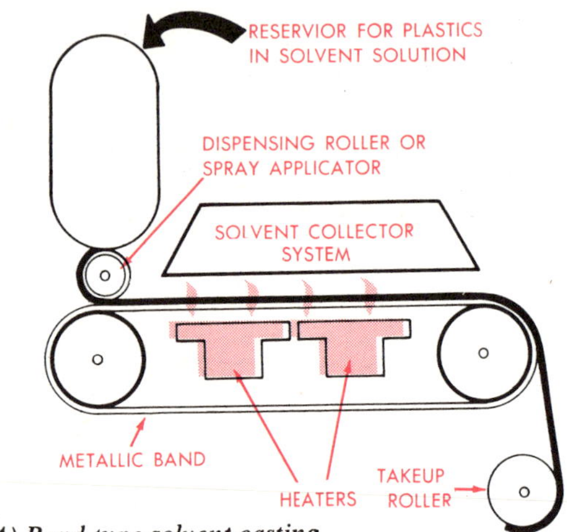

(A) Band type solvent casting.

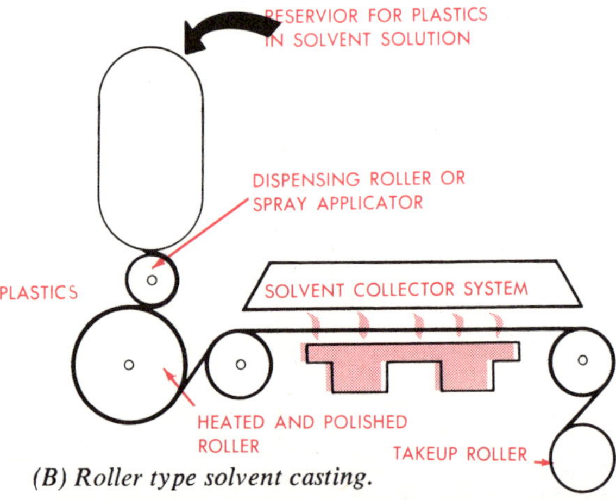

(B) Roller type solvent casting.

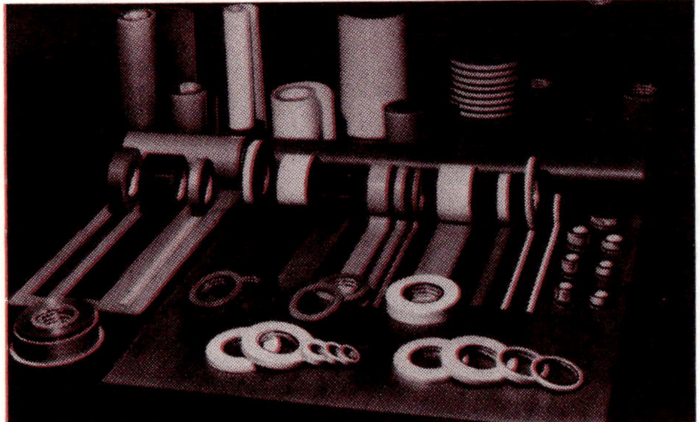

(C) Polytetrafluoroethylene films may be cast as aqueous dispersions on heated belts. The fabricated film may then be cut to other sizes. (Chemplast, Inc.)

Fig. 16-4. Film casting.

Films uniform in thickness, optically clear, and without orientation or stress are possible with this method. In order to be economically feasible a solvent recovery system is necessary in solvent casting of film. Cellulose acetate, cellulose butyrate, cellulose propionate, ethyl cellulose, polyvinyl chloride, polymethyl methacrylate, polycarbonate, polyvinyl alcohol, and other copolymers may be solvent cast. Casting of liquid plastics latexes on Teflon coated surfaces may also be used to produce special films.

Aqueous dispersions of polytetrafluoroethylene and polyvinyl fluoride are cast on heated belts at temperatures below their melting points. The method provides a convenient means to fabricate films and sheets of materials which are very difficult to process by any other means. These films are used as nonstick coatings, gasket material, and as sealing components for pipes and joints.

Other cast film applications include water soluble packaging for bleaches, detergents, etc., and skin and blister packaging of various products.

Hot-Melt Casting

Hot-melt plastics were used for casting during World War II. Today, hot-melt formulations may be based on ethyl cellulose, cellulose acetate butyrate, polyamide, butyl methacrylate, polyethylene, and various other mixtures. The largest use is for stripable coatings and adhesives which will be discussed later. Hot-melt resins, however, are used in potting and encapsulation (Fig. 16-5). Not all potting compounds are thermoplastic and hot-melting. The use of silicone for coating, sealing, and casting is commonly used. Epoxy and polyester resins are also used.

Electrical parts may be placed in molds and hot resin poured over the components. When cool, the plastics provides a protective "housing" for wires and vital parts. The encapsulated or potted components may then be placed with other assemblies to complete the finished product. Some encapsulations and pottings are not cast in separate molds. They are produced by pouring the molten compound directly over the components while inside the case of the finished product. The insulation of parts in a radio chassis or motor is a familiar example. If the components are "cast" in place and not removed from a mold shape, they must be classed as coatings.

These compounds are used to seal various electrical components from hostile environments; how-

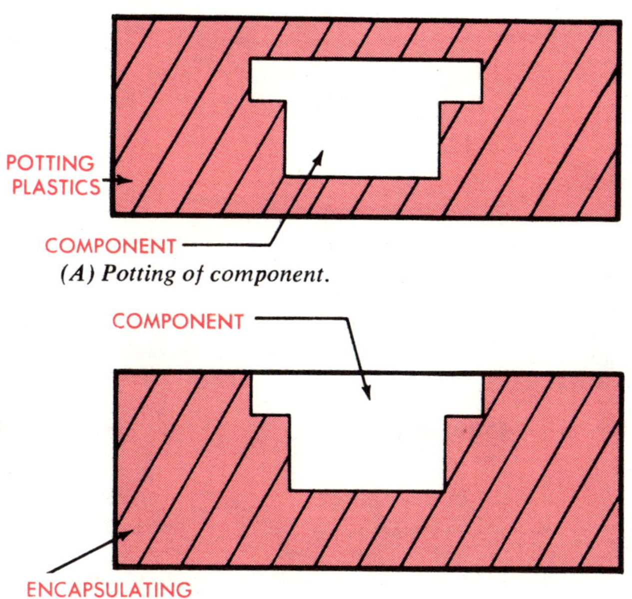

(A) Potting of component.

Fig. 16-5. Hot-melt potting and encapsulation.

ever, hot-melts may be used for making molds into which other materials are cast.

Slush Casting

The process of slush casting involves pouring dispersions of polyvinyl chloride or other plastics into a heated, hollow, but open mold. As the material hits the walls of the mold, it begins to solidify. (See Fig. 16-6A.) The wall thickness continues to build up as long as the mold remains hot and the solution is left in the mold. When the desired wall thickness is reached, the excess plastisol dispersion is poured from the mold. The mold, with the remaining solidified material, is then placed in an oven until the plastics melts (fuses together) and evaporation of solvents is complete. After water cooling, the mold is opened (may be one-piece mold) and the product removed. Commercial molds are generally of aluminum for rapid cycling and lower tooling costs; however, ceramic, steel, or even plaster and plastics molds may be used. To eliminate air bubbles in the plastisol product, vibrating, spinning, or vacuum chambers are sometimes necessary.

Dry thermoplastic powders are also used in a similar process sometimes called *static casting*. The metal mold is filled with powdered plastics and placed in a hot oven (Fig. 16-6B). As the heat penetrates the mold, the powder melts and fuses to the mold wall. When the desired wall thickness is obtained the excess powder is removed from the mold. The melted powder remaining in the mold is then returned to the oven until all particles have completely fused together. Huge storage tanks and containers with heavy walls are examples of this type of casting. Cellular polystyrene or polyurethanes may be placed in the remaining space and used as tough-skinned flotation devices.

Slush-cast articles include hollow toys, syringe bulbs, doll parts, and special containers.

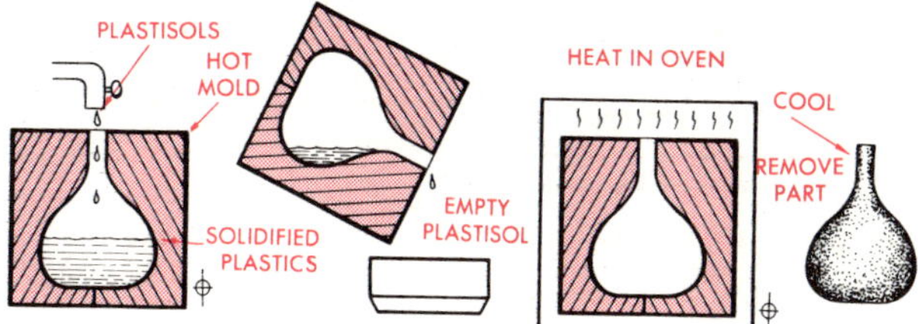

(A) Basic slush casting with plastisols.

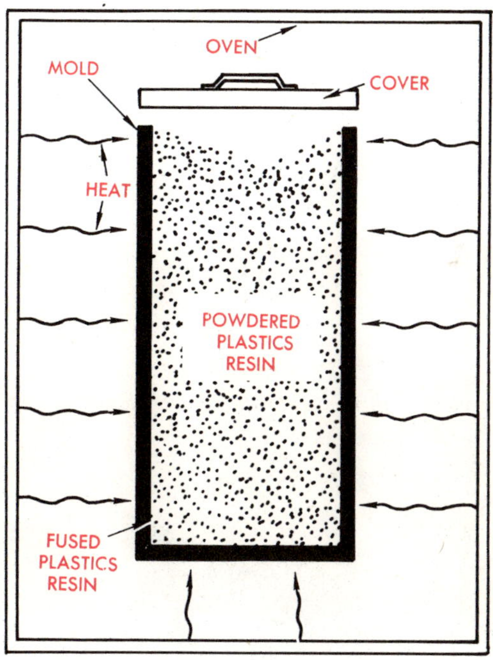

(B) Static casting with dry thermoplastic powders.

Fig. 16-6. Principles of slush casting.

Rotational Casting

Rotational casting is similar to slush casting except it may be used for completely hollow objects such as balls, toys, various containers, and indus-

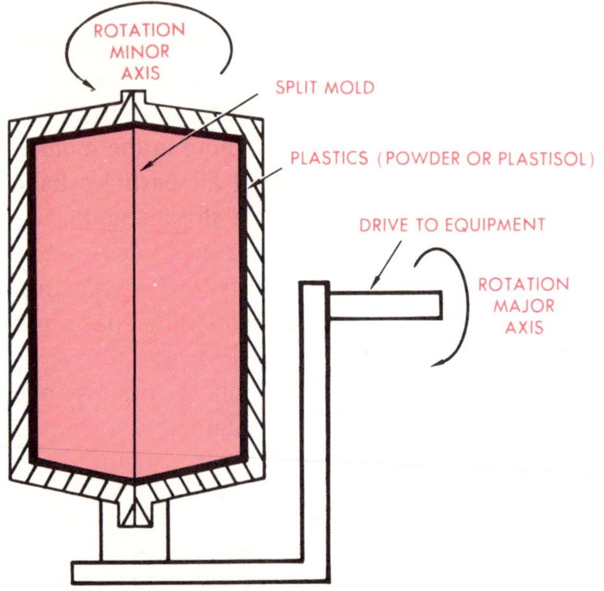

(A) Basic principle of rotation.

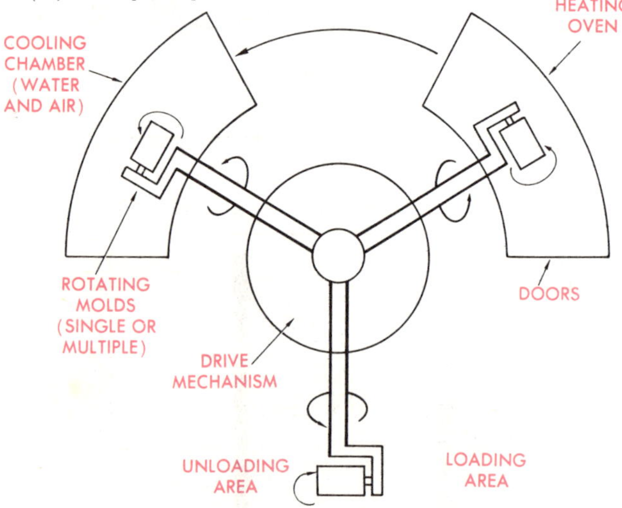

(B) Top view of three-mold rotational molder.

(C) Rotational-cast horse from three-piece mold. (McNeil Femco Corp.)

(D) Rotational-cast 300-gal garbage container. (Phillips Petroleum Co.)

(E) Rotational-cast one-piece chair seat and back. (McNeil Femco Corp.)

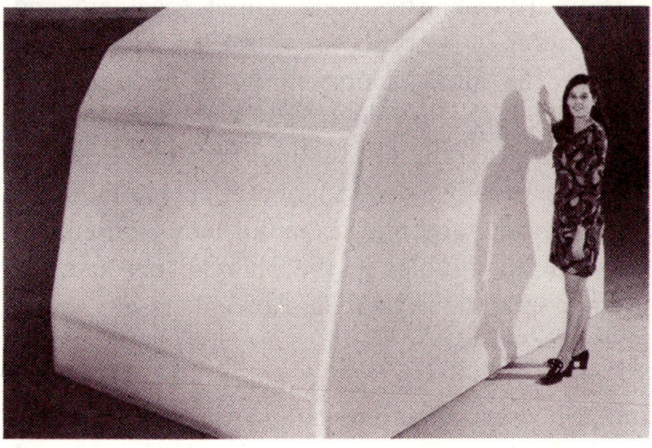

(F) Large rotational-cast tank. (Lunn Laminates, Inc.)

Fig. 16-7. Rotational casting and its products.

trial parts including rests, sunvisors, fuel tanks, and floats.

Plastics powders or dispersions are measured and placed in (usually aluminum) two or more piece molds. The mold with its measured charge of plastics is then placed in an oven where it is rotated in two planes (axes) at the same time (Fig. 16-7A). This action evenly distributes the material on the walls of the hot mold. The material solidifies as it contacts the hot mold surfaces providing a one-piece coating. When all the powders or dispersions have melted and fused together, the heating cycle is complete. The mold then enters a cooling chamber as it continues to rotate. The cooled plastics product is removed when cool enough to handle.

Nearly all thermoplastic powders may be used in rotational casting. The most common dispersion is polyvinyl chloride.

Rotational cast parts may be produced as large commercial containers, tubing (ends cut open), ice chests, pans, and boxes of all shapes. Luggage shapes, for example, may be cast as one piece and then this single piece may be cut at the seam to form two perfectly fitting halves. Figs. 16-7C through 16-7F show some rotational cast items.

By programming the major and minor axis speed, the wall thickness in various areas may be controlled. If for example, it was desirable to have a thick wall section around the parting line of a ball, the minor axis would be programmed to turn at a faster speed, thus placing more powdered material against the hot mold in that area. Fig. 16-8A shows how this thickness is obtained in the casting of a plastics ball.

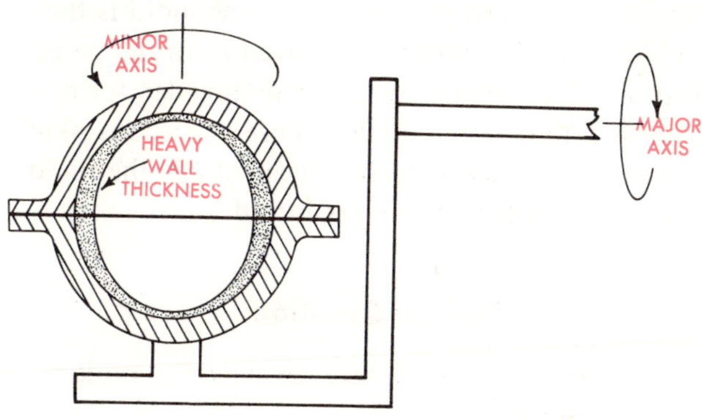

(A) Wall thickness in ball when minor axis is rotated at a faster speed than major axis.

(B) Rotational casting of three parts at once. (McNeil Femco Corp.)

(C) Rotational casting of large tank with spray water cooling. (McNeil Femco Corp.)

(D) Rotational casting of eight parts at once. (McNeil Femco Corp.)

(E) This tube is cast in a special centrifugal casting machine.

Fig. 16-8. Rotational casting machines.

Dip Casting

Dip casting should not be confused with dip coating, which is not removed from the substrate. In dip casting, a preheated mandrel the shape and size of the *inside* of the product is dipped into a plastisol dispersion. (See Fig. 16-9.) As the resin hits the hot mold surface it begins to solidify. The thickness of the piece continues to increase as it remains in the solution. If additional thickness is desired, reheating of the coated piece may be required. After the desired thickness is obtained, the mold is removed from the oven and cooled in a water quench. The part is then stripped from the mold.

Plastics gloves, overshoes, coin purses, spark plug covers, and toys may be examples of dip-cast products.

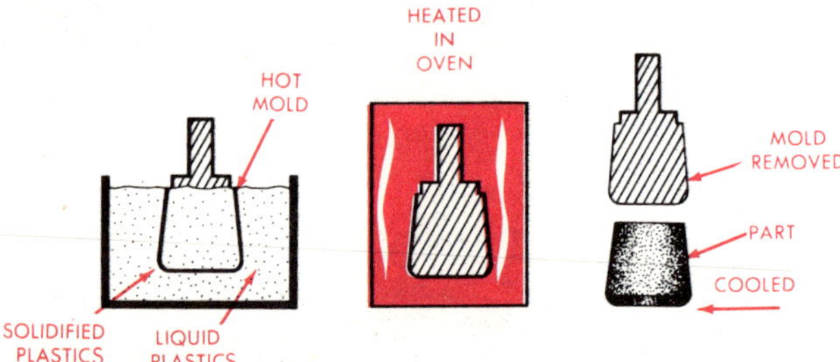

(A) Scheme of dip casting.

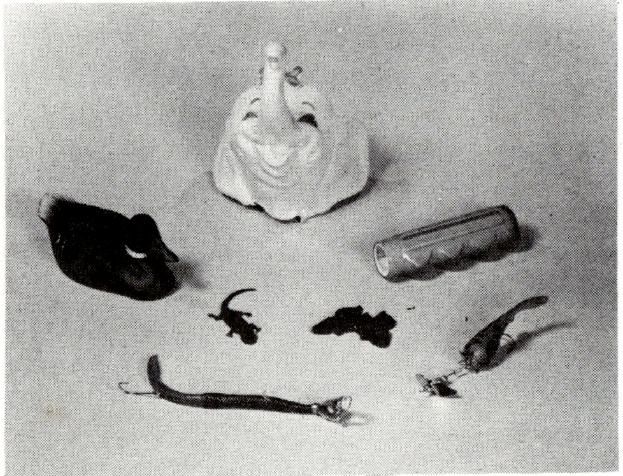

(B) Recreational dip-cast parts. (B.F. Goodrich Chemical Co.)

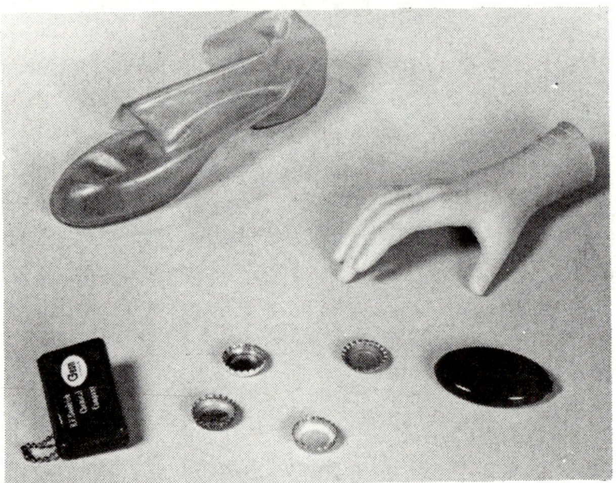

(C) Assorted dip-cast parts. (B.F. Goodrich Chemical Co.)

Fig. 16-9. Dip casting and dip-cast products.

Glossary of New Terms

Casting—The process of placing plastics or resins in a mold and polymerizing or heating until cured or molten and then allowing to form into this shape. No pressure is required.

Dip casting—The process of submerging a hot mold into a resin. After cooling the product is removed from the mold.

Encapsulating—Enclosing an article in a closed envelope of plastics by immersing in resin and allowing the resin to polymerize or, if hot, to cool.

Gel coat—A thin, outer layer of resin applied to a low-pressure laminate. Reinforced layers may then be built up.

Potting—Similar to encapsulating, except the object may be simply covered and not surrounded by an envelope of plastics.

Rotational casting—A method used to make hollow objects from plastisols or powders. The mold is charged and rotated in one or more planes. The hot mold fuses the material into a gel during rotation, covering all surfaces. The mold is then chilled and the product removed from the mold.

Slush casting—Resin in liquid or powder form is poured into a hot mold where a viscous skin forms. The excess slush is drained off, the mold is cooled and the casting removed.

Review Questions

16-1. What is the difference between molding and casting?

16-2. What kind of molds may be used for casting?

16-3. What is the difference between dip casting and dip coating?

16-4. What is potting? Encapsulation?

16-5. What are plastisols? What causes them to solidify?

16-6. What processes are likely to utilize plastisols? Why?

16-7. What is slush casting? Rotational casting? Dip casting?

16-8. What materials and products are made by casting?

16-9. What characteristics of the casting process make it less expensive than molding? More expensive?

16-10. How do rotational, slush, and solid casting differ?

16-11. What is a vinyl dispersion?

16-12. Why are molds for rotational casting less expensive than molds for injection molding of the same hollow article?

16-13. Why are two heating cycles necessary with a slush casting?

16-14. What conditions determine wall thickness of the part in rotational casting?

17

Thermoforming Processes

Thermoforming processes are made possible by the ability of thermoplastic sheet or film stock to soften and be reshaped when heated and to retain the new shape when cooled. Most thermoplastic materials may be formed by this process. However, acetals, polyamides, and fluorocarbons are not normally thermoformed. Extruded, calendered, laminated, cast, and blown films or sheet forms may be thermoformed.

The process of heating a thermoplastic material and forcing it to take a mold shape by mechanical, air, or vacuum pressure is popular because tooling costs are usually low and parts with large surface areas may be produced economically. Prototypes and short runs are also practical. Although dimensional accuracy is good, material thinning in some part designs is a problem.

In thermoforming, tooling can run from inexpensive plaster molds to water cooled steel for extremely long production runs. Although cast aluminum is probably the most common tooling material, gypsum, hardboard, pressed wood, cast phenolic resins, filled or unfilled polyester or epoxy resins, "sprayed" metal, and steel may be used.

According to one source, thermoforming processes date back to the ancient Egyptians, who discovered that horns and tortoise shells could be heated and formed into a variety of vessels and shapes. In the United States, John Hyatt thermoformed Celluloid sheets over wooden cores for keys on the piano.

Today, sheets and films may be thermoformed by eleven basic techniques: (1) straight vacuum forming, (2) drape forming, (3) matched-mold forming, (4) pressure bubble-plug assist vacuum forming, (5) plug assist vacuum forming, (6) plug assist pressure forming, (7) vacuum snap-back forming, (8) pressure bubble vacuum snap-back forming, (9) trapped sheet, constant heat, pressure forming, (10) free forming, and (11) mechanical forming.

Examples of articles fabricated by thermoforming processes include signs, light fixtures, ice-cube trays, blister packages, ducts, drawers, instrument panels, tote trays, housewares, toys, refrigerator panels, transparent aircraft enclosures, and boat windshields (see Fig. 17-1). Fig. 17-2 shows some modern industrial thermoforming machines.

(A) Sports car made of panels which were vacuum formed, glued together, and then painted to produce an all-plastics body. (U.S. Gypsum)

(B) Blister package with a recloseable sliding door for display rack visibility and convenient dispensing. (Celanese Plastics Co.)

(C) Package frame is thermoformed of propionate sheet and is overwrapped with a clear shrink film. (Celanese Plastics Co.)

(D) Clear plastics frames for a display package are thermoformed on a continuous "web" of plastics sheet. (Celanese Plastics Co.)

(E) Vacuum formed clear covers protect and help display these bakery products.

Fig. 17-1. Some articles fabricated by thermoforming processes.

(A) High-speed pressure/vacuum thermoformer operates from either roll stock or inline with an extruder.

(B) Rotary style unit used for large industrial components with a comparatively high production rate.

(C) Twin-sheet thermoforming machine with separate, independent clamping frames.

Fig. 17-2. Modern industrial thermoforming machines. (Brown Machine Co.)

Straight Vacuum Forming

Vacuum forming is the most versatile and widely used thermoforming process. Vacuum equipment generally costs less than pressure or mechanical type processing equipment.

In straight vacuum forming a plastics sheet is clamped in a frame and heated. While the hot sheet is rubbery or in an elastic state it is placed over a female mold cavity and the air removed from this cavity by vacuum (see Fig. 17-3). This causes the atmospheric pressure (14.7 psi) to force the hot sheet against the walls and contours of the mold. When the plastics has cooled, the formed part is removed for final finishing and decorating if necessary. Blowers or fans are usually used to speed cooling. One disadvantage of thermoforming is that formed pieces generally must be trimmed and the scrap reprocessed.

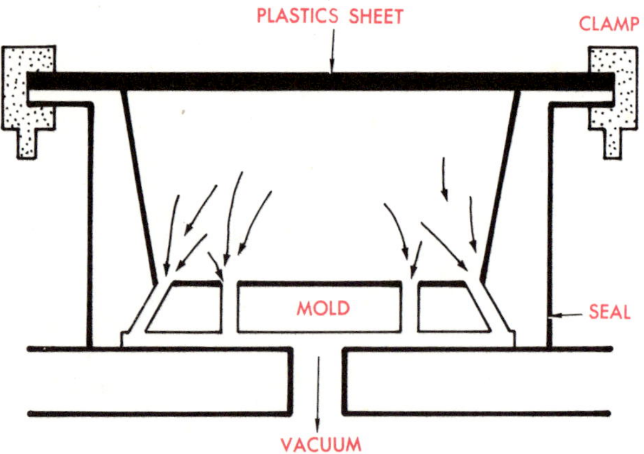

(A) Clamped, heated plastics sheet is forced down into the mold by a vacuum. (Atlas Vac Machine Co.)

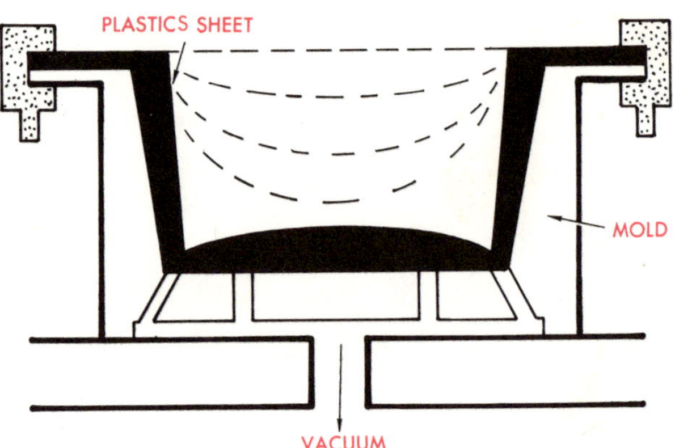

(B) Plastics sheet cools as it contacts the mold. (Atlas Vac Machine Co.)

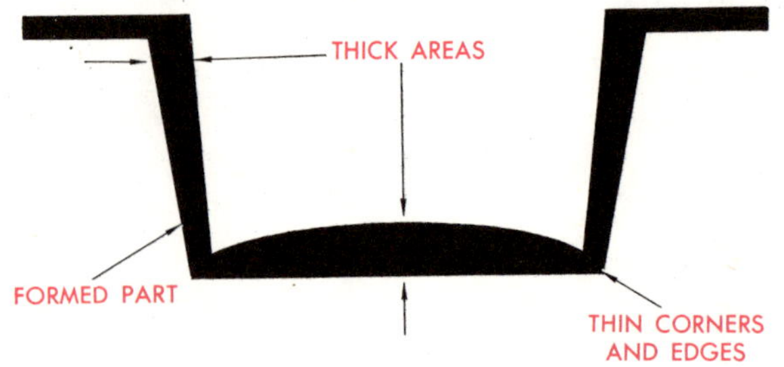

(C) Areas of the sheet which reach the mold last are the thinnest. (Atlas Vac Machine Co.)

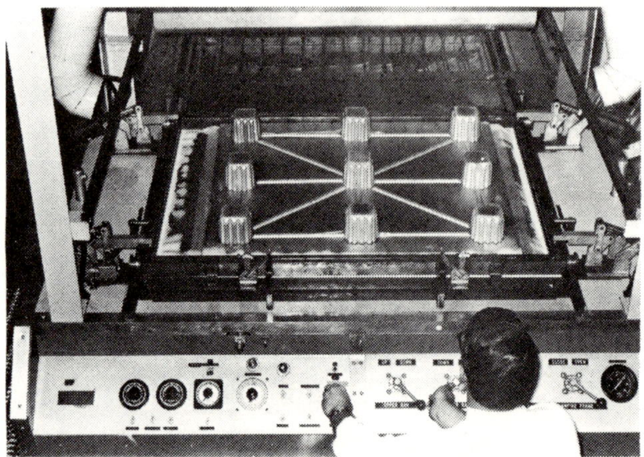

(D) Note frame holding hot thermoformed plastics sheet being drawn over the mold. (Chemplex Co.)

Fig. 17-3. Straight vacuum forming.

Most vacuum systems have a surge tank to ensure a constant 20 to 30 inches of mercury for the vacuum. Superior parts are formed by rapidly applying the vacuum before any portion of the sheet has cooled. It has been proved that slots rather than holes are more desirable and efficient in providing the vacuum air escape. However, slots or holes should be kept smaller than 0.025 inch in diameter to avoid possible surface blemishes on the formed part. As a general rule a hole or slot must be placed in all low or unconnected portions of the mold. If this is not done, air may be trapped under the hot sheet with no way for escape. Unless a collapsible mold is provided, molds should include a 2 to 7 degree angle (draft) for easy part removal. Thinning in deep female molds is a disadvantage. This is caused by the fact that hot plastics is normally drawn to the center of the mold, but the sheeting at the edges of the mold must stretch the most. This becomes the thinnest portion of the formed item.

The "draw" or *draw ratio* of a female mold is the ratio of the maximum cavity depth to the minimum span across the top opening. For high-density polyethylene sheet the best results have been achieved when this ratio did not exceed 0.7:1. If preprinted flat sheets are formed, thinning must be considered to compensate for distortion during forming.

Drape Forming

Drape forming (sometimes incorrectly called "mechanical forming") is similar to straight vacuum forming except that after the plastics is framed and heated, it is mechanically stretched over a male mold and a vacuum (actually a pressure differential) is applied, pulling the hot plastics against all portions of the mold (Fig. 17-4). By draping the sheet over the mold, that part of the sheet touching the mold remains close to the original thickness of the sheet. Side walls are formed from the material draped between the top edges of the mold and the bottom seal area at the base. When the plastics has cooled, it is removed for trimming or postprocessing if necessary. Mark-off (marks from the mold) is on the inside of the product. In straight vacuum forming it appears on the outside of the part.

It is possible to drape form items with a depth to diameter ratios of nearly 4:1. Although high draw ratios are possible with drape forming, this technique is also more complex. Male molds are easy to make and generally cost less than female ones. Male molds are more easily damaged.

Drape forming has also been applied to forming a hot plastics sheet over male or female molds by gravitational forces only.

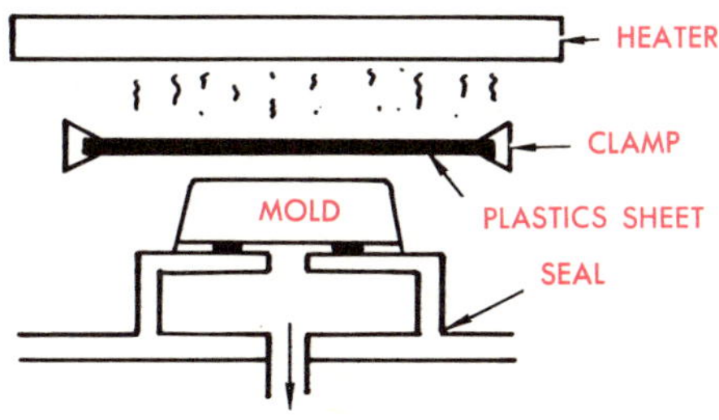

(A) Clamped, heated plastics sheet is drawn over mold either by pulling it over the mold or by forcing the mold into the sheet.

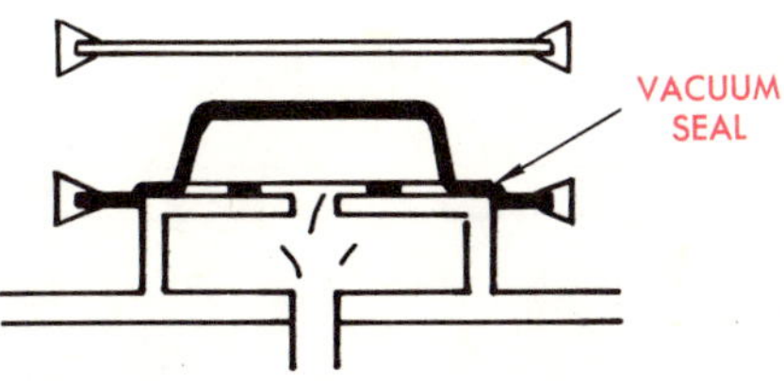

(B) When the mold has been forced into the sheet and a seal created, the vacuum applied beneath the mold forces the sheet over the male mold.

(C) Final wall thickness distribution of the molded part.

Fig. 17-4. Principle of drape forming of plastics. (Atlas Vac Machine Co.)

Matched-Mold Forming

Matched-mold forming is similar to compression molding except a heated sheet is trapped and formed between male and female dies, which may be wood, plaster, epoxy, etc. (see Fig. 17-5). Ac-

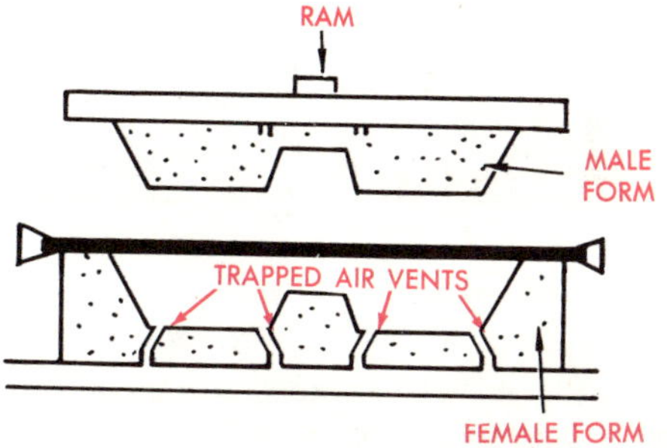

(A) Heated plastics sheet may be clamped over female die as shown, or can be draped over the mold force.

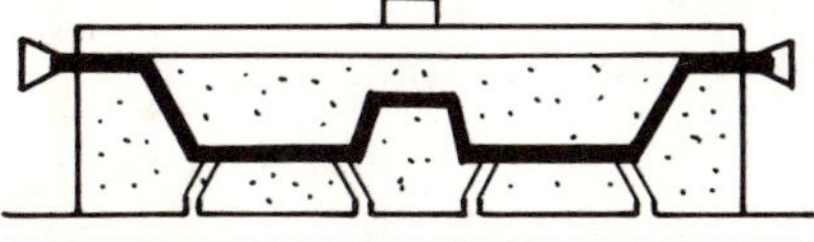

(B) As mold closes, it forms the sheet, with vents allowing trapped air to escape.

(C) Material distribution of formed part depends on the shapes of the two forms.

Fig. 17-5. Principle of matched-mold forming. (Atlas Vac Machine Co.)

curate, close-tolerance parts may be rapidly produced by this method in expensive, water cooled molds. Excellent reproduction of mold detail and dimensional accuracy can be obtained, including lettering and grained surfaces. There is mark-off on both sides of the finished product; consequently, mold dies *must* be protected from scratches or damage.* As a rule thermoplastic materials will form to the surface appearance of the mold. A smooth surface mold should not be used with polyolefins because air may be trapped between the hot plastics and the highly polished mold. Sandblasted mold surfaces are normally used for these materials.

Fig. 17-6 shows a matched mold operation where the blank is smaller than the finished part. Under the ram it flows out to the periphery without forming trim or flash. The total mold cycle here is 10-20 seconds.

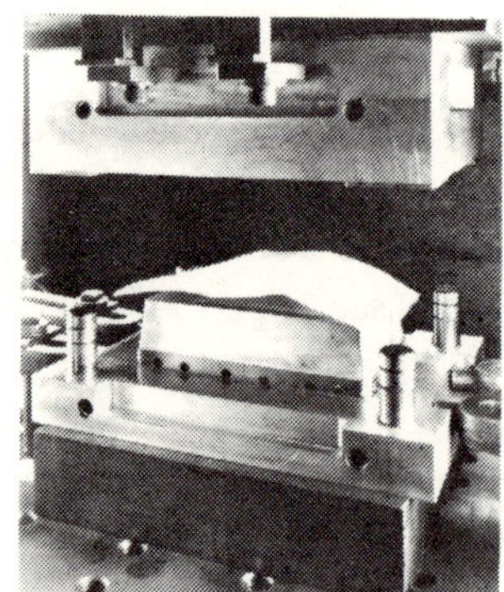

(A) Preheated sheet placed between cooled matched metal dies.

(B) The finished and formed part is larger than original sheet.

Fig. 17-6. Actual matched-mold operation with conventional metal-stamping press. (G.R.T.L. Co.)

Pressure Bubble-Plug Assist Vacuum Forming

For deep thermoforming, pressure bubble-plug assist vacuum forming is an important process. By this process it is possible to control the thickness of the formed article. The article may have a uniform thickness or the thickness may be varied.

Once the sheet has been placed in the frame and heated, controlled air pressure creates a bubble (see Fig. 17-7). This bubble stretches the material to a predetermined height, usually controlled by a photocell. The male plug assist is then lowered, forcing the stretched stock down into the cavity. The male plug is normally heated so as not to prematurely chill the plastics. This plug is made as large as possible so the plastics is stretched close to the final shape of the finished product. Plug penetration should be from 70 to 80 percent of the mold cavity depth. Air pressure is then applied from the plug side and vacuum is drawn on the cavity to help form the hot sheet. For many products, only vacuum is drawn on the cavity to complete formation of the sheet. In Fig. 17-7 both vacuum and pressure is applied during the forming process. The female mold must be vented to allow trapped air to escape from between the plastics and the female mold.

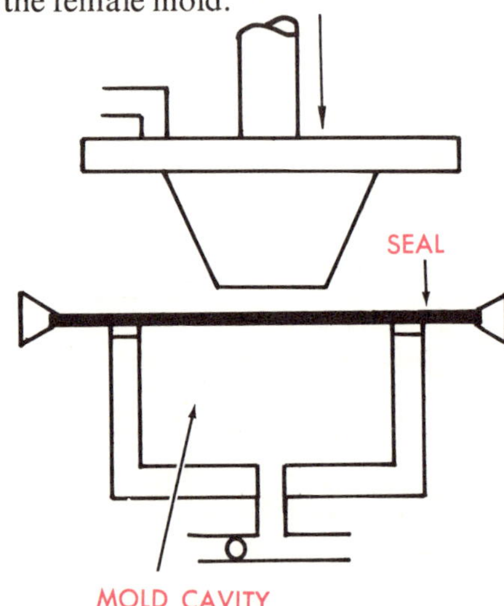

(A) The plastics sheet is heated and sealed across the mold cavity.

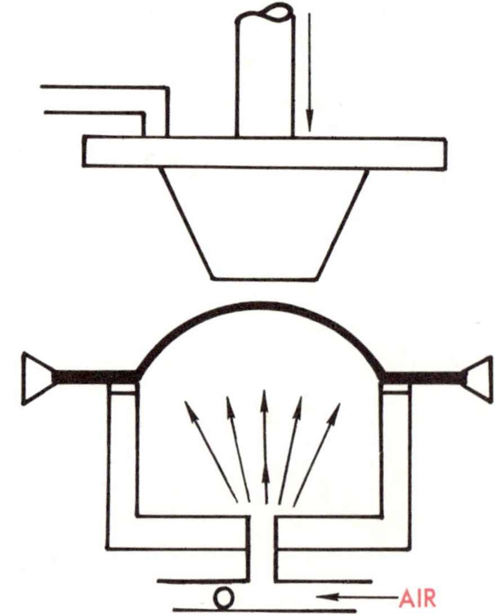

(B) Air is introduced into the mold cavity and blows the sheet upward into a bubble, stretching it evenly.

*See Mechanical Forming, p. 298.

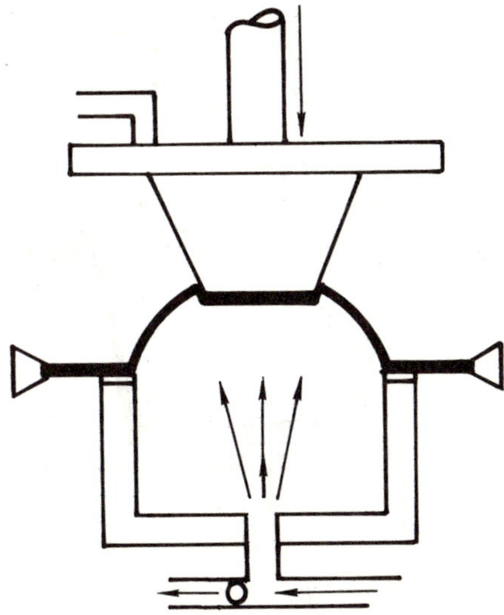

(C) A plug shaped roughly to the contour of the cavity plunges into the plastics sheet.

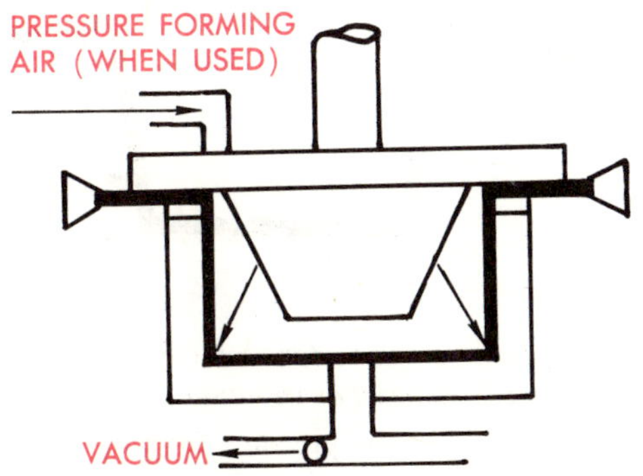

(D) When the plug reaches its lowest position a vacuum is drawn on the cavity to complete formation of the sheet.

Fig. 17-7. Pressure bubble-plug assist vacuum forming. (Atlas Vac Machine Co.)

Plug Assist Vacuum Forming

To help prevent corner or periphery thinning of cup or box shaped articles, a plug assist is used to mechanically stretch and pull additional plastics stock into the female cavity (Fig. 17-8). The plug is normally heated to a temperature just below the forming temperature of the sheet stock. The plug should be from 10 to 20 percent smaller in length and width than the female dimensions. Once the plug has forced the hot sheet into the female cavity, air is drawn from the mold to complete formation of the part. The plug design or shape determines the wall thickness, as is shown in cross section in Fig. 17-8D.

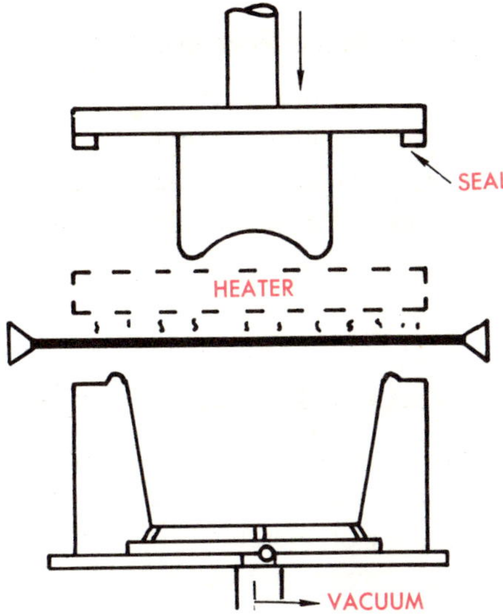

(A) Heated, clamped plastics sheet is positioned over mold cavity.

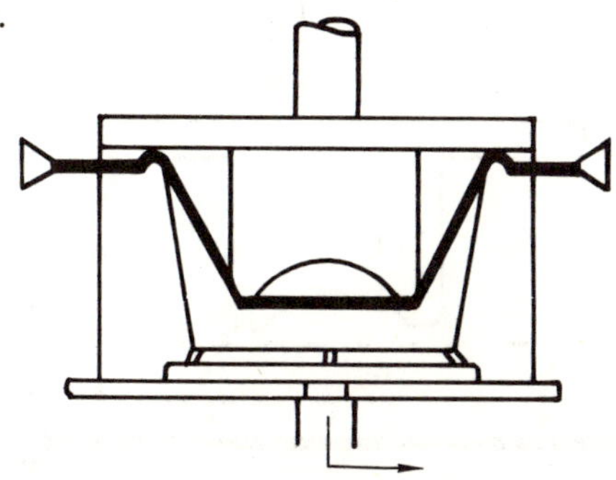

(B) Plug, shaped roughly like the mold cavity but smaller, is plunged into the plastics sheet and prestretches it.

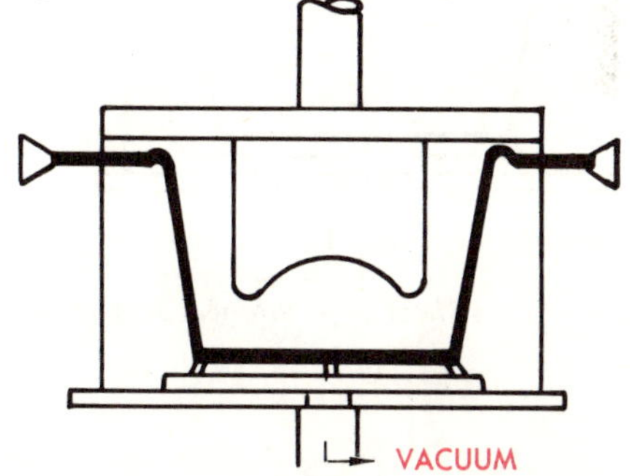

(C) When plug platen reaches its closed position, a vacuum is drawn on the mold cavity.

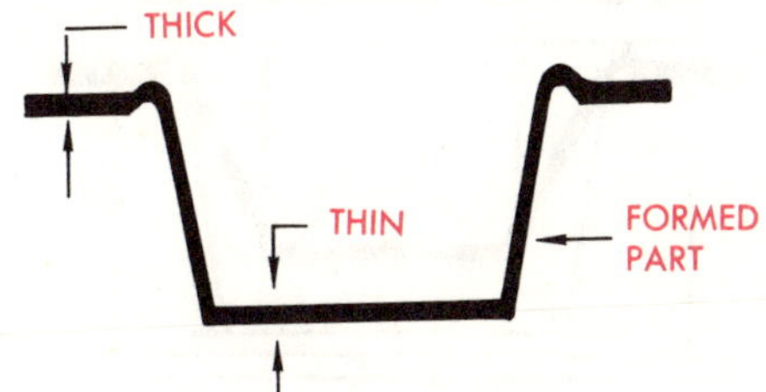

(D) Areas of the plug touching the sheet first create thicker areas due to chilling effect.

Fig. 17-8. Plug assist vacuum forming. (Atlas Vac Machine Co.)

Plug Assist Pressure Forming

Plug assist pressure forming is similar to plug assist vacuum forming except that as the plug forces the hot plastics into the female cavity, air pressure is applied from the plug, forcing the plastics sheet against the walls of the mold. This is shown in Fig. 17-9.

Vacuum Snap-Back Forming

In vacuum snap-back forming the hot plastics sheet is placed over a box and a vacuum (measured in inches of mercury) is drawn, causing a bubble to be forced into the box (Fig. 17-10). A

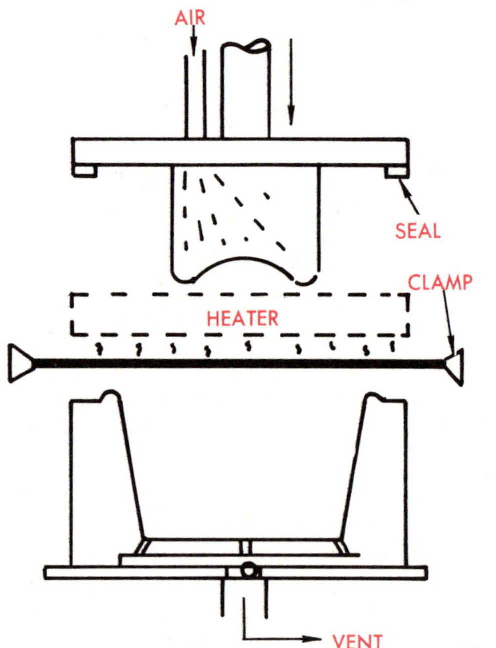

(A) Heated, clamped sheet is positioned over mold cavity.

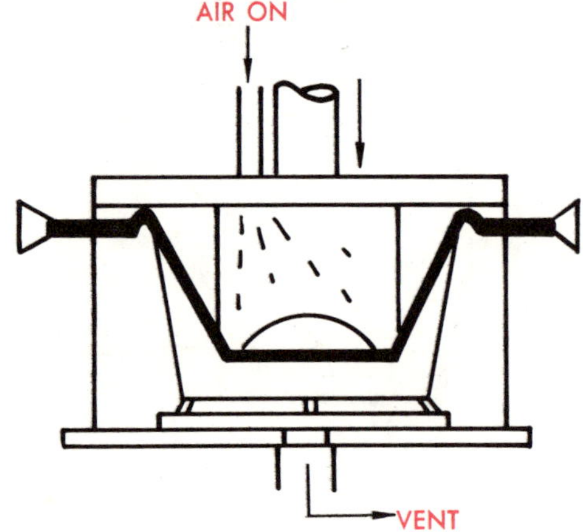

(B) As plug enters sheet air under sheet is vented to the atmosphere.

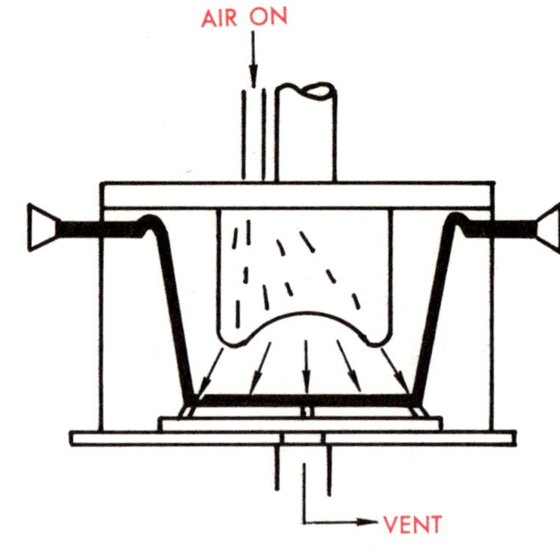

(C) When plug completes its stroke and seals the mold, air pressure is applied from the plug side.

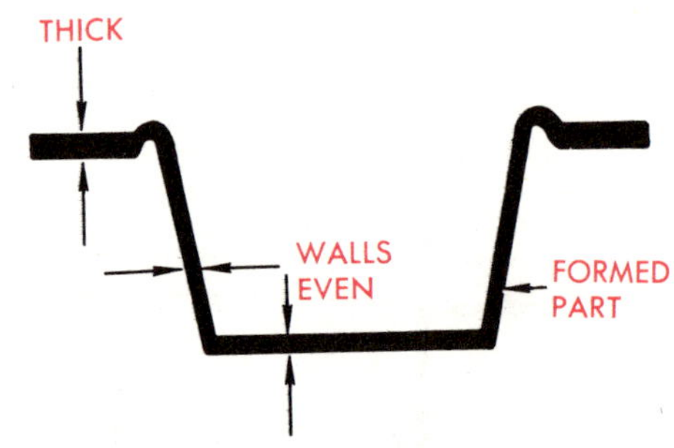

(D) Plug assist pressure forming can be controlled to produce uniform material distribution.

Fig. 17-9. Plug assist pressure forming (Atlas Vac Machine Co.)

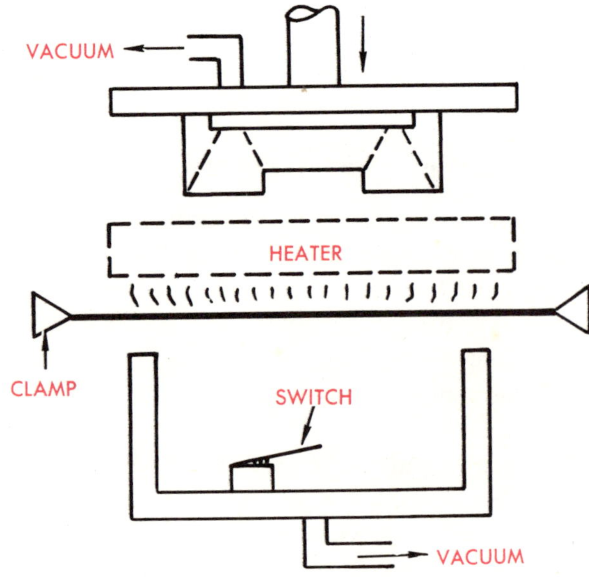

(A) Plastics sheet is heated and sealed over the top of the female vacuum box. (Atlas Vac Machine Co.)

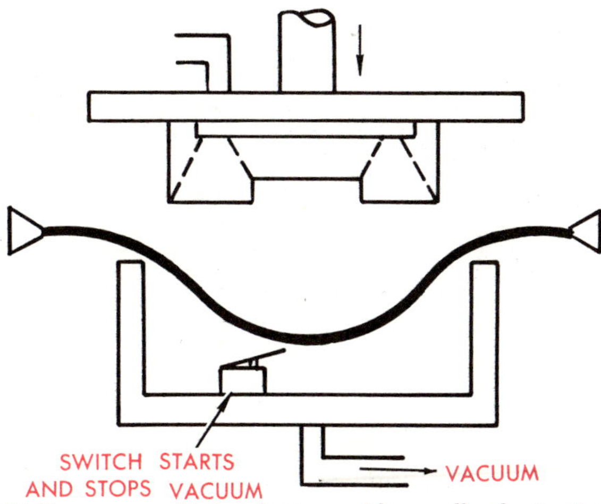

(B) Vacuum applied at bottom of box pulls plastics into a concave shape. (Atlas Vac Machine Co.)

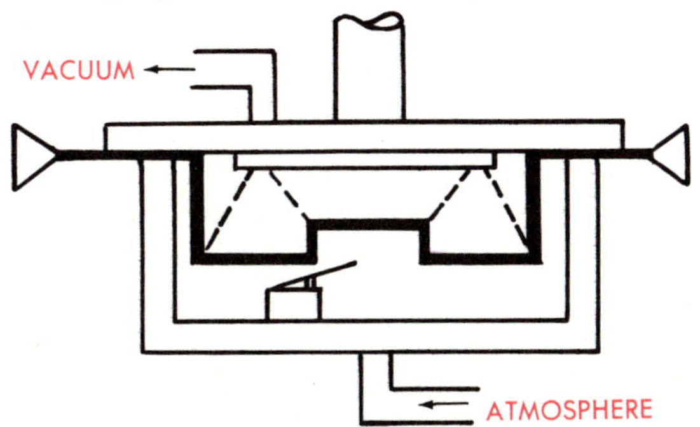

(C) Male plug enters the sheet and a vacuum is drawn through the male plug, while vacuum under the sheet is vented. (Atlas Vac Machine Co.)

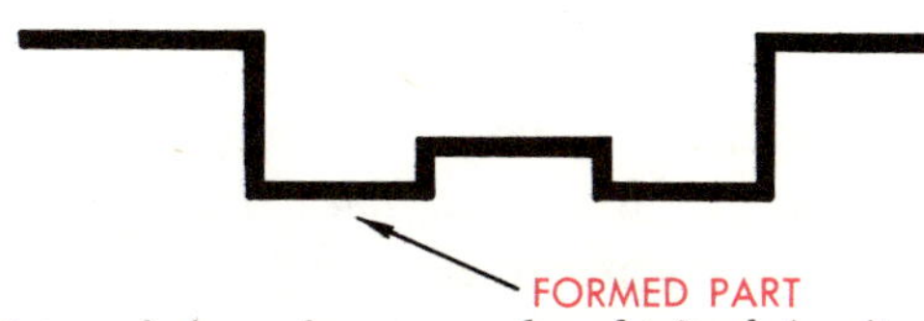

(D) External deep draws can be obtained for items like luggage, auto parts, etc. (Atlas Vac Machine Co.)

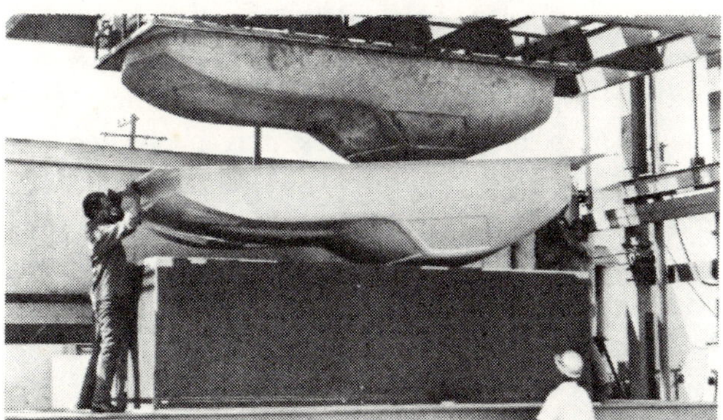

(E) ABS-polycarbonate ¼-in sheets are used to vacuum form this complete automobile body in a 20-min cycle. (Marbon Div., Borg-Warner Corp.)

Fig. 17-10. Vacuum snap-back forming. (Atlas Vac Machine Co.)

male mold is then lowered and the vacuum in the box is released, causing the plastics to "snap back" around the male mold. For some molds a vacuum may also be placed in the male mold to assist in pulling the plastics into place.

Pressure-Bubble Vacuum Snap-Back Forming

As the name implies, the sheet is heated and then stretched into a bubble shape by air pressure (see Fig. 17-11). The sheet prestretches about

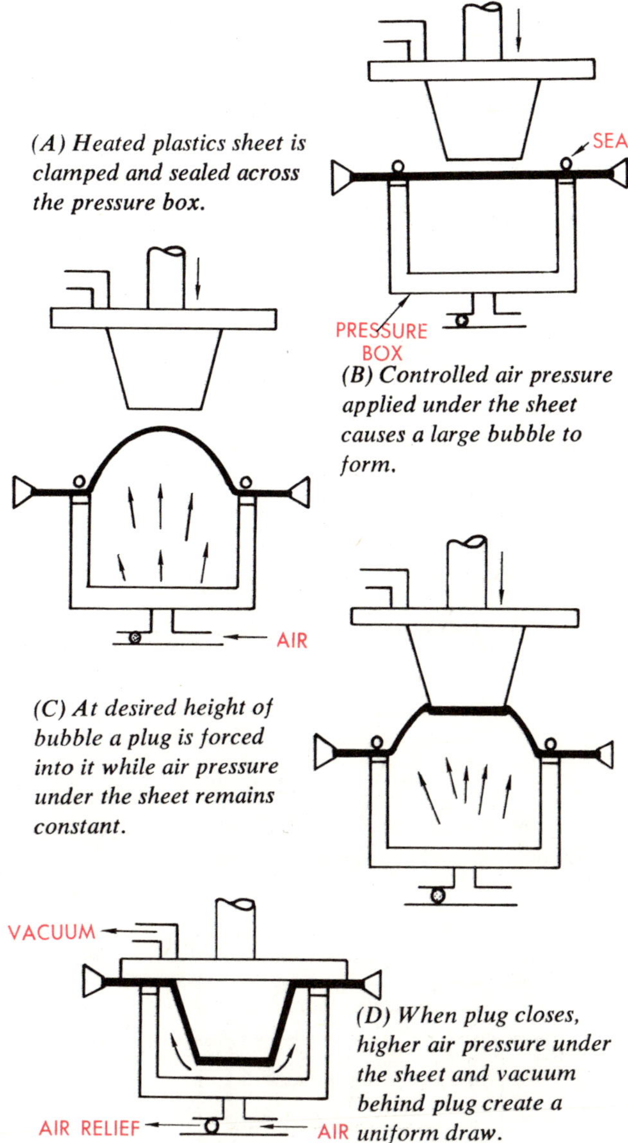

(A) Heated plastics sheet is clamped and sealed across the pressure box.

(B) Controlled air pressure applied under the sheet causes a large bubble to form.

(C) At desired height of bubble a plug is forced into it while air pressure under the sheet remains constant.

(D) When plug closes, higher air pressure under the sheet and vacuum behind plug create a uniform draw.

Fig. 17-11. Pressure-bubble vacuum snap-back forming. (Atlas Vac Machine Co.)

35 to 40 percent. The male mold is then lowered, and a vacuum is applied to the male mold while air pressure is forced into the female cavity. This causes the hot sheet to "snap-back" on the male mold. Mark-off is on the male mold side.

Trapped-Sheet Contact-Heat Pressure Forming

This process is similar to straight vacuum forming except air pressure and a vacuum assist may be used to force the hot plastics into a female mold (see Fig. 17-12).

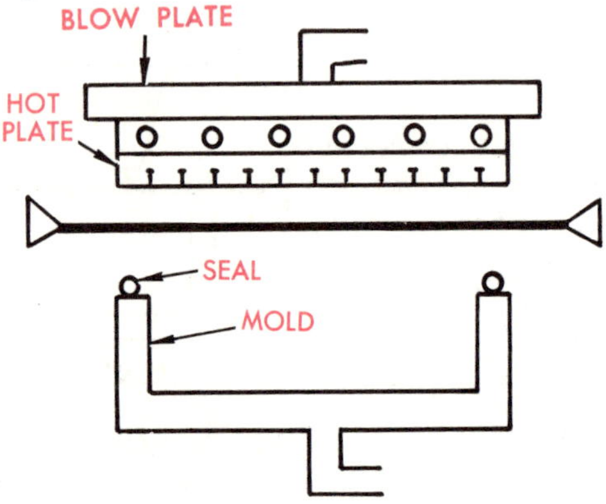
BLOW PLATE
HOT PLATE
SEAL
MOLD

(A) Flat, porous plate allows air to be blown through its face.

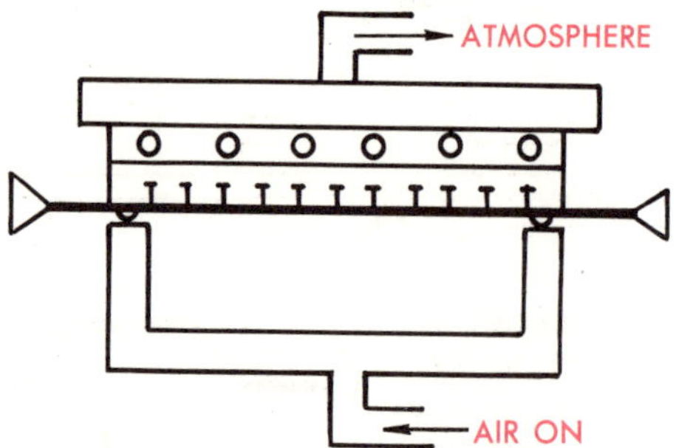
ATMOSPHERE
AIR ON

(B) Air pressure in female mold and often a vacuum on the hot plate are used to bring sheet totally in contact with hot plate.

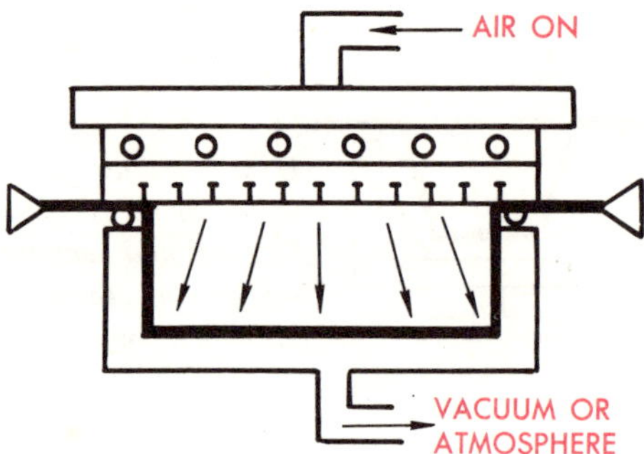
AIR ON
VACUUM OR
ATMOSPHERE

(C) Air pressure through hot plate forms the sheet into the female mold.

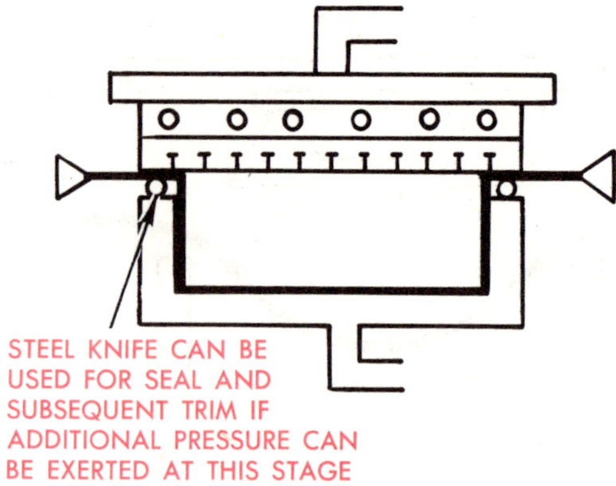
STEEL KNIFE CAN BE
USED FOR SEAL AND
SUBSEQUENT TRIM IF
ADDITIONAL PRESSURE CAN
BE EXERTED AT THIS STAGE

(D) After forming, additional pressure can be exerted.

Fig. 17-12. Trapped-sheet contact-heat pressure forming. (Atlas Vac Machine Co.)

Free Forming

In free forming, air pressures of over 400 pounds per square inch may be used to blow a hot plastics sheet through the silhouette of a female mold (Fig. 17-13). The air pressure causes the sheet to form a smooth bubble-shaped article. A stop may be provided and used to form special contours in the bubble. Skylight panels and aircraft canopies are familiar examples.

Mechanical Forming

In mechanical forming no vacuum or air pressure is used to form the part. It is similar to matched molding except close-fitting matched male and female molds are *not* used. The process is sometimes classed as a fabrication or postforming operation. Ovens, strip heaters, or heat guns may be used as a heat source. The forming process may utilize simple wooden forming jigs to help form the desired shape. Flat stock may be heated and wrapped around cylindrical shapes or heated in a narrow strip and bent at right angles. Tubes, rods, and other profile shapes may also be mechanically formed (see Fig. 17-14).

Plug and ring forming (Fig. 17-15) is sometimes classed as a separate forming process. Because no vacuum or air pressure is used, however, it may be classed as a mechanical forming operation. The process consists of a male mold shape and a similarly shaped female silhouette mold (not a matched mold). Normally, the hot plastics is

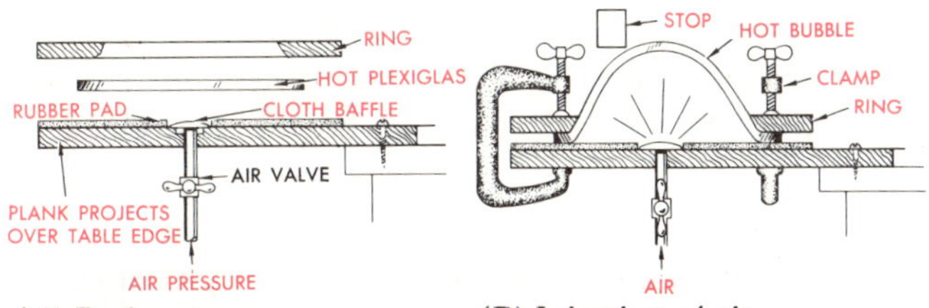

(A) Basic setup. *(B) Injection of air.*

(C) Removing free formed acrylic bubble-shaped product. (Rohm & Haas Co.)

Fig. 17-13. Free forming of plastics bubbles.

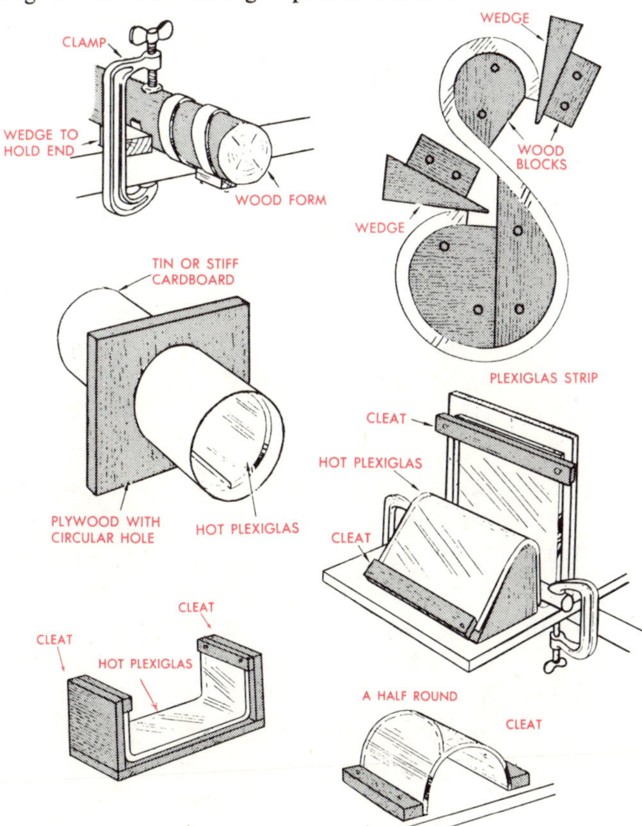

Fig. 17-14. Examples of mechanical forming. (Rohm & Haas Co.)

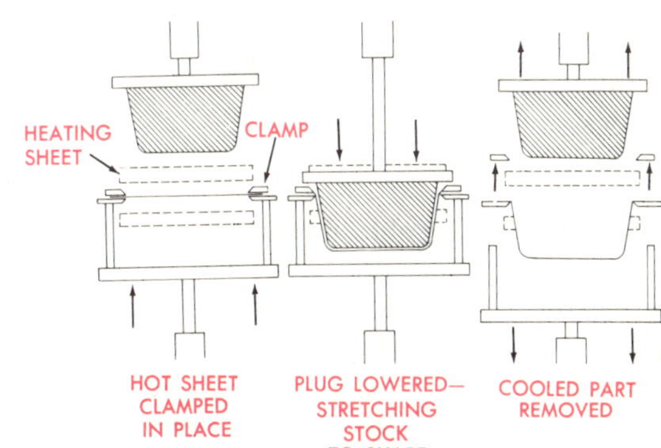

(A) Basic principle of plug and ring forming.

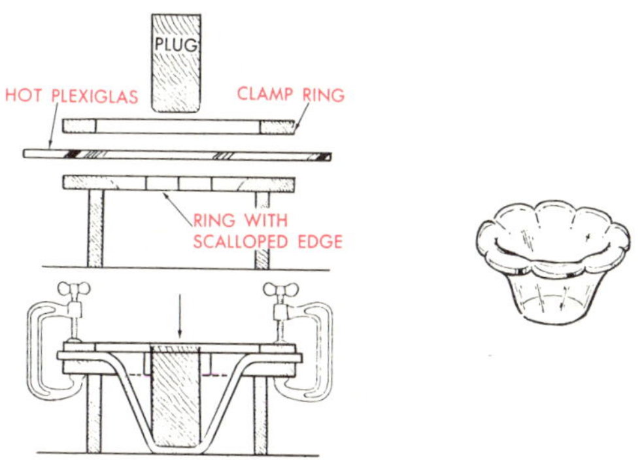

(B) Vase. (Rohm & Haas Co.)

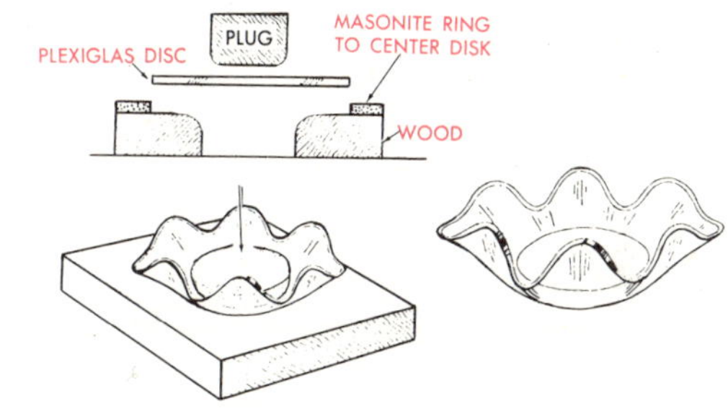

(C) Decorative bowl. (Rohm & Haas Co.)

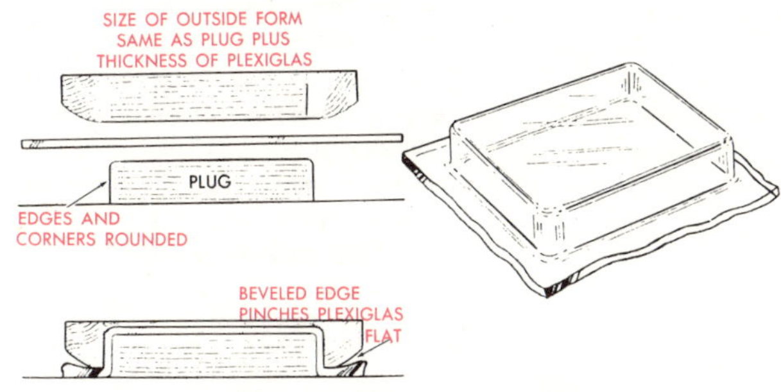

(D) Plastics pan. (Rohm & Haas Co.)

Fig. 17-15. Examples of plug and ring forming.

forced through the "ring" (not necessarily a rough shape) shape of the female mold by the male. The cooling plastics take the shape of the male mold that it touches.

Table 17-1 is a general-use table giving some common problems encountered in thermoforming plastics.

Table 17-1. Troubleshooting Thermoforming.

Defect	Possible Remedy
Pinholes or ruptures	Vacuum holes too large, too much vacuum or uneven heating Attach baffles to the top clamping frame
Webbing or bridging	Sharp corners on deep draw, change design or mold layout Use mechanical drape or plug assists or add vacuum holes Check vacuum system and shorten heating cycle
Markoff	Slow draping action may trap air Clean mold or remove high surface gloss from mold Remove all tool marks or wood grain patterns from mold Mold may be chilling plastic sheet too quickly
Excessive post shrinkage	Rotate sheet in relation to mold Increase cooling time
Blisters or bubbles	Overheating sheet: lower heater temperature Ingredients of sheet formulation incorrect or hygroscopic
Sticking to mold	Smooth mold or increase taper and draft Use mechanical releasing tools, air pressure, or mold release Mold may be too warm or increase cooling cycle
Incompletely formed pieces	Lengthen heating cycle and increase vacuum Add vacuum holes
Distorted pieces	Poor mold design: check tapers and ribs Increase cooling cycle or cool molds Sheet is removed too quickly while still hot
Change in color intensity	Use proper mold design and allow for thinning of piece Lengthen heating cycle and warm mold and assists Use heavier gauge sheet and add vacuum holes

Glossary of New Terms

Drape forming—Method of forming thermoplastic sheet in a movable frame, heated and draped over high points of a male mold. Vacuum is then pulled to complete the forming.

Free forming—Air pressure is used to blow a heated sheet of plastics being held in a frame until the desired shape or height is attained.

Matched-mold forming—Hot sheets are formed between matched male and female molds.

Mechanical forming—Heated sheets of plastics are shaped or formed by hand or with the aid of jigs and fixtures. No mold is used.

Shrink wrapping—A technique of packaging in which the strains in a plastics film are released by raising the temperature of the film, thus causing it to shrink over the package.

Snap-back forming—A technique in which an extended plastics sheet is allowed to contract over a male form shaped to the desired contours.

Thermoforming—Any process of forming thermoplastic sheet which consists of heating the sheet and pulling it down onto a mold surface.

Vacuum forming—Method of sheet forming in which the plastics is clamped in a stationary frame, heated, and drawn down by a vacuum into a mold.

Review Questions

17-1. What is thermoforming?

17-2. Name six common thermoformed items.

17-3. Give two advantages of thermoforming.

17-4. Give two disadvantages of thermoforming.

17-5. What is straight vacuum thermoforming?

17-6. Is a vacuum used in drape forming?

17-7. What is the major advantage in matched-mold thermoforming? The major disadvantage?

17-8. How is the product thickness controlled in pressure bubble-plug assist vacuum forming?

17-9. What determines product wall thickness in pressure assist vacuum thermoforming?

17-10. How does plug assist pressure forming differ from plug assist vacuum forming?

17-11. What do the words "snap-back" refer to in vacuum snap-back forming?

17-12. What is free forming?

17-13. Describe mechanical forming. Plug and ring forming.

18

Expanding Processes

Expanded plastics (sometimes called frothed, cellular, blown, foamed, or bubble plastics) may be classified by cell structure, density, type of plastics, and degree of flexibility. Nearly all plastics materials, including thermoplastic and thermosets have been expanded. Expanded plastics are a distinctly cellular plastic material.

Resins are made into expanded plastics by six basic methods: (1) thermal decomposition of a chemical blowing agent liberating a gas, (2) dissolving a gas in the resin which expands at room temperature, (3) mixing a liquid component which vaporizes when heated, (4) whipping air into the resin and then rapidly curing or cooling the resin, (5) adding components that react (liberate gas) within the resin by chemical reaction, and (6) volatilizing moisture (steam left in resins by the heat generated from exothermic chemical reaction.) Syntactic type "expanded" plastics are sometimes included as a separate process. Syntactic plastics are produced by blending microscopically small hollow balls of glass or plastics in a resin binder. This produces a lightweight closed-cell material (the hollow balls may have been expanded). The puttylike mixture may be molded or applied by hand into spaces not easily reached by other means.

Expanded plastics may be flexible or rigid, have cells that are closed or interconnected (open-cell), and have densities from that of the solid parent resin down to less than one pound per cubic foot. Expanded plastics are used for insulation, packaging, cushioning, and flotation. Some act as acoustical insulation as well as thermal insulation. Others are used as moisture barriers in construction. Epoxy expanded materials are used in lightweight tooling fixtures and models. They may also be used as noncorroding, lightweight, shock-absorbing materials for automobiles, aircraft, furniture, boats, honeycomb structures, etc. In the textile industry, expanded materials are used as padding and insulation, and give garments a special texture and "feel."

During World War II the Dow Chemical Company introduced expanded polystyrene products in the United States. General Electric produced expanded phenolic products.

The two main expanded plastics today are probably polystyrene and polyurethane. Polystyrene products are rigid cellular structures, while polyurethane products may be rigid or flexible. Expanded products familiar to the consumer are ceiling tile, Christmas decorations, flotation materials, toys, package liners for fragile articles, mattresses, pillows, carpet backing, sponges, and various disposable containers (see Fig. 18-1).

Numerous plastics are expanded and formed into products by several different processes. These processes may be divided into four basic methods: (1) molding, (2) casting, (3) expand-in-place, and (4) spray.

Molding

Various methods have been developed for molding expandable plastics, including injection molding, compression molding, extrusion molding,

(A) Polystyrene (cellular) containers.

(B) Injection-molded polyethylene bowl. (Phillips Petroleum Co.)

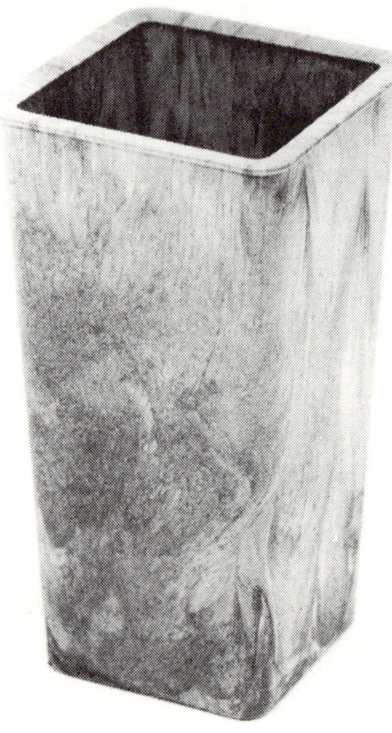

(C) Injection-molded polyethylene wastebasket. (Phillips Petroleum Co.)

(D) Polyethylene display sign with woodlike texture and grain pattern. (Phillips Petroleum Co.)

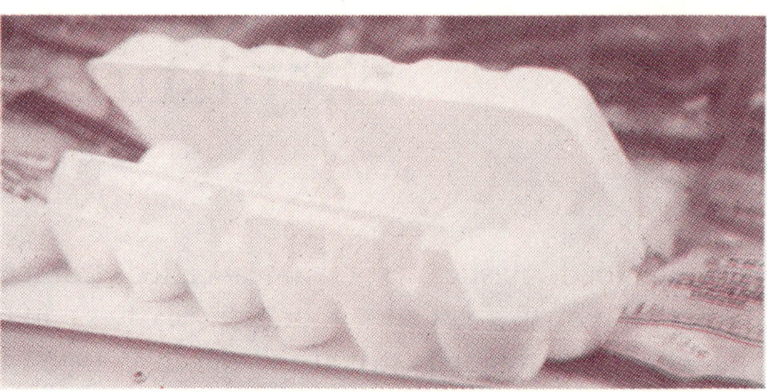

(E) Vacuum formed polystyrene egg container.

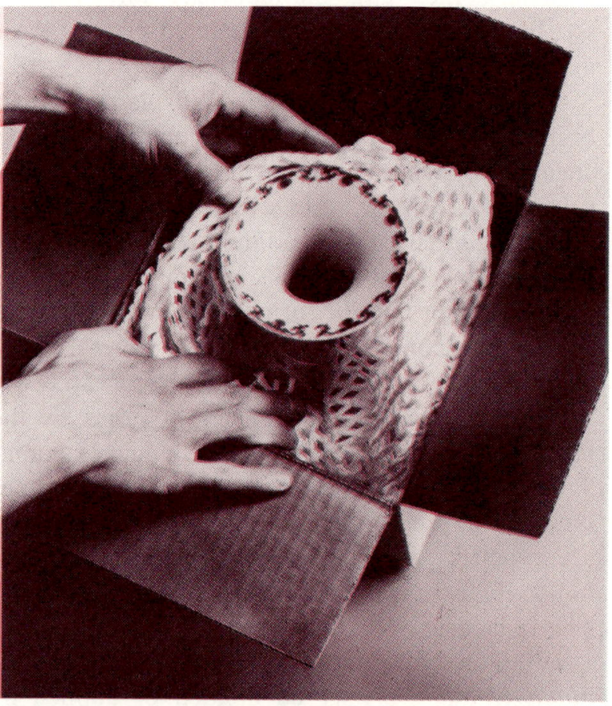

(F) Polystyrene mesh for cushion-in-transit material. (Foster Grant Co.)

Fig. 18-1. Various expanded polystyrene and polyethylene products.

dri-electric (high-frequency expanding method) molding, and steam chamber or probe molding.*

In one injection and extrusion method the expanding agent is added directly to the hot mix.

*See Molding Processes, Chap. 15.

The pressure of the plunger or extruder does not allow this material to expand until it is forced into a mold (see Fig. 18-2).

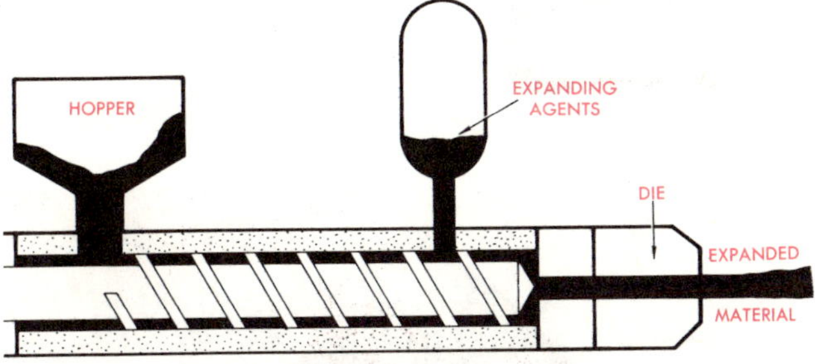

(A) Screw type method.

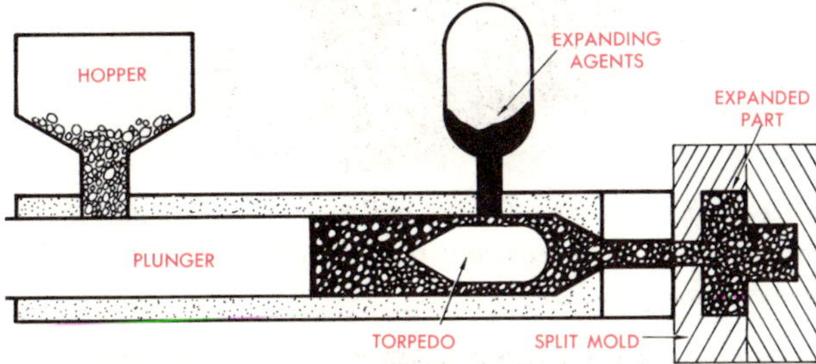

(B) Plunger type method.

Fig. 18-2. Adding expanding agents directly to the hot mix.

In other injection and extrusion methods the blowing or expanding agent, colorants, and other additives are metered directly into the molten plastics just before it enters the mold (see Fig. 18-3). The expansion of the plastics material takes place in the molding cavity.

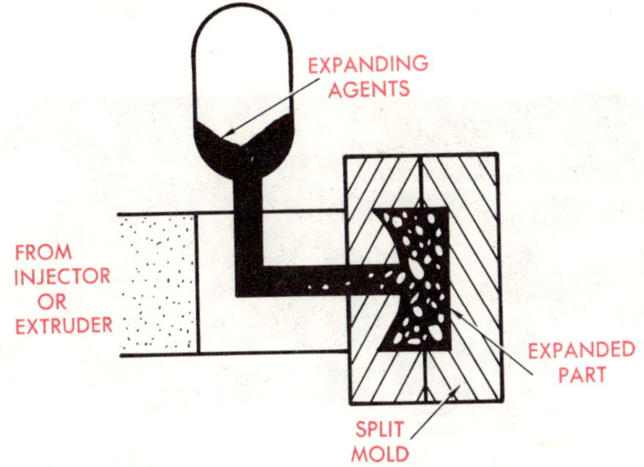

Fig. 18-3. Molding method in which expanding agent is added to molten plastics just before it enters the mold.

In yet another process the extruded material with expanding agents are fed into an accumulator (see Fig. 18-4). When the predetermined charge is reached the plunger of the accumulator forces

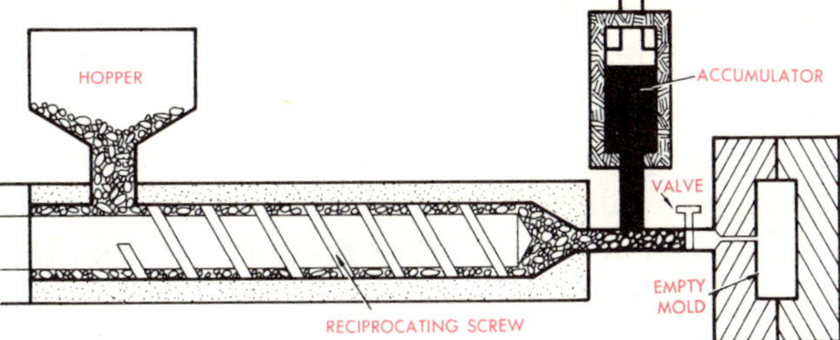

(A) Material enters accumulator.

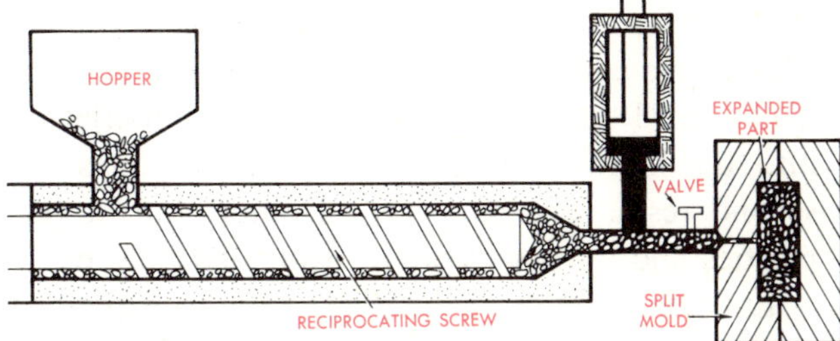

(B) Accumulator forces material into mold cavity.

Fig. 18-4. Molding method in which an accumulator is used.

the material into the mold cavity, where expansion takes place.

An expanded plastics with a skin (unexpanded layer) layer is produced by forcing the hot mixture around a fixed torpedo (Fig. 18-5). The extruded shape is actually a hollow form the shape of the sizing die. The extrudate then expands, filling this

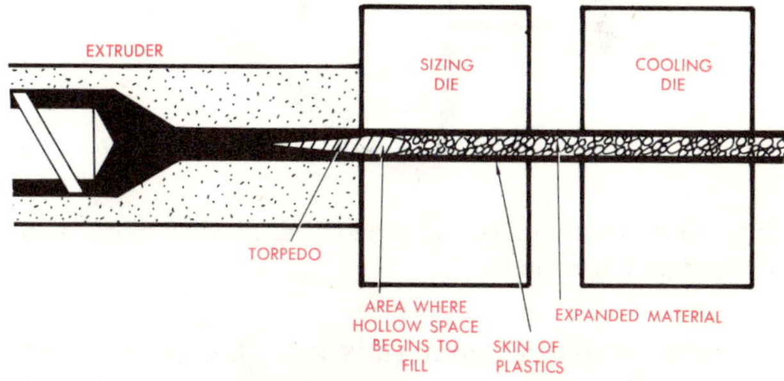

(A) A method of production.

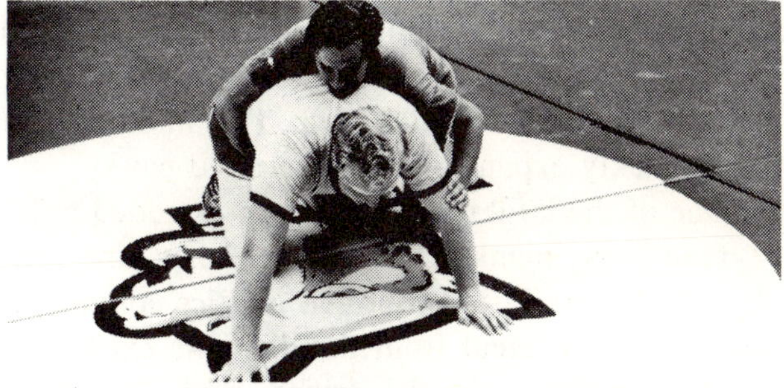

(B) Vinyl formed mat with a "skin" on each side.

Fig. 18-5. Forming a "skin" on expanded plastics.

hollow space. The skin is formed by the cooling action of the sizing and cooling dies. Structural profiles are produced by this method.

In another process two plastics of the same formulation or different families are injected one after the other into a mold. The first plastics does not contain expanding agents and is partially injected into the mold. The second plastics, containing the expanding agents, is then injected against the first plastics, forcing it against the edges of the mold. This forms a shell around the expandable plastics. To close off the shell the first resin is again injected into the mold. This completely encapsulates the second resin. The part has an outer skin of one plastics and an inner core of expanded plastics.

Extruded polyvinyl materials may be expanded as they emerge from the die or stored for future expansion. They are used in garment industries as single components or given a cloth backing.

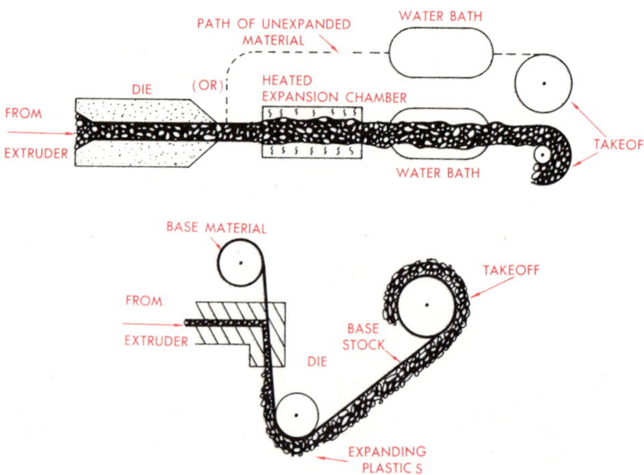

Fig. 18-6. Two methods of expanding plastics material as it emerges from the die.

Various foamed materials are placed on carpet or other flooring materials by the processing techniques indicated in Fig. 18-7.

In compression molding of expandable plastics the resin formulation is extruded into the molding chamber and the mold closed. The molten resin then quickly expands, filling the mold cavity.

One of the largest markets is for extruded polystyrene logs, planks, and sheets. They are produced by extruding the molten plastics containing the expanding agent from the die. The expansion occurs rapidly at the die orifice. Rods, tubes, or other shapes may be produced in this manner.

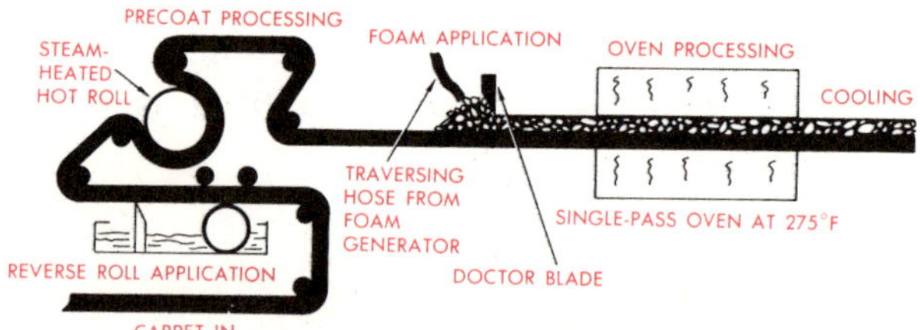

(A) Backing a carpet. (Union Carbide Corp.)

(B) Tough polyamide "grass" will not wrinkle or pull loose, and players may wear any type of shoe including spikes and cleats. (3M Co.)

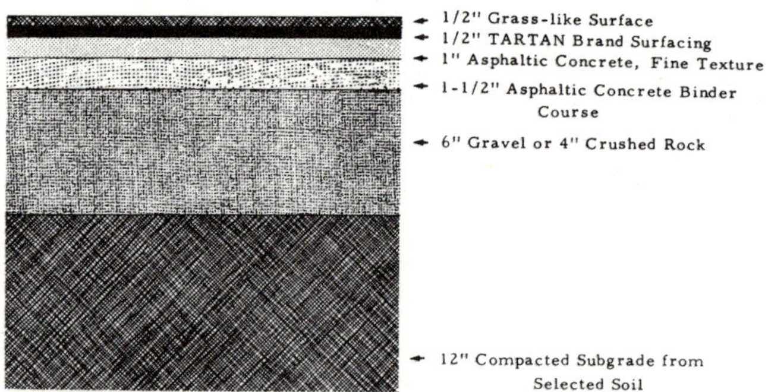

(C) Suggested makeup for a playing field having the surface shown in (B). (3M Co.)

(D) Workers wet-pouring the ½-in cushion layer of surfacing material over asphalt. A thin resin layer is then troweled on and a top layer of "grass" immediately laid. (3M Co.)

Fig. 18-7. This is how many foams are applied to flooring materials.

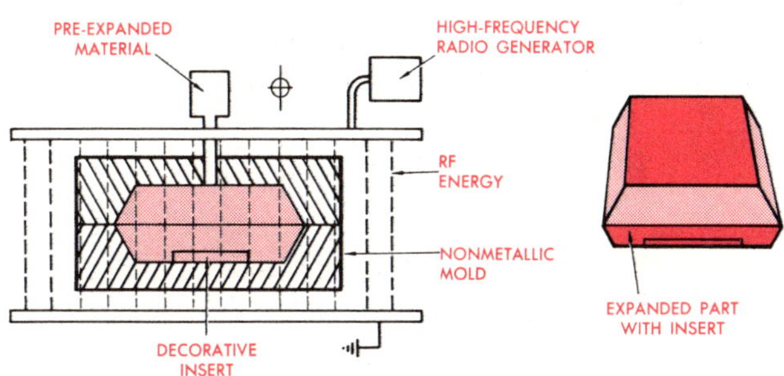

(A) Basic principle using radio-frequency energy.

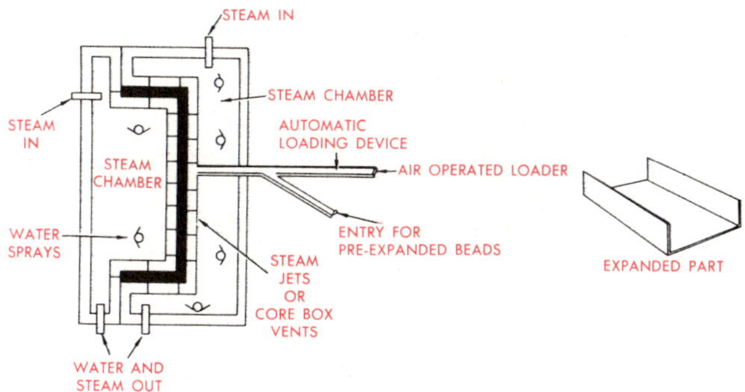

(B) Typical bead-expanding mold using steam.

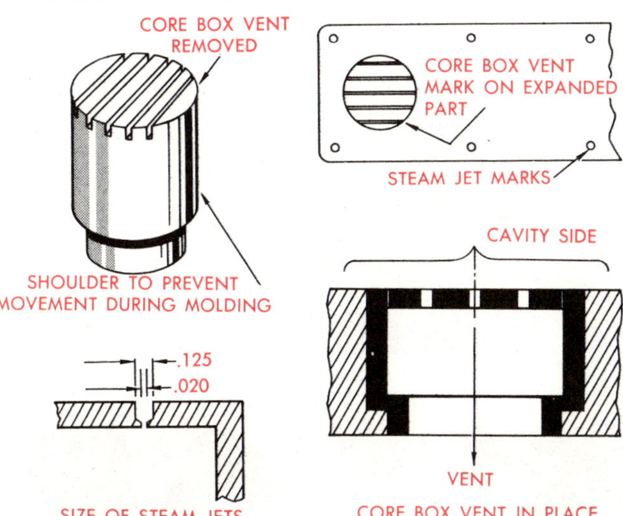

(C) Part showing core box vent and steam jet marks.

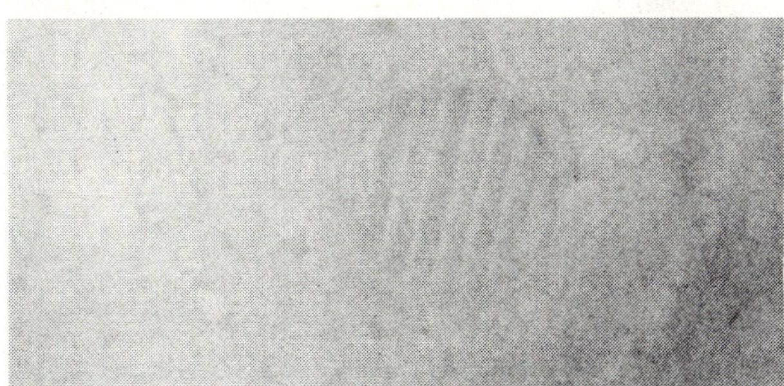

(D) Closeup of core box vent mark left on expanded product.

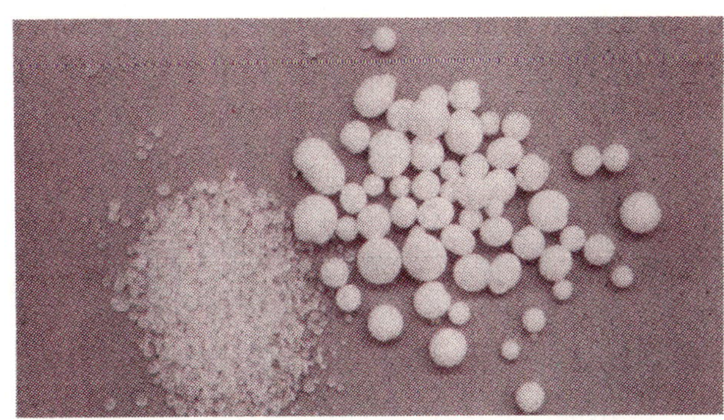

(E) Unexpanded (left) and expanded (right) polystyrene beads.

(F) Expandable polystyrene molding machine capable of molding ice chests. (Koehring Co.)

(G) Expandable polystyrene molding machine for producing 12-oz. cups. (Springfield Cast Products, Inc.)

Fig. 18-8. Expanding plastics without using a hot-melt method.

Not all plastics are expanded utilizing a hot-melt method. Polystyrene is produced in the form of very small beads containing an expanding agent. These beads may be pre-expanded by heat or radiation and placed in a mold cavity. They are again heated, causing further expansion. The expanding pressure packs the beads into a closed cellular structure, some expanding up to forty times their original size. The actual heating of the beads in the mold cavity may be accomplished by the thermal heat of steam or by thermal excitation of molecules by high radio-frequency energy. The latter method, sometimes called dri-electric molding, does not require steam lines, moisture and steam vents, or metallic molds. Inlays or decorative substrates of paper, fabric, or plastics may also be molded in place by this method.

Pre-expansion is accomplished by dry thermal heat, radio-frequency radiation, steam, or boiling water. Pre-expanded beads are used within a few days. Stored beads may loose their volatile blowing or expanding agent. They should be kept in a cool airtight container until ready for use.

Insulated cups, ice chests, holiday decorations, novelty items, toys and various flotation and thermal insulating products are familiar products of this method.

Casting

In casting expandable plastics materials the resin mix containing catalysts and chemical expanding agents are placed in a mold where it expands into a cellular structure (see Fig. 18-9). Polyurethanes, polyethers, urea-formaldehyde, polyvinyls, and phenolics are commonly cast-expanded plastics. Flotation devices, sponges, duck decoys, mattress-

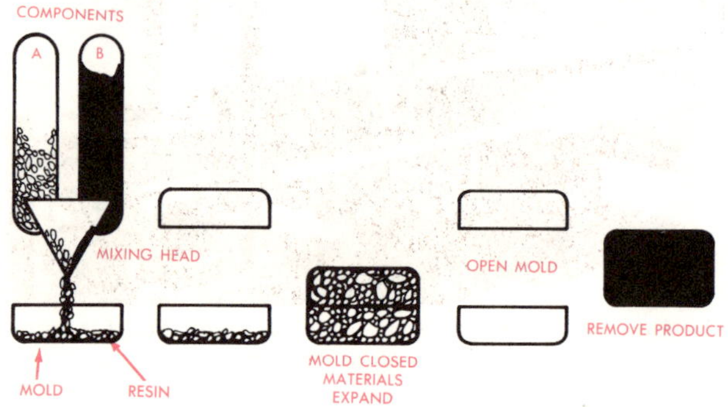

(A) Production scheme of casting plastics materials.

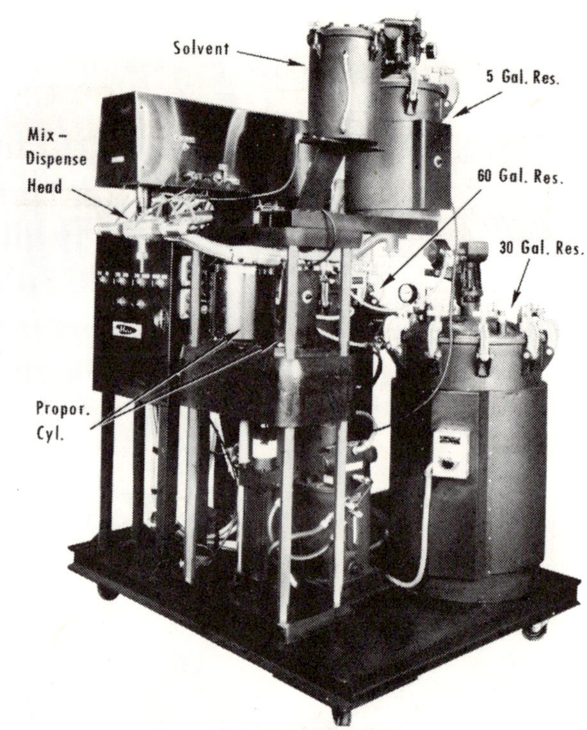

(B) Machine blends liquid resins with one or two curing agents to produce polyurethane foam. (Hull Corp.)

(C) Cellular plastics cast in an open container is being sliced into sheets. (McNeil Femco Corp.)

(D) Example of foam shaping. (McNeil Femco Corp.)

Fig. 18-9. The production of cast plastics materials.

es, and numerous safety cushioning materials are cast. Large slabs or blocks of flexible polyurethanes are cast in open and closed molds. These slabs or blocks are cut into mattress stock or shredded for cushioning. Crash pads and pillow products may be cast in closed molds.

Expand-In-Place

Expand-in-place is similar to casting except the expanded plastics and the mold become the finished product. Insulation in truck trailers, rail cars, refrigerator doors, flotation material in boats, and coatings on fabrics are familiar examples.

In this process the resin, catalyst, expanding agents, or other ingredients are mixed and poured into the cavity (see Fig. 18-10). The expanding takes place at room temperature or the mixture may be heated for a more volatile expanding reaction. This method is said to be done *in situ* (in place or position). Syntactic forms of plastics may be placed in this category.

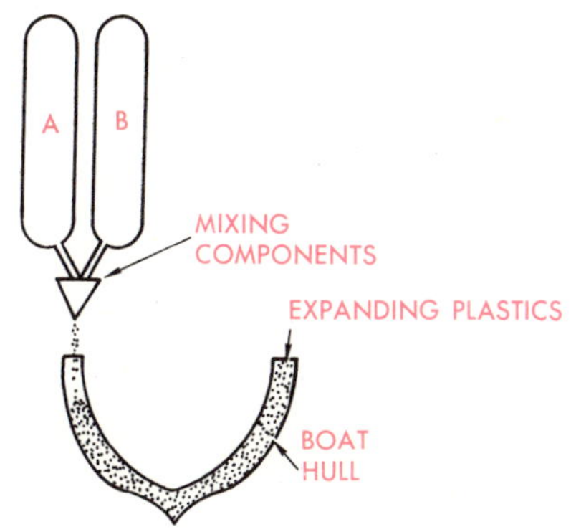

(A) Between inner and outer boat hulls.

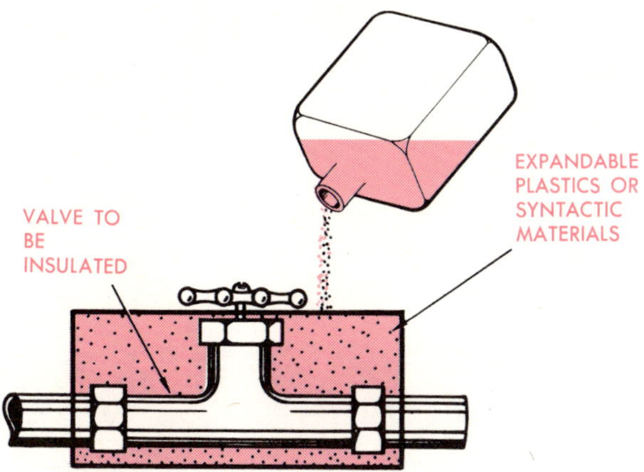

(B) Insulating a valve.

Fig. 18-10. Expanded-in-place forming.

Spray

A special spraying device is used to place expandable plastics on mold surfaces or (*in situ*) on walls and roofs for insulation. Fig. 18-11 shows two examples of this type of forming.

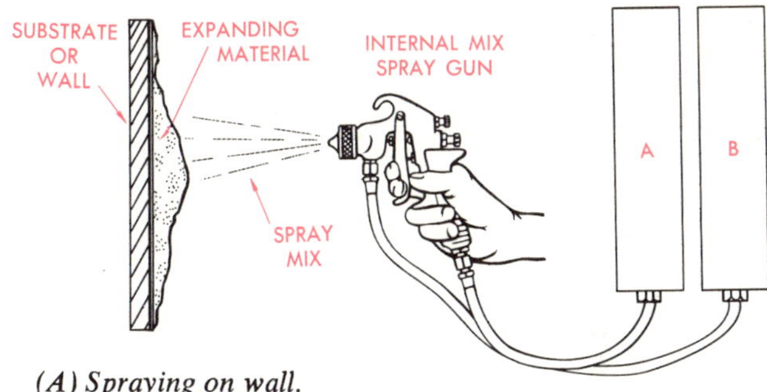

(A) Spraying on wall.

(B) Making a shell house.

Fig. 18-11. Examples of spray forming.

Glossary of New Terms

Expanded (foamed) plastics—Plastics that are cellular or spongelike.

Foaming agents—Chemicals added to plastics that generate inert gases on heating, causing the resin to assume a cellular structure.

Foam-in-place—Refers to the deposition of foams which requires that the foaming machine be brought to the work.

In-situ foaming—The technique of depositing a foamable plastics (prior to foaming) into the place where it is intended that foaming shall take place.

Review Questions

18-1. What other names are given to expanded plastics?

18-2. What are some general uses of expanded plastics?

18-3. Describe two ways expanding agents are added to the base material in the molding of expanded plastics.

18-4. Describe one way of producing an expanded plastics with a "skin."

18-5. What causes polystyrene beads to expand?

18-6. Why must pre-expanded beads be used quickly in forming expanded products?

18-7. How is the closed cellular structure obtained in polystyrene beads?

18-8. What useful items of expanded plastics may be cast?

18-9. What does the term *in situ* mean?

18-10. What is the difference between expanding-in-place and casting expanded plastics?

18-11. What is meant by *spray forming* of expanded plastics?

19

Coating Processes

For a process to be classed as a coating the plastics material must remain on the substrate. In dip casting and film casting, however, the plastics is removed from the substrate or mold. Sometimes the actual coating process is confused with other processes because similar equipment is utilized or variations of processing are shown. For example, in extrusion processing wire coatings may be used to illustrate how more than one material may be put through an extrusion die. In extrusion or calendering of films it is common to show how hot films are placed on other substrates. The casting of liquid dispersions or solvent solutions is a casting method if the film is removed, but it is a coating process if it remains on a substrate.

Coatings are applied to substrates to enhance the properties of the product. They may protect, insulate, lubricate, or add durable beauty. They may have a combination of properties no other material can match. Coatings may be flexible, textured, colored, and yet be transparent.

There are eight broad techniques in which plastics are placed on substrates: (1) extrusion coating, (2) calender coating, (3) powder coating, (4) transfer coating, (5) knife or roller coating, (6) dip coating, (7) spray coating, and (8) metal coating.

Extrusion Coating

Extrusion coating is the technique where a hot film of plastics is placed on a substrate and allowed to cool. For best adhesion the hot film should strike the preheated and dried substrate just before it reaches the nip of the pressure roller (see Fig. 19-1). The chill roller is water cooled to speed cooling of the hot film. It is usually chrome-plated for durability and high gloss transfer; however, it may be embossed to produce special textures on the film surface. The thickness of the film is controlled by the orifice of the die and by the surface speed of the chill roller. Because the substrate is traveling faster than the hot extrudate as it emerges from the extruder, it is drawn out to the desired thickness just before it reaches the nip of the pressure and chill rollers.

Expanded plastics are also extruded onto various substrates. The substrate may also be drawn through the extrusion die as in the coating of wire, cable, rods, and some textiles.

Calender Coating

Calendered films may be placed as a coating on numerous substrates in a technique similar to extrusion. The hot film is squeezed onto the substrate by the pressure of the heated gauging rollers (see Fig. 19-2).

Powder Coating

Although there are ten known techniques for applying plastics powder coatings, fluidized bed, electrostatic bed, and electrostatic powder gun coatings are the three major processes used today. The process of coating a substrate with a dry

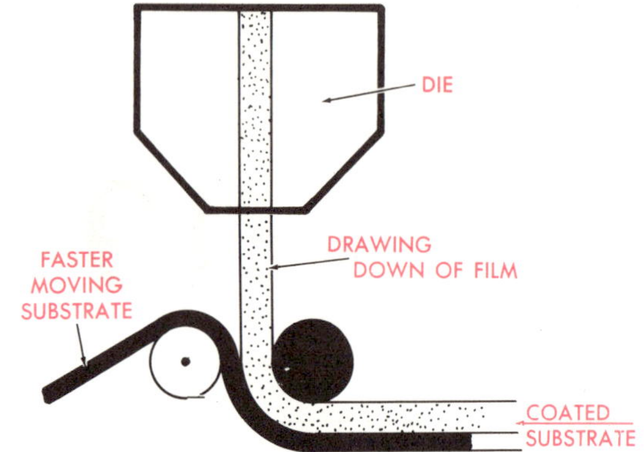

(A) Detail of die and pressure and chill rollers.

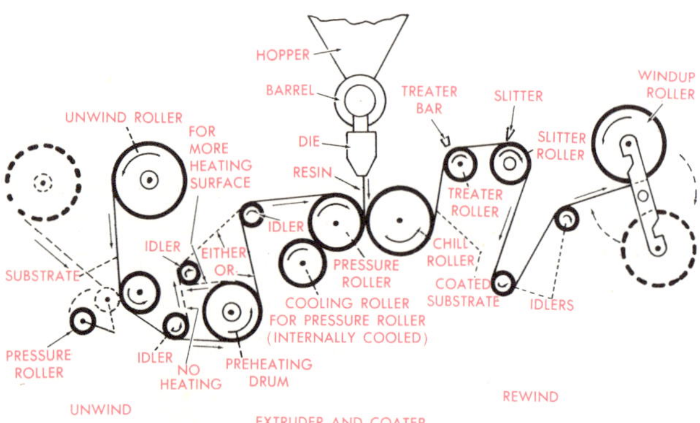

(B) Entire extrusion coating setup, with unwind and wind equipment. (USI)

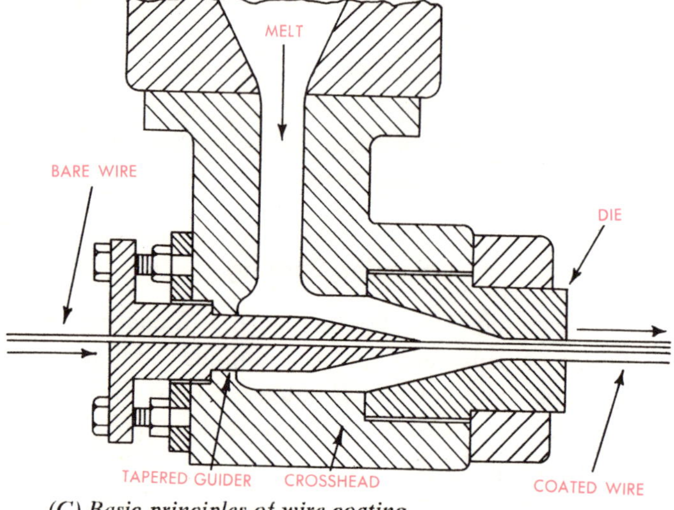

(C) Basic principles of wire coating.

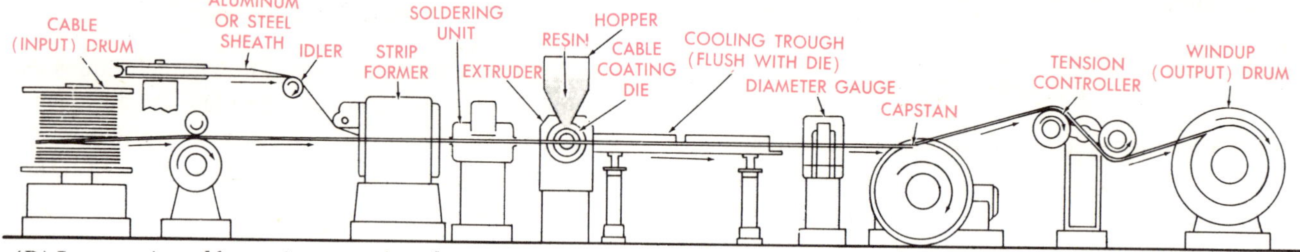

(D) Layout of a cable coating extrusion plant.

(E) An extruder coats cable with polyethylene plastics. (Western Electric Co.)

(F) Polyethylene shielded cable is buried here. (Western Electric Co.)

(G) Polyethylene and polyvinyl coated electrical wires are being spliced by workman. (Western Electric Co.)

Fig. 19-1. Method and products of extrusion coating.

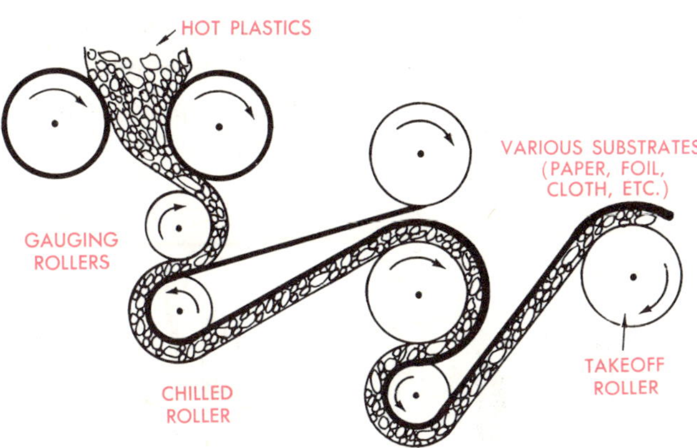

Fig. 19-2. Calender coating.

plastics powder has sometimes been called *dry painting*.

Fluidized Bed Coating

In fluidized bed coating a heated part is suspended in a tank of fine powder (see Fig. 19-3A). The bottom of the tank has a porous base membrane allowing air (or inert gas) to atomize the powdered plastics into a cloudlike dust storm. Perhaps "fog cloud" would be more descriptive because the air velocity is carefully controlled throughout the coating operation. This air-solid phase appears and acts as a boiling liquid—hence the term "fluidized bed."

When the aerated powder hits the hot part it melts and clings to the surfaces of the part. The part is then removed from the coating tank and placed in a heated oven. Part size is limited by the size of the fluidized tank. Heating fuses or cures the powder coating. Epoxy, polyesters, polyethylene, polyamides, polyvinyls, cellulosics, fluoroplastics, polyurethanes, and acrylics are used in powder coating.

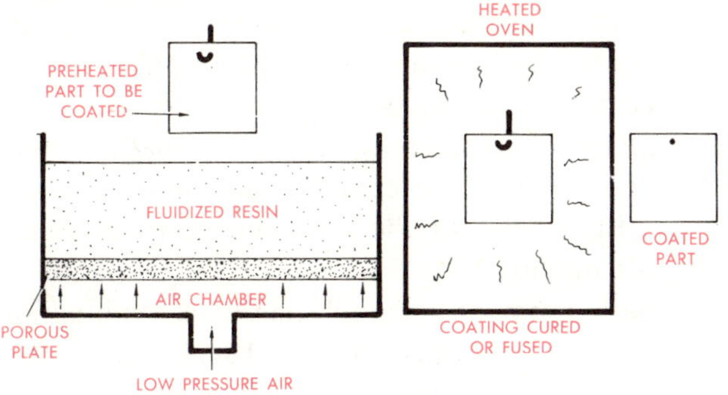

(A) Principle of operation. (W. S. Rockwell Co.)

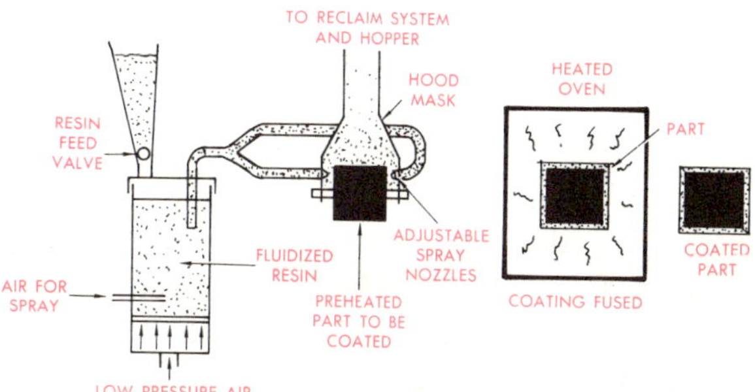

(B) Fluidized bed spray coating technique. (W. S. Rockwell Co.)

(C) Large fluidized bed coating operation showing metal dip preparation at left. (Michigan Oven Co.)

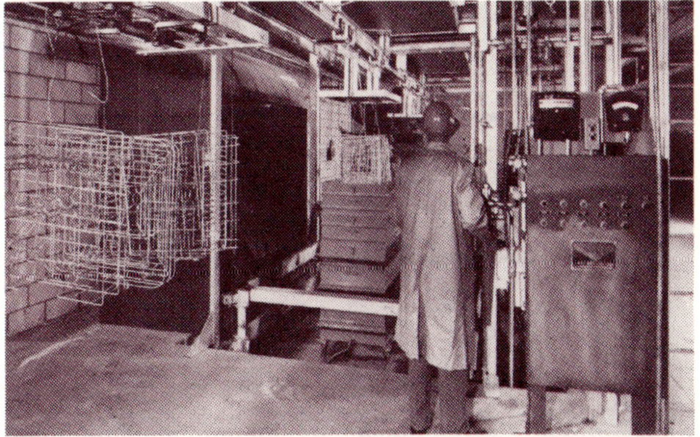

(D) These parts are being cleaned, heated, and coated with plastic coatings by fluidized bed techniques. (Michigan Oven Co.)

(E) Transformer can tops (right) are receiving a primer application prior to heating and coating by fluidized bed techniques. (Michigan Oven Co.)

Fig. 19-3. Fluidized bed coating technique.

The process of fluidized bed originated in Germany in 1953 but has grown into an important plastics process in the United States.

In a variation of the fluidized bed operation the fluidized powder is sprayed onto preheated parts in a separate chamber. The oversprayed powder is collected and can be reused. (See Fig. 19-3B.) (This process is sometimes called *fluidized bed spray coating.*) The coating on the part is then fused in a heated oven.

Electrostatic Bed Coating

In electrostatic bed coating a fine cloud of negatively charged plastics powders is atomized and deposited on a positively charged object to be coated (Fig. 19-4). These parts may or may not require preheating. If the powder is not preheated, the curing or fusing must take place before the plastics powder loses its charge. The curing is done in a heated oven, as shown in Fig. 19-4. Thin foils, screens, pipes, parts for dishwashers, refrigerators, washing machines, automobiles, and marine and farm equipment are electrostatic bed coated.

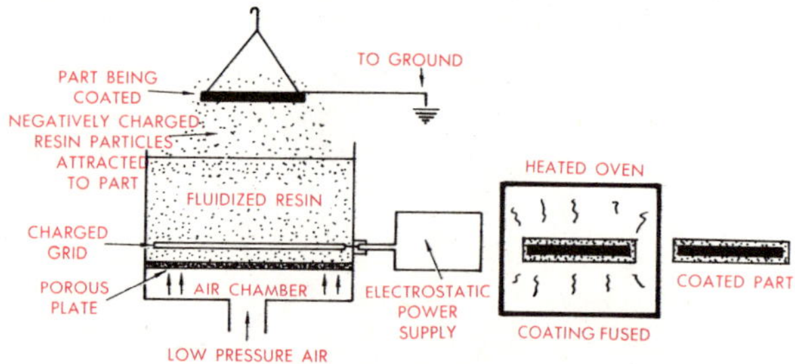

Fig. 19-4. Electrostatic bed coating technique. (W. S. Rockwell Co.)

Electrostatic Powder Gun Coating

Electrostatic powder gun applications are similar to painting with atomized spray gun equipment. In the electrostatic powder gun process dry powdered plastics is given a negative electrical charge as it is discharged in a spray pattern against the grounded object to be coated (Fig. 19-5). Fusion or curing must take place in ovens before the powder particles lose their electrical charge and fall from the part. Electrostatic gun operations have the advantage of being able to coat intricate shapes of nearly any size. The fusing oven is the only limiting factor in relation to size. Tanks, washing machine parts, and construction or manufacturing industries utilize the advantages of electrostatic powder coatings. The American Automobile Manufacturers are looking toward the replacement of liquid finishing processes with the powder coating methods.

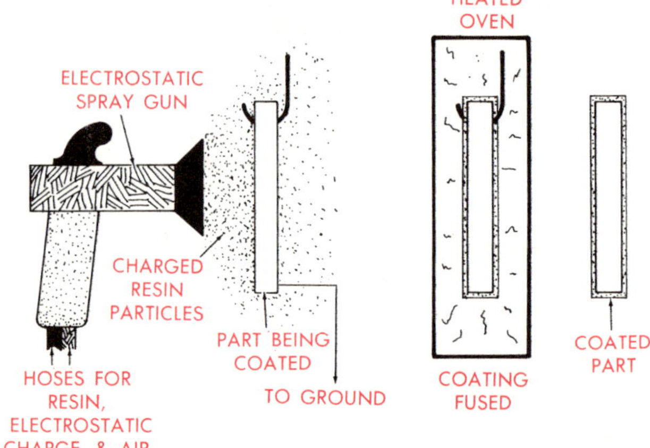

Fig. 19-5. Electrostatic powder gun coating technique. (W. S. Rockwell Co.)

Transfer Coating

In transfer coating a release paper is coated with plastics solution and dried in an oven. A second coat of plastics solution is then applied over the first coat, and a fabric layer is placed on the wet layer. The coated textile then passes through nip rollers and a drying oven. The release paper is then stripped away from the coated fabric. This method produces a tough, skinlike layer on fabric for use as leatherlike materials. See Fig. 19-6.

Knife or Roller Coating

Knife and roller coating methods are other means to spread a dispersion or solvent mixture of plastics on a substrate. The curing or drying of the plastics coating on the substrate may be accomplished by heating ovens, evaporating systems, heated rollers, catalysts, or irradiation.

The knife method may be a simple blade scraper or a narrow jet of air called an *air knife* (Fig. 19-7A). Both sides of the substrate may be coated by this method.

Coating may be accomplished by a combination of rollers as illustrated in Figs. 19-7B and 19-7F. Paper and fabric are commonly coated by this method.

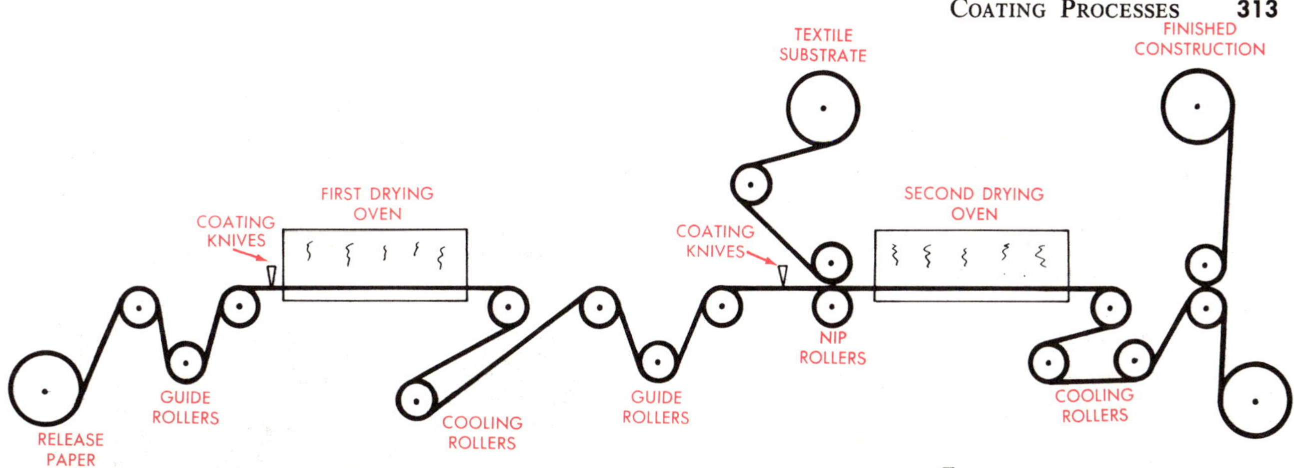

(A) Diagram of a transfer coating line.

(B) Plastics coatings are placed on paper substrates on all of these paperback books.

Fig. 19-6. Transfer coating production and products.

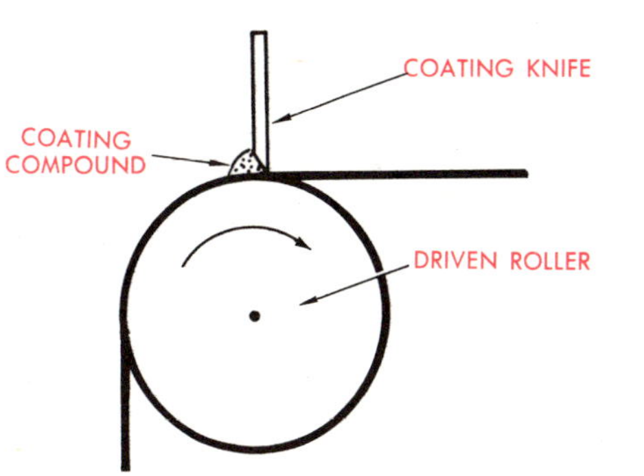

(C) Knife-over-roller coating head. (John Waldron Corp.)

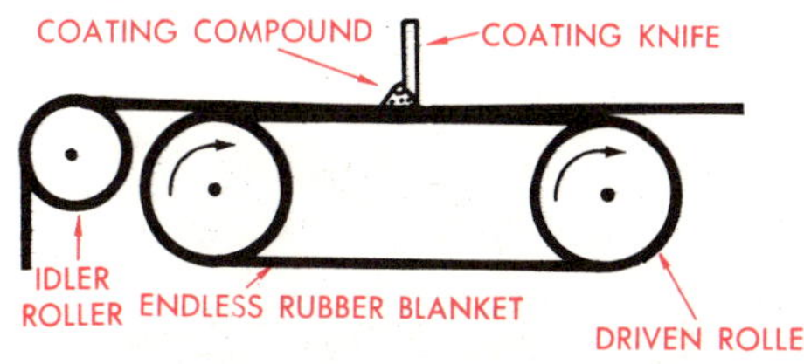

(D) Continuous blanket knife coater. (John Waldron Corp.)

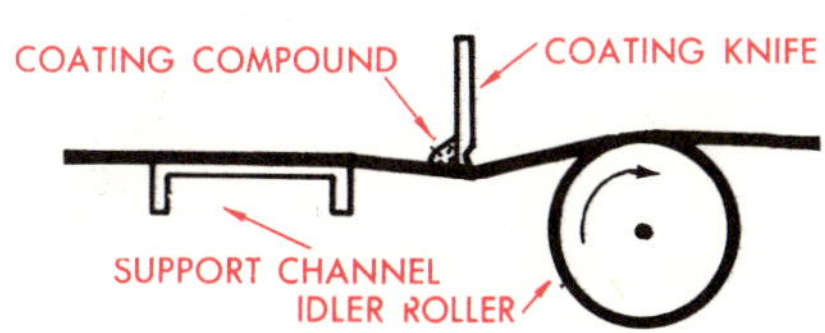

(E) Floating doctor knife. (John Waldron Corp.)

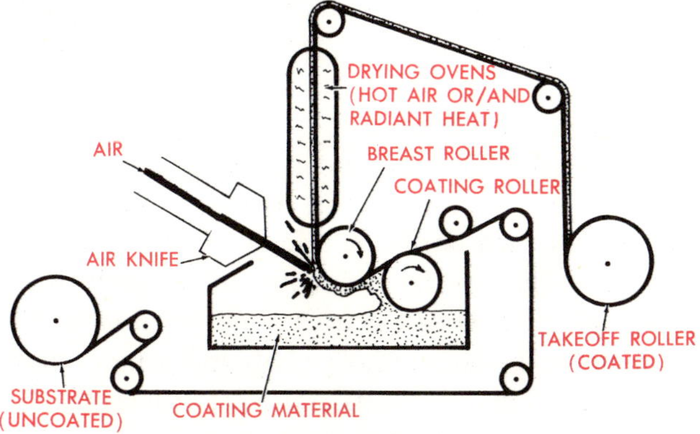

(A) Typical air-knife coating line.

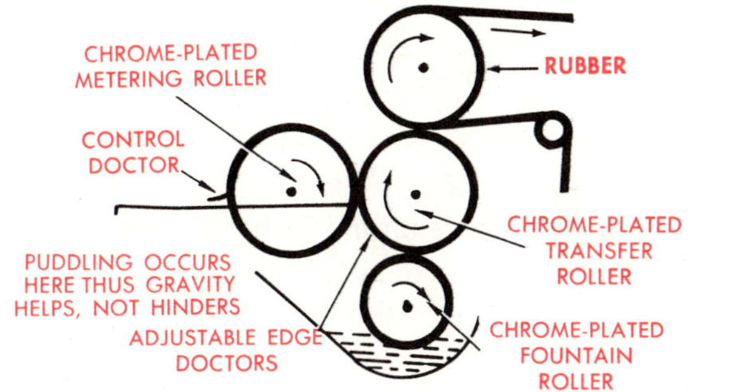

(B) Contracoater method. (Black-Clawson Co., Inc., Dilts Division)

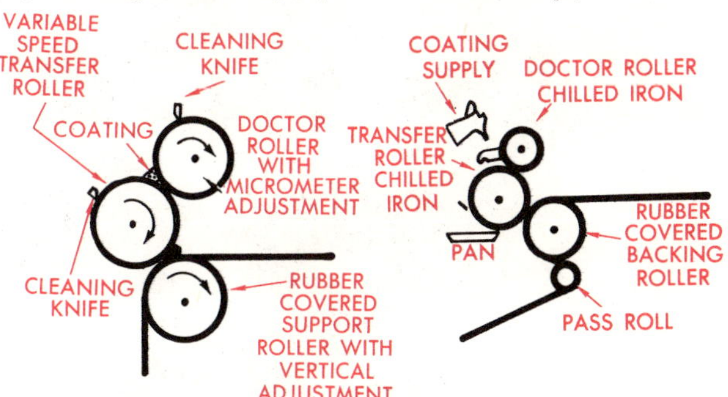

(F) Reverse roll coaters.

Fig. 19-7. Knife or roller coating.

Dip Coating

Dip coatings are applied by dipping a heated object in liquid dispersion or solvent mixtures of plastics. The most common plastics used is polyvinyl chloride. For dispersions a heating cycle is required to fuse or cure the plastics on the coated object. Tool handles and dish-drain racks are probably the most common dip coated products. Some dip coatings may harden by simple evaporation of solvents. Objects are generally limited by the size of the dipping tank.

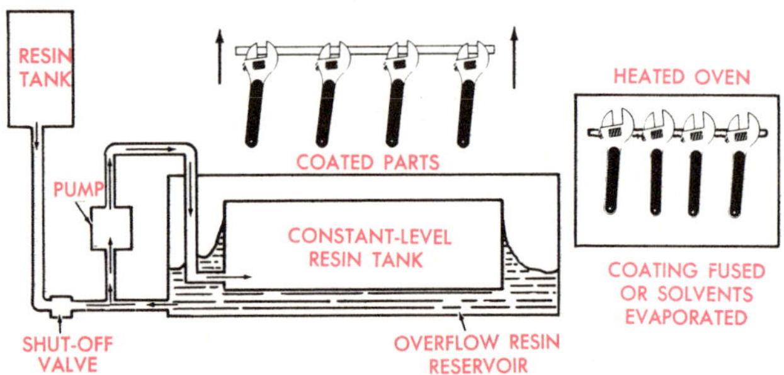

(A) Dip coating technique. (W. S. Rockwell Co.)

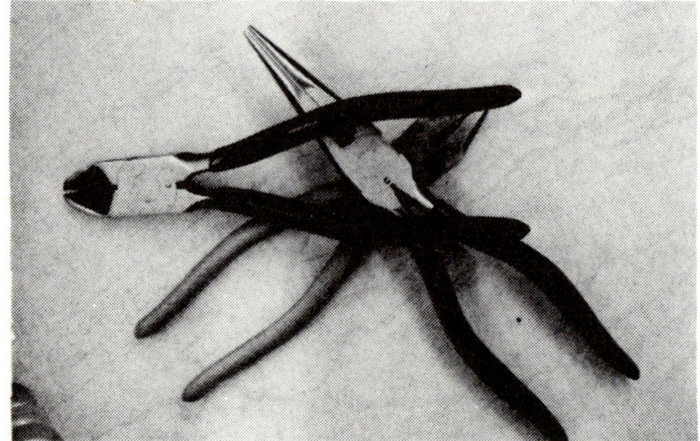

(B) PVC dip coated tool handles.

(C) Spray and dip PTFE-coated household utensils. (Chemplast, Inc.)

(D) Spray and dip PTFE-coated tools. (Chemplast, Inc.)

Fig. 19-8. Dip coating production and products.

Spray Coating

In this method dispersions, solvent solutions, or molten powders are atomized by the action of air (or inert gas) or the pressure of the solution itself, and are deposited on the substrate. The coating of furniture, houses, and automobiles with various plastics paints or varnishes is a familiar solvent-solution spray coating process. Dispersions of polyvinyl chloride (plastisol) have been spray coated on the inside and out of railroad cars utilizing this spray process. (See Fig. 19-9.)

In a process sometimes called *flame coating* finely ground powders are blown through a specially designed burner nozzle of a spray gun (Fig. 19-10). The powder is quickly melted as it passes through this gas or electrically heated nozzle. The hot molten plastics quickly cools and adheres to the substrate. This process is useful for articles that are too large for other coating operations.

Metal Coating

Metal coatings are also included as a decorating and finishing process. They are also done as a special coating application. Perhaps metal coating should not be classed as a basic process of the plastics industry! Besides providing a decorative finish, metal coatings may provide an electrical conducting surface, a wear and corrosion resistant surface, or added heat deflection.

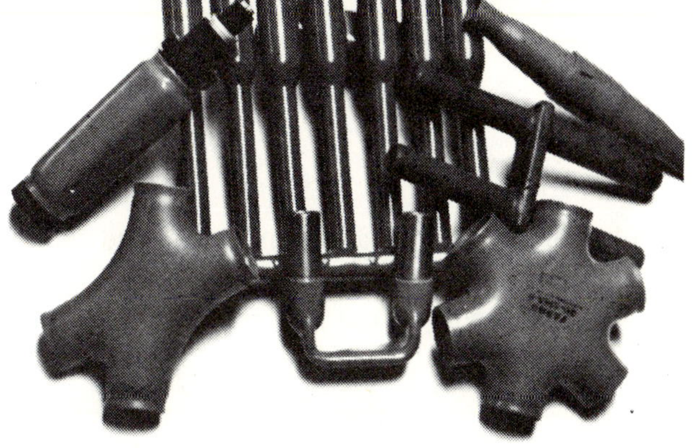

(A) Various plastisol coated parts for the electrical industry. (Quelcor, Inc.)

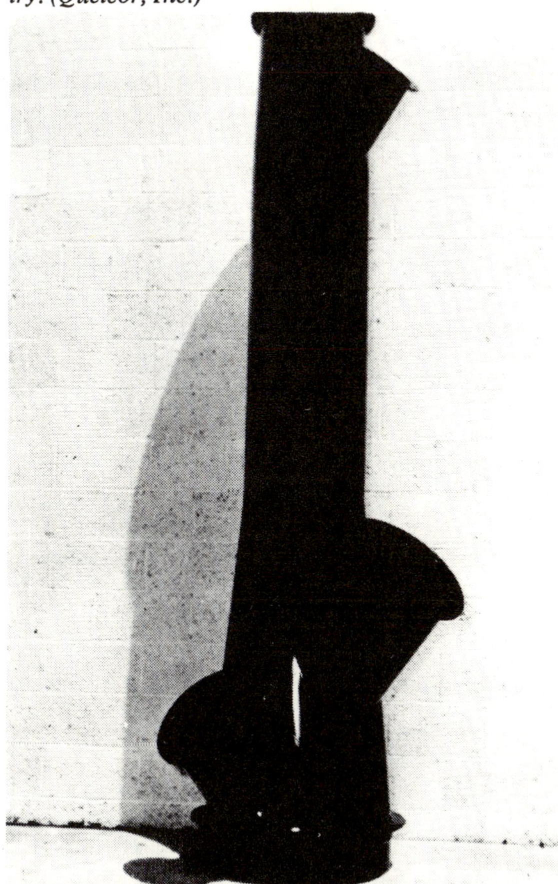

(B) Plastisol coated duct. (Michigan Chrome & Chemical Company)

(C) Plastisol coated hood (Michigan Chrome & Chemical Company)

(D) Aerosol spray cans of plastics materials for coatings.

Fig. 19-9. Sprays and spray coated products.

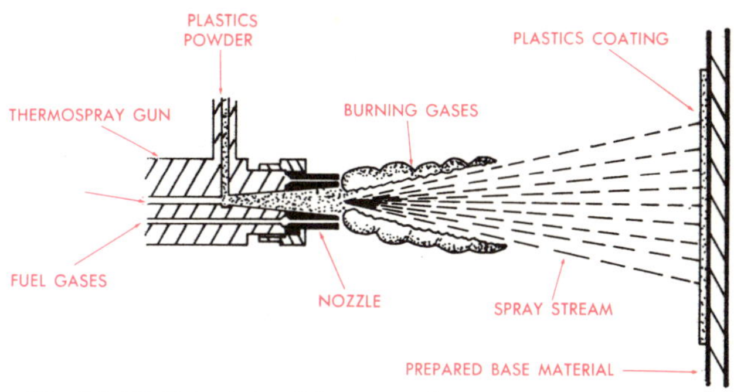

Fig. 19-10. Principle of flame coating.

There are three major methods for applying a metal coating on a substrate: (1) adhesives, (2) electroplating, and (3) vacuum metallizing.

Adhesives are used to apply foils on many surfaces. Intricate or irregular parts, however, are difficult to coat. The carpet and textile industries have used this method to adhere metal foils to special garment designs. Polyethylene, fluoroplastics, and polyamides are difficult to adhere to.

Electroplating is accomplished on many plastics. Both the resin and the mold design must be carefully considered in producing a metal coating on plastics parts. Ribs, fins, slots, or indentations should be rounded or given tapers. (See Fig. 19-11.) Phenolic, urea, acetal, ABS, polycarbonate, polyphenylene oxide, acrylics, and polysulfone articles are commonly plated.

The electrolysis preplating process is accomplished by carefully cleaning the plastics part and etching the surface to ensure adhesion. (See Fig. 19-12.) The etched part is again cleaned and the surfaces "seeded" with an inactive noble metal catalyst. An accelerator is added to activate the noble metal, and the ion solution of metal reacts

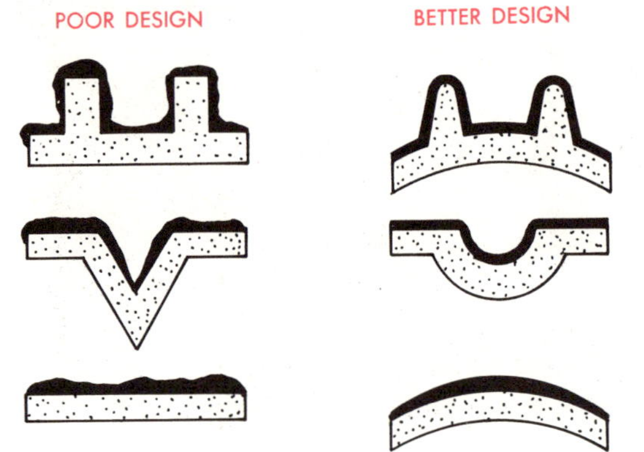

Fig. 19-11. Large-radius fillets and bends are desirable when parts are to be plated.

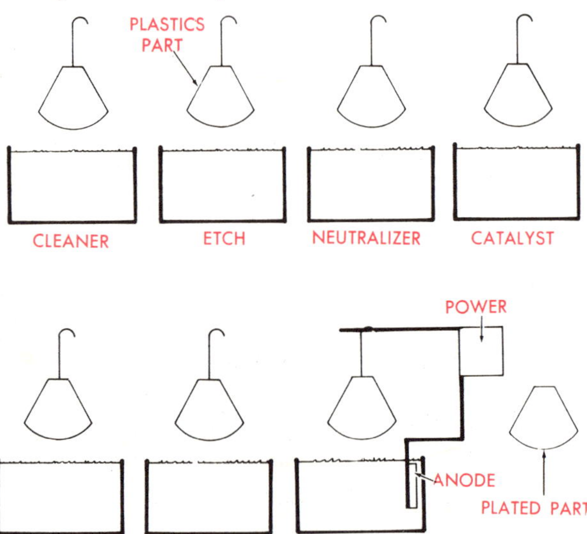

Fig. 19-12. Sequence in electroplating plastics.

autocatalytically in the electrolysis solution. Copper, silver, and nickel electrolysis solutions are used in preplating. These preplating deposits are normally from 10-30 millionths of an inch in thickness. Once a conductive surface has been established, commercial plating solutions may be used; chrome, nickel, brass, gold, copper, and zinc are well established. The majority of the plated plastics requires a chromelike finish.

In vacuum metallizing, plastics parts or films are thoroughly cleaned and given a base coat of lacquer to eliminate surface defects and to seal the pores of the plastics. Polyolefins and polyamides are chemically etched to ensure good adhesion of the metal coating. (See Fig. 19-13.) The carefully cleaned plastics are then placed in a vacuum chamber. Small pieces or strips of coating metal such as chromium, gold, silver, zinc, or aluminum are placed on special heating filaments. The chamber is sealed and the vacuum cycle is started. When the desired vacuum is reached (0.5 micron or 5×10^{-4}mm Hg) the heating filaments are heated until the metal coating strips melt and vaporize into the vacuum chamber. This vaporized metal coats everything it contacts in the chamber, condensing or solidifying on cooler surfaces. Parts must be rotated for full coverage since the vaporized metal travels in a line-of-sight path. Once the plating is accomplished the vacuum is released and

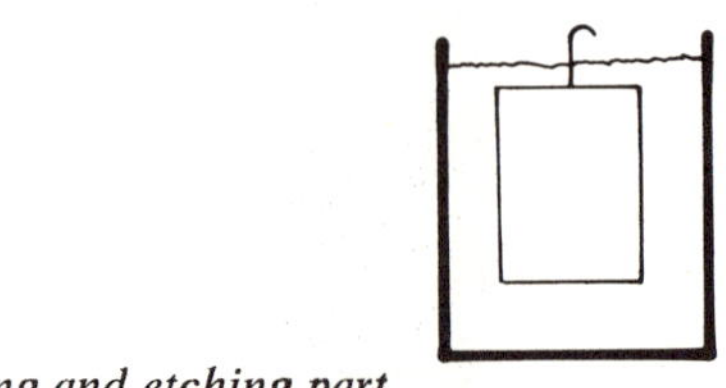

(A) Cleaning and etching part.

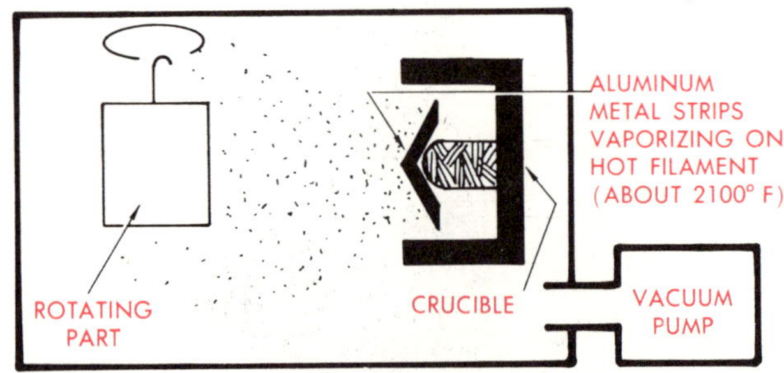

(B) Vaporizing aluminum to coat plastics.

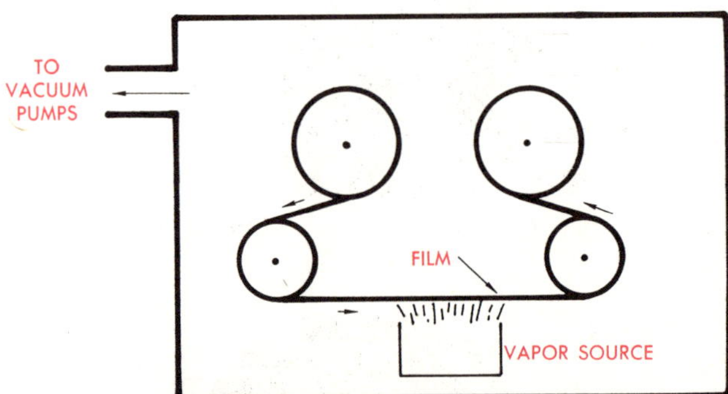

(C) Vacuum metallizing on plastics film.

(D) "Selectric" typewriter head is metal coated plastics.

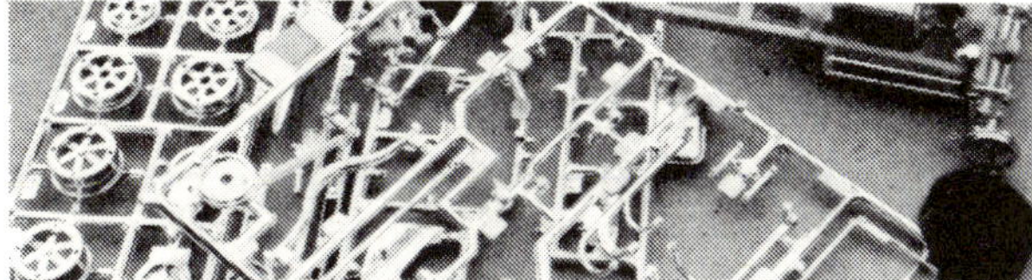

(E) Vacuum metallized parts for toy model.

(F) Closeup of part in (E), showing where clamp held molding and no metal was deposited.

Fig. 19-13. Vacuum metallizing of plastics parts.

(A) Toy horns are loaded onto holding fixtures. While this full rack is being processed, another will be loaded. Lacquered and baked products are loaded on rotating fixtures and placed in the vacuum chamber. Fixture rotation is controlled from outside through vacuum seals.

(B) Base coat of lacquer, to smooth out minor surface blemishes and provide initial glossiness, is applied by dipping. Lacquer can also be applied by spraying or flow coating. The base coat of lacquer is then baked in oven.

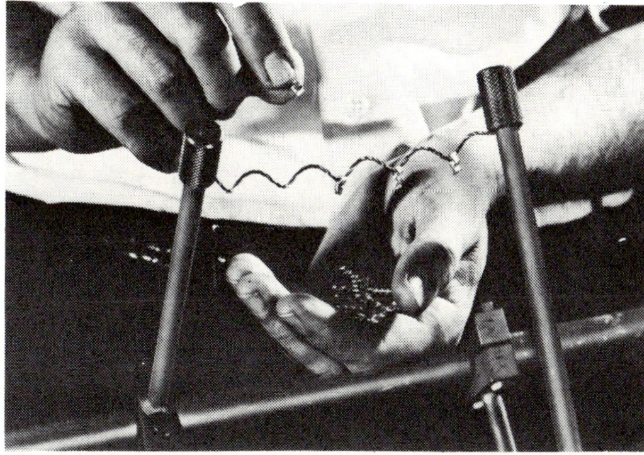

(C) Small staples of plating material—in this case aluminum—are loaded on coils of stranded tungsten wire filaments. Chamber is then closed, proper vacuum is drawn (about 0.5 micron for aluminum), and the filaments are heated by electric current to incandescence (1200°F).

(D) The aluminum melts, spreads thinly over filaments and vaporizes as filament temperature is raised quickly to 2100°F. Vaporization, or "flashing the filaments," takes only 5 to 10 seconds. The metallized products are then removed from vacuum chamber and from fixtures. The products are dipped in topcoat lacquer. A wide range of colors can be produced by dipping products in dye solutions, which will color the topcoat. The topcoat protects the metal film from abrasion and wear.

Fig. 19-14. An eight-step technique for metallizing toy horns. (Stokes-Pennwalt Corp.)

the parts removed. To help protect the plated surface from oxidization and abrasion, a lacquer coating is applied.

Glossary of New Terms

Dip coating—Applying a coating by dipping the article to be coated into a tank of melted resin or plastisol and chilling. The article may be heated and powders used as a coating. The powders melt as they strike the hot object.

Extrusion coating—The resin coating placed on a substrate by extruding a thin film of molten resin and pressing it into or onto the substrate or both, without the use of adhesives.

Flame spraying—Method of applying a plastics coating in which finely powdered plastics together with suitable fluxes are projected through a cone of flame onto a surface.

Fluidized bed—A method of coating articles that are heated by immersion in a dense-phase fluidized bed of powdered resin and reheated in an oven to provide a smooth coating.

Knife coating—A method of coating a substrate by an adjustable knife or bar set at a suitable angle to the substrate.

Plastisols—Dispersion mixtures of plasticizers and resin.

Vacuum metallizing—Process in which surfaces are thinly coated by metal by exposing them to the vapor of metal that has been evaporated under vacuum.

Review Questions

19-1. What is plastics coating?

19-2. What functions do coatings have?

19-3. What is extrusion coating?

19-4. How is the hot film applied to the substrate in calender coating?

19-5. What are the three major processes of dry powder coating?

19-6. Describe the fluidized bed coating process.

19-7. In electrostatic bed coating, how is the dry powder deposited on the part?

19-8. What is electrostatic powder gun coating? How does it differ from electrostatic bed coating?

19-9. What materials are transfer coated?

19-10. What is an *air knife*? Where is it used?

19-11. Briefly describe the dip coating process.

19-12. What kind of objects are spray coated?

19-13. What are the three major methods of metallizing a substrate?

19-14. Briefly describe the electroplating process. Which plastics materials are well suited to this process?

19-15. Why must most plastics materials be lacquered before vacuum metallizing?

19-16. Why is it necessary to rotate the plastics part during the vacuum metallizing cycle?

20

Decorating Processes

Various decorating processes are utilized in the production of plastics parts. These processes may be done during molding, directly afterward, or before final assembly and packaging. The most inexpensive method to produce decorative designs on the product is to incorporate the desired design in the cavity of the mold. These designs may be textures, raised or depressed contours, or informative messages such as trademarks, patent numbers, symbols, letters, numbers, or directions.

In any decorating of plastics articles, in the mold or out, surface treatment and cleanliness is of primary importance. Not only do the molds have to remain clean and mark-free but the molded articles must be properly prepared to ensure good decorating results. For example, blushing is the result of applying coatings over articles which have not been properly dried of surface moisture. Crazing is due to solvent cutting along lines of strain in the molded plastics. These fine cracks (crazing) may extend on or under the surface or through a layer of the plastics material. A change in mold design may be needed to eliminate this problem and produce a strain-free molding.

Prior to decoration the surface of plastics must be cleaned of mold release, internal plastics lubricants, and plasticizers. Plastics parts become electrostatically charged and attract dust and disrupt the even flow of coating. Solvent or electronic destaticizers may be used to clean and destatic plastics articles prior to decorating.

Polyolefins, polyacetals, and polyamides must be treated by one of the following methods to ensure satisfactory adhesion of the decorating media. (1) Flame treatment consists of passing the part through a hot oxidizing flame of 1100°C to 2800°C. This momentary flame treatment does not cause distortion of the plastics but makes it receptive to decorating methods. (2) A chemical treatment consists of submerging the part or portions of the part to be decorated in an acid bath. On polyacetals and polymethylpentene polymers the acid bath results in an etched surface which is receptive to decorating. For many thermoplastics solvent vapors or baths may be used for the etching treatment. (3) Corona discharge is a process where the surface of the plastics may be oxidized by electron discharge (corona). The part or film is oxidized when passed between two discharging conductor electrodes. (4) In a process called *plasma treating*, plastics are subjected to an electrical discharge in a closed vacuum chamber. Atoms on the surface of the plastics are physically changed and rearranged in such a manner that excellent adhesion is possible.

The most widely used decorating processes used in the plastics industry are: (1) coloring, (2) painting, (3) hot-leaf stamping, (4) plating, (5) engraving, (6) printing, (7) in-mold decorating, (8) heat transfer, and (9) pressure sensitive labels, decalcomanias ("decals" for short), flocking, etc.

Coloring

The most economical way to color plastics is to blend the color pigments into the base resin. Color matching may be a problem because successive batches of colored plastics may vary slightly. Most producers of colored resins and plastics, encourage the use of stock or standard colors. When plastics parts of an assembly are produced in different plant locations and at different times, it becomes necessary that color "standards" be carefully considered. Plasticizers, fillers, and the molding process may affect the final product color.

Colorants in the form of dry powders, paste concentrates, organic chemicals, and metallic flakes are usually blended with a given resin mix. Banbury, two-roll, and continuous mixers are used to disperse the pigments thoroughly in the resin. The colored resin may then be cast or extruded into familiar molding forms. Water and chemical solvent dyes have been successfully used on many plastics. The procedure consists merely of dipping the parts in the dye bath and air drying.

Painting

Painting of plastics is a popular, inexpensive way to decorate parts or provide flexibility in product color design. Transparent, clear, or colored plastics may be painted on the back surface for a striking contrast, variety, or appearance not possible by other methods. Painting processes used in decorating plastics include: (1) spray painting, (2) electrostatic spraying, (3) dip coating, (4) fill-in marking, (5) screen painting, and (6) roller coating. In all painting processes the solvents or curing systems used in the paint must be carefully selected and controlled. As a rule thermosetting plastics are less susceptible to swelling, etching, crazing, and deteriorating effects of solvents than thermoplastics. Temperature may also be a limiting factor for the curing or heat baking of paints on many plastics. Radiation curing is one method of curing coatings on plastics.

The most versatile and frequently used method of decorating all sizes of plastics articles in spray painting. It is an economical rapid method of applying coatings. The spray guns may utilize air pressure or hydraulic pressure of the paint itself to atomize the paint.

Masking is required when areas of the part are not to be painted. These masking materials may consist of the familiar paper-backed masking tape or they may be durable form-fitting metal masks. Polyvinyl alcohol masks may be sprayed over areas and later removed by stripping or solvents. Electroformed metal masks are preferred because they conform to the contour of the article and are durable. Four basic types of electroformed masks are shown in Fig. 20-1.

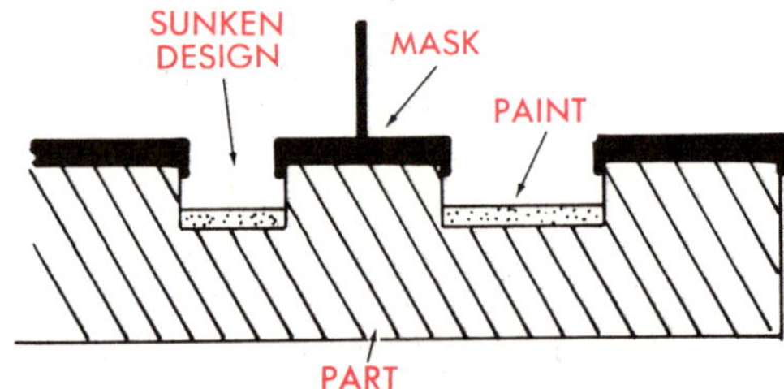

(A) Lip mask on sunken design.

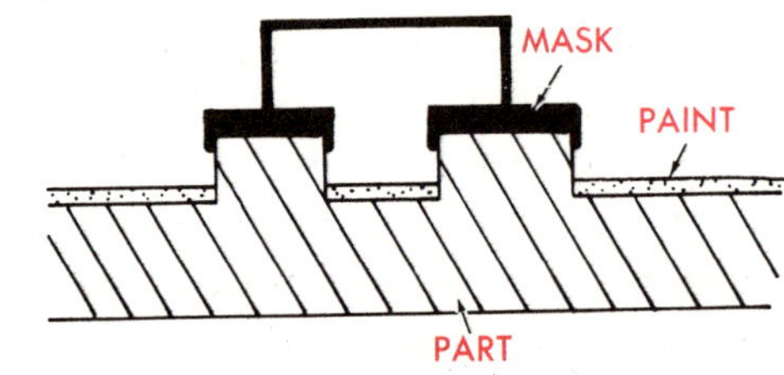

(B) Cap mask on raised design.

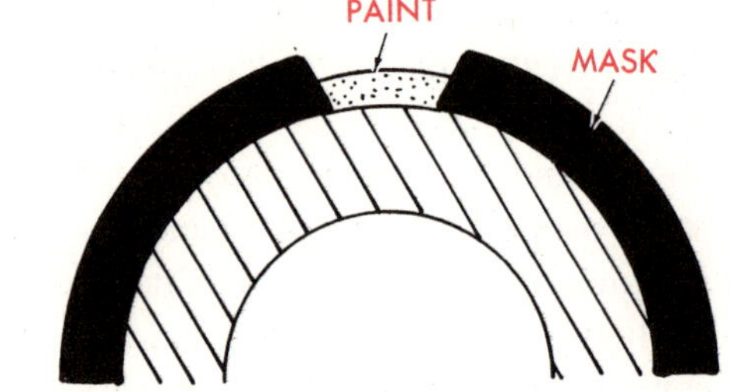

(C) Surface cutout mask.

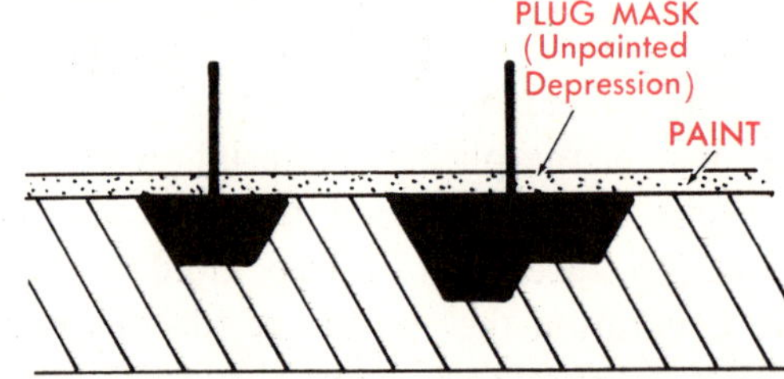

(D) Plug mask for unpainted depressions.

Fig. 20-1. Basic types of electroformed masks.

In electrostatic painting the plastics surface must be treated to take an electrical charge. In electrostatic painting the surface must conduct an electrical charge as it passes through the oppositely charged and atomized paint. The paint may be atomized by air or hydraulic pressure or by centrifugal force (Fig. 20-2). Nearly 95 percent of the paint that was atomized is attracted to the charged surface; thus it is a highly efficient method of applying paint. Narrow recesses are difficult to coat and metal masks are not practical.

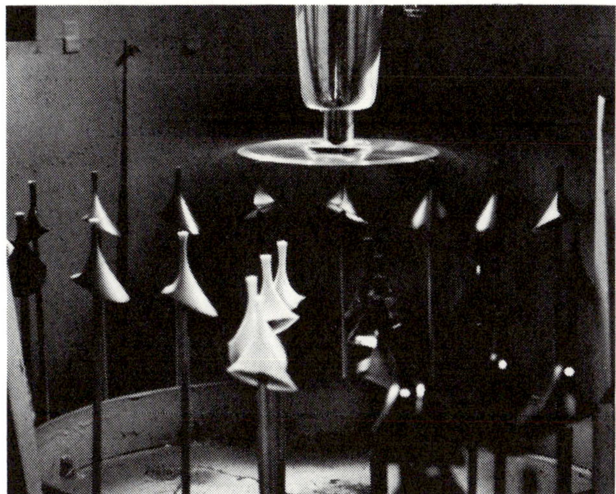

(A) Centrifugal force atomization.

(B) Compressed air atomization.

(C) Hydraulic atomization.

Fig. 20-2. Methods of paint atomization in electrostatic painting. (Ransburg Corp.)

Dip painting (Fig. 20-3) is useful when only one color or a base color is required. A uniform coating may be applied if the part is withdrawn from the paint very slowly and sufficient time is allowed for drainage. The excess (tear) drops of paint may be removed by spinning the part, hand wiping, or by electrostatic methods.

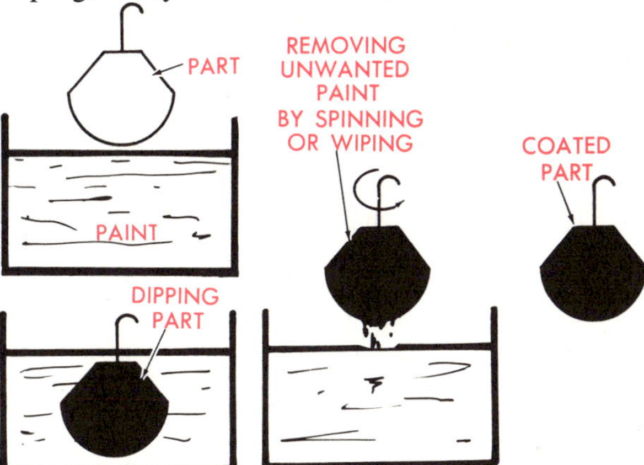

Fig. 20-3. Dip painting.

Screen painting is a versatile and attractive method for decorating plastics articles. The process consists of transferring a special ink or paint through the small openings of a stenciled screen onto the product surface. The process is sometimes referred to as *silk screen* painting or printing because earlier screens were made of silk. Today the screens may be made of metal mesh or finely woven polyamide, polyester, or other plastics. The screen stencil may be prepared by blocking out the areas where no paint is wanted. For intricate designs or lettering, photographic stencils are used and applied to the screen. When exposed and immersed in a developer bath, the exposed areas wash away. It is through these openings that paint will be transferred onto the plastics surface.*

In a process just the opposite of roller coating, paint is placed in low or indented portions of the article (see Fig. 20-4). Letters, figures, or designs are placed on the mold surface to produce a depression in the molded part. These decorations are filled by spraying or wiping paint in the depressions. To ensure a sharp image, the depression should be deep and narrow. If the depression or design is too wide, the buffing or wiping action may remove the decorative paint. Excess paint

*See Printing, p. 325.

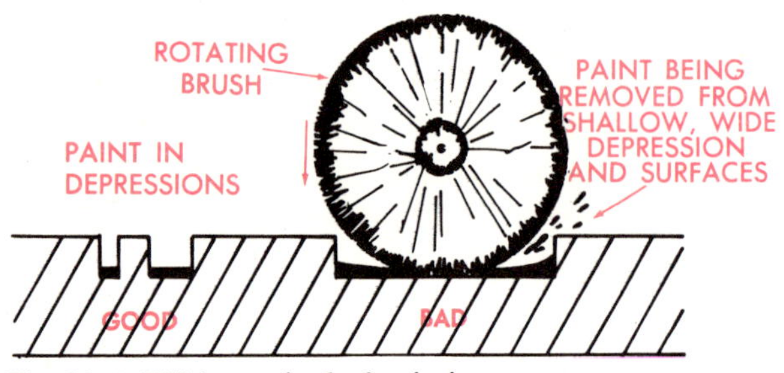

Fig. 20-4. Fill-in method of painting.

around the design may be removed by wiping or buffing operations.

Raised portions, letters, figures, or other designs may be painted by passing a coating roller over the raised portions (Fig. 20-5). In some cases masking out portions of the article may be required; however, if edges and corners are sharp and highly raised, a good coating detail will be obtained. The roller coating operation may be automated or on small runs done by hand with a brayer (hand roller).

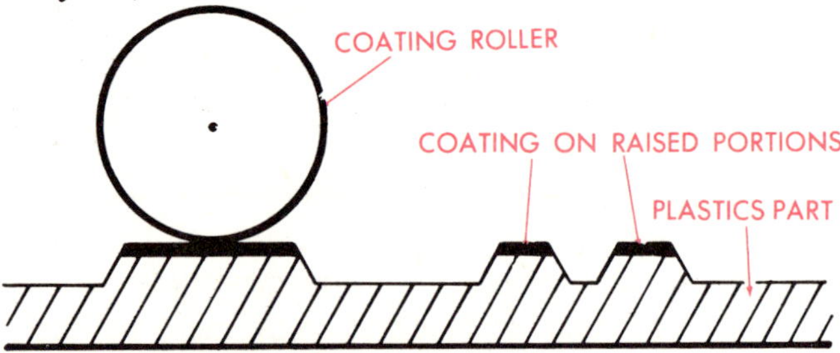

Fig. 20-5. Roller coating of raised portions.

Hot-Leaf Stamping (Hot Stamping)

Hot-leaf stamping (sometimes called *roll-leaf* or simply *hot stamping*) offers an economical, simple method for producing a durable decorations on plastics. Letters, designs, trademarks, and messages may be hot-leaf stamped. The process consists of placing the part to be decorated under a hot stamping die. The hot die then strikes the surface of the part through a metallized or painted roll-leaf carrier. The paint carried on the roll is fused into the impression made by the stamp, providing a durable, clear decoration. The hot-stamping dies may be made of machine engraved or chemically etched metal or of tough, heat resistant, flexible silicone (see Fig. 20-6). For textured, uneven, or large surfaces, silicone dies may be preferred. Roller dyes are used to transfer the design on large areas. Hot-leaf stamping can be done on all thermoplastics and some thermosets. Thermosetting materials are not easily hot stamped because of the heat and pressure required. The process is similar to a branding operation on thermosets. Melamines are never hot stamped, and urea based resins are rarely decorated by this method.

Genuine gold, silver, or other metal foils (leaf) as well as paint pigments may be placed on plastics. A typical bright metallized hot-stamping foil is shown in Fig. 20-6A. Because these foils and pigments are dry, they are easy to handle and may be placed over painted surfaces. No masking is required, and the process may be automatic or hand operated. The carrier film for hot-leaf stamping which supports the decorative coatings until they are pressed on the plastics part is made of cellophane, acetate, or polyester. A thin layer of heat sensitive material is then placed on the carrier as a releasing agent. A lacquer coating is then applied over the releasing layer to provide protection for the metal foil. If paints are to be used instead of metal foils, the lacquer and pigmented colors are combined into one layer. The bottom layer functions as a heat- and pressure-sensitive hot-melt adhesive. Heat and pressure must have time to penetrate (dwell) the various film coatings and layers to bring the adhesive to a liquid state. Before the carrier film is stripped away, a short cooling time is allowed to ensure that the adhesive has solidified.

Fig. 20-7A shows a hot-stamping press for applying multiple color decoration to a molded plastics cannister in one operation. The designs are preprinted on carrier and transferred and fused to the part with application of heat, pressure, and dwell. This process is dry, as is the case with all hot-stamping transfers, so that the newly decorated parts can be handled, assembled, or packaged. This particular press setup will also apply the conventional hot-stamping method using regular transfer dies and hot-stamping foil. Flat and shaped decorating areas can be accommodated, and with a special attachment the complete circumference of cylindrical parts can be marked by this machine. Fig. 20-7C shows a machine for hot-stamping a squeezable plastics tube with a highly decorative design. The tooling in this case consists of a rotary dial table assembly permitting continuous operation of the press, requiring only the loading of the part to the nests. Ejection after marking is automatic.

Plating

Plating and vacuum metallizing have been discussed under the topic of metal coatings. Although there are many functional applications of coating plastics with metal the decorative applications outnumber functional ones. Metallized foils for dielectrics, semiconductors, and resistors are functional applications. Flexible mirrors and plating with greater corrosion resistance are also very functional. The mirrorlike finish on countless automotive items, appliances, jewelry, and toy parts are familiar decorative applications.

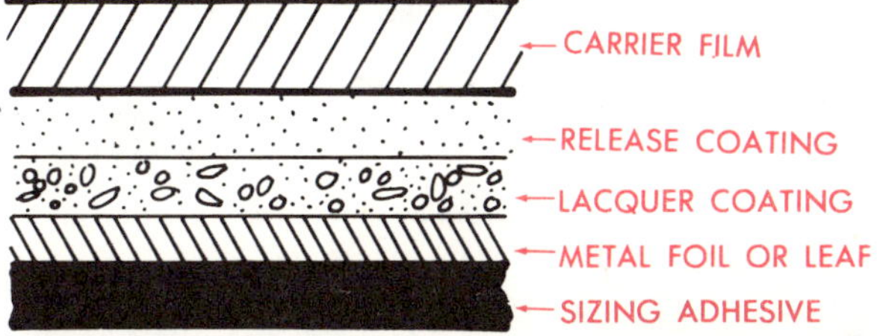

(A) Diagram of a typical metallized hot-stamping foil.

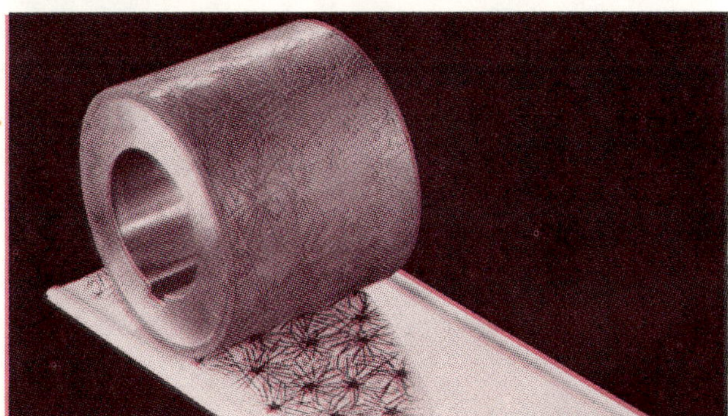

(B) Textured silicone hot-stamping rolls produce continuous designs on flat products. (Gladen Enterprises, Inc.)

(C) Ring which was hot stamped by the foil shown in (B). (Gladen Enterprises, Inc.)

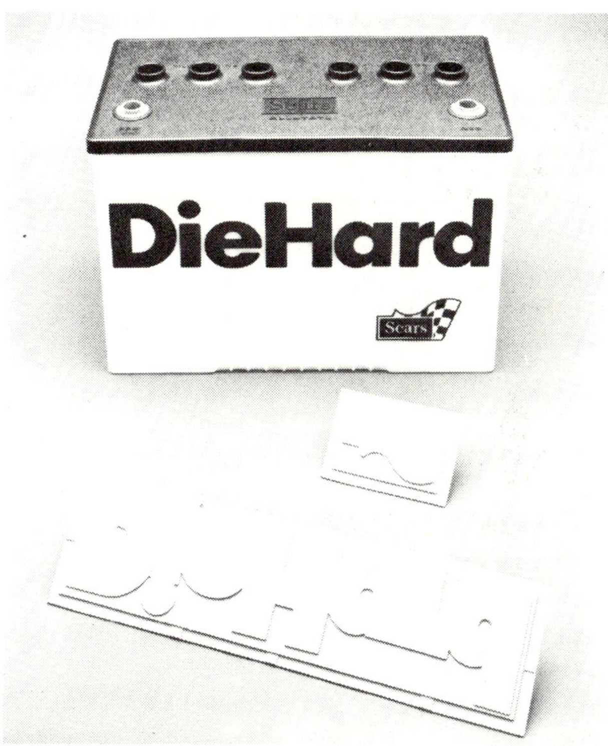

(D) Example of hot stamping with a silicone die. (Gladen Enterprises, Inc.)

(E) Molded silicone rubber die used to stamp glass. The glass at the bottom of the mug was coated with silk screening epoxy and forced-air dried prior to hot stamping. (Gladen Enterprises, Inc.)

Fig. 20-6. Hot stamping and hot-stamped products.

(A) *Hot-stamping machine which applies multiple colors. (The Acromark Co.)*

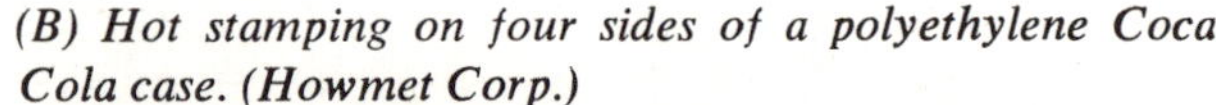

(B) *Hot stamping on four sides of a polyethylene Coca Cola case. (Howmet Corp.)*

(C) *Machine for hot stamping squeezable plastics tubes. (The Acromark Co.)*

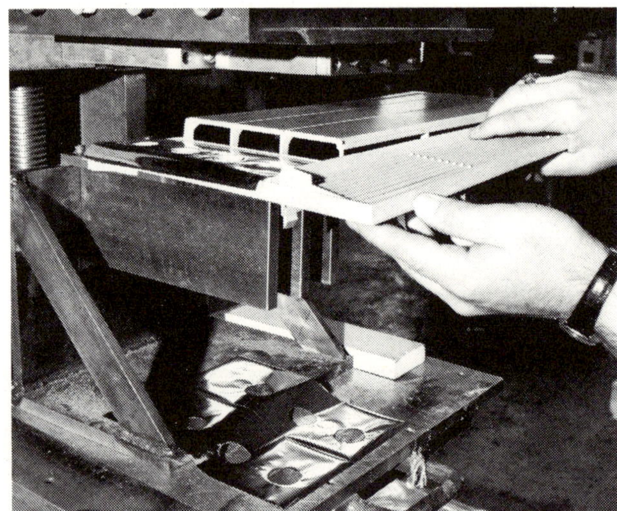

(D) *Hot-stamping of bright aluminum finishes on an air conditioner. (Howmet Corp.)*

(E) *Roll-on hot stamping of woodgrain design on top and sides of TV cabinet. (Howmet Corp.)*

(F) Decorative hot stamping of plastics drinking tumbler. (The Acromark Co.)

Fig. 20-7. Hot-stamping machines.

Engraving

Engraving is seldom used on production scale. It does, however, provide a durable means of marking and decorating plastics. Pantographic engraving machines may be automatic or manual and are commonly used to engrave on laminated nametags, door signs, directories, and equipment, and to place identifying names and marks on bowling balls, golf clubs, and other parts. Laminated engraving sheets usually contain two or more color layers of plastics. Engraving cuts through the top layer, exposing the contrasting second-color layer.

Printing

There are over eleven different methods and many combinations for printing on plastics. (1) *Letterpress* is a method where raised, rigid printing plates are "inked" and pressed against the plastics part. The raised portion of the plate transfers the image. (2) *Letterflex* is similar to letterpress except that flexible printing plates are used. Flexible plates may transfer their designs to irregular surfaces (Fig. 20-8A). (3) *Flexographic* printing is similar to letterflex except that a liquid ink rather than a paste ink is used. The plate is commonly a rotary one transferring inks that set or dry rapidly by solvent evaporation. (4) *Dry offset* is a method where a raised, rigid printing plate transfers a paste-type ink image on a special roller called an "offset blanket." This roller blanket then places the ink image on the plastics part. If multicolor is required, a series of offset heads can be used to apply different colors to the blanket roller. The multicolored image is then transferred (offset) to the plastics part in a single printing process (see Fig. 20-8B). (5) *Offset lithography* is similar to dry offset except the impression on the roller drum or plate is not raised or sunken. The process is based on the principle that oil and water do not mix. The image or message to be printed is placed on the plate by a photographic-chemical process. Images may be placed directly on the plate by special "grease" typewriter ribbons or pencils. The greasy or treated images will be receptive to ink while those areas not treated will be receptive to water, but will repel ink. A water roller must first pass over the offset plate and then the ink roller will deposit ink on the receptive areas. The image is transferred from the printing plate to a rubber offset cylinder (blanket roller) which places the image on the plastics part. The process is called "offset" because the printing image is not printed directly on the plastics surface but is transferred to an offset cylinder and then to the plastics. (6) In *rotogravure* or *intaglio* printing the image is depressed or sunken into the printing plate. Ink is applied to the entire surface and a (doctor) blade is used to scrape or remove all excess ink except in the sunken areas. The ink which is left in these depressed areas is transferred directly to the product. (7) *Silk screen* printing is a process where ink or paint is forced by a rubber squeegee through a fine metallic or fabric screen onto the product. The screen is blank or blocked off in areas where no ink is wanted. (8) *Stenciling* is similar to silk screen printing except the open areas or the area to be printed does not have a connecting mesh. Stencils may be positive or negative. In positive stencil printing the image is open and spray or rollers are utilized to transfer the ink through these open areas onto the product. In negative stencil printing the image is blocked out and the background is inked, leaving no ink in the stencil area. Stencil printing may be considered a masking operation. (9) Electrostatic printing has been adapted to several well-known printing techniques; however, all employ the principle of electrostatics. In this process dry inks are attracted to the areas to be printed by a difference in electrical potential. In electrostatic printing there is no direct contact between the printing

(A) One-color direct-printing letterflex press is ideal for printing on irregular surfaces because of the rubber plate used.

(B) This machine may print one or more colors by dry offset or direct letterflex printing methods.

Fig. 20-8. Letterflex and dry offset machines. (Apex Machine Co.)

plate or screen and the product. There are several methods where a screen is made conductive in the areas of the image and nonconductive in other areas. Dry, charged particles are held in these open areas until discharged toward an oppositely charged back plate. The object to be printed is placed between the screen and the backup plate. When the ink is discharged toward the backup plate, it strikes the substrate surface. A fixing agent is then applied to provide a permanent image. The image is faithfully reproduced regardless of the surface configuration of the substrate. Images can be printed on the yolk of an uncooked egg or on other food products by this method. Edible inks are used to identify, decorate, and supply messages on fruits and vegetables. (10) *Heat transfer printing* is used as a decorating process and as an important printing method. The process is similar to hot-leaf stamping in that a carrier film (or paper) supports the release layer and the ink image. The thermoplastic ink is heated and transferred to the product by means of a heated rubber roll. (11) *Hot-leaf stamping,* the process of transferring a colorant or a decorative material from a dry carrier film to a product by heat and pressure, is sometimes used as a printing method.

In-Mold Decorating

During in-mold decorating, an overlay or a coated film called a "foil" becomes part of the mold product. The decorative image and, if possible, the film carrier are made of the same material as the part to be molded. With thermosetting products the film may be a clear cellulose sheet covered with a partially cured resinlike molding material. The in-mold overlay is placed in the molding cavity while the thermosetting material is only partially cured. The molding cycle is then completed with the decoration becoming an integral part of the product. With thermoplastic materials the overlay may be placed directly in the mold cavity before any molten material is forced into the mold. The overlay material should be the same as the product. As the molten thermoplastic material flows into the mold cavity the overlay becomes fully bonded and part of the final product. It should be obvious that the placement of gates is important to prevent wrinkled or "washed" overlays. In both thermosetting and thermoplastic molding the overlay may be held in place in the mold by cutting the overlay physically so that it fits snugly in the cavity, or, where physical methods are not desirable, by electrostatic means.

In a process similar to heat transfer, blow molded parts may be decorated in-mold. The ink or paint image is placed on a carrier film or paper. As the hot plastics expands, filling the mold cavity, the image is transferred from the carrier to the hot plastics material.

Heat Transfer

In heat transfer decorating the image is transferred from a carrier film onto the plastics part. The structure of heat transfer decorating stock is shown in Fig. 20-9A.

The preheated carrier stock is transferred to the product by means of a heated rubber roller as shown in Fig. 20-9C.

A decorating or printing process which resembles a combination of engraving and offset printing is the "Tampo-Print." In this process a flexible transfer pad is used to pick up the impression from the inked engraving plate (Fig. 20-10A) and transfers it to the item to be printed (Fig. 20-10B). The entire ink supply carried by the transfer pad is deposited on the part, leaving the pad clean of ink residue. The flexible pad adapts to rough and uneven surfaces while maintaining absolute reproduction sharpness. Variously shaped printing heads can accommodate a wide variety of objects and textures. Multicolor wet-on-wet printing, including halftones, can be accomplished. Virtually any type of printing ink or paint may be used by this simple process. Depending on the type of product, up to 20,000 parts per hour can be automatically decorated.

Pressure Sensitive Labels and Miscellaneous Methods

There are numerous miscellaneous decorating methods, including (1) pressure sensitive labels, (2) decalcomanias, (3) flocking, and (4) decorative coatings or clads.

Pressure sensitive labels are easy to use and apply. The designs or messages are generally printed on the adhesive backed foil or film label. The labels are then placed on the finished product by hand or mechanical means.

Decalcomanias ("decals" for short) are a means of transferring a picture or design to plastics. They are generally placed as a decorative film on a paper backing. The decalcomania is moistened in water and the adhesive backed film is "slipped" onto the plastics surface. This process is not widely used because the decalcomanias are difficult to place accurately and rapidly on the plastics surface.

Flocking by mechanical or electrostatic means is an important method of placing a velvetlike finish on virtually any surface. The process consists of coating the product with an adhesive and placing plastics fibers on the adhesive areas. The velvetlike coatings or designs on walls, wallpaper, cases, toys, and furniture are familiar examples.

There are various woodgraining decorative processes. Some are accomplished by rolling engraved or etched woodgraining plates over a contrasting background color. This is actually a printing process adaptation where woodgrain is being reproduced on various substrates. Various decorative laminates and clad coatings are also used to decorate substrates. Polyvinyl clad (coated) metal products are used in store fixtures, partitions, room dividers, furniture, automobiles, kitchen equipment, and bus interiors. They are durable as well as decorative applications.

Glossary of New Terms

Corona discharge—A method of oxidizing a film of plastics to render it printable by passing the film between the electrodes and subjecting it to a high-voltage discharge.

Electrostatic printing—The deposit of ink on a plastics surface where electrostatic potential is utilized to attract the dry ink through an open area defined by opaquing.

Engraving—The act of cutting figures, letters, etc. into a surface. A plastics web is often printed or decorated by interposing a resilient offset roll between the engraved roll and the web.

(A) Structure of heat transfer decorating stock.

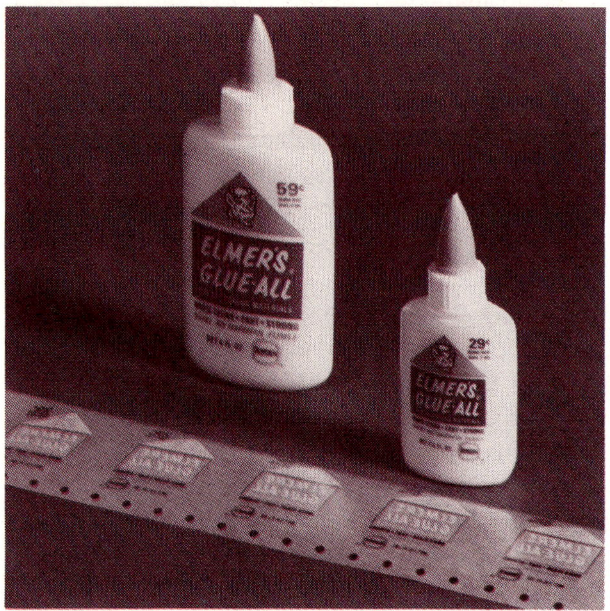

(B) Familiar example of container decoration by heat transfer process. (Therimage® Products Group, Dennison Mfg. Co.)

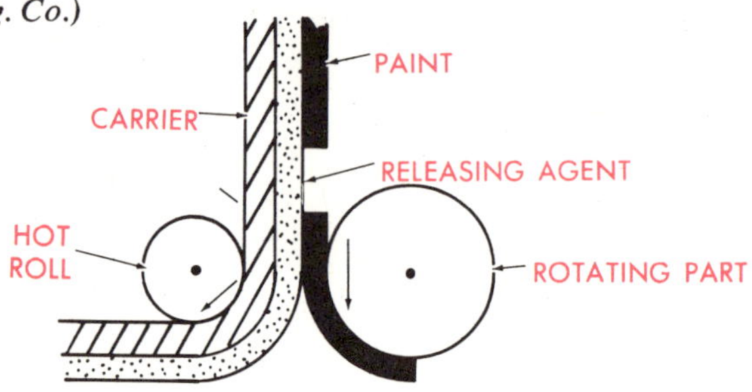

(C) Roller method for transferring design.

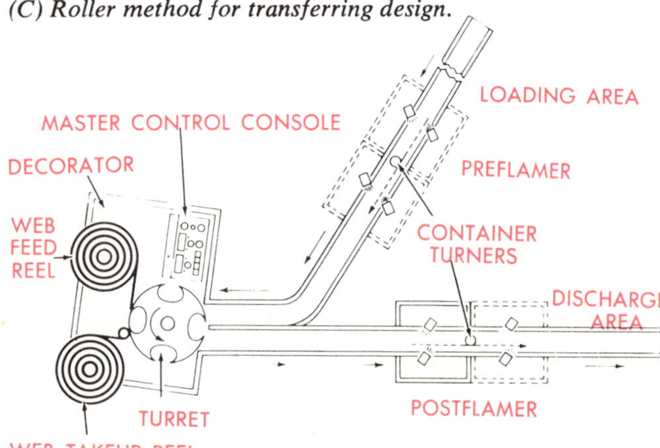

(D) Scheme for heat transferring decorations to containers. (Therimage® Products Group, Dennison Mfg. Co.)

(E) Heat transferring decorations according to the scheme shown in (D). (Therimage® Products Group, Dennison Mfg. Co.)

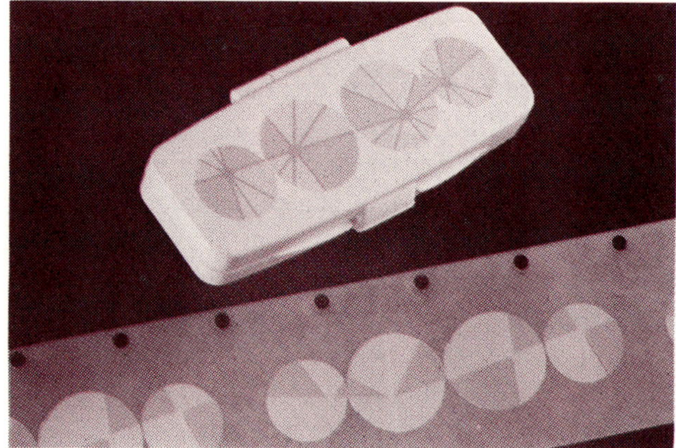

(F) Flat 1/8-in thick silicone rubber pad is used to apply transfer to cosmetic compact. (Gladen Enterprises, Inc.)

(G) Assortment of containers decorated by heat transfer process. (Therimage® Products Group, Dennison Mfg. Co.)

Fig. 20-9. The method of heat transfer decorating.

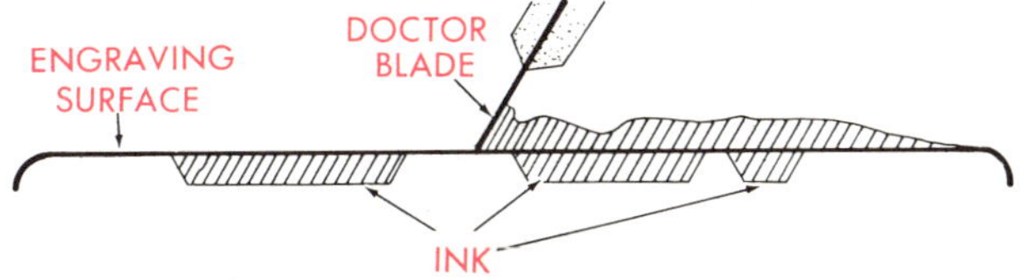

(A) Etched plate used with pigment held in recessed design area. Engraving surface is cleaned by means of doctor blade so that ink remains only in the recesses.

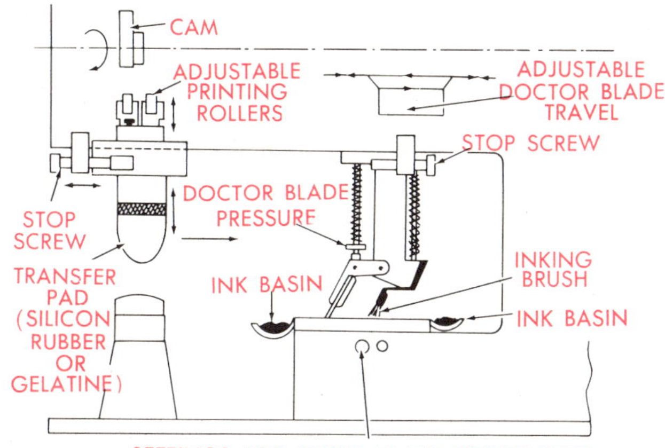

(B) Movement of the transfer pad and inking brush or squeegee.

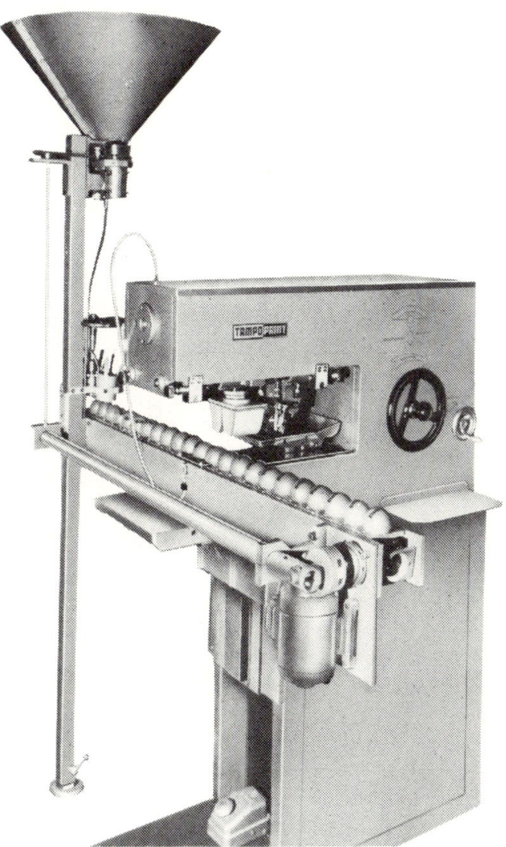

(C) Entire printing machine setup.

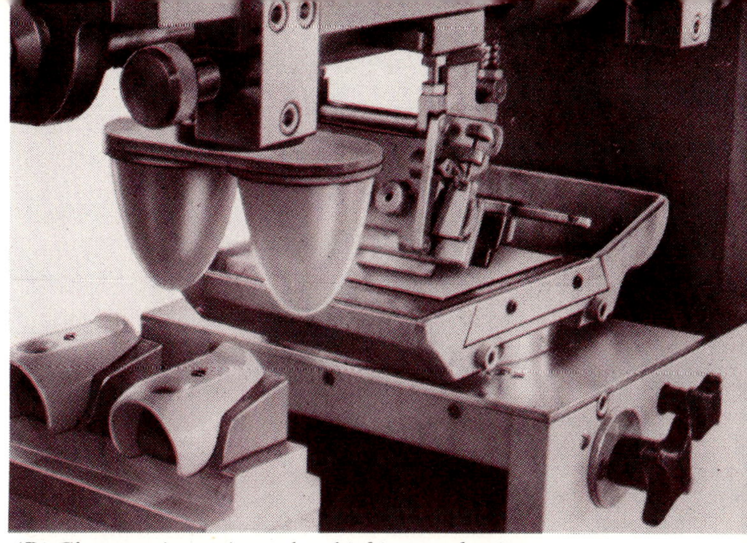

(D) Closeup of transfer pad and inking mechanism.

Fig. 20-10. Tampo-Print process is used to print or decorate various products by use of transfer pad. (Dependable Machine Co.)

Heat transfer decorating—A process in which the image is transferred from the carrier film to the product by stamping with rigid or flexible shapes and with heat and pressure.

Hot-leaf stamping—Decorating operation for marking plastics in which a roll leaf is stamped with heated metal dies onto the face of the plastics. Ink compounds can also be used. *See* Heat transfer decorating.

In-mold decorating—Producing decorations or patterns on molded products by placing the pattern or image in the mold cavity before the actual molding cycle. The pattern becomes part of the plastics article as it is fused by heat and pressure of the plastics material.

Offset printing—A printing technique in which ink is transferred from a bath onto the raised surface of the printing plate by rollers. Subsequently the printing plates transfer the ink to the object to be printed.

Therimage—A trademark of a decorating process of plastics which transfers the image of a label or decoration to the object under the influence of heat and light pressure.

Tampo-Print—A process of transferring ink from an engraved ink-filled surface by the use of a flexible printing (transfer) pad to a product surface.

Review Questions

20-1. What is the most inexpensive way to produce decorative designs on a molded plastics object?

20-2. What must be done to plastics objects before they are decorated?

20-3. What is the most economical way to color plastics objects? What precautions should be observed in this process?

20-4. Which method of painting plastics objects is the most popular? Why?

20-5. What is masking? Why is it used in painting processes?

20-6. Name the three ways of atomizing paint.

20-7. How is excess paint removed in the dip coating process?

20-8. What is screen painting?

20-9. What is the major design restriction in the fill-in method of decorating plastics objects?

20-10. What type of design may be roller coated?

20-11. Briefly describe hot-leaf stamping.

20-12. Give three decorative applications of plating.

20-13. How is engraving used to produce multicolor decorations on plastics objects?

20-14. Describe five methods of printing on plastics objects.

20-15. What is the basic principle of in-mold decorating?

20-16. Why is the Tampo-Print process especially useful for decorating irregular surfaces?

20-17. What is flocking? Clading?

20-18. What is the difference in application between a pressure sensitive label and a decal?

21

Machining and Finishing

The vast number of machines and processes used in the machining and finishing of plastics does not allow detailed discussion. There are, however, certain fundamentals which may apply to all machining and finishing processes. Additives, fillers, and the plastics of each family add to the individual machining and finishing characteristics. Few plastics articles are made in their entirety by machining; many molded parts must be finished or fabricated into useful articles.

The techniques for processing plastics have come from the machining and finishing of wood and metal. Nearly all plastics may be machined. As a rule thermosets are more abrasive to cutting tools than thermoplastics. High-pressure laminates and many filled plastics may require the use of diamond or carbide tipped tools. The lower thermal conductivity and the low modulus of elasticity (softness, flexibility) of most thermoplastics suggest that tools should be properly sharpened to cut clean without burning, clogging, or generating frictional heat from the cutting operation. The low melting points of some thermoplastic materials cause a tendency of these materials to gum, melt, or craze when machined. When heat builds up on plastics they expand more than most materials. The coefficient of thermal expansion for plastics is roughly ten times greater than that of metals. Liquid or air cooling agents may be used to keep the cutting tool clean and free of chips while cooling the cutting edge and the plastics surface. With cooling, cutting speeds increase, smoother cuts result, tool life is longer, and dust is largely eliminated.

The operations of machining and finishing include sawing, filing, drilling, tapping, turning, planing, milling, shaping, routing, sanding, shearing, punching, laser cutting, tumbling, grinding, ashing, buffing, polishing, transparent coating, polishing by solvents, annealing, and postcuring.

Sawing

Nearly all types of saws have been adapted to cutting plastics. Backsaws, coping saws, hacksaws, saber saws, hand saws, and jeweler's saws are normally used for hobbycraft or short run cutting. The shape of the cutting tooth is important in order to cut plastics satisfactorily.

Circular blades should have plenty of set or be hollow ground. Blades should have a deep, well-rounded gullet (see Fig. 21-1A). The rake (or hook) angle is zero (or slightly negative) and the back clearance should be about thirty degrees. The number of teeth per inch will vary, depending on the thickness of the material to be cut. More than six teeth per inch should be used for cutting thin materials, while fewer teeth per inch are required for plastics over one inch in thickness.

A skip-tooth bandsaw blade is preferred (Fig. 21-2A). The wide gullet in this blade provides ample space for plastics chips to be carried from the kerf (cut made by the saw). Best results are obtained if the teeth have a zero rake and some set. Table 21-1 gives recommended teeth per inch for various circular saw and band saw speeds, with materials less than ¼ inch in thickness and materials greater than ¼ inch in thickness.

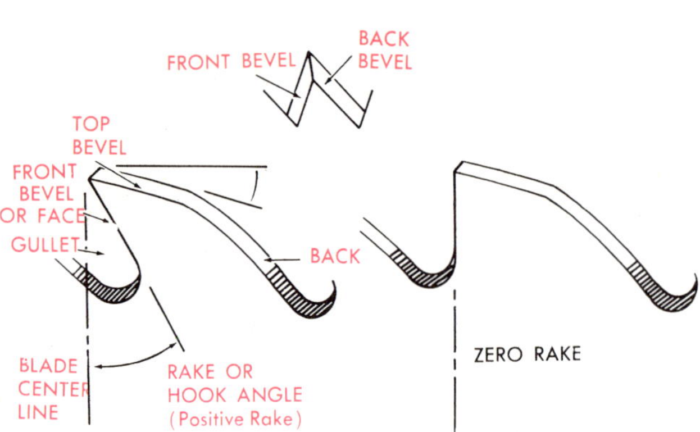

(A) Parts of a circular saw tooth.

(B) Zero rake of circular saw tooth.

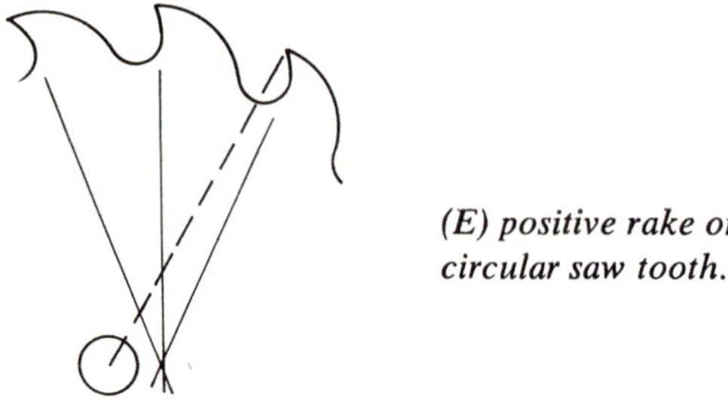

(E) positive rake on circular saw tooth.

Fig. 21-1. Circular saw blade characteristics.

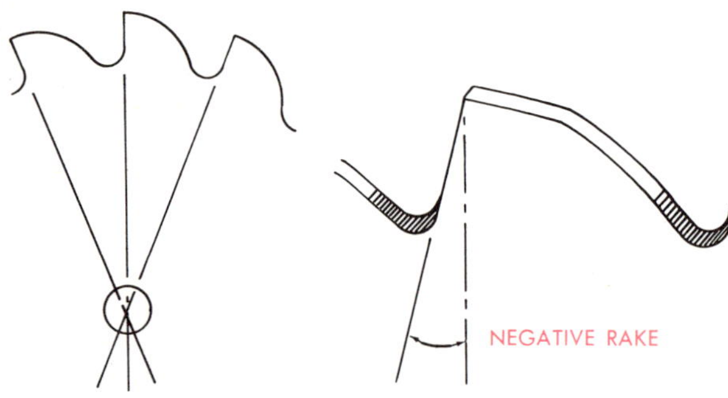

(C) Zero rake where line of tooth face crosses center of blade.

(D) Negative rake on circular saw tooth.

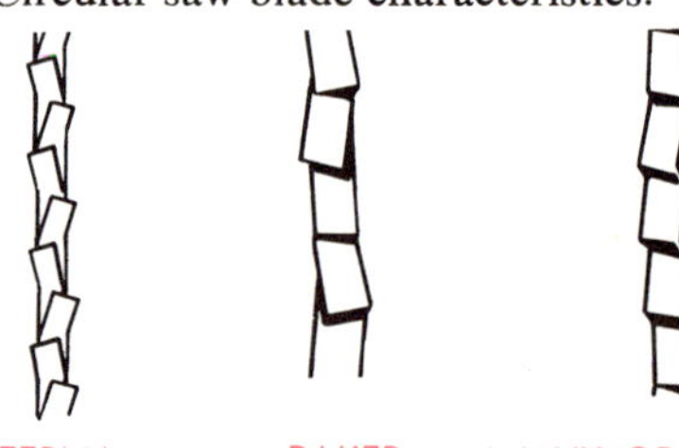

(A) Skip tooth provides large gullet and good chip clearance, and hook tooth is sometimes preferred for cutting glass filled thermosetting materials.

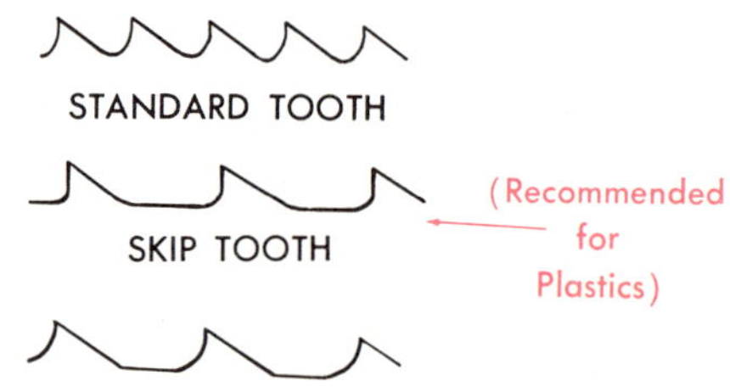

(B) Common bandsaw blade teeth (cutting edge view).

Fig. 21-2. Bandsaw blades.

Table 21-1. Circular and Band Power Sawing.

Plastics	Teeth Per Inch (<¼ In)	Teeth Per Inch (>¼ In)	Speed (FPM) Circular	Teeth Per Inch (<¼ In)	Teeth Per Inch (>¼ In)	Speed (FPM) Band (>¼ In)
Acetal	6	4	8000	16	10	1800-1500
Acrylic	4	3	3000	12	5	4000-2000
ABS	6	4	4000	9	5	3000-1000
Cellulose Acetate	6	4	3000	8	4	3000-1500
Diallyl Phthalate	10	8	2500	18	10	2500-2000
Epoxy	10	8	3000	18	10	2000-1500
Ionomer	8	6	6000	8	6	2000-1500
Melamine Formaldehyde	10	8	5000	18	10	4500-2500
Phenol Formaldehyde	10	8	3000	18	10	3000-1500
Polyallomer	6	4	9000	6	4	1500-1000
Polyamide	10	8	5000	6	4	1500-1000
Polycarbonate	6	4	8000	6	4	2000-1500
Polyester	10	8	5000	18	10	4000-3000
Polyethylene	10	8	9000	6	4	2000-1500
Polyphenylene Oxide	8	6	5000	6	4	3000-2000
Polypropylene	10	8	9000	6	4	2000-1500
Polystyrene	6	4	2000	18	10	2500-2000
Polysulfone	6	4	3000	10	6	3000-2000
Polyurethane	6	4	4000	6	4	2000-1500
Polyvinyl Chloride	6	4	3000	10	6	3000-2000
Tetrafluoroethylene	6	4	8000	8	6	2000-1500

NOTE: Fewer teeth per inch are required for cutting plastics over ¼ inch in thickness. Thin or flexible plastics may be cut on shears or blanking dies. Foams require cutting speeds in excess of 8000 fpm.

For frequent cutting of reinforced, filled or many thermosetting plastics, carbide tipped blades will give accurate cuts and reasonably long blade life. Abrasive or diamond tipped blades may be used; however, a liquid coolant is recommended to prevent clogging or overheating (see Fig. 21-3). All cutting tools should have protective shields and safety devices.

Many thin sheets of plastics may be cut by shearing on metal or specially designed shears. A white crazing line is often produced at the cutting line after shearing operations. Thermoplastics may be heated to aid the shearing action in cutting brittle materials.

(A) Diamond cutoff which is used to cut boron-epoxy reinforced tube. (Whittaker Corp.)

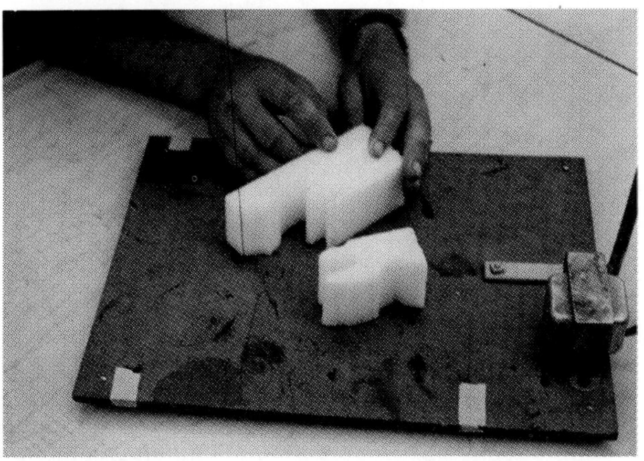

(B) Expanded polystyrene may be easily cut by using a hot, Nichrome wirecutter which "melts" a path through the cellular plastics.

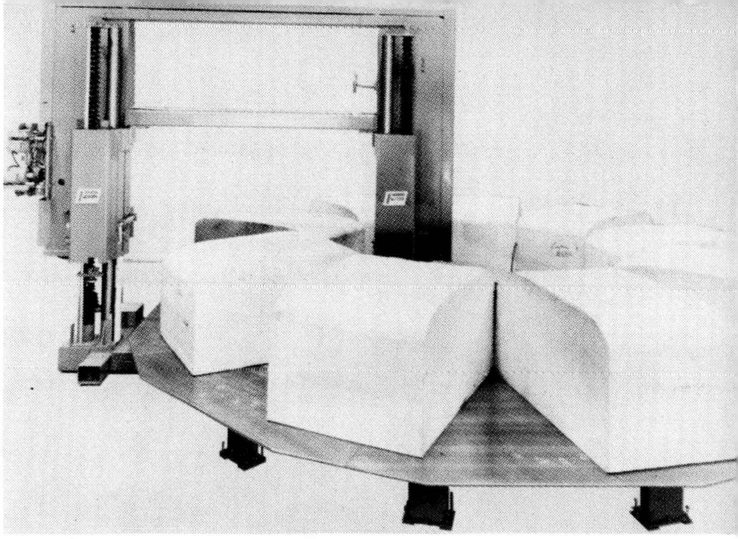

(C) Large blocks of cast foam and cellular plastics are trimmed and cut to shape on this machine. (McNeil Femco Corp.)

(D) This machine is designed to skive or pare sheets of plastics from cylindrical stock. (McNeil Femco Corp.)

Fig. 21-3. Other plastics cutting methods.

Filing

Thermosetting plastics are relatively hard and brittle. Filing removes material in the form of a light powder. Aluminum type A, shear-tooth, or other files that have coarse, single-cut teeth with an angle of 45 degrees are preferred (see Fig. 21-4). The deep-angled file teeth enable the file to clear itself of plastics chips. Many thermoplastics have a tendency to clog files. Curved-tooth files like those used in auto body shops are good because they clear themselves of plastics chips. Specially designed files for plastics should be kept clean and not used for filing metals.

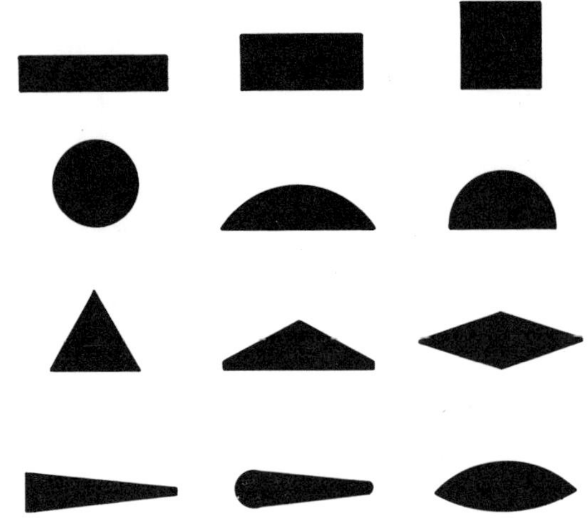

(A) Profile shapes of various files.

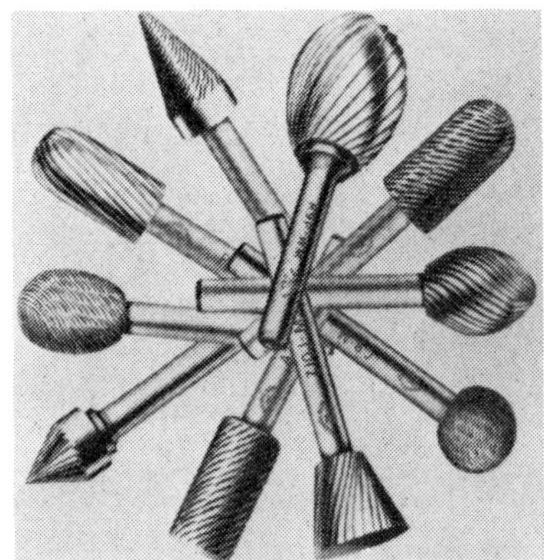

(B) Rotary files may be used on plastics.

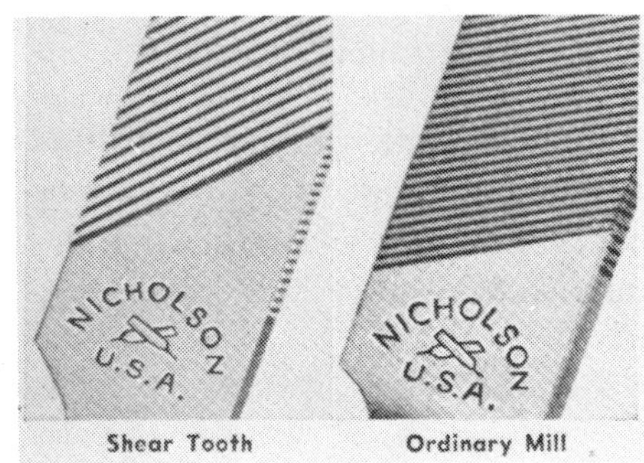

(C) Comparison of shear-tooth and ordinary mill files.

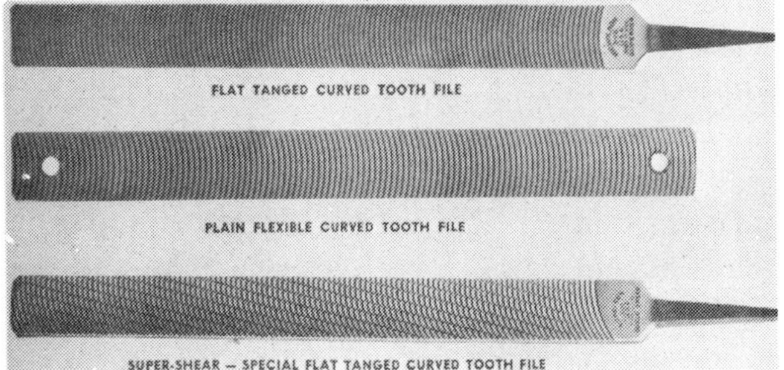

(D) Various curved-tooth files used on plastics.

Fig. 21-4. Files which are used on plastics.

Drilling

Thermoplastic and thermosetting materials may be drilled with any standard twist drill; however, specially designed drills are commercially available which produce better results. Carbide-tipped drills will give long tool life. Holes drilled in most thermoplastics and some thermosets are generally 0.002 to 0.004 inch undersized. Thus a ¼-inch drill will not produce a hole large enough to admit a ¼-inch rod. Thermoplastics may require cooling to reduce frictional heat and gumming.

For drilling most plastics materials drills should be ground with a 70 to 120 degree point angle and a 10 to 25 degree lip clearance angle (see Fig. 21-5A). The rake angle on the cutting edge (Fig. 21-5B) should be zero or several degrees negative (opposite of positive rake angle). Table 21-2 gives rake and point angles for various plastics materials. Highly polished, large, slowly twisting flutes (large helix or helix rake angle) are most desirable for good chip removal.

Table 21-2. Drill Geometry.

Material	Rake Angle	Point Angle	Clearance	Rake
Thermoplastic				
Polyethylene	10°-20°	70°- 90°	9°-15°	0°
Rigid polyvinyl chloride	25°	120°	9°-15°	0°
Acrylic (polymethyl methacrylate)	25°	120°	12°-20°	0°
Polystyrene	40°-50°	60°- 90°	12°-15°	0°-neg. 5°
Polyamide resin	17°	70°- 90°	9°-15°	0°
Polycarbonate	25°	80°- 90°	9°-15°	0°
Acetal resin	10°-20°	60°- 90°	10°-15°	0°
Fluorocarbon TFE	10°-20°	70°- 90°	9°-15°	0°
Thermosetting				
Paper or cotton base	25°	90°-120°	10°-15°	0°
Fibrous glass or other fillers	25°	90°-120°	10°-15°	0°

There are a number of influential factors which affect the cutting speed of plastics: (1) type of plastics, (2) tool geometry, (3) lubricant or coolant, and (4) the feed and depth of cut.

The cutting speed of plastics is given in surface feet per minute (fpm). "Fpm" does not refer to the revolutions per minute (rpm) of the drill, but

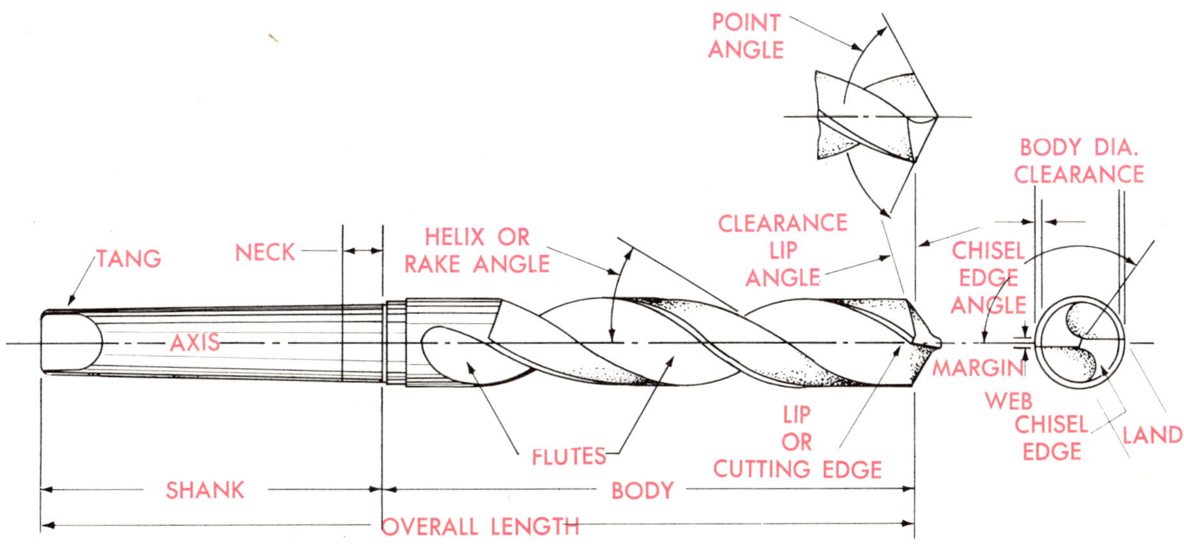

(A) Selected drill nomenclature for twist drill with tapered shank. Drills ½ inch and under usually have straight shanks. (Morse Twist Drill Machine Co.)

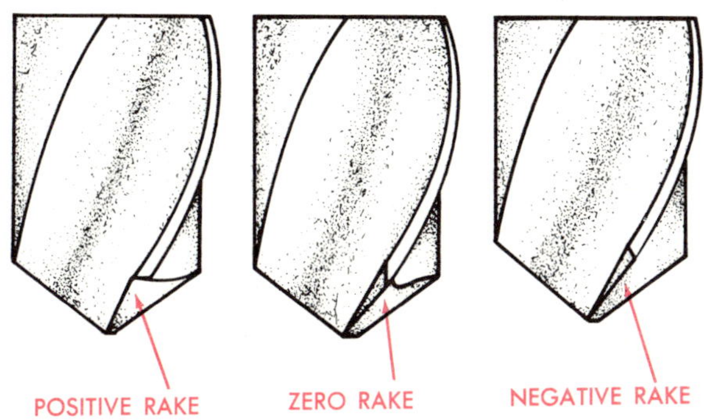

(B) Various rake angles cut on lip or cutting edge. Zero rake on cutting edge is recommended for most plastics. A negative rake on the cutting edge is sometimes used in cutting polystyrene.

Fig. 21-5. Drill nomenclature.

to the distance the cutting edge(s) travels when measured on the circumference of the cutting tool in one minute. The following formula is generally used to determine surface feet per minute, although fpm is usually obtained from handbooks:

$$\text{fpm} = \frac{\pi D \times \text{rpm}}{12}$$

$$\text{rpm} = \frac{\text{fpm} \times 12}{\pi D}$$

where rpm = revolutions per minute,
fpm = surface feet per minute,
D = diameter of cutting tool in inches,
π = 3.1416.

As a rule of thumb, plastics have a machining speed of approximately 200 fpm. A guide for drilling thermoplastics and thermosets is shown in Tables 21-3 and 21-4.

Table 21-3. Guide to Speeds for Drilling Plastics.

Drill Size	Speed for Thermoplastics (rpm)	Speed for Thermosets (rpm)
No. 33 and smaller	5000	5000
No. 17 through 32	3000	2500
No. 1 through 16	2500	1700
1/16 inch	5000	5000
1/8 inch	3000	3000
3/16 inch	2500	2500
1/4 inch	1700	1700
5/16 inch	1700	1300
3/8 inch	1300	1000
7/16 inch	1000	600
1/2 inch	1000	600
A-C	2500	1700
D-O	1300	1300
P-Z	1300	1000

The rate at which the drill or cutting tool advances into the plastics is also important. The distance the tool is fed into the work each revolution is called the *feed*. Feed is measured in decimal fractions of an inch. Drill feed ranges from 0.001 to 0.030 inch for most plastics, depending on thickness of the material.

Many of the same principles apply to reaming, countersinking, spotfacing, and counterboring (Fig. 21-6). Holes may be punched in many thermoplastics with hollow punches. Warming the stock may help the punching operation.

Table 21-4. Drilling Feeds of Plastics.

Material	Speed (fpm)	Feed (in/revolution) Nominal Hole Diameter (in) 1/16	1/8	1/4	1/2	3/4	1
Thermoplastics							
Polyethylene	150-200	0.002	0.003	0.005	0.010	0.015	0.020
Polypropylene							
TFE fluorocarbon							
Butyrate							
High impact styrene	150-200	0.002	0.004	0.005	0.006	0.006	0.008
Acrylonitrile-butadiene-styrene							
Modified acrylic							
Nylon	150-200	0.002	0.003	0.005	0.008	0.010	0.012
Acetals							
Polycarbonate							
Acrylics	150-200	0.001	0.002	0.004	0.008	0.010	0.012
Polystyrenes	150-200	0.001	0.002	0.003	0.004	0.005	0.006
Thermosets							
Paper or cotton base	200-400	0.002	0.003	0.005	0.006	0.010	0.012
Homopolymers	150-300	0.002	0.003	0.004	0.006	0.010	0.012
Fiber glass, graphitized, & asbestos base	200-250	0.002	0.003	0.005	0.008	0.010	0.012

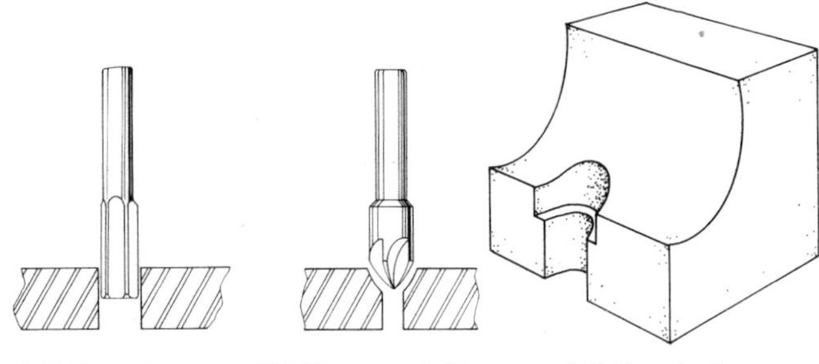

(A) *Reaming.* (B) *Countersinking.* (C) *Spotfacing.*

Stamping, Blanking, or Diecutting

Many thermoplastics and thin pieces of thermosets may be cut using rule dies blanking dies or matched molding dies. It is normally done on flat parts less than 0.250 inch thick. Holes may be drilled or diecut. Heating plastics stock may aid in cutting operation.

Tapping and Threading

Standard machine shop tools and procedures may be used for tapping and threading. To prevent overheating, taps should be finish ground and have polished flutes. Lubricants may also facilitate clearing the chips from the hole. If transparency is required a wax stick may be inserted in the drilled hole before tapping. The wax lubricates, helps expel chips, and produces a more transparent thread.

Because of the elastic recovery of most plastics oversized taps should be used. Oversized taps are commercially available and designated:

H1: Basic-basic + 0.0005 inch
H2: Basic + 0.0005 basic + 0.0010 inch
H3: Basic + 0.0010 basic + 0.0015 inch
H4: Basic + 0.0015 basic + 0.0020 inch

The cutting speed for machine tapping (fpm) should be less than 50 fpm and the taps should be

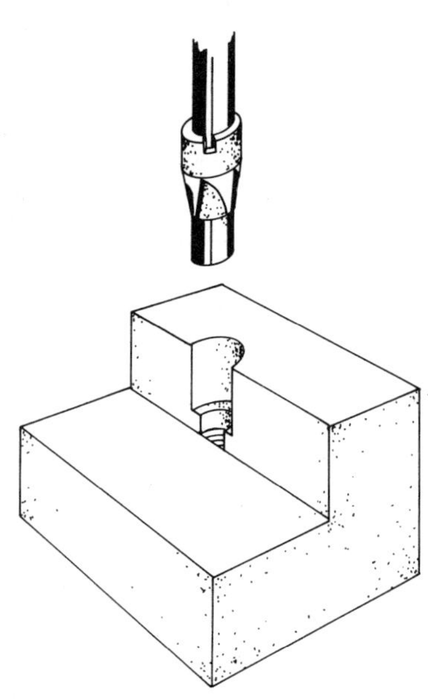

(D) *Counterboring.*
Fig. 21-6. Drilling operations in plastics.

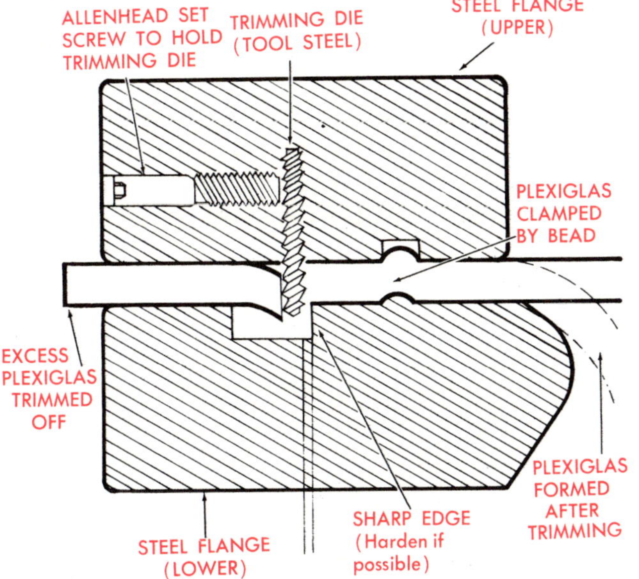

(A) Blank die and clamp for cutting Plexiglas.

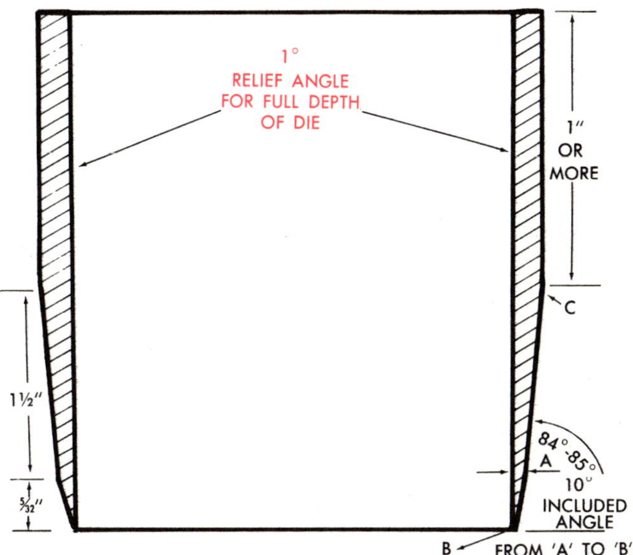

(B) Modified shoemaker die.

Fig. 21-7. Dies for cutting plastics. (Rohm & Haas Co.)

backed out often to clear the plastics chips. As a rule not more than 75 percent of the full thread is cut into the plastics. Sharp V-threads are not recommended. Acme type threads, Unified National Coarse Series (UNC), United States Standard (USS), and National Pipe Thread Series (NPT) are preferred. See Fig. 21-8 and Table 21-5.

Plastics may also be tapped and threaded on conventional lathes and screw machines.

Table 21-5. Coarse-Thread Series, UNC and NC— Basic Dimensions (Unified Screw Threads).

Sizes	Basic Major Diam. Inches	Thds. per Inch	Basic Pitch Diam. Inches	Minor Diameter Ext. Thds. Inches	Minor Diameter Int. Thds. Inches	Lead Angle at Basic P.D. Deg.	Lead Angle at Basic P.D. Min.	Area of Minor Diam. Sq. In.	Tensile Stress Area Sq. In.	Tap Drill Size
1 (.073)*	0.0730	64	0.0629	0.0538	0.0561	4	31	0.00218	0.00263	52
2 (.086)	0.0860	56	0.0744	0.0641	0.0667	4	22	0.00310	0.00370	48
3 (.099)*	0.0990	48	0.0855	0.0734	0.0764	4	26	0.00406	0.00487	44
4 (.112)	0.1120	40	0.0958	0.0813	0.0849	4	45	0.00496	0.00604	41
5 (.125)	0.1250	40	0.1088	0.0943	0.0979	4	11	0.00672	0.00796	35
6 (.138)	0.1380	32	0.1177	0.0997	0.1042	4	50	0.00745	0.00909	31
8 (.164)	0.1640	32	0.1437	0.1257	0.1302	3	58	0.01196	0.0140	28
10 (.190)	0.1900	24	0.1629	0.1389	0.1449	4	39	0.01450	0.0175	22
12 (.216)*	0.2160	24	0.1889	0.1649	0.1709	4	1	0.0206	0.0242	13
¼	0.2500	20	0.2175	0.1887	0.1959	4	11	0.0269	0.0318	5
5/16	0.3125	18	0.2764	0.2443	0.2524	3	40	0.0454	0.0524	G
⅜	0.3750	16	0.3344	0.2983	0.3073	3	24	0.0678	0.0775	O
7/16	0.4375	14	0.3911	0.3499	0.3602	3	20	0.0933	0.1063	⅜
½	0.5000	13	0.4500	0.4056	0.4167	3	7	0.1257	0.1419	7/16
9/16	0.5625	12	0.5084	0.4603	0.4723	2	59	0.162	0.182	31/64
⅝	0.6250	11	0.5660	0.5135	0.5266	2	56	0.202	0.226	17/32
¾	0.7500	10	0.6850	0.6273	0.6417	2	40	0.302	0.334	21/32
⅞	0.8750	9	0.8028	0.7387	0.7547	2	31	0.419	0.462	49/64
1	1.0000	8	0.9188	0.8466	0.8647	2	29	0.551	0.606	⅞
1⅛	1.1250	7	1.0322	0.9497	0.9704	2	31	0.693	0.763	1.000
1¼	1.2500	7	1.1572	1.0747	1.0954	2	15	0.890	0.969	1.125
1⅜	1.3750	6	1.2667	1.1705	1.1946	2	24	1.054	1.155	1.218
1½	1.5000	6	1.3917	1.2955	1.3196	2	11	1.294	1.405	1.343
1¾	1.7500	5	1.6201	1.5046	1.5335	2	15	1.74	1.90	

*Secondary sizes.

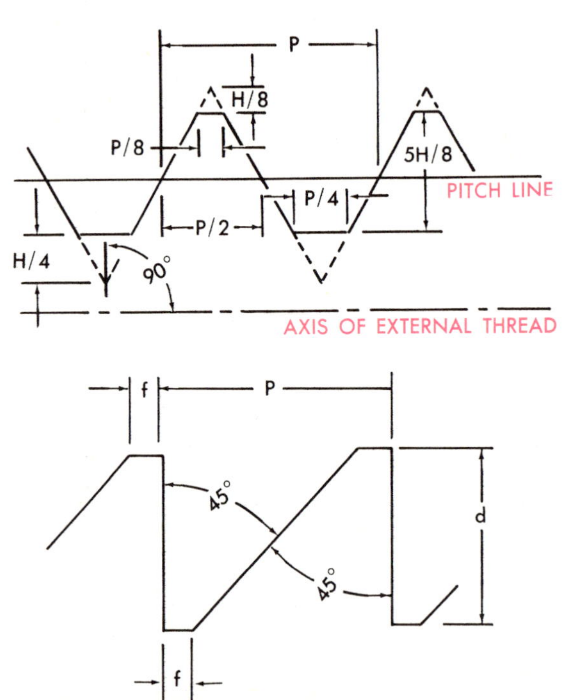

(A) Butress thread form. (Machinery's Handbook, *The Industrial Press)*

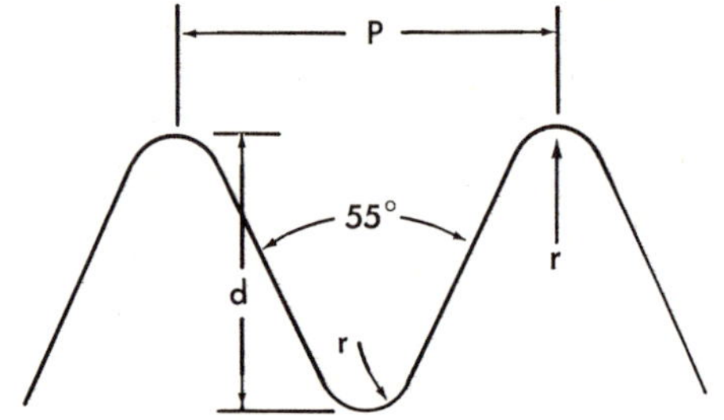

(B) Whitworth thread form. (Machinery's Handbook, *The Industrial Press)*

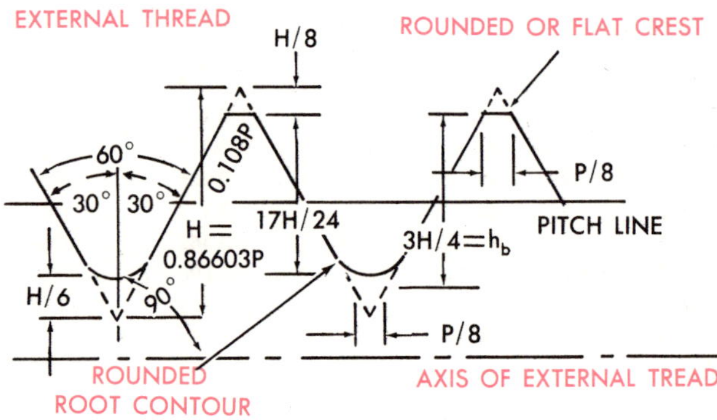

(C) Basic form of Unified screen threads.

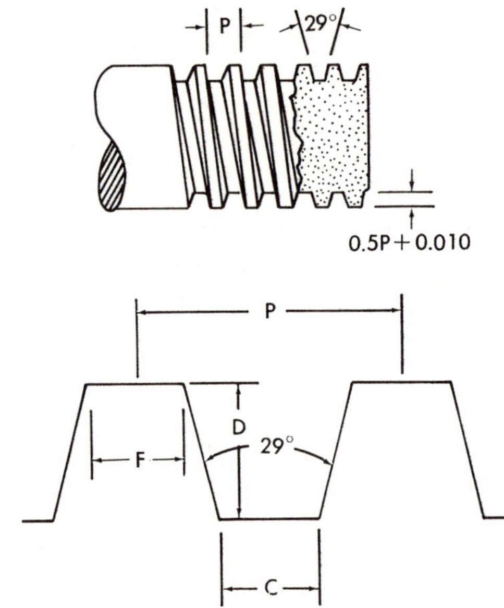

(D) Acme thread form.

Fig. 21-8. Various types of thread forms.

Turning, Milling, Planing, Shaping, and Routing

High-speed steel or carbide cutting tools used for machining brass and aluminum are recommended for machining plastics. The feeds and speeds should also match those used for brass or aluminum. For many plastics a surface speed of 500 feet per minute (fpm) with feeds (depth of cut) of 0.002 to 0.005 inch per revolution (ipr) will produce good results. Depth of cut refers to the distance the cutter has been fed into the plastics. On cylindrical stock a 0.050-inch cut will reduce the diameter by 0.100 inch.

Climb-cutting, milling operations with lubrication will give a good machined finish (Fig. 21-9A). The feed rate on multiple-edged milling cutters is usually expressed as inches of cut per cutting edge per minute. The feed of a milling machine is expressed in inches per minute of table movement rather than inches per spindle rotation. The following formula is used to determine the amount of feed in inches per minute:

$$\text{ipm} = t \times \text{fpt} \times \text{rpm}$$

where

t = teeth (number),
ipm = feed in inches per minute,
fpt = feed per tooth (chip load),
rpm = revolutions per minute (spindle or work).

Table 21-6 gives turning and milling data for various plastics materials. Table 21-7 gives side and end relief angles and back rake angles for cutting tools used on various plastics.

For all operations of milling, planing, shaping and routing, carbide-tipped cutters are recommended. The feed and speeds should be similar to machining brass or aluminum. Conventional, high-speed-steel, woodworking shapers, planers, and router tools may be used if tools are properly sharpened. Routers and shapers are useful for cutting beads, rabbets, flutes, and trimming edges.

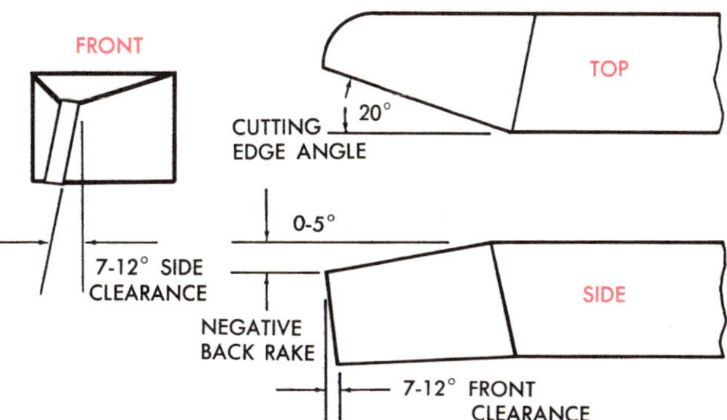

(B) Cutting tool rake and clearance angles for general-purpose turning of plastics. Note 0-5° negative back rake angle. A +20°-rake, sharp-pointed tool is used in turning polyamides.

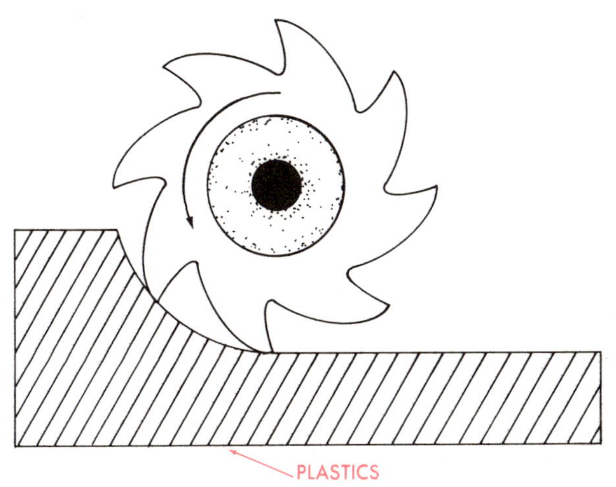

(A) Climb milling, where the work moves in the same direction as the rotating cutter.

(C) Machined polyamide block polymer articles which can be produced within small tolerances in either small or large production runs. (BASF Corp.)

Fig. 21-9. Machining plastics.

Table 21-6. Turning and Milling Plastics.

	Turning Single Point (H-S Steel)			Milling Tool Per Tooth (H-S Steel)		
Material	Depth of cut (Inches)	Speed (FPM)	Feed (IPR)	Depth of cut (Inches)	Speed (FPM)	Feed (Inches Per Tooth)
Thermoplastics						
Polyethylene	0.150	250-350	0.010	0.150	500-750	0.016
Polypropylene	0.025	300-400	0.002	0.150	500-750	0.016
TFE-fluorocarbon				0.060	750-1000	0.004
Butyrates				0.150	500-750	0.016
ABS	0.150	250-350	0.015	0.150	500-750	0.016
Polyamides	0.150	300-400	0.010	0.150	500-750	0.016
Polycarbonate	0.025	400-500	0.002	0.060	750-1000	0.004
Acrylics	0.150	250-300	0.002	0.060	750-1000	0.004
Polystyrenes, low	0.150	75-100	0.005	0.150	500-750	0.016
& medium impact	0.025	150-200	0.001	0.150	500-750	0.016
Thermosets						
Paper &	0.150	500-1000	0.012	0.060	400-500	0.005
cotton base	0.025	1000-2000	0.005	0.060	400-500	0.005
Fiber glass &	0.150	200-500	0.012	0.060	400-500	0.005
graphite base	0.025	500-1000	0.005	0.060	400-500	0.005
Asbestos base	0.150	650-750	0.012	0.060	400-500	0.005

Table 21-7. Design for Turning Cutting Tool.

Work Material	Side Relief Angle (Deg.)	End Relief Angle (Deg.)	Back Rake Angle (Deg.)
Polycarbonate	3	3	0-5
Acetal	4-6	4-6	0-5
Polyamide	5-20	15-25	neg. 5-0
TFE	5-20	0.5-10	0-10
Polyethylene	5-20	0.5-10	0-10
Polypropylene	5-20	0.5-10	0-10
Acrylic	5-10	5-10	10-20
Styrene	0-5	0-5	0
Thermosets			
Paper or Cloth	13	30-60	neg. 5-0
Glass	13	33	0

Laser Cutting

A CO_2 laser can deliver powerful radiation at a wavelength of 10.6 microns. It is currently being used to punch intricate holes and cut delicate patterns from plastics. (See Fig. 21-10.) The laser power can be directed to etch the plastics surface barely or actually vaporize and melt it. There is no physical contact between the plastics and the laser equipment and no dust or drill chips are produced.

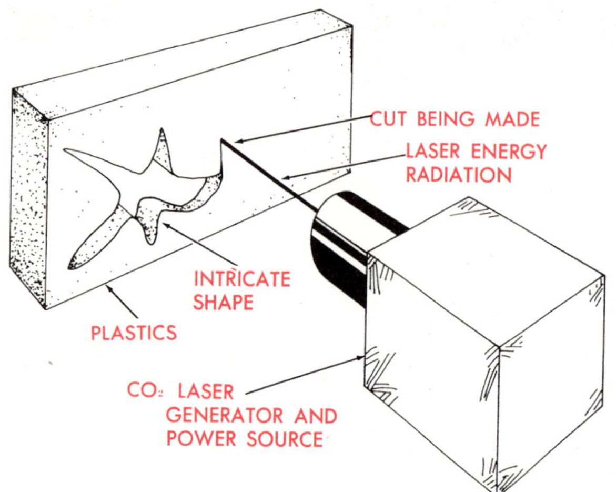

Fig. 21-10. Light energy from laser can be used to cut intricate shapes in plastics.

Smoothing and Polishing

The smoothing and polishing techniques for finishing plastics are similar to those used on woods, metals, and glass. Grinding is not recommended unless open grit wheels with a coolant are used. Hand and machine sanding is an important operation. Open grit sandpaper on machine operations is used to prevent clogging. A number 80 grit silicon-carbide abrasive is recommended for rough sanding. In any machine sanding light pressure is used to prevent overheating of the plastics. Disk sanders operating at 1750 rpm and belt sanders operating at a surface speed of 3600 feet per minute are used for dry sanding. If water coolants are used the abrasive lasts longer and cutting action is increased. Progressively finer abrasives are used in finishing. The initial rough sanding should be followed by finer grades of abrasive paper. After the 80-grit sanding, 280-grit silicon-carbide wet or dry sandpaper is used. The final sanding may be with 400-grit or 600-grit sandpaper. After the sanding is completed and the abrasives removed further finishing operations are made.

Ashing, buffing, and polishing are usually done on abrasive charged wheels. These wheels may be made of cloth, leather, or bristles, and a separate wheel is used for each abrasive grit. Finishing wheels should not exceed 2000 surface feet per minute. When coolants are used surface speed may be increased.

Never finish plastics on wheels used for metal. Small metal particles may be left in the wheel and damage the plastics surface. Machines should be grounded because static electricity is readily generated by the frictional movement of the wheels.

Ashing is a finishing operation in which a wet abrasive, usually number 00 pumice is applied to a loose muslin wheel. (See Fig. 21-11.) Because the operation is wet a hood or shield must be used over the wheel. Surface speeds of over 4000 feet per minute may be used for ashing. Overheating is avoided in this process and the loose muslin wheel is fast cutting on irregular surfaces.

Buffing is an operation in which grease or wax filled abrasive cakes are applied to a loose or sewn muslin wheel. The buffing wheels are charged by holding the cake of abrasive compound against the wheel while it is running. This produces frictional heat, leaving the wax filled abrasive on the wheel (charging). The most common buffing abrasives are tripoli, rouge, or other fine silica powders in a cake (wax) form.

Polishing, sometimes called *luster buffing* or *burnishing*, employs wax compounds composed of

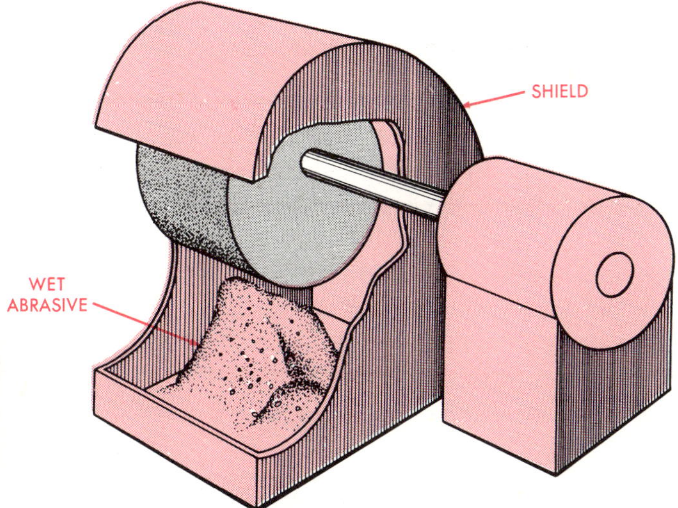

Fig. 21-11. Ashing with wet pumice cuts more rapidly than grease or wax based compounds and has a better cooling action.

(A) Charging flannel wheel with whiting.

(B) Abrasive wheel passing over an edge. About half of each side is done as the part is pulled toward the operator.

Fig. 21-12. Details on buffing plastics materials.

the finest abrasives, such as levigated alumina or whiting. Polishing wheels are generally made of loose flannel or chamois. A final polishing is sometimes done with clean, abrasive-free waxes on a flannel or chamois wheel. The wax fills many imperfections and protects the polished surface.

In operating finishing wheels never let the wheel rotate to the edge of a part because it may be jerked from your hands. The wheel may pass over an edge but never to it. It is better to do about half of the surface and then turn it around and finish the remaining portion. The part should be moved or pulled toward the operator in rapid, even strokes. Do not spend very much time at the finishing wheels. Remove the coarse tool marks before the finishing wheels are used.

Solvent-dip polishing of cellulosic and acrylic plastics may be used to dissolve surface irregularities (Fig. 21-13A). The parts are either dipped or sprayed with selected solvents for approximately one minute. Solvents are sometimes used to polish edges or drilled holes. All parts solvent polished should be annealed to prevent crazing.

Surface coatings may be used on most plastics to produce a high surface gloss. This operation may be less costly than other finishing operations.

Flame polishing with an oxygen-hydrogen flame may be used to polish several plastics (see Fig. 21-13B).

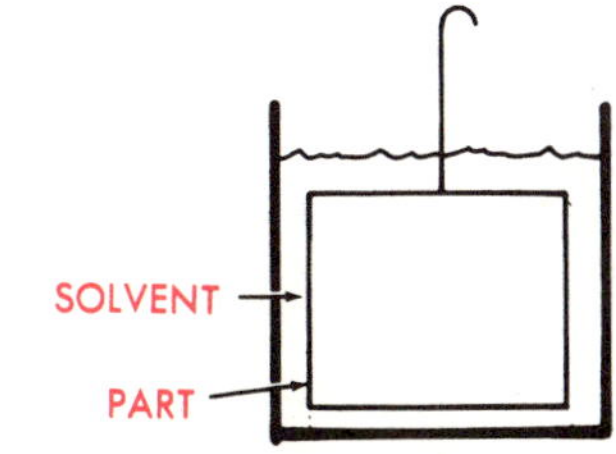

(A) Solvent-dip polishing.

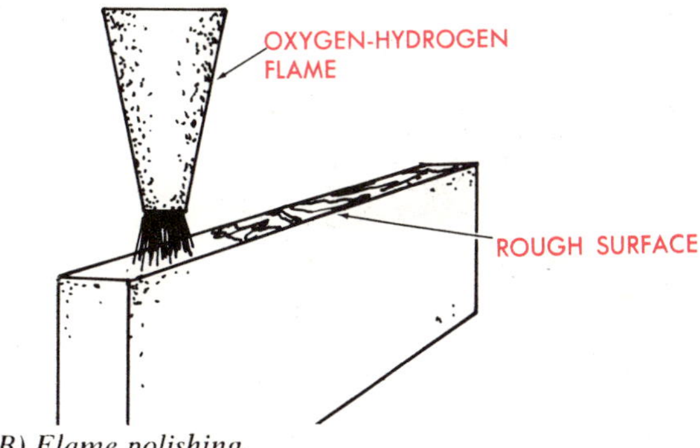

(B) Flame polishing.

Fig. 21-13. Two polishing methods.

Tumbling

The tumbling barrel process is one of the most economical methods for rapidly finishing plastics molded parts. It may be used to produce a smooth finish on plastics by rotating them in a drum with selected abrasives and lubricants (Fig. 21-14A). The rotation of the drum causes the parts and abrasives to rub against each other with a polishing effect. The amount of material to be removed is dependent on the speed of tumbling, the abrasive grit, and the length or period of the tumbling cycle.

In another tumbling process, abrasive grit is sprayed over the parts as they tumble on an endless rubber belt. Fig. 21-14B shows parts being tumbled while being doused with abrasive grit.

Dry ice is sometimes used in tumbling processes to remove molding flash. The dry ice thoroughly chills the thin flash while the tumbling breaks it free in a very short time.

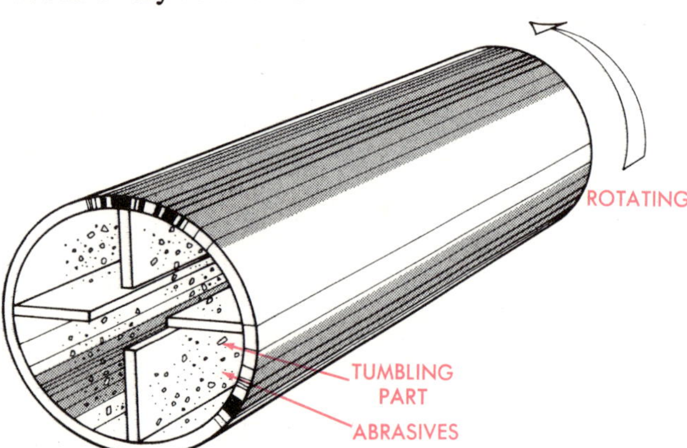

(A) Parts being tumbled in a revolving drum.

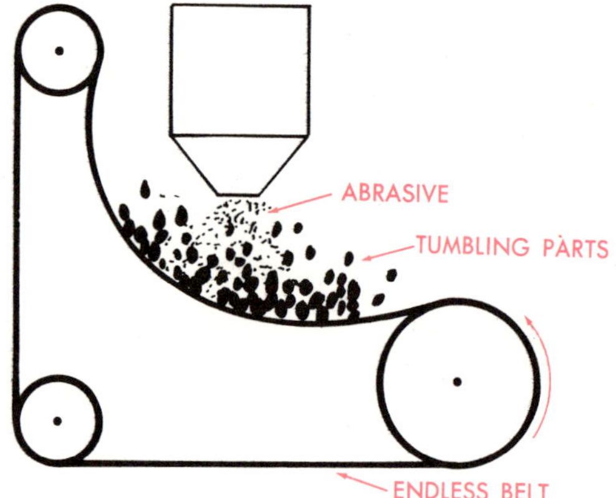

(B) Parts being tumbled on an endless revolving belt.

Fig. 21-14. Two tumbling methods.

Annealing and Postcuring

During the molding operation, various finishing and fabrication processes, the plastics part may develop internal stresses. Chemicals may sensitize the plastics and cause crazing. Annealing consists of prolonged heating of the plastics part at temperatures lower than molding temperatures. The parts are then slowly cooled. The internal stresses set up during molding or fabrication are reduced or eliminated by this treatment. All machined parts should be annealed before cementing.

Fig. 21-15. This large oven may be used for annealing and postcuring of plastics products. (Precision Quincey Corp.)

Tables 21-8 and 21-9 give heating and cooling times for the annealing of Plexiglas. Fig. 21-15 shows a large oven which may be used in the process.

Glossary of New Terms

Annealing—Slow cooling of plastics after heating, to remove internal stress or improve physical properties and workability.

Table 21-8. Heating Times for Annealing of Plexiglas.

Thickness (Inches)	Hours† in a Forced-Circulation Oven at the Indicated Temperature									
	Plexiglas G, II and 55					Plexiglas I-A				
	230°F*	210°F*	195°F*	175°F	160°F**	195°F*	175°F*	160°F*	140°F	120°F
0.060 to 0.150	2	3	5	10	24	2	3	5	10	24
0.189 to 0.375	2½	3½	5½	10½	24	2½	3½	5½	10½	24
0.500 to 0.750	3	4	6	11	24	3	4	6	11	24
0.875 to 1.125	3½	4½	6½	11½	24	3½	4½	6½	11½	24
1.250 to 1.500	4	5	7	12	24	4	5	7	12	24

(Rohm & Haas Co.)

†Includes period of time required to bring part up to annealing temperature, but not cooling time. See Table 21-9.
*Formed parts may show objectionable deformation when annealed at these temperatures.
**For Plexiglas G and Plexiglas II only. Minimum annealing temperature for Plexiglas 55 is 175°F.

Table 21-9. Cooling Times for Annealing of Plexiglas.

Thickness (Inches)	Hours to Cool from Annealing Temperature to Maximum Removal Temperature*								
	Rate	Plexiglas G, II and 55			Plexiglas I-A				
	°F/Hr	230°F	210°F	195°F	175°F	195°F	175°F	160°F	140°F
0.060 to 0.150	120	¾	½	½	¼	¾	½	½	¼
0.187 to 0.375	49	1½	1¼	¾	½	1½	1¼	¾	½
0.500 to 0.750	25	3¼	2¼	1½	¾	3	2¼	1½	¾
0.875 to 1.125	18	4¼	3	2	1	4	3	2¼	1
1.250 to 1.500	13	5¾	4½	3	1½	5¾	4½	3	1½

(Rohm & Haas Co.)

*Removal temperature is 160°F for Plexiglas G II, 175°F for Plexiglas 55, and 120°F for Plexiglas I-A.

Blanking—The cutting of flat sheet stock to shape by striking it sharply with a punch while it is supported on a mating die. Punch presses are used.

Charging—In molding, the measurement or weight of the material placed in a mold. In polishing, the depositing of abrasive on a revolving wheel, generally cloth.

Diecutting—Blanking or cutting shapes from sheet stock by striking it sharply with a shaped knife edge known as a steel-rule die.

Feed—The distance the cutting tool is fed into the work each revolution.

Fpm—Feet per minute.

Hollow ground—Referring to a blade that has been ground thin to allow clear cutting in the kerf. The cutting teeth are the thickest portion.

Kerf—The slit or notch made by a saw or cutting tool.

Laser cutting—Means of cutting materials by laser energy.

Negative rake—Applies to tools with cutting teeth with a rake of less than zero.

Open grit sandpaper—Coarse sandpaper, with a number of 80 or less.

Rpm—Revolutions per minute.

Tripoli—A silica abrasive.

Tumbling—Finishing operation for small plastics articles by which gates, flash, and fins are removed and/or surfaces are polished by rotating them in a barrel or on a belt together with wooden pegs, sawdust, and some polishing compounds.

Whiting—Calcium-carbonate powder abrasive.

Review Questions

21-1. In machining and finishing, what precautions must be taken due to the high thermal coefficient of plastics?

21-2. How many teeth per inch should a circular saw have for cutting thin plastics materials? For thick materials?

21-3. What type of bandsaw blade and what cutting speed should you use to cut ⅛-inch acrylic plastics?

21-4. What is a skip-tooth blade?

21-5. Name various saws that may be used to cut plastics and indicate the kinds of jobs for which each is suitable.

21-6. What is the difference between a negative rake and a positive rake on a cutting tool?

21-7. What types of files are selected for cutting plastics? Why?

21-8. What is drill feed? What is its range in inches for most plastics?

21-9. What factors affect the cutting speed of drills in plastics materials?

21-10. What precaution must be observed in drilling a hole for a given size rod in plastics material?

21-11. Why are wax sticks often inserted in drilled holes before tapping plastics materials?

21-12. Why are oversized taps necessary for plastics materials?

21-13. Name the preferred thread forms for taps used on plastics.

21-14. Cutting tools for which metals are recommended for plastics?

21-15. What is climb milling? Why is it used?

21-16. What is laser cutting of plastics? Where is it used?

21-17. What number grit sandpaper should be used on final finishing of plastics?

21-18. What is ashing? Buffing? Charging a wheel?

21-19. Which abrasives are used in buffing? In burnishing?

21-20. Briefly describe solvent-dip polishing and flame polishing.

21-21. What is tumbling? Why is it so named?

21-22. What does the smoothing in the tumbling operation?

21-23. What is annealing or postcuring of machined or molded parts? Why is this process done?

22

Assembly or Fabrication

There are only four broad methods by which plastics are fabricated or ways various assemblies are joined together: (1) adhesion, (2) cohesion, (3) mechanical linkage, and (4) friction fits. Under each of these broad categories are several distinct assembly methods.

Adhesion

Adhesive bonding is a high-speed, effective means of bonding plastics assemblies. The adhesive does not cause intermingling of molecules between pieces but "adheres" the pieces together. Gluing of wood, paper or metals is a common example of adhesion. Hot-melts are generally adhesive materials that set by cooling.

Cohesion

In cohesive bonds there is an intermingling of the molecules between the parts. (See Fig. 22-1.) Cementing, spin welding, hot gas welding, heat joining, dielectric joining, and ultrasonic and induction bonding are methods of cohesive assembly.

Cementing

There are two kinds of cements in general use: solvent cements and dope cements. Solvent cements are solvents or blends of solvents which melt the selected plastics joints together. Dope cements are sometimes called *laminating cements* or *solvent mixes* and are composed of solvents and a small quantity of the plastics to be joined. This cement is a viscous, syrupy material and leaves a thin film of the parent plastics on the joint when dried.

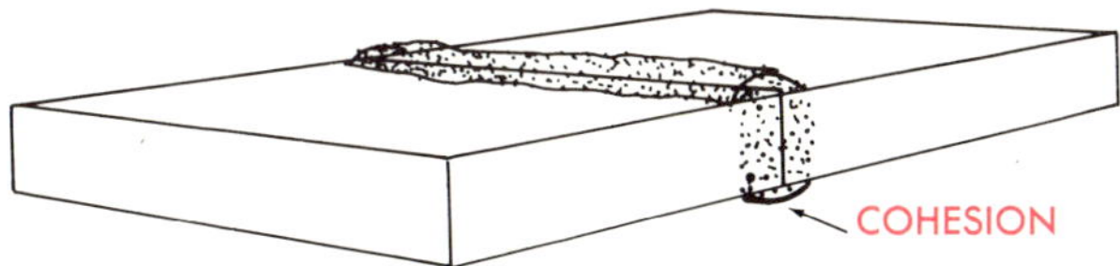

(A) Cohesive bonding.

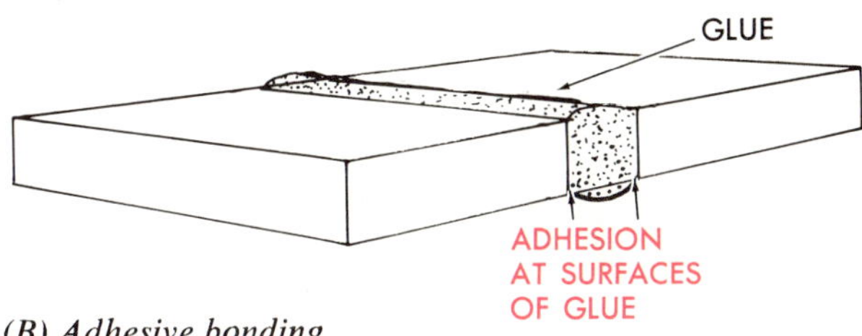

(B) Adhesive bonding.

Fig. 22-1. Comparison of cohesive and adhesive bonding.

In Table 22-1 the solvents with low boiling points evaporate rapidly. Methylene chloride has a boiling point of 104°F; the joint must be positioned rapidly before all the solvent evaporates.

Solvent cements may be applied to the plastics joints by several methods. All joints should be clean and sanded smooth. A V joint is preferred for making butt joints by many manufacturers and fabricators. See Fig. 22-2.

Joints may simply be soaked in a solvent until a soft surface is obtained. The pieces are then immediately placed together under slight pressure until all solvents evaporate. If too much pressure is applied, the soft portion may be squeezed out of the joint, resulting in a poor bond.

Table 22-1. Common Solvent Cements for Thermoplastics.

Plastics	Solvent	Boiling Point (°F)
Acrylic	Methylene chloride	104
	Vinyl trichloride	189
	Ethylene dichloride	183
ABS	Methyl ethyl ketone	176
	Methylene chloride	104
Cellulose acetate,	Methylene dichloride	106
butyrate, propionate	Ethylene dichloride	183
ethyl cellulose	Acetone	135
Polycarbonate	Methylene chloride	104
Polyphenylene oxide	Toluene	232
	Chloroform	142
	Ethylene dichloride	183
Polystyrene	Methyl ethyl ketone	176
	Methylene chloride	104
	Ethylene dichloride	183
	Toluene	232
Polyvinyl chloride	Acetone	135
and copolymers	Methyl ethyl ketone	176
	Tetrahydrofuran	149

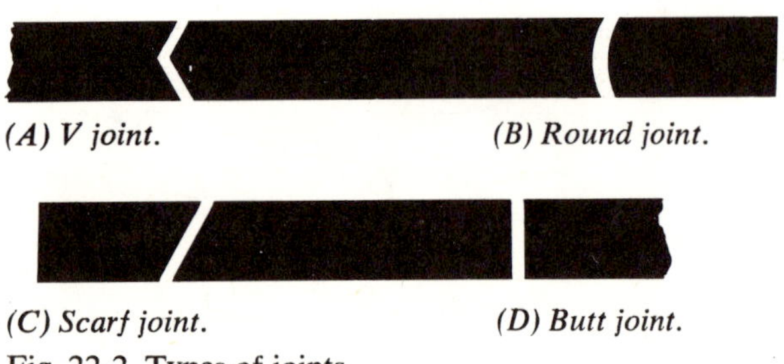

(A) V joint. *(B) Round joint.*

(C) Scarf joint. *(D) Butt joint.*

Fig. 22-2. Types of joints.

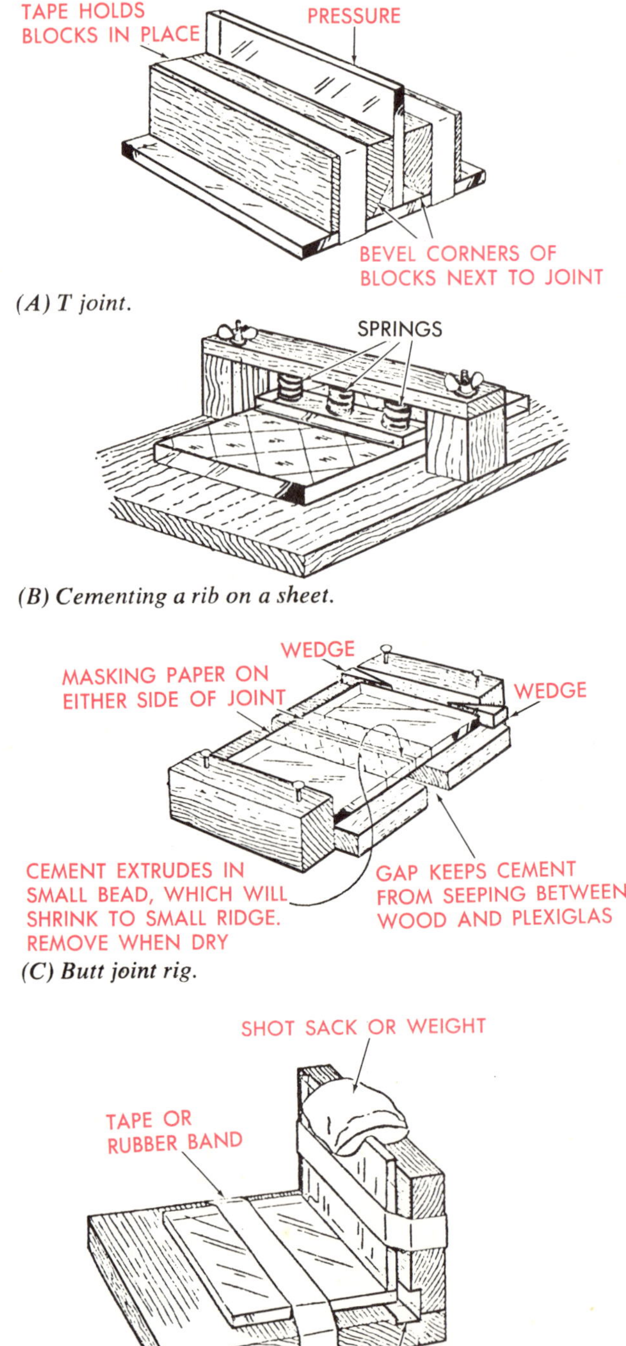

(A) T joint.

(B) Cementing a rib on a sheet.

(C) Butt joint rig.

(D) Corner cementing.

Fig. 22-3. Methods of cementing. (Cadillac Plastics Co.)

Large surfaces may be dipped or sprayed with solvent cements. Cohesive bonds may be made by allowing the solvent to flow into crack joints by capillary action. Small paint brushes and hypodermic syringes are convenient cementing techniques. Fig. 22-3 shows several methods of cementing.

Spin Welding

Spin welding is a method of frictionally joining circular thermoplastic parts. When one or both parts are rotated against each other, frictional heat causes a cohesive melt at that point. (See Fig. 22-4.) Depending on the diameter and the material, joints must spin at about 20 feet per second with less than 20 pounds per square inch of contact area. When the melt has occurred, the parts are stopped and the melt solidifies under pressure.

Joints may also be spin welded by rapidly rotating a filler rod on the joint. A heavy rod of the parent materials is rotated at about 5000 rpm and simply moved along the joint as it melts (Fig. 22-4B). The plastics weld looks similar to arc welds on metals.

(A) Plastics rod spin welded to plastics sheet.

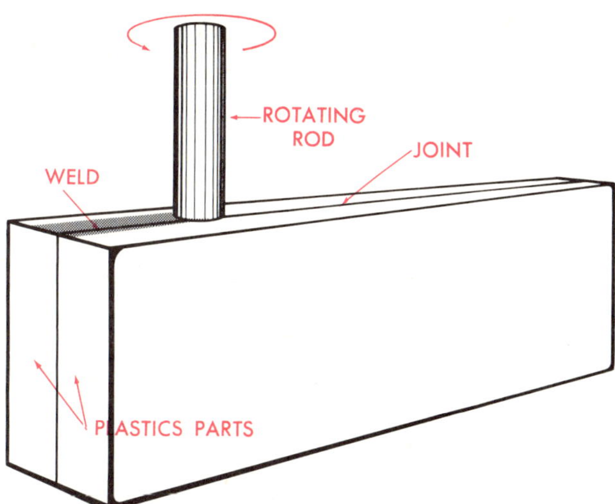

(B) A method of spin welding joints.

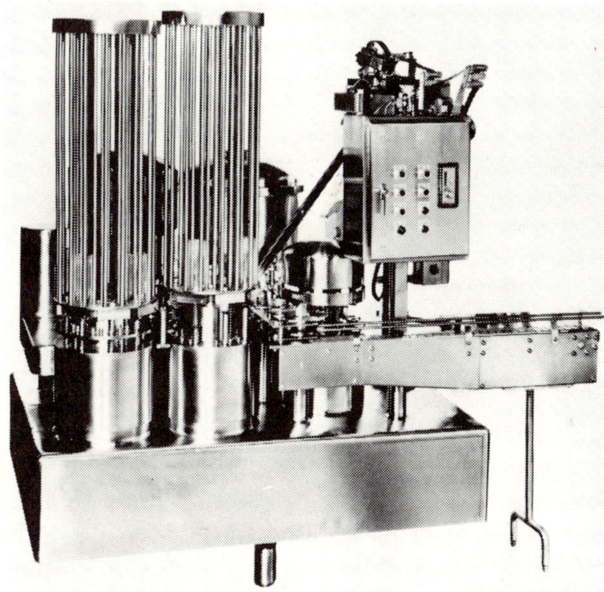
(C) This unit spin welds, fills, and caps preform thermoplastic container halves. (Brown Machine Co.)

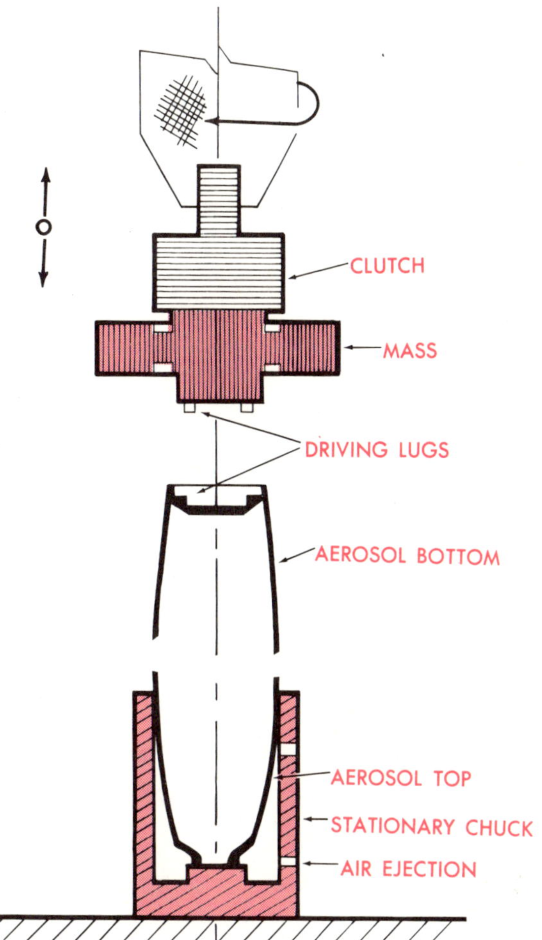

(D) Spin welding aerosol bottle halves. (E. I. du Pont de Nemours & Co., Inc.)

Fig. 22-4. Principle of spin welding.

Hot Gas Welding

Hot gas welding consists of directing a heated gas (usually nitrogen) at temperatures of 400°F to 800°F onto the joints to be melted together. The process is similar to open flame welding of metals. Filler rods or materials similar to the parent plastics are used to build up the welded area. Welds may exceed 85 percent of the tensile strength of the parent material strength (see Table 22-2). As in any welding technique the joint area must be properly cleaned and prepared. Butt joints should be beveled to 60 degrees (see Fig. 22-5D) and the area filled with a filler rod of the parent material. The temperature of the "flameless" hot gas torch is controlled by regulating the gas flow or the heating source. Electric heating elements are preferred with a nitrogen or air pressure of 2 to 4 pounds.

Table 22-2. Plastics Weldability of Selected Materials.

Material	Percent Weld Strength	Spot Weld	Staking and Inserting	Swaging	Welding
Polystyrene unfilled	95-100+	E	E	F	E
ABS	95-100+	E	E	G	E
Polycarbonate	95-100+	E	E	G-F	E
Polyamide	90-100	E	E	F-P	G
Polysulfone	95-100+	E	E	F	G
Acetal	65-70	G	E	P	G
Polyimide	80-90	F	G	P	G
Acrylics	95-100+	G	E	P	E
Polyphenylene	95-100+	E	G	F-P	G
Phenoxy	90-100	G	E	G	G
Polypropylene	90-100	E	E	G	G-P
Polyethylene	90-100	E	E	G	G-P
Butyrates	90-100	G	G-F	G	P
Cellulosics	90-100	G	G-F	G	P
Vinyls	40-100	G	G-F	G	F-P

CODE: E = excellent, G = good, F = fair, P = poor.

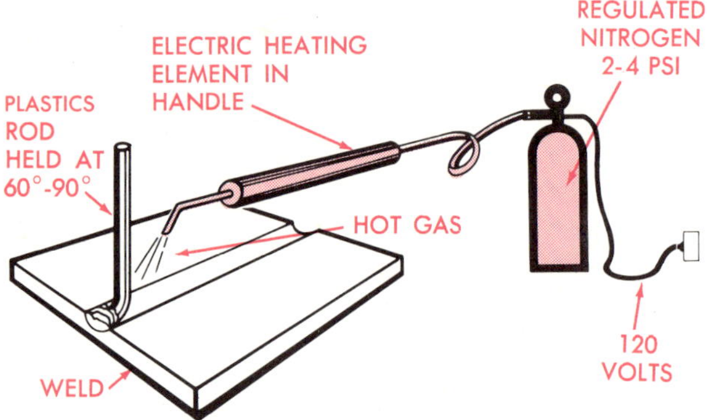

(A) Principle of hot gas welding.

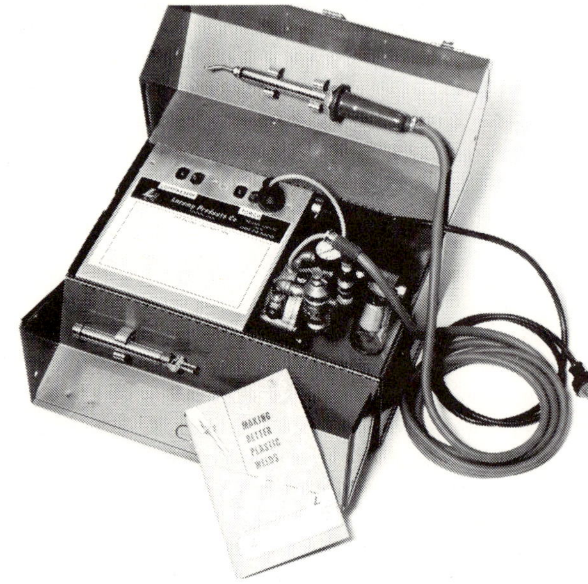

(B) Typical plastics welding unit. (Laramy Products Co.)

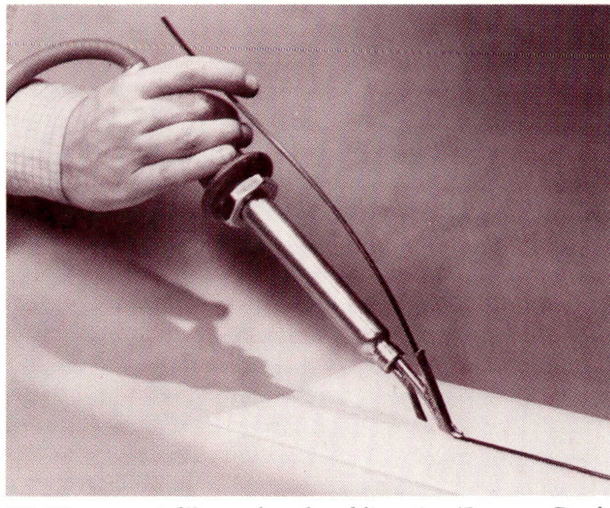

(C) Closeup of filler rod and welding tip. (Laramy Products Co.)

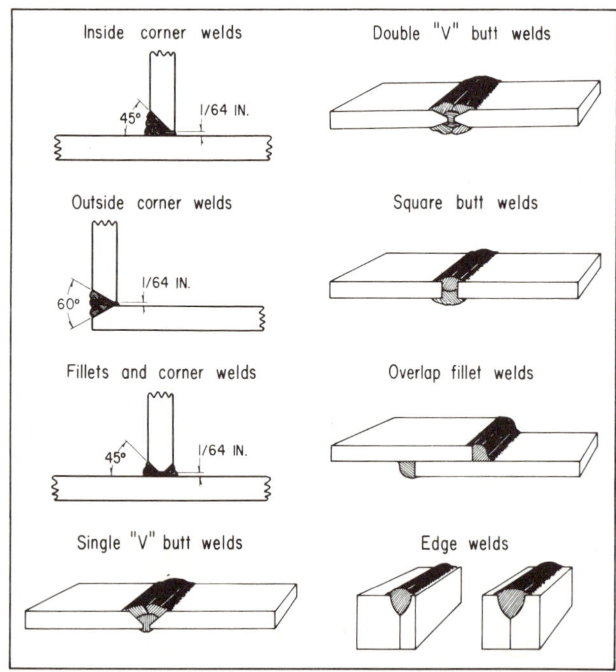

(D) Various types of joints produced by hot gas welding of thermoplastics. (Modern Plastics Magazine)

Fig. 22-5. Hot gas welding.

Heat Joining

Heat joining is a method where like materials are heated by various means and the joints brought together while in the molten stage. Fusion welding is another name often used to describe the technique of "fusing" plastics together using heated tools. Electric strip heaters, hot plates, soldering irons, or other specially designed heated tools are used to melt the plastics surfaces. The heated areas are then quickly brought together under pressure until cooled (see Fig. 22-6A). Various pipes and pipe fittings may be joined in this manner. The surfaces of the heating tools may be Teflon coated but the use of lubricants or other materials to prevent sticking of the plastics to the hot metal is not recommended. These materials contaminate and weaken the weld.

One of the most common applications of heat joining is employed in the fusing of films (see Fig. 22-6B). Not all thermoplastics may be heat sealed. Many, however, are given a coating layer of plastics that may be heat sealed. Various electrically heated rollers, jaws, plates, or metal bands are used to melt and fuse the film layers together.

Dielectric Joining

Dielectric heat sealing is used to join plastics films, fabrics, and foams. Only those plastics that have a high dielectric loss characteristic (dissipation factor) may be joined by this method. Cellulose acetate, ABS, polyvinyl chloride epoxy, polyether, polyester, polyamide, polyurethane, and others have sufficiently high dissipation factors to allow dielectric sealing. Polyethylene, polystyrene, and fluoroplastics have very low dissipation factors and cannot be heat sealed electronically. The actual fusion is caused by the presence of high-frequency (radio-frequency) voltage. In the areas where the high frequency is directed, molecules are rapidly trying to realign themselves with the oscillations (Fig. 22-7). This molecular movement causes frictional heat and the areas become molten. The Federal Communications Commission regulates the high-frequency energy because the generated signals are similar to those produced by TV and FM transmitters. The transmitter or generators are available in various kilowatt sizes and operate at frequencies between 20 and 40 megahertz (millions of cycles per second).

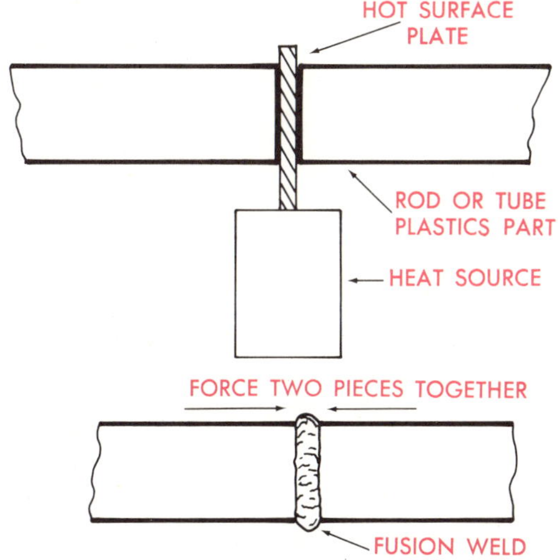

(A) Heat joining of plastics parts by fusion welding.

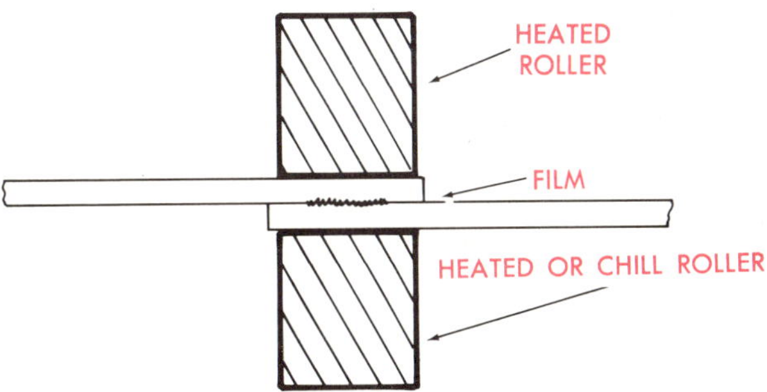

(B) Heat joining of plastics film by heat sealing method using rollers.

(C) Heat sealing of plastics films on a hot platen.

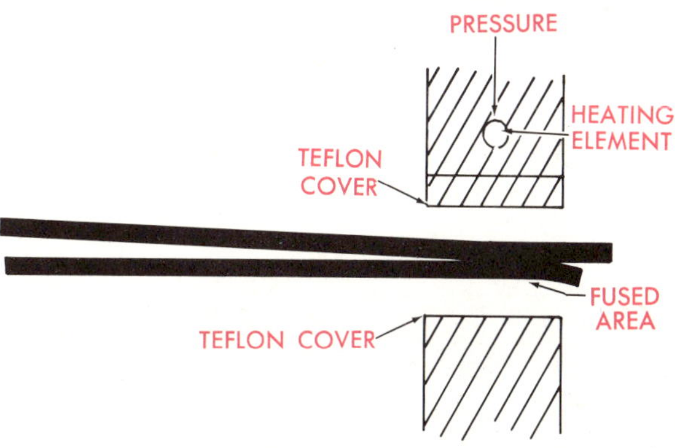

(D) Heat joining of plastics film by heat sealing method using press.

Fig. 22-6. Heat joining methods.

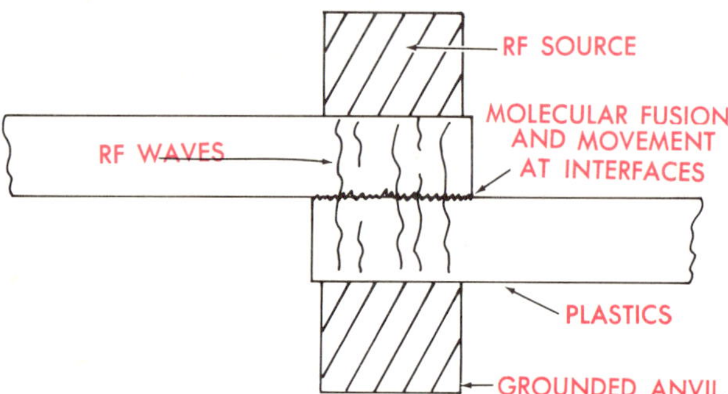

Fig. 22-7. Dielectric heat sealing with radio-frequency voltage.

Ultrasonic Bonding

Ultrasonic bonding is used to vibrate plastics together mechanically. High-frequency mechanical vibrations at 20 kilohertz to 40 kilohertz are directed to the plastics part by a tool called a horn (Fig. 22-8A). The high frequency causes the plastics molecules to vibrate, thus creating sufficient frictional heat to melt the thermoplastic. An electronic transducer converts 60 hertz energy to the 20-40 kilohertz frequencies.

Ultrasonic techniques are used to weld and assemble metal and plastics parts, spot weld, activate adhesives to a molten state, and sew or stitch films and fabrics together, eliminating needles and threads. Simple joints may be ultrasonic welded in 0.2 to 5.0 seconds, depending on the material and the areas to be joined.

Staking is a term used to describe the ultrasonic forming of a locking head on plastics studs. It is similar to forming a head on a metal rivet. (See Fig 22-8B.) Plastics parts with studs may be assembled by this technique.

Metallic inserts are placed in plastics by ultrasonic means (Fig. 22-8C). The horn is used to hold the insert and direct the high-frequency vibrations into an undersized hole. As the plastics melts, the pressure of the horn permits the insert to slip into the hole. Upon cooling, the plastics reforms itself around the insert.

A permanent molecular bond is made when small areas are caused to become molten by ultrasonic spot welding. This process is similar to metal spot welding.

Many adhesives may be made molten and caused to become cured or molten by ultrasonic vibrations (Fig. 22-8D).

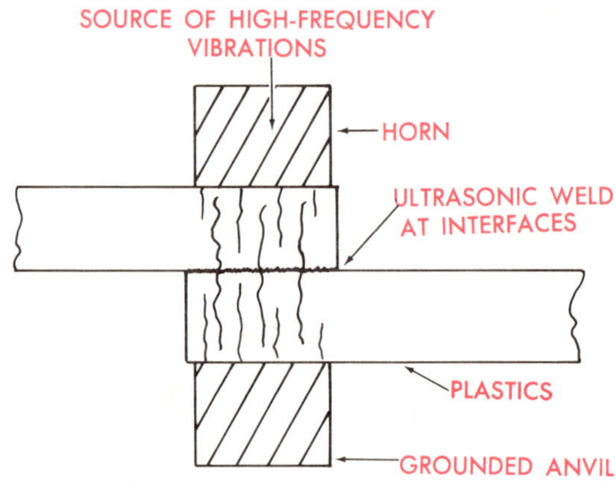

(A) Welding of plastics parts.

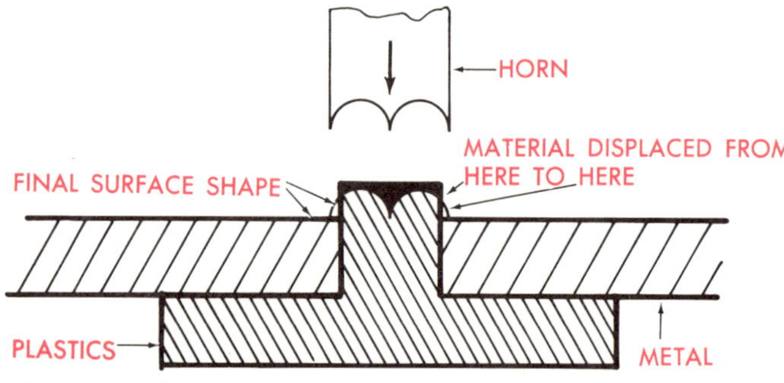

(B) Assembly of metal and plastics. (Branson Sonic Co.)

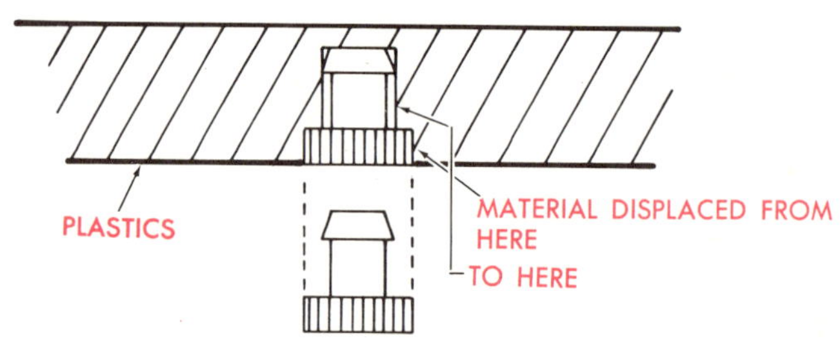

(C) Staking a plastics stud. (Branson Sonic Co.)

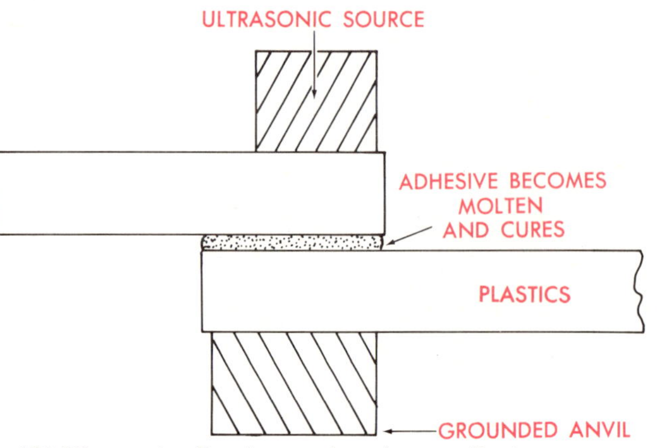

(D) Ultrasonic vibrations melt and cure adhesives.

Fig. 22-8. Ultrasonic bonding.

Induction welding is an electromagnetic technique for bonding thermoplastics. The heat is produced from an induction generator with power from 1 to 5 kilowatts and frequencies from 4 to 47 megahertz. The thermoplastic cannot be heated directly by induction. Consequently metal powders or inserts must be placed in the plastics interface joints. As the metallic particles or inserts are excited by the high-frequency induction source they become hot, melting the surrounding plastics. The metallic inserts or powders must remain in the final weld. Only a few seconds and little pressure is required for this rapid method of assembly.

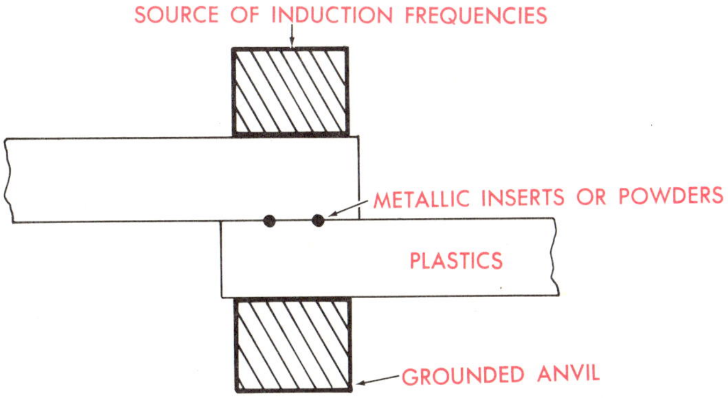

SOURCE OF INDUCTION FREQUENCIES

METALLIC INSERTS OR POWDERS

PLASTICS

GROUNDED ANVIL

Fig. 22-9. Induction welding by electromagnetic technique.

Mechanical Linkage

There is a wide variety of mechanical fasteners for use with plastics. Self-threading screws are used if the fastener is not to be removed very often. When frequent disassembly is anticipated, threaded metal inserts are placed in the plastics. They may be molded in place or ultrasonically installed. Drive screws and metallic or plastics rivets provide permanent assembly. Standard nuts, bolts, and machine screws are available in metal or plastics and are used in conventional assembly methods. Speed nuts are used with sheet metal screws for rapid, reinforced assembly. Hinges, knobs, catches, clips, dowels, and other devices are also used for assembly of plastics. (See Fig. 22-10.)

Friction Fits

"Press fits," "shrink fits," and "snap fits" are terms used to indicate a pressure-tight joint of like or dissimilar materials without any mechanical linkage.

Press fitting may be used to insert plastics or metallic shafts or parts into other plastics components. The parts may be joined while the plastics parts are still warm. When the shaft is press fitted into the bearing or sleeve, the outside diameter as well as the inside diameter may be expanded (see Fig. 22-11A).

"Shrink fitting" refers to placing inserts in the plastics immediately after molding and allowing the plastics to cool (Figs. 22-11B and 22-11C). It also refers to the process of placing various plastics parts over substrates and then heating until the plastics shrinks to its original shape by a property called "memory."

Snap fitting is a means of assembly in which parts are snap-fit into place. The plastics is simply forced over a lip or into an undercut retaining ring. The simple locks or catches on plastics boxes or the light covers on many parts including automotive dome lenses, flashbulb covers, and instrument panels are familiar examples.

Glossary of New Terms

Adhesion—The physical ability of a substance which, applied as an intermediate, is capable of holding materials together by surface attachment.

Cohesion—The action involved when materials are being joined and an actual intermingling of molecules holds the materials together.

Heat joining—A method of joining plastics by simultaneous application of heat and pressure to areas in contact. Heat may be supplied conductively or dielectrically.

Hot gas welding—A technique of joining thermoplastics materials whereby the materials are softened by a jet of hot air from a welding torch and joined together at the softened points. Generally a thin rod of the same material is used to fill and consolidate the gap.

Induction welding—High-frequency electromagnetic fields are used to excite the molecules of the metallic insert placed in the plastics or in the interfaces, thus fusing the plastics. The metallic insert remains in the joint.

Mechanical linkage—Mechanical means of joining plastics, including machine screws, self-tapping

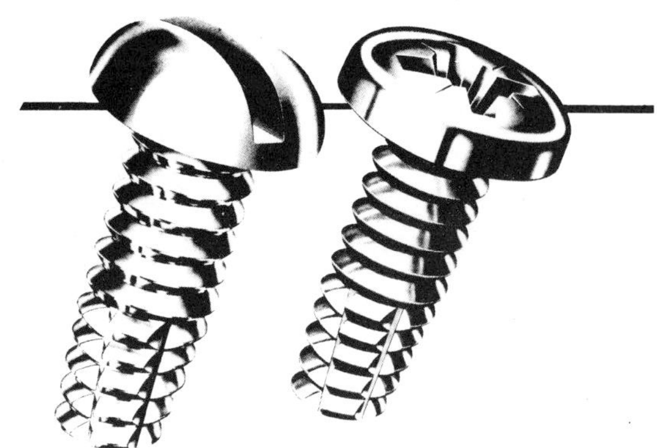

(A) "Thread cutting" thread screws intended primarily for plastics. (Parker-Kalon Fastener Corp.)

Pull Mandrel—Nonbreak

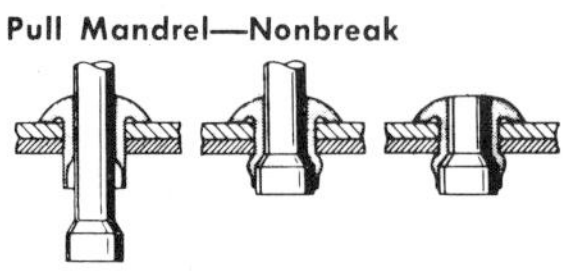

Pull Mandrel—Break Mandrel Closed-End

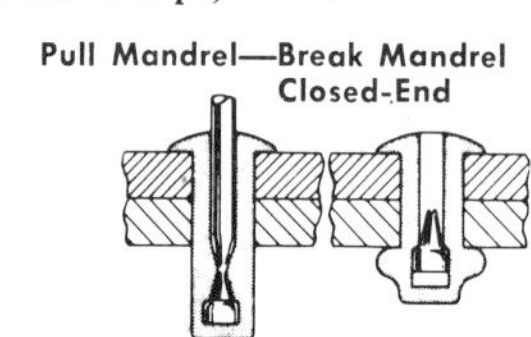

Pull Mandrel—Break Mandrel Open-End

Drive Pin

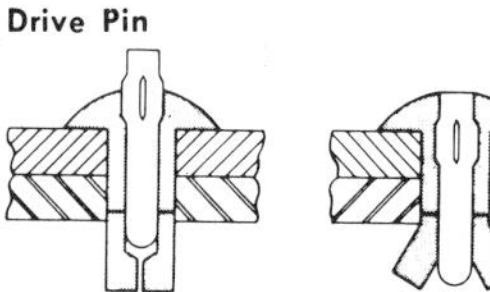

Pull Mandrel—Pull-Through

Chemically Expanded—Open-End

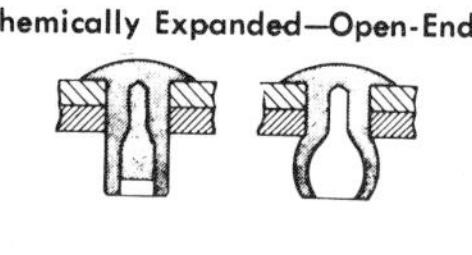

Threaded

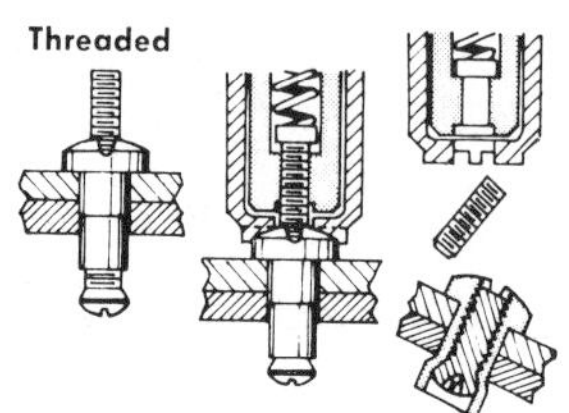

(B) Blind rivets (Fastener Div., USM Corp.)

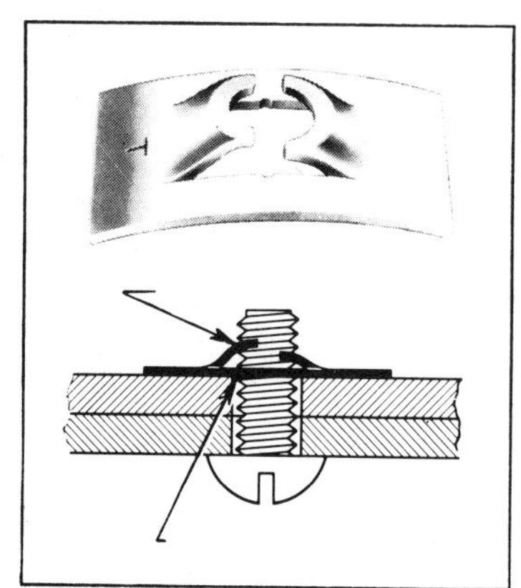

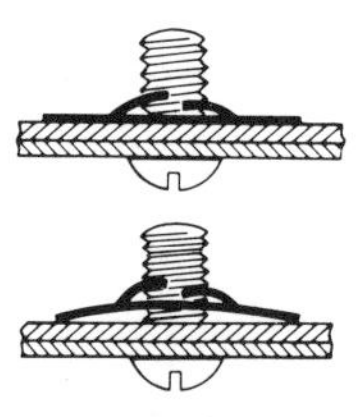

Single-thread locknut, which is speedily applied, locks by grip of arched prongs when bolt or screw is tightened.

(C) Single-thread locknut. (Tinnerman Products, Inc.)

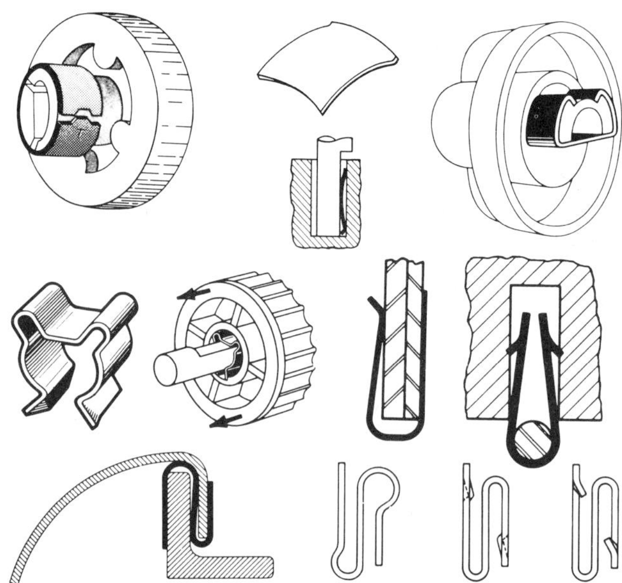

(D) Mechanical fasteners and devices for assembly of plastics parts.

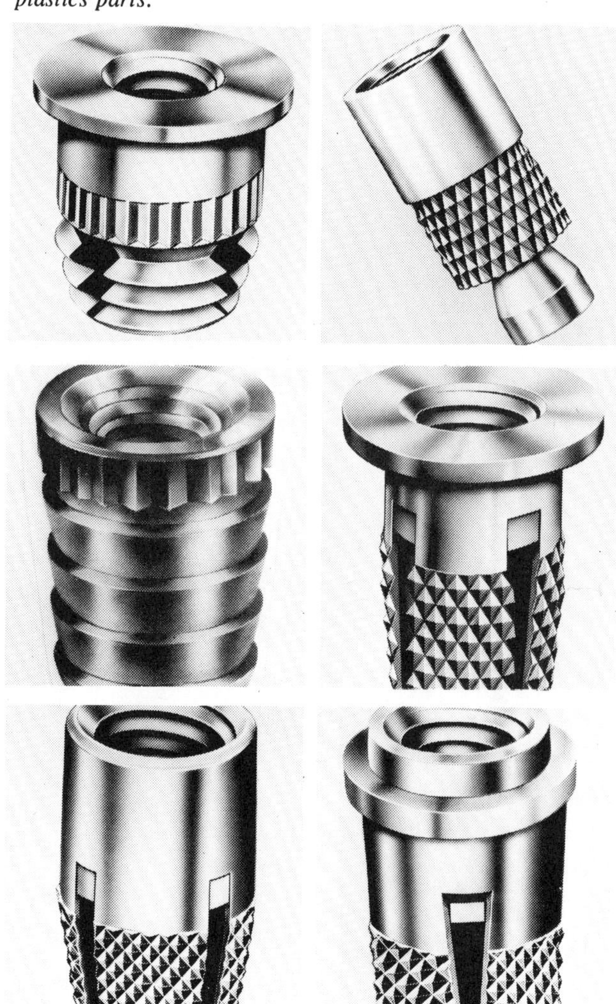

(E) Inserts designed for use in plastics are installed quickly and easily after molding. (Heli-Coil Products)

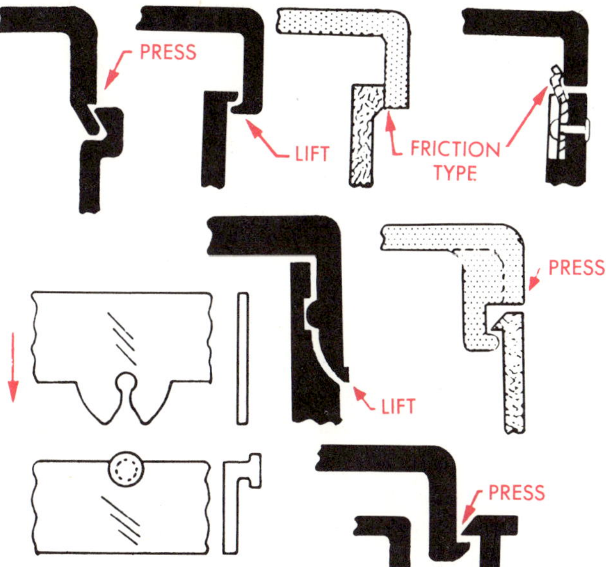

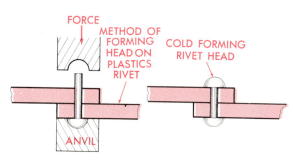

(F) *Cold forming of plastics rivets. Heads may be made by mechanical or explosive means.*

(G) *Possible types of clasps for containers.*

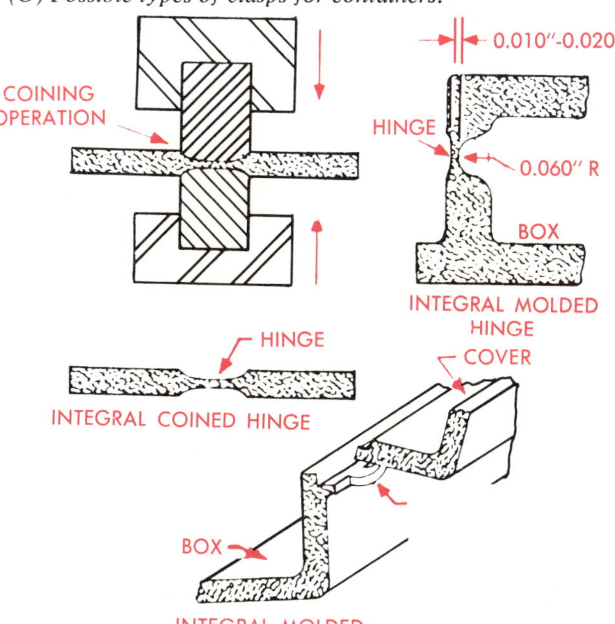

(H) *Examples of integral molded and coined hinges.*

Fig. 22-10. Screws, rivets, and fasteners used with plastics.

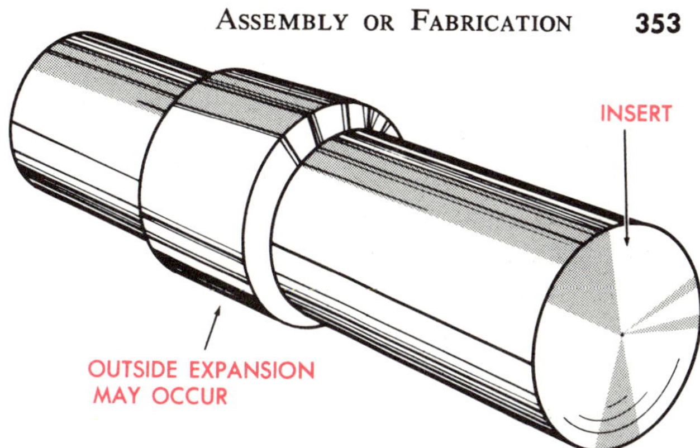

(A) *Press fit showing expansion of outside diameter of plastics part.*

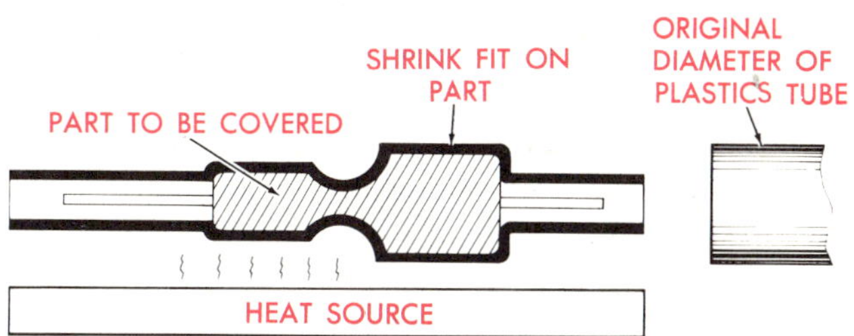

(B) *Shrink fit of plastics tube over electronic substrate (part). Note original size of tube before heating.*

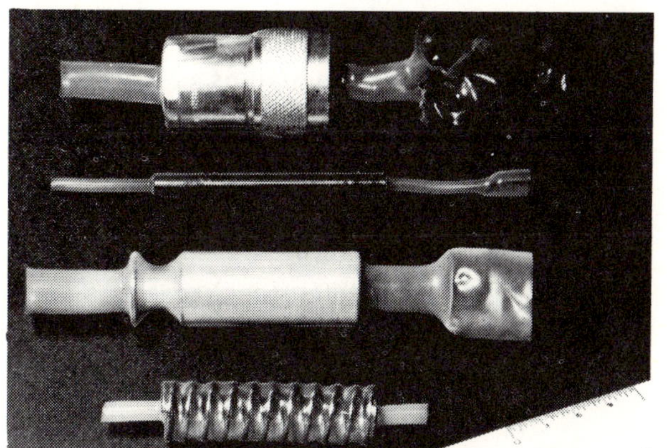

(C) *Shrink fitting plastics over parts for protection or antistick properties. (Chemplast, Inc.)*

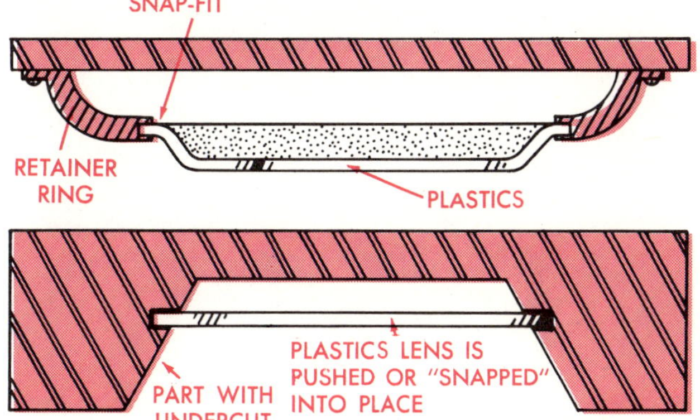

(D) *Two examples of snap fitting.*

Fig. 22-11. Press, shrink, and snap fitting.

screws, drive screws, rivets, spring clips, clips, dowels, catches, etc.

Shrink fit—Joining plastics or other parts by inserting the plastics onto (or the insert into) a part while the plastics part is hot. Shrink fitting takes advantage of the fact that plastics expand when heated and shrink when cooled. The plastics is normally heated and the insert placed in an undersized hole. On cooling, the plastics shrinks around the insert.

Spin welding—A process of fusing two objects together by forcing them together while one of the pair is spinning, until frictional heat melts the interface. Spinning is then stopped and pressure held until the parts are frozen together.

Ultrasonic bonding—A method of joining accomplished through the application of vibratory mechanical pressure at ultrasonic frequencies. Electrical energy is converted to ultrasonic vibrations through the use of either a magnetostrictive or piezoelectric transducer.

Review Questions

22-1. What are the four basic ways plastics are joined together?

22-2. What is the difference between adhesive and cohesive binding?

22-3. What is the difference between solvent cement and dope cement?

22-4. How are joints solvent cemented?

22-5. What is spin welding? Of the four basic joining methods, which is spin welding?

22-6. Describe hot gas welding of thermoplastic materials.

22-7. Is nitrocellulose a heat sealable product?

22-8. "Heat sealing" usually refers to the welding of what thickness of plastics?

22-9. By which method are plastics films fused together?

22-10. Name three methods of heating sheets and films in heat sealing.

22-11. What type of plastics may be sealed by dielectric joining?

22-12. What is the basic physical principle of dielectric joining of plastics?

22-13. Briefly describe ultrasonic bonding.

22-14. What are some of the uses of dielectric bonding?

22-15. What is staking? How is it done?

22-16. What is used to create the heat in induction bonding?

22-17. What is meant by the term "mechanical linkage" of plastics parts? Name some mechanical linkages.

22-18. What is meant by the term "friction fit?"

22-19. What is press fitting?

22-20. Briefly describe shrink fitting.

22-21. What is snap fitting? Where is snap fitting used?

23

Tooling

Tools for long run production work are usually made of metal. However, there are numerous types of materials used in tooling, including (1) gypsum plasters, (2) plastics, (3) wood, and (4) metals.

Gypsum Plasters

The United States Gypsum Company has developed a number of high-strength plasters with sufficient strength to produce prototype models, die models, transfer (takeoff) tools, patterns, and die molds for forming plastics. The tradenames Ultracal, Hydrocal, and Hydro-stone refer to plasters used for tooling. Many patterns are produced by this method and numerous vacuum-formed pattern molds are made of low-cost plasters. Hydro-stone has an average compressive strength of nearly 11,000 pounds per square inch. (See Fig. 23-1.)

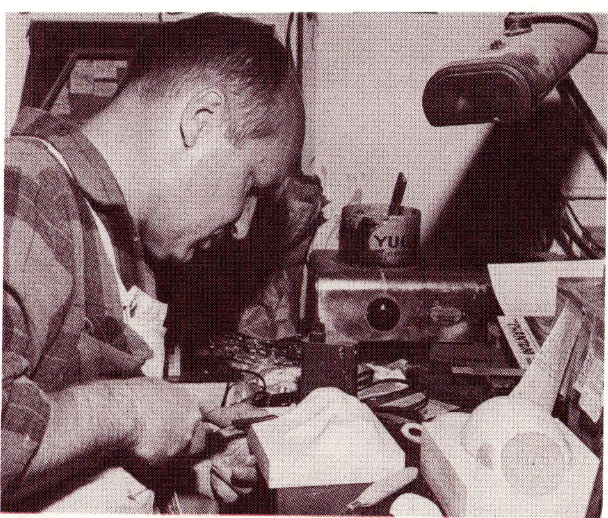

(A) This man is working on a gypsum plaster pattern for forming plastics. (Revell, Inc.)

(B) An epoxy resin master model of the pilot's enclosure for a military jet plane after removal from the plaster mold. (U.S. Gypsum Co.)

Fig. 23-1. Gypsum plaster pattern and plaster mold.

Plastics

Plastics models, dies, and patterns are replacing many wooden and plaster toolings. (See Fig. 23-2.) They are used mainly for making dies, jigs, and foundry patterns (Fig. 23-3). Laminated, reinforced, and filled plastics are normally used for tooling. The use of plastics in die fabrication is one of the most rapidly advancing areas of tooling. Metal filled and glass reinforced phenolics, ureas, melamines, polyesters, epoxies, silicones, and polyurethanes are used in the plastics and metal tooling industry. These materials are strong, lightweight, and easy to machine. Many may be foam

(A) Basic shape and form.

(B) Filling with syntactic foam.

(C) Machining to final shape and tolerance.

Fig. 23-2. Lightweight, strong tooling may be fabricated of honeycomb structures with bonding and fill of syntactic forms of extrudable epoxy. (Ren Plastics, Inc.)

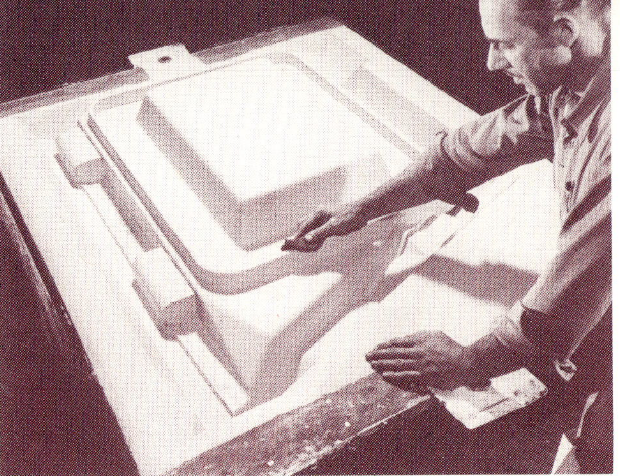

Fig. 23-3. A large foundry pattern made by laminating glass cloth and epoxy resins. The basic patterns are being used in place of metal patterns because of economy and speed of fabrication. (U.S. Gypsum Co.)

filled. Plastics tools have been used as bending dies, stretch forming dies, and drop hammer dies. Acetals, polycarbonates, high-density polyethylene, fluoroplastics, and polyamides are also used as toolings. They are used primarily as matched stamping dies, jigs, and fixtures. The acceptability of plastics tools is verified by the many applications used in the aerospace, aircraft, and automotive industries.

Plastics tooling has several advantages over tools of metal or wood. Plastics tools may be cast in inexpensive molds with less labor, duplicate easily, and allow frequent changes in design. Plastics tooling is also light in weight and corrosive resistant. Hot-melt compounds are replacing many of the wood and steel dies, hammers, mockups, prototypes, and numerous other jig fixtures used by industry. See Fig. 23-4.

Fig. 23-4. Sheet metal parts being formed with plastics tool made of polyamide resin. (General Mills, Inc.)

For prototype or low-run work low-melting metals may be used. Zinc, lead, tin, bismuth, cadmium, and aluminum are used in thermoforming dies, casting patterns, and duplicate models. Aluminum is a popular metal for many molding processes because it is lightweight, easy to machine, and a good thermal conductor.

Berylium copper is used in some injection molds and blow molding processes (see Fig. 23-5). If a long-run production is required, various steels must be used. In applications where high compressive strength and wear resistance at elevated temperatures are required an AISI type H21 steel may be used.

Type H21 Analysis:		
	Carbon	0.35%
	Manganese	0.25%
	Silicon	0.50%
	Chromium	3.25%
	Tungsten	9.00%
	Vanadium	0.40%

AISI type W1 is a high-quality, straight-carbon, water-hardening tool steel used in various tooling:

Type W1 Analysis:		
	Carbon	1.05%
	Manganese	0.20%
	Silicon	0.20%
	Alloys	None

Toolmakers have long expressed their desire for a nondeforming die steel that would combine the deep-hardening characteristics of air-hardening steels with the simplicity of low-temperature heat treatment possible in many oil-hardened steels. AISI type A6 is used to meet this need:

Type A6 Analysis:		
	Carbon	0.70%
	Manganese	2.25%
	Silicon	0.30%
	Chromium	1.00%
	Molybdenum	1.35%
	Plus alloy sulphides	

There are a number of processes employed in the manufacture of tools and dies from steels, including milling, turning, drilling, boring, grinding, hobbing, casting, planing, etching, electroforming, electrical discharge machining, plating, welding, and heat treating.

Milling, turning, drilling, boring, and grinding are processes of mechanically removing pieces of metal by the use of a cutting tool. Shapers, planers, lathes, drilling machines, grinding machines, milling machines, and various pantograph duplicating machines are only representative of the many types of equipment used for cutting metal in the making of molds and dies. In Fig. 23-6 a machinist is carefully removing steel with the aid of a vertical milling machine and cutting tool.

(A) Steps necessary to produce simulated wood furniture components. From left to right: hand-carved wooden model, silicone impression, gypsum model, and the beryllium-copper injection molds which have the fine grain detail as that of the original wooden model.

(B) Application of simulated wood furniture components. Note chair backs.

Fig. 23-5. Simulated wood furniture made with beryllium-copper injection molds. (Shell Oil Co.)

In Fig. 23-6B a specially modified vertical milling machine is used to cut and duplicate in steel from the master pattern. The tracing head on the right is a control over the movements of the work table and cutter spindle. This pattern is made of metal and the duplicator setup is machining a cavity for a blow mold. The pattern to workpiece ratio is 1:1.

Pantograph machines are similar and function like duplicating machines, except that they operate on a ratio as high as 20:1. In Fig. 23-6C a plaster master is shown much larger than the workpiece of steel. The large ratio reduction from the plaster pattern to the steel will be machined with very delicate detail by the coordinated movements of the table and cutting tool.

(A) Machinist carefully removes steel in making mold of steel. (Revell, Inc.)

(B) Duplicating steel mold from pattern at right. (Cincinnati Milacron)

(C) Pantograph (duplicator) machine with a plastics master model which will be used to machine a metal piece. (U.S. Gypsum Co.)

(D) Epoxy resin mold is the master pattern from which the machine cuts into steel each detail at one-fourth scale. (Revell, Inc.)

Fig. 23-6. Making tools and dies from steel and plastics.

Hobbing, etching, electroforming, and electrical discharge machining are generally classed as a *metal-displacement* process of making molds where no cutting tools are involved.

Cold hobbing (no heat is used) consists of pushing a piece of very hard steel into a blank of unhardened steel (Fig. 23-7). The process is performed at room temperatures with pressures, depending on the hobbing metals and blanking material, from 100 to 200 tons psi. This may require press pressures as high as 3000 tons in capacity. Standardization of the die blank may be beneficial so that the die cavities are interchangeable from mold to mold. Hobs are often made from oil-hardened tool steels with a high percentage of chromium. Multicavity molds may be numbered to permit instant location of any molding troubles. Although it may be economical to hob single die cavities the process is usually used for producing large numbers of impressions for multicavity molds. A slight draft must be provided in order to remove the hob from the forming blank. It is also imperative that the hob is clean. Even a pencil mark on the hob may be transferred to the cavity during the hobbing operation.

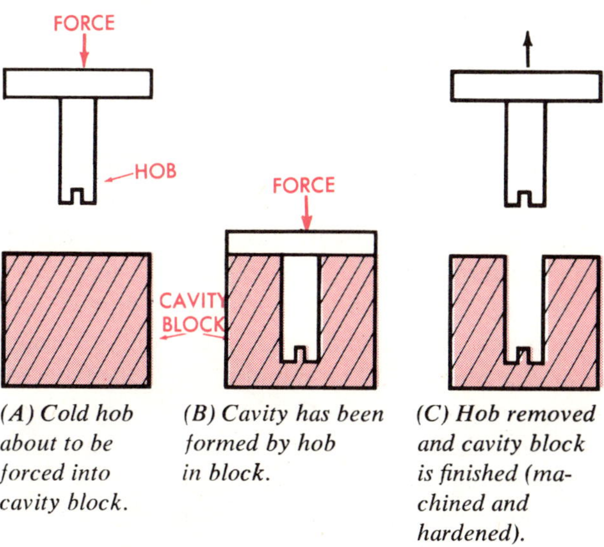

(A) Cold hob about to be forced into cavity block.

(B) Cavity has been formed by hob in block.

(C) Hob removed and cavity block is finished (machined and hardened).

Fig. 23-7. Diagram of metal displacement or hobbing process.

After the hobbing is completed the blank is machined and hardened before placement in mold frames. In Fig. 23-8A a finished hob (right) has formed the cavity (center) into the blank of steel. The hobbed cavity will then be machined and hardened. Shown on the left is the force or male portion of the compression mold. A finished, compression molded part is shown in Fig. 23-8B after it was molded in the finished mold cavity.

(A) Hob (at right) was pressed into steel block (center).

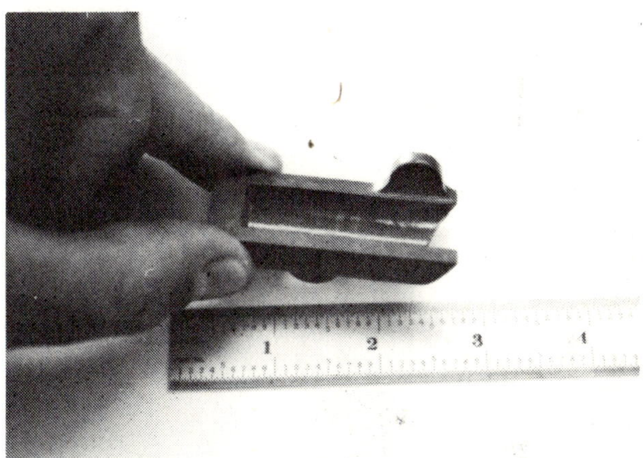

(B) Part produced on hobbed compression mold shown in (A).

Fig. 23-8. Example of hobbing and finishing process.

Electrical erosion or *electrical discharge machining* (EDM) is a comparatively slow method of removing metal compared with mechanical methods. Steel is removed at about 0.016 cubic inch per minute. The workpiece may be hardened before the cavity is formed, thus eliminating the dangers of heat treatment after machining or forming. The actual machining process consists of making a master pattern of copper, zinc, or graphite and placing the pattern approximately 0.001 inch from the workpiece. Both the workpiece and the master are submerged in a very poor dielectric fluid such as kerosene or light oil. As the current (on the order of 10^6 amperes per square inch) is forced across the gap between the master workpiece, each discharge removes minute amounts of material from both. The loss of material from the tool master must be compensated for in obtaining accurate cavities. For inexpensive tool masters made of carbon or zinc the ratio of material removed from the work to that removed from the tool may be more than 20:1. Accuracy may be within $\pm$ 0.001 of an inch with a finish cut of less than 30 microinches. The principle of electrical discharge machining is shown in Fig. 23-9.

In chemical erosion or etching, an acid or alkaline solution is used to create a depression or cavity. The process usually involves the use of photosensitive resistants or a chemical resistant maskant such as wax or many plastics based paints and films. Photosensitive resists are commonly used in the printing industry. In all processes the maskant is removed from those areas where the metal is to be chemically removed. Shallow cavities or designs are commonly reproduced with textures duplicating fabrics and leather.

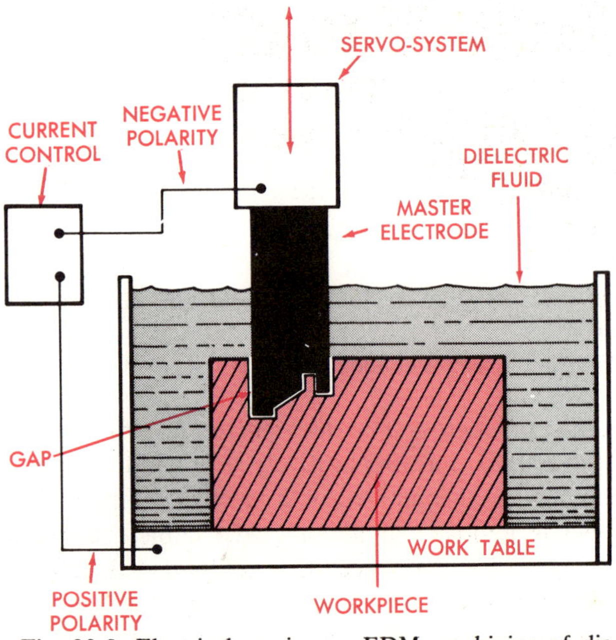

Fig. 23-9. Electrical erosion or EDM machining of die. Note that the gap between workpiece and master tool is very uniform.

Allowance must be made to compensate for effects of "etching radius" or "etch factor." As the etchant acts on the workpiece there is a tendency to undercut the maskant pattern, leaving a small undercut. In deep cuts this may be a serious problem. Fig. 23-10 depicts the effects of the "etch factor" in chemical erosion.

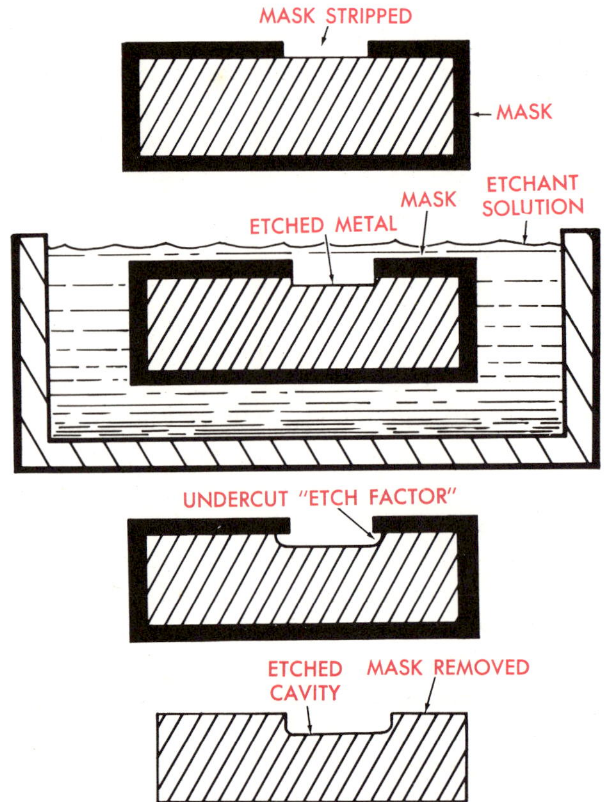

Fig. 23-10. Chemical erosion method for producing a die cavity.

Casting and electroforming are sometimes called *metal-deposition* processes involving the deposit of a metallic (sometimes ceramic and plastics) coating on a master form.

In Fig. 23-11A a steel master is dipped into molten lead compounds until a coating is formed over the mandrel. The mandrel may then be removed and used again. Casting resins may be cast into the shell and removed when polymerized. Low-melting materials such as wax may be used and simply melted from the cast shape.

The hot casting of metals by lost wax, sand, or permanent metal molds may be used to produce precision molds. Molten metal may be poured over a hardened steel master to form a cavity as shown in Fig. 23-11B. This process is sometimes called *hot hobbing* because the molten metal is cast over a hob and pressure is applied during the cooling.

Electroforming is an electroplating process where an accurate mandrel is made electrically conductive (plastics, wax, and glass) and a metal deposit is plated on it. Plastics, glass, wax and various metals are used as masters to electrically

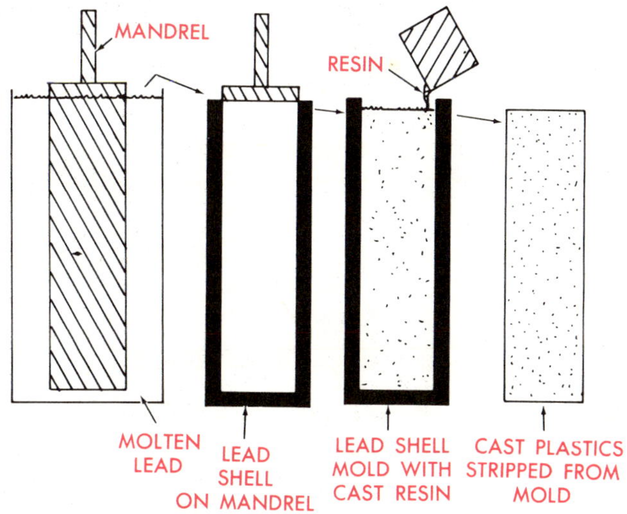

(A) Steps in casting of plastics parts using cast molds.

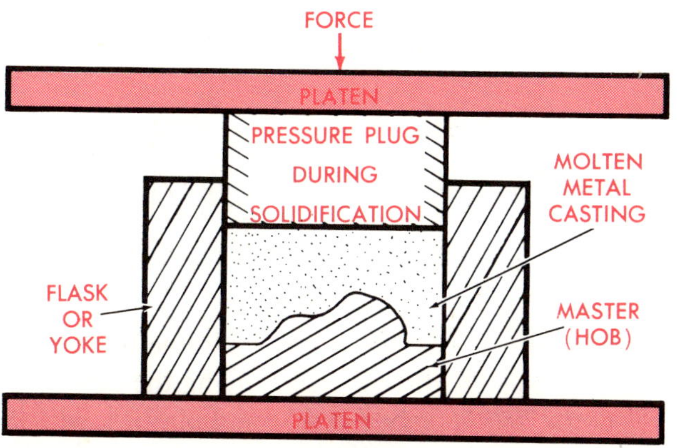

(B) Molten metal is cast on master, and pressure plug forces casting into a dense sound casting.

Fig. 23-11. Casting of plastics and metal.

deposit the metallic ions from a chemical solution (Fig. 23-12). The electrical deposit of metallic ions from a chemical solution may be made on plastics, glass, wax, and various metal masters, provided they are made electrically conductive. These *thin* shelled molds may have severe undercuts and a highly polished finish. The cavity may then be strengthened by copper plating the back of the shell. Further strength may be provided by placing the die cavity into filled epoxy. The cavities may then be used for thermoforming, blow molding, or injection molding.

There are several other metal depositing techniques that may be used for mold making, including flame spraying of metals and vacuum metallizing. (See Chap. 20.)

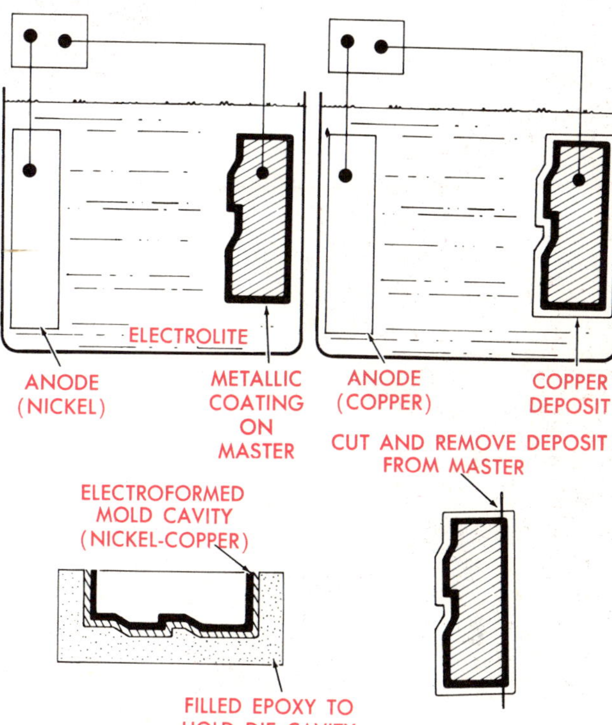

Fig. 23-12. Electroformed mold cavity. Copper coating and filled epoxy strengthen the nickel (shell) cavity.

Electroplating, welding, and heat treating are additional operations used in the production of molds. Many of the finishing operations are accomplished by hand. In Fig. 23-13 a toolmaker is working on a steel mold. It may then be electroplated (given a metallic coating) to protect the die cavity from corrosion and provide the desired finish on the plastics product.

Mold frames are equally important to the toolmaker. Frames are the means of holding the cavities in place and are made with sufficient thickness to provide heating and cooling for the cavities. The mold frames are made from mild steel and are never hardened. Mold frames have been made into standard sizes and may be purchased to accommodate most custom or proprietary molds.

Alignment pins are used to ensure proper alignment of cavities when the molding frame is brought together during the molding operation. If the final assembly is not properly aligned and the parting lines not in registry, or the molded part continues to hold to the die cavity, last-minute stoning, hand grinding, and polishing may be required for a proper finish.

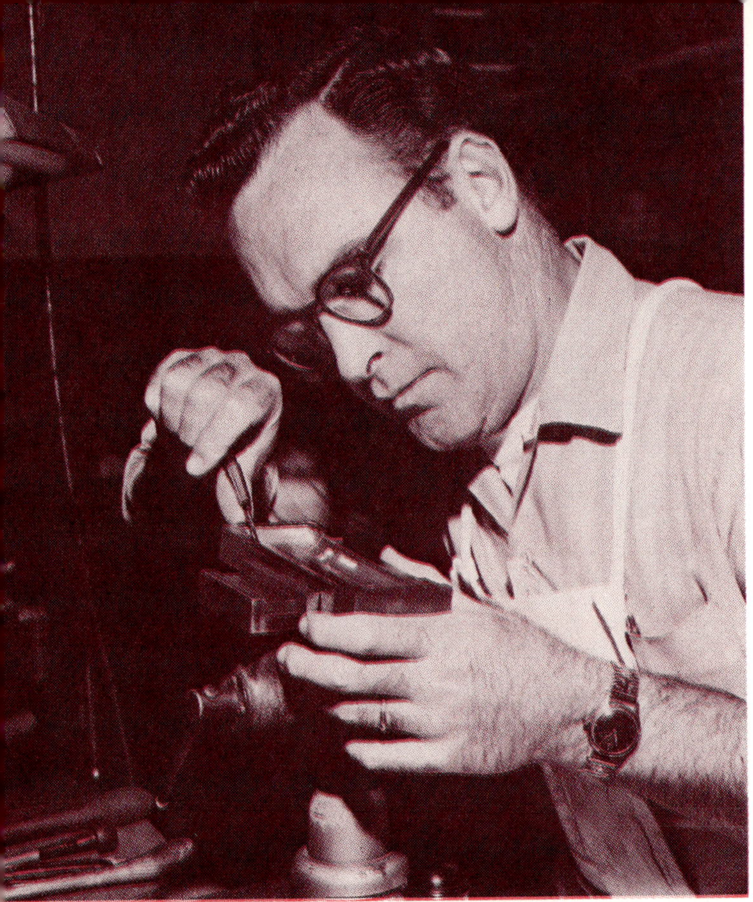

Fig. 23-13. A toolmaker working on a steel mold. (Revell, Inc.)

Sometimes it is desirable to have the molded piece cling slightly to one half of the mold, depending on the knockout mechanism. Excessive sticking may be caused by dents or undercuts in the mold or by a dirty cavity surface. In cleaning the dies and cavities a wax, lubricant, or silicone spray is often used. For stubborn spots a wooden scraper or a brass brush may be used. Never use steel scrapers in cleaning cavities. Steel instruments may scratch or damage the polished finish in the cavity. (See Fig. 23-14.)

(A) Tool and die maker puts finishing touches on injection mold of steel. Note intricate parts. (Revell, Inc.)

(B) Tool and pattern maker is carefully inspecting steel mold. (Revell, Inc.)

Fig. 23-14. Care must be taken with molds.

Glossary of New Terms

Air-hardening—Refers to steel which is cooled in air.

Alignment pins—Devices that maintain proper alignment of cavity as mold closes.

Chemical erosion—Method of removing metal by chemical means.

Deep-hardening—Refers to the depth of hardening in a piece of steel.

EDM—Electrical discharge machining is a process where a high-frequency intermittent electrical spark is used to erode the workpiece.

Electroforming—A process of depositing a thin metallic coating on a form by electroplating on the reverse pattern. Molten steel may be then sprayed on the back of the mold to increase strength.

Electroplating—A method of applying metallic coatings to a substrate.

Gypsum—Crystalline hydrated sulphate of calcium ($CaSO_4 \cdot 2H_2O$), used for making plaster of Paris and Portland cement.

Hobbing—Forming multiple mold cavities by forcing a hob into soft steel or beryllium-copper cavity blanks.

Jig—An appliance for accurately guiding and locating tools during the operations involved in producing interchangeable parts.

Oil-hardened steel—Steel which is cooled by an oil bath.

Plastics tooling—Tools, dies, jigs, fixtures, etc. for the metalforming trades constructed of plastics —generally laminates or casting materials.

Review Questions

23-1. Why are gypsum plasters used for experimental work with plastics?

23-2. Why are plastics tools and dies being used to replace other kinds?

23-3. Why is aluminum used for many molding operations? Beryllium copper? Steel?

23-4. What is a pantograph machine?

23-5. What is a metal-displacement process?

23-6. What is cold hobbing? Hot hobbing?

23-7. Briefly describe electrical erosion machining.

23-8. What is chemical erosion etching?

23-9. Briefly describe the electroforming process.

23-10. What precautions must be taken in cleaning mold cavities?

Bibliography

Books

Akin, Russell B. *Acetal Resins.* New York: Van Nostrand Reinhold Co., 1962.

American Society of Tool & Manufacturing Engineers. *Plastic Tooling and Manufacturing Handbook.* Dearborn, Michigan: American Society of Tool & Manufacturing, 1965.

Arnold, L. K. *Plastics as Materials of Construction.* Ames, Iowa: Iowa State University Press, 1958.

Baskekis, C. H. *ABS Plastics.* New York: Van Nostrand Reinhold Co., 1964.

Bebb, R. H. *Plastic Mould Design.* New York: Iliffe-NTP, Inc., 1962.

Benning, Calvin J. *Plastic Foams.* 2 vols. New York: John Wiley & Sons, Inc., 1969.

Blais, John F. *Amino Resins.* New York: Van Nostrand Reinhold Co., 1959.

Boenig, H. V. *Polyolefins: Structure and Properties.* New York: American Elsevier Publishing Co., Inc., 1966.

Brenner, Walter. *High-Temperature Plastics.* New York: Van Nostrand Reinhold Co., 1962.

Bruins, Paul F. *Polyurethane Technology.* New York: John Wiley & Sons, Inc., 1969.

——— *Plastics for Industrial Insulation.* New York: John Wiley & Sons, Inc., 1968.

Buehr, Walter. *Plastics, the Man-Made Miracle.* New York: Morrow, Williams & Co., Inc., 1967.

Butler, J. *Compression and Transfer Moulding of Plastics.* New York: John Wiley & Sons, 1960.

Buttrey, D. N. *Cellulose Plastics.* New York: Interscience Publishers, Inc.

Christopher, William F. and Daniel W. Fox. *Polycarbonates.* New York: Van Nostrand Reinhold Co., 1962.

Dietz, Albert G. *Plastics for Architects and Building.* Cambridge, Massachusetts: MIT Press, 1969.

Dovaly, Ken. *Handbook of Plastic Furniture Manufacturing.* New York: Academic Press, Inc., 1970.

Dubois, J. G. and W. I. Pribble. *Plastics Mold Engineering.* New York: Van Nostrand Reinhold Co., 1965.

DuBois, J. Harry. *Plastics History, U.S.A.* Boston, Massachusetts: Cahners Publishing Company, Inc., 1972.

Duffin, D. J. *Laminated Plastics.* New York: Van Nostrand Reinhold Co., 1958.

Farkas, Robert. *Heat Sealing.* New York: Van Nostrand Reinhold Co., 1964.

Ferrigno, T. M. *Rigid Plastic Foams.* New York: Van Nostrand Reinhold Co., 1967.

Floyd, D. E. *Polyamide Resins.* New York: Van Nostrand Reinhold Co., 1966.

Freeman, G. G. *Silicones.* New York: Iliffe-NTP, Inc., 1962.

Gaylord, Norman G. *Polyethers.* New York: Van Nostrand Reinhold Co., Inc.

Gould, D. F. *Phenolic Resins.* New York: Van Nostrand Reinhold Co., 1959.

Griff, A. L. *Plastics Extrusion Technology.* New York: Van Nostrand Reinhold Co., 1967.

Haim, George. *Manual for Plastic Welding.* New York: Chemical Publishing Co., 1960.

Harsock, John A. *Design of Foam-Filled Structures*. Stamford, Connecticut: Technomic Publishing Co., 1969.

Haslam, J. and H. A. Willis. *Identification and Analysis of Plastics*. New York: Crane-Russak & Co., Inc., 1972.

Horn, Milton B. *Acrylic Resins*. New York: Van Nostrand Reinhold Co., 1960.

Ian, O. D. *Principles of Polymerization*. New York: McGraw-Hill Book Co., 1969.

International Plastic Directory. 2 vols., 3rd ed. New York: International Publications Service, 1968-1970.

Jacobi, H. R. *Screw Extrusion of Plastics*. New York: Iliffe-NTP, Inc., 1963.

Jones, David and Thomas W. Mullen. *Blow Molding*. New York: Van Nostrand Reinhold Co., 1939.

Kaufman, Morris. *Giant Molecules*. New York: Doubleday and Company, Inc., 1969.

Kresser, T. J. *Polyolefin Plastics*. New York: Van Nostrand Reinhold Co., 1969.

——— *Polypropylene*. New York: Van Nostrand Reinhold Co., 1960.

Lawrence, John R. *Polyester Resins*. New York: Van Nostrand Reinhold Co., 1960.

Lenk, R. S. *Plastics Rheology*. New York: John Wiley & Sons, Inc., 1968.

Mark, H. F., et al. *Man-Made Fibers*. New York: John Wiley & Sons, Inc., 1968.

Mark, Hermann and L. Sandel. *Giant Molecules*. Morriston, New Jersey: Silver Burdett Co., 1966.

Martens, C. R. *Alkyd Resins*. New York: Van Norstrand Heinhold Co.

Melville, Harry W. *Big Molecules*. Riverside, New Jersey: MacMillan Publishing Co., 1958.

Moacanin J., et al. *Block Copolymers*. New York: John Wiley & Sons, Inc., 1969.

Modern Plastics Encyclopedia. (annual). New York. McGraw-Hill Book Co.

Moseyev, A. A., et al. *Expanded Plastics*. New York: Pergamon Press, Inc., 1963.

Morgerison, D. and G. C. East. *Introduction to Polymer Chemistry*. New York: Pergamon Press, Inc., 1966.

Mosle, Ernest. *Runnerless Molding*. New York: Van Nostrand Reinhold Co., 1960.

Mullen, Thomas. *Cellular Plastics*. 1966.

Narcus, Harold. *Metallizing of Plastics*. New York: Van Nostrand Reinhold Co., 1960.

Noll, Walter. *Chemistry and Technology of Silicones*. New York: Academic Press, Inc., 1968.

Oleesday, Samuel S. and Gilbert J. Mohr. *Handbook of Reinforced Plastics of SPI*. New York: Van Nostrand Reinhold Co., 1963.

Paist, W. D. *Cellulosics*. New York: Van Nostrand Reinhold Co., 1958.

Palin, G. R. *Plastics for Engineers*. New York: Pergamon Press, Inc., 1966.

Park, William R. *Plastic Film Technology*. New York: Van Nostrand Reinhold Co., 1970.

Parkyn, B. *Polyesters*. New York: American Elsevier Publishing Co., Inc., 1967.

Patton, T. C. *Alkyd Resin Technology*. New York: John Wiley & Sons, Inc., 1962.

Plastics Engineering Handbook, 3rd. ed. New York: Van Nostrand Reinhold Co., 1960.

Properties and Testing of Plastics Materials, 3rd ed. Cleveland, Ohio: Chemical Rubber Co., 1962.

Raech, K., Jr. *Allylic Resins and Monomers*. New York: Van Nostrand Reinhold Co., 1965.

Redfarn, C. A. *Experimental Plastics*. New York: John Wiley & Sons, Inc., 1960.

Rosato, D. V. and R. T. Schwartz, eds. *Environmental Effects on Polymeric Materials*. 2 vols. Plainfield, New Jersey: Textile Book Service, 1968.

Rudner, M. A. *Fluorocarbons*. New York: Van Nostrand Reinhold Co., 1958.

Saunders, K. J. *Identification of Plastics and Rubber*. New York: Halsted Press, 1966.

Schmitz, J. V. *Testing of Polymers*. 2 vols. New York: John Wiley & Sons, Inc., 1965-1966.

Schwartz, R. T. and H. S. Schwartz. *Plastic Composites*. New York: John Wiley & Sons, Inc., 1968.

Simonds, H. R. *Concise Guide to Plastics*. 2nd ed. New York: Van Nostrand Reinhold Co., 1963.

Sittig, Marshall. *Fluorinated Hydrocarbons and Polymers*. New York: McGraw-Hill Book Co., 1966.

———. *Polyacetal Resins*. Houston, Texas: Gulf Publishing Co., 1963.

Skeist, Irving. *Handbook of Adhesives*. New York: Van Nostrand Reinhold Co., 1962.

———. *Plastics in Building*. New York: Van Nostrand Reinhold Co., 1966.

Smith, W. Maye. *Vinyl Resins*. New York: Van Nostrand Reinhold Co., 1958.

Soderberg, George A. *Finishing Technology*. Bloomington, Illinois: McKnight Publishing Company, 1969.

Sutermeister, F. *Casein and Its Application*. New York: Van Nostrand Reinhold Co., 1939.

Swanson, Robert S. *Plastics Technology, Basic Materials and Processes*. Bloomington, Illinois: McKnight Publishing Company, 1965.

Teach, W. C. and G. C. Kessling. *Polystyrene*. New York: Van Nostrand Reinhold Co., 1960.

Wohlrabe, Raymond. *Exploring Giant Molecules*. Cleveland, Ohio: World Publishing Co., 1969.

Yarsley, W. E. *Cellulosic Plastics*. New York: Iliffe-NTP, Inc., 1964.

Yescombe, E. R. *Sources of Information on Rubber, Plastics and Allied Industries*. New York: Pergamon Press, Inc., 1968.

Periodicals

Adhesives Age, Palmerton Publishing Company, New York.

Bakelite Review, Union Carbide Corporation, New York.

British Plastics, Thomas Skinner and Company, Ltd., New York.

Creative Plastics, du Pont of Canada, Ltd., Canada.

Durez Foundry News, Durez Plastic Division, Hooker Chemical Corporation, New York.

Materials Engineering, Van Nostrand Reinhold Co., New York.

Modern Plastics, McGraw-Hill Book Co., New York.

Plastics Review, Union Carbide International, New York.

Plastics World, Cahners Publishing Company, Denver, Colorado.

SPE Journal, Society of Plastics Engineers, Connecticut.

Appendix A

Abbreviations for Selected Plastics*

Term	Abbreviation	Thermoplastic	Thermosetting
Acrylonitrile-butadiene-styrene plastics	ABS	•	
Carboxymethyl cellulose	CMC	•	
Casein	CS	•	
Cellulose acetate	CA	•	
Cellulose acetate-butyrate	CAB	•	
Cellulose acetate propionate	CAP	•	
Cellulose nitrate	CN	•	
Cellulose propionate	CP	•	
Cresol-formaldehyde	CF		•
Diallyl phthalate	PDAP or DAD		•
Epoxy, epoxide	EP		•
Ethyl cellulose	EC	•	
Melamine formaldehyde	MF		•
Perfluoro (ethylene-propylene) copolymer	FEP	•	
Phenol-formaldehyde	PF		•
Poly(acrylic acid)	PAA	•	
Polyacrylonitrile	PAN	•	
Polyamide (nylon)	PA	•	
Polybutadiene-acrylonitrile	PBAN	•	
Polybutadiene-styrene	PBS	•	
Polycarbonate	PC	•	
Poly(diallyl phthalate)	PDAP		•
Polyethylene	PE	•	
Polyethylene terephthalate	PETP	•	
Poly(methylchloroacrylate)	PMCA	•	
Poly(methyl methacrylate)	PMMA	•	
Polymonochlorotrifluoroethylene	PCTFE	•	
Polyoxmethylene, polyacetal	POM	•	
Polypropylene	PP	•	
Polyphenylene oxide	PPO	•	
Polystyrene	PS	•	
Polytetrafluoroethylene	PTFE	•	
Poly(vinyl acetate)	PVAc	•	
Poly(vinyl alcohol)	PVAL	•	
Poly(vinyl butyral)	PVB	•	
Poly(vinyl chloride)	PVC	•	
Poly(vinyl chloride-acetate)	PVCAc	•	
Poly(vinyl fluoride)	PVF	•	
Poly(vinyl formal)	PVFM	•	
Silicone plastics	SI		•
Styrene-acrylonitrile	SAN	•	
Styrene-butadiene plastics	SBP	•	
Styrene-rubber plastics	SRP		•
Urea-formaldehyde	UF		•
Urethane plastics	UP		•

*ASTM Standards, Vol. 27

Appendix B

Tradenames and Manufacturers

Tradename	Polymer	Manufacturer
Absaglas	Fiber glass reinforced ABS	Fiberfill Div., Dart Industries, Inc.
Absinol	ABS	Allied Resinous Products, Inc.
Abson	ABS resins & compounds	B. F. Goodrich Chemical Co.
Acelon	Cellulose acetate film	May & Baker, Ltd.
Acetophane	Cellulose acetate film	UCB-Sidac
Aclar	CTFE fluorohalocarbon films	Allied Chemical Corp., Fabricated Products Div., Plastic Film Dept.
Acralen	Ethylene-vinyl acetate polymer	Verona Dyestuffs, Div. Verona Corp.
Acrilan	Acrylic (acrylonitrile-vinyl chloride)	Monsanto Co.
Acroleaf	Hot stamping foil	Acromark Co.
Acrylaglas	Fiber glass reinforced styrene-acrylonitrile	Fiberfill Div., Dart Industries, Inc.
Acrylicomb	Acrylic sheet faced honeycomb	Dimensional Plastics Corp.
Acrylite	Acrylic molding compounds; cast acrylic sheets	American Cyanamid Co., Industrial Chemicals and Plastics Div.
Acryloid	Acrylic modifiers for PVC; coating resins	Rohm & Haas Co.
Acrylux	Acrylic	Westlake Plastics Co.
Aeroflex	Polyethylene extrusion	Anchor Plastics Co.
Aeron	Plastic-coated nylon	Flexfilm Products, Inc.
Aerotuf	Polypropylene extrusions	Anchor Plastics Co.
Afcolene	Polystyrene and SAN copolymers	Pechiney-Saint-Gobain
Afcoryl	ABS copolymers	Pechiney-Saint-Gobain
Alathon	Polyethylene resins	E. I. du Pont de Nemours & Co.
Alfane	Thermosetting epoxy resin cement	Atlas Minerals & Chemicals Div., ESB Inc.
Alpha-Clan	Reactive monomer	Marbon Div., Borg-Warner Corp.
Alphalux	PPO	Marbon Chemical Co.
Alsynite	Reinforced plastic panels	Reichhold Chemicals, Inc.
Amberlac	Modified alkyd resins	Rohm & Haas Co.
Amberol	Phenolic and maleic resins	Rohm & Haas Co.
Amer-Plate	PVC sheet material	Ameron Corrosion Control Div.
Ampol	Cellulose acetates	American Polymers, Inc.
Amres	Thermosetting liquid resins	Pacific Resins & Chemicals, Inc.
Ancorex	ABS extrusions	Anchor Plastics Co.
Anvyl	Vinyl extrusions	Anchor Plastics Co.
Apogen	Epoxy resin series	Apogee Chemical, Inc.
Araclor	Polychlorinated polyphenyls	Monsanto Co.
Araldite	Epoxy resins and hardeners	CIBA Products Co., Ciba-Geigy Corp.
Armorite	Vinyl coating	John L. Armitage & Co.
Arnel	Cellulose (triacetate fiber)	Celanese Corp.
Arochem	Modified phenolic resins	Ashland Chemical Co., Div. Ashland Oil, Inc.
Arodure	Urea resins	Ashland Chemical Co., Div. Ashland Oil, Inc.
Arofene	Phenolic resins	Ashland Chemical Co., Div. Ashland Oil, Inc.
Aroplaz	Alkyd resins	Ashland Chemical Co., Div. Ashland Oil, Inc.
Aroset	Acrylic resins	Ashland Chemical Co., Div. Ashland Oil, Inc.
Arothane	Polyester resin	Ashland Chemical Co., Div. Ashland Oil, Inc.
Artfoam	Rigid urethane foam	Strux Corp.
Arylon	Polyarl ether compounds	Uniroyal, Inc.

Tradename	Polymer	Manufacturer
Ascot	Coated spunbonded polyolefin sheet	Appleton Coated Paper Co.
Astralit	Vinyl copolymer sheets	Dynamit Nobel of America Inc.
Atlac	Polyester resin	Atlas Chemical Industries, Inc.
Avisco	PVC films	American Viscose Div., FMC Corp.
Avistar	Polyester film	American Viscose Div., FMC Corp.
Avisun	Polypropylene	Avisun Corp.
Bakelite	Polyethylene, ethylene copolymers, epoxy, phenolic, polystyrene, phenoxy, ABS and vinyl resins and compounds	Union Carbide Corp., Chemicals and Plastics
Beetle	Urea molding compounds	American Cyanamid Co., Industrial Chemicals and Plastics Div.
Betalux	TFE-filled acetal	Westlake Plastics Co.
Blanex	Crosslinked polyethylene compounds	Reichhold Chemicals, Inc.
Blapol	Polyethylene compounds & color concentrates	Reichhold Chemicals, Inc.
Blapol	Polyethylene molding & extrusion compounds	Blane Chemical Div., Reichhold Chemicals, Inc.
Blendex	ABS resin	Marbon Div., Borg-Warner Corp.
Bolta Flex	Vinyl sheeting & film	General Tire & Rubber Co., Chemical/Plastics Div.
Bolta Thene	Rigid olefin sheets	General Tire & Rubber Co., Chemical Plastics Div.
Boltaron	ABS or PVC rigid plastic sheets	General Tire & Rubber Co., Chemical Plastics Div.
Boronol	Polyolefins with boron	Allied Resinous Products, Inc.
Bronco	Supported vinyl or pyroxylin	General Tire & Rubber Co., Chemical Plastics Div.
Budene	Polybutadiene	Goodyear Tire & Rubber Co., Chemical Division
Butaprene	Styrene-butadiene latexes	Firestone Plastics Co., Div. Firestone Tire & Rubber Co.
Cadco	Plastic rod, sheet, tubing, film; resins, glass fibers, glass strands	Cadillac Plastic & Chemical Co.
Capran	Nylon 6 film	Allied Chemical Corp., Plastics Div.
Capran	Nylon films and sheet	Allied Chemical Corp., Fabricated Products Div., Plastic Film Dept.
Carbaglas	Fiber glass reinforced polycarbonate	Fiberfil Div., Dart Industries, Inc.
Carolux	Filled urethane foam, flexible	North Carolina Foam Industries, Inc.
Carstan	Urethane foam catalysts	Cincinnati Milacron Chemicals, Inc.
Castcor	Cast polyolefin films	Mobil Chemical Co., Films Dept.
Castethane	Castable molding urethane elastomer system	Upjohn Co., CPR Div.
Castomer	Urethane elastomer systems	Baxenden Chemical Co.
Castomer	Urethane elastomer and coatings	Isocyanate Products Div., Witco Chemical Corp.
Celanar	Polyester film	Celanese Plastics Co.
Celcon	Acetal copolymer resins	Celanese Plastics Co.
Cellasto	Microcellular urethane elastomer parts	North American Urethanes, Inc.
Celluliner	Resilient expanded polystyrene foam	Gilman Brothers Co.
Cellulite	Expanded polystyrene foam	Gilman Brothers Co.
Celpak	Rigid polyurethane foam	Dacar Chemical Products Co.
Celthane	Rigid polyurethane foam	Dacar Chemical Products Co.
Chem-o-sol	PVC plastisol	Chemical Products Corp.
Chem-o-thane	Polyurethane elastomer casting compounds	Chemical Products Corp.
Chemfluor	Fluorocarbon plastics	Chemplast, Inc.
Chemglaze	Polyurethane based coating materials	Hughson Chemical Co., Div. Lord Corp.
Chemgrip	Epoxy adhesives for TFE	Chemplast, Inc.
Cimglas	Fiber glass reinforced polyester moldings	Cincinnati Milacron, Molded Plastics Div.
Clocel	Rigid urethane foam systems	Baxenden Chemical Co.
Clopane	PVC film and tubing	Clopay Corp.
Cloudfoam	Polyurethane foam	International Foam Div., Holiday Inns of America, Inc.
Co-Rezyn	Polyester resins and gel coats; pigment pastes	Interplastic Corp., Commercial Resins Div.
Cobocell	Cellulose acetate butyrate tubing	Cobon Plastics Corp.
Coboflon	Teflon tubing	Cobon Plastics Corp.
Cobothane	Ethylene-vinyl acetate tubing	Cobon Plastics Corp.
Colorail	Polyvinyl chloride handrails	Blum, Julius & Co.
Colovin	Calendered vinyl sheeting	Columbus Coated Fabrics
Conathane	Polyurethane casting, potting, tooling and adhesive compounds	Conap, Inc.
Conolite	Polyester laminate	Woodall Industries, Inc.
Cordo	PVC foam and films	Ferro Corp., Composites Div.

Tradename	Polymer	Manufacturer
Cordoflex	Polyvinylidene fluoride solutions, etc.	Ferro Corp., Composites Div.
Coror-Foam	Urethane foam systems	Cook Paint & Varnish Co.
Coverlight HTV	Vinyl coated nylon fabric	Reeves Brothers, Inc.
Crystic	Unsaturated polyester resins	Scott Bader Co.
Cumar	Coumarone-indene resins	Neville Chemical Co.
Curithane (Series)	Polyaniline polyamine; organo mercury catalyst	Upjohn Co., Polymer Chemicals Div.
Curon	Polyurethane foam	Reeves Brothers, Inc.
Cycolac	ABS resins	Marbon Div., Borg-Warner Corp.
Cycolon	Synthetic resinous compositions	Marbon Div., Borg-Warner Corp.
Cycoloy	Alloys of synthetic polymers with ABS resins	Marbon Div., Borg-Warner Corp.
Cycovin	Self-extinguishing ABS graft polymer blends	Marbon Div., Borg-Warner Corp.
Cyglas	Glass filled polyester molding compound	American Cyanamid Co., Industrial Chemicals & Plastics Div.
Cymel	Melamine molding compounds	American Cyanamid Co., Industrial Chemicals & Plastics Div.
Dacovin	PVC compounds	Diamond Shamrock Chemical Co., Plastic Div.
Dapon	Diallyl phthalate resins	FMC Corp., Organic Chemicals Div.
Daran	Polyvinylidene chloride emulsion coatings	Grace, W. R. & Co., Polymers & Chemicals Div.
Daratak	Polyvinyl acetate homopolymer emulsions	Grace, W. R. & Co., Polymers & Chemicals Div.
Darex	Styrene-butadiene latices	Grace, W. R. & Co., Polymers & Chemicals Div.
Davon	TFE resins and reinforced compounds	Davies Nitrate Co.
Delrin	Acetal resin	E. I. du Pont de Nemours & Co.
Densite	Molded flexible urethane foam	General Foam Div., Tenneco Chemical, Inc.
Derakane	Vinyl ester resins	Dow Chemical Co.
Diaron	Melamine resins	Reichhold Chemicals, Inc.
Dielux	Acetal	Westlake Plastics Co.
Dion-Iso	Isophthalic polyesters	Diamond Shamrock Chemical Co., Resinous Products Div.
Dolphon	Epoxy resin and compounds, polyester resins	John C. Dolph Co.
Dorvon	Molded polystyrene foam	Dow Chemical Co.
Dow Corning	Silicones	Dow Corning Corporation
Dri-Lite	Expanded polystyrene	Poly Foam, Inc.
Duco	Lacquers	E. I. du Pont de Nemours & Co.
Duracel	Lacquers for cellulose acetate and other plastics	Maas & Waldstein Co.
Duracon	Acetal copolymer	Polyplastics Co.
Dural	Acrylic modified semirigid PVC	Alpha Chemical & Plastics Corp.
Duramac	Oil modified alkyds	Commercial Solvents Corp.
Duraplex	Alkyd resins	Rohm & Haas Co.
Durelene	PVC flexible tubing	Auburn Plastic Engineering, Div. Plastic Warehousing Corp.
Durethene	Polyethylene film	Sinclair-Koppers Co.
Duron	Phenolic resins and molding compounds	Firestone Foam Products Co.
Dyal	Alkyd and styrenated alkyd resins	Sherwin Williams Chemicals, Div. Sherwin Williams Co.
Dyalon	Urethane elastomer material	Thombert, Inc.
Dyfoam	Expanded polystyrene	Construction Products Div., W. R. Grace & Co.
Dylan	Low- and medium-density polyethylene	Sinclair-Koppers Co.
Dylel	ABS plastics	Sinclair-Koppers Co.
Dylene	Polystyrene resin and oriented sheet	Sinclair-Koppers Co.
Dylite	Expandable polystyrene beads, extruded sheets, etc.	Sinclair-Koppers Co.
E-Form	Epoxy molding compounds	Epoxy Products Co., Div. Allied Products Corp.
Easy-Kote	Fluorocarbon release compound	Borco Chemicals, Inc.
Easypoxy	Epoxy adhesive kits	Conap, Inc.
Eccosil	Silicone resins	Emerson & Cuming, Inc.
El Rexene	Polyethylene, polypropylene, polystyrene and ABS resins	Rexene Polymers Co., Div. Dart Industries, Inc., Chemical Group
Elastolit	Urethane engineering thermoplastic	North American Urethanes, Inc.
Elastollyx	Urethane engineering thermoplastics	North American Urethanes, Inc.
Elastolur	Urethane coatings	North American Urethanes, Inc.
Eslastonate	Urethane isocyanate prepolymers	North American Urethanes, Inc.
Elastonol	Urethane polyester polyols	North American Urethanes, Inc.
Elastopel	Urethane engineering thermoplastics	North American Urethanes, Inc.
Elvace	Acetate-ethylene copolymers	E. I. du Pont de Nemours & Co.
Elvacet	Polyvinyl acetate emulsions	E. I. du Pont de Nemours & Co.

Tradename	Polymer	Manufacturer
Elvacite	Acrylic resins	E. I. du Pont de Nemours & Co.
Elvamide	Nylon resins	E. I. du Pont de Nemours & Co.
Elvanol	Polyvinyl alcohols	E. I. du Pont de Nemours & Co.
Elvax	Vinyl resins; acid terpolymer resins	E. I. du Pont de Nemours & Co.
Ensocote	PVC lacquer coating	Uniroyal, Inc.
Ensolex	Cellular plastic sheet material	Uniroyal, Inc.
Ensolite	Cellular plastic sheet material	Uniroyal, Inc.
Epi-Rez	Basic epoxy resins	Celanese Resins, Div. Celanese Coatings Co.
Epi-Tex	Epoxy ester resins	Celanese Resins, Div. Celanese Coatings Co.
Epikote	Epoxy resin	Shell Chemical Co.
Epocap	Two-part epoxy compounds	Hardman, Inc.
Epocast	Epoxies	Furane Plastics, Inc.
Epocrete	Two-part epoxy materials	Hardman, Inc.
Epocryl	Epoxy acrylate resin	Shell Chemical Co.
Epocure	Epoxy curing agents	Hardman, Inc.
Epolast	Two-part epoxy compounds	Hardman, Inc.
Epolite	Epoxy compounds	Rezolin, Div. Hexcel Corp.
Epomarine	Two-part epoxy compounds	Hardman, Inc.
Epon	Epoxy resin; hardener	Shell Chemical Co.
Eponol	Linear polyether resin	Shell Chemical Co.
Eposet	Two-part epoxy compounds	Hardman, Inc.
Epotuf	Epoxy resins	Reichold Chemicals, Inc.
Estane	Polyurethane resins and compounds	B. F. Goodrich Chemical Co.
Ethafoam	Polyethylene foam	Dow Chemical Co.
Ethocel	Ethyl cellulose resin	Dow Chemical Co.
Ethofil	Fiber glass reinforced polyethylene	Fiberfil Div., Dart Industries, Inc.
Ethoglas	Fiber glass reinforced polyethylene	Fiberfil Div., Dart Industries, Inc.
Ethosar	Fiber glass reinforced polyethylene	Fiberfil Div., Dart Industries, Inc.
Ethylux	Polyethylene	Westlake Plastics Co.
Evenglo	Polystyrene resins	Sinclair-Koppers Co.
Everflex	Polyvinyl acetate copolymer emulsion	Grace, W. R., & Co., Polymers & Chemicals Div.
Everlon	Urethane foam	Stauffer Chemical Co., Plastics Div.
Excelite	Polyethylene tubing	Thermoplastic Processes, Inc.
Exon	PVC resins, compounds and latexes	Firestone Plastics Co., Div. Firestone Tire & Rubber Co.
Extane	Polyurethane tubing	Extron Div., Pipe Line Service Co.
Extrel	Polyethylene and polypropylene films	Extrudo Film Corp., Aff. Enjay Chemical Co.
Extren	Fiber glass reinforced polyester shapes	Koppers Co., Reinforced Plastics Div.
Fabrikoid	Pyroxylin-coated fabric	Stauffer Chemical Co., Plastics Div.
Fassgard	Vinyl coating on nylon	M. J. Fassler & Co.
Fasslon	Vinyl coating	M. J. Fassler & Co.
Felor	Nylon filaments	E. I. du Pont de Nemours & Co.
Flamolin	Flame retarded polyolefin	Raychem Corp.
Flexane	Urethanes	Devcon Corp.
Flexocel	Urethane foam systems	Baxenden Chemical Co.
Floranier	Cellulose for cellulose esters	ITT Rayonier, Inc.
Fluokem	Teflon spray	Bel-Art Products
Fluon	TFE resin	ICI American, Inc.
Fluorglas	PTFE-coated and impregnated woven glass fabric, laminates, belting	Dodge Industries, Inc., Sub. Oak Electro/Netics Corp.
Fluorocord	Fluorocarbon material	Raybestos Manhattan, Inc.
Fluorofilm	Cast Teflon films	Dilectrix Corp.
Fluoroglide	Dry-film lubricant of TFE	Chemplast, Inc.
Fluororay	Filled fluorocarbon	Raybestos Manhattan, Inc.
Fluorored (series code)	Compounds of TFE	John L. Dore Co.
Fluorosint	TFE-fluorocarbon base composition	Polymer Corp.
Foamthane	Rigid polyurethane foam	Pittsburgh Corning Corp.
Formadall	Polyester premix compound	Woodall Industries, Inc.
Formaldafil	Fiber glass reinforced acetal	Fiberfil Div., Dart Industries, Inc.
Formaldaglas	Fiber glass reinforced acetal	Fiberfil Div., Dart Industries, Inc.
Formaldasar	Fiber glass reinforced acetal	Fiberfil Div., Dart Industries, Inc.
Formica	High-pressure laminate	American Cyanamid Company

Tradename	Polymer	Manufacturer
Formrez	Urethane elastomer chemicals	Witco Chemical Corp., Organics Div.
Forticel	Cellulose propionate flake, resins	Celanese Plastics Co.
Fortiflex	Polyethylene resins	Celanese Plastics Co.
Fosta-Net	Polystyrene foam extruded mesh	Foster Grant Co.
Fosta Tuf-Flex	High-impact polystyrene	Foster Grant Co.
Fostacryl	Thermoplastic polystyrene resins	Foster Grant Co.
Fostafoam	Expandable polystyrene beads	Foster Grant Co.
Fostalite	Light-stable polystyrene molding powder	Foster Grant Co.
Fostarene	Polystyrene molding powder	Foster Grant Co.
Futron	Polyethylene powder	Fusion Rubbermaid Co.
Genthane	Polyurethane rubber	General Tire & Rubber Co., Chemical Plastics Div.
Gentro	Styrene butadiene rubber	General Tire & Rubber Co., Chemical Plastics Div.
Geon	Vinyl resins, compounds, latexes	B. F. Goodrich Chemical Co.
Gil-Fold	Polyethylene sheet	Gilman Brothers Co.
Glaskyd	Alkyd molding compounds	American Cyanamid Co., Industrial Chemicals & Plastics Div.
Gordon Superdense	Polystyrene in pellet form	Hammond Plastics, Inc.
Gordon Superflow	Polystyrene in granular or pellet form	Hammond Plastics, Inc.
Gracon	PVC compounds	Grace, W. R., & Com. Elm Coated Fabrics Div.
GravoFLEX	ABS sheets	Hermes Plastics, Inc.
GravoPLY	Acrylic sheets	Hermes Plastics, Inc.
Halon	TFE molding compounds	Allied Chemical Corp., Plastics Div.
Haysite	Polyester laminates	Synthane-Taylor Corp., an Alco Standard Co.
Herox	Nylon filaments	E. I. du Pont de Nemours & Co.
Hetrofoam	Fire retardant urethane foam systems	Durez Div., Hooker Chemical Corp.
Hetron	Fire retardant polyester resins	Durez Div., Hooker Chemical Corp.
Hex-One	High-density polyethylene	Gulf Oil Co.
Hi-fax	Polyethylene	Hercules, Inc.
Hi-Styrolux	High-impact polystyrene	Westlake Plastics Co.
Hydrepoxy	Water-based epoxies	Acme Chemicals Div., Allied Products Corp.
Hydro Foam	Expanded phenol-formaldehyde	Smithers Co.
Implex	Acrylic molding powder	Rohm & Haas Co.
Intamix	Rigid PVC compounds	Diamond Shamrock Chemical Co., Plastic Div.
Interpol	Copolymeric resinous systems	Freeman Chemical Corp., Div. H. H. Robertson Co.
Irvinil	PVC resins and compounds	Great American Chemical Corp.
Isoderm	Urethane rigid and flexible integral skinning foam	Upjohn Co., CPR Div.
Isofoam	Urethane foam systems	Isocyanate Products Div., Witco Chemical Corp.
Isonate	Diisocyanates and urethane systems	Upjohn Co., CPR Div.
Isonate (series)	Isocyanates, diisocyanates	Upjohn Co., Polymer Chemicals Div.
Isoteraglas	Isocyanate elastomer-coated Dacron glass fabric	Natvar Corp.
Isothane	Flexible polyurethane foams	Bernel Foam Products Co.
Jetfoam	Polyurethane foam	International Foam Div., Holiday Inns of America, Inc.
K-Prene	Urethane cast material	Di-Acro Kaufman
Kalex	Two-part polyurethane elastomers	Hardman, Inc.
Kalspray	Rigid urethane foam systems	Baxenden Chemical Co.
Keltrol	Vinyl toluene copolymer	Spencer Kellogg Div. Textron Inc.
Ken-U-Thane	Polyurethanes: urethane foam ingredients	Kenrich Petrochemicals, Inc.
Kencolor	Silicone/pigments dispersion	Kenrich Petrochemicals, Inc.
Kodacel	Cellulosic film and sheeting	Eastman Chemical Products, Inc., Sub. Eastman Kodak Co.
Kohinor	Vinyl resins and compounds	Pantasote Co.
Korad	Acrylic film	Rohm & Haas Co.
Koroseal	Vinyl films	B. F. Goodrich Chemical Co.
Kralastic	ABS high-impact resin	Uniroyal, Inc.
Kralon	High-impact styrene and ABS resins	Uniroyal, Inc.
Kraton	Styrene-butadiene polymers	Shell Chemical Co.
Krene	Plastic film and sheeting	Union Carbide Corp., Chemicals & Plastics
Krystal	PVC sheet	Allied Chemical Corp., Fabricated Products Div., Plastic Film Dept.
Krystaltite	PVC shrink films	Allied Chemical Corp., Fabricated Products Div., Plastic Film Dept.
Kydene	Acrylic/PVC molding powder	Rohm & Haas Co.

Tradename	Polymer	Manufacturer
Kydex	Acrylic/PVC sheets	Rohm & Haas Co.
Kynar (Series)	Polyvinylidene fluoride	Pennwalt Corp.
Lamabond	Reinforced polyethylene	Lamex, Columbian Carbon Co.
LAmar	Mylar vinyl laminate	Morgan Adhesives Co.
Laminac	Polyester resins	American Cyanamid Co., Industrial Chemicals & Plastics Div.
Last-A-foam	Plastic foam	General Plastics Mfg. Co.
Lexan	Polycarbonate resins, film, sheet	General Electric Co., Plastics Dept.
Lucite	Acrylic resins	E. I. du Pont de Neumours & Co.
Lucolite	Press polished clear flexible PVC	Tenneco Advanced Materials, Inc.
Lumasite	Acrylic sheet	American Acrylic Corp.
Lustran	SAN and ABS molding and extrusion resins	Monsanto Co.
Lustrex	Polystyrene molding and extrusion resins	Monsanto Co.
Macal	Cast vinyl film	Morgan Adhesives Co.
Marafoam	Polyurethane foam resin	Marblette Co.
Maraglas	Epoxy casting resin	Marblette Co.
Maraset	Epoxy resin	Marblette Co.
Maraweld	Epoxy resin	Marblette Co.
Marlex	Polyethylenes, polypropylenes, other polyolefin plastics	Phillips Petroleum Co., Chemical Department, Plastics Div.
Marvinol	Vinyl resins and compounds	Uniroyal, Inc.
Meldin	Polyimide and reinforced polyimide	Dixon Corp.
Merlon	Polycarbonate	Mobay Chemical Co.
Metallex	Cast acrylic sheets	Hermes Plastics, Inc.
Meticone	Silicone rubber dies and sheet	Gladen Enterprises, Inc.
Metre-Set	Epoxy adhesives	Metachem Resins Corp., Mereco Products Corp. Div.
Micarta	Thermosetting laminates	Westinghouse Electric Corp., Industrial Plastics Div.
Microsol	Vinyl plastisol	Michigan Chrome & Chemical Co.
Micro-Matte	Extruded acrylic sheet with matte finish	Extrudaline, Inc.
Micropel	Nylon powders	Nypel, Inc.
Microthene	Powdered polyolefins	U. S. Industrial Chemicals Co., Div National Distillers & Chemical Co.
Milmar	Polyester	Morgan Adhesives Co.
Mini-Vaps	Expanded polyethylene	Nalge Co., Agile Div.
Minit Grip	Epoxy adhesives	Schramm Fiberglass Products Div., High-Strength Plastics Corp.
Minit Man	Epoxy adhesive	Kristal Draft, Inc.
Mipoplast	Flexible PVC sheets	Dynamit Nobel of America, Inc.
Mirasol	Alkyd resins: epoxy esters	C. J. Osborn Chemicals, Inc.
Mirbane	Amino resin	Showa Highpolymer Co.
Mirrex	Calendered rigid PVC	Tenneco Chemicals, Inc., Tenneco Plastics Div.
Mista Foam	Urethane foam systems	M. R. Plastics & Coatings, Inc.
Mod-Epox	Epoxy resin modifier	Monsanto Co.
Molycor	Glass fiber reinforced epoxy tubing	A. O. Smith, Inland, Inc., Reinforced Plastics Div.
Mondur	Isocyanates	Mobay Chemical Co.
Monocast	Direct polymerized nylon	Polymer Corp.
Moplen	Isotactic polypropylene	Montecatini Edison S. p. A.
Multrathane	Urethane elastomer chemicals	Mobay Chemical Co.
Multron	Polyesters	Mobay Chemical Co.
Mylar	Polyester film	E. I. du Pont de Nemours & Co., Inc.
Nalgon	Plasticized PVC tubing	Nalge Co., Div. Sybron Corp.
Napryl	Polypropylene	Pechiney-Saint-Gobain
Natene	High-density polyethylene	Pechiney-Saint-Gobain
Naugahyde	Vinyl coated fabrics	Uniroyal, Inc.
NeoCryl	Acrylic resins and resin emulsions	Polyvinyl Chemicals, Inc.
NeoRez	Styrene emulsions and urethane solutions	Polyvinyl Chemicals, Inc.
NeoVac	PVA emulsions	Polyvinyl Chemicals, Inc.
Nevillac	Modified coumarone-indene resin	Neville Chemical Co.
Nimbus	Polyurethane foam	General Tire & Rubber Co., Chemical Plastics Div.
Nitrocol	Nitrocellulose base pigment dispersion	C. J. Osborn Chemicals, Inc.
Nob-Lock	PVC sheet material	Ameron Corrosion Control Div.

Tradename	Polymer	Manufacturer
Nopcofoam	Urethane foam systems	Diamond Shamrock Chemical Co., Resinous Products Div.
Norchem	Low-density polyethylene resin	Northern Petrochemical Co.
Noryl	Modified polyphenylene oxide	General Electric Co., Plastics Dept.
Nyglathane	Glass filled polyurethane	Nypel, Inc.
Nylafil	Fiber glass reinforced nylon	Fiberfil Div., Dart Industries, Inc.
Nylaglas	Fiber glass reinforced nylon	Fiberfil Div., Dart Industries, Inc.
Nylasar	Fiber glass reinforced nylon	Fiberfil Div., Dart Industries, Inc.
Nylasint	Sintered nylon parts	Polymer Corp.
Nylatron	Filled nylons	Polymer Corp.
Nylo-Seal	Nylon 11 tubing	Imperial-Eastman
Nylux	Nylon	Westlake Plastics Co.
Nypelube	TFE-filled nylons	Nypel, Inc.
Nyreg	Glass reinforced nylon molding compounds	Nypel, Inc.
Oasis	Expanded phenol-formaldehyde	Smithers Co.
Oilon Pv 80	Acetal based resin sheets, rods, tubing, profiles	Cadillac Plastic & Chemical Co.
Olefane	Polypropylene film	Amoco Chemicals Corp.
Olefil	Filled polypropylene resin	Amoco Chemicals Corp.
Oleflo	Polypropylene resin	Amoco Chemicals Corp.
Olemer	Copolymer polypropylene	Amoco Chemicals Corp.
Oletac	Amorphous polypropylene	Amoco Chemicals Corp.
Opalon	Flexible PVC materials	Monsanto Co.
Orgalacqe	Epoxy and PVC powders	Aquitaine-Organico
Orgamide R	Nylon 6	Aquitaine-Organico
Orlon	Acrylic fiber	E. I. du Pont de Nemours & Co., Inc.
Papi	Polymethylene polyphenylisocyanate	Upjohn Co., Polymer Chemicals Div.
Paradene	Dark coumarone-indene resins	Neville Chemical Co.
Paraplex	Polyester resins and plasticizers	Rohm & Haas Co.
Pelaspan	Expandable polystyrene	Dow Chemical Co.
Pelaspan-Pac	Expandable polystyrene	Dow Chemical Co.
Pellethane	Thermoplastic urethane	Upjohn Co., Polymer Chemicals Div.
Pellon Aire	Nonwoven textile	Pellon Corp., Industrial Div.
Penton	Chlorinated polyether	Hercules Inc.
PermaRez	Cast epoxy	Permali, Inc.
Permelite	Melamine molding compounds	Melamine Plastics, Inc., Div. Fiberite Corp.
Petrothene	Low-, medium-, and high-density polyethylene	U. S. Industrial Chemical Co., Div. National Distillers & Chemical Co.
Petrothene XL	Crosslinkable polyethylene compounds	U. S. Industrial Chemical Co., Div. National Distillers & Chemical Co.
Phenoweld	Phenolic adhesives for nylon, etc.,	Hardman, Inc.
Philjo	Polyolefin films	Phillips-Joana Co., Div. Joana Western Mills Co.
Piccoflex	Acrylontrile-styrene resins	Pennysylvania Industrial Chemical Corp.
Piccolastic	Polystyrene resins	Pennsylvania Industrial Chemical Corp.
Piccotex	Vinyl toluene copolymer	Pennsylvania Industrial Chemical Corp.
Piccoumaron	Coumarone-indene resins	Pennsylvania Industrial Chemical Corp.
Piccovar	Alkyl-aromatic resins	Pennsylvania Industrial Chemical Corp.
Pienco	Polyester resins	Mol-Rex Div., American Petrochemical Corp.
Pinpoly	Reinforced polyurethane foam	International Foam Div., Holiday Inn of America, Inc.
Plaskon	Plastic molding compounds	Allied Chemical Corp., Plastics Div.
Plastic Steel	Epoxy tooling and repair materials	Devcon Corp.
Pleogen	Polyester resins and gel coats; Polyurethane systems	Mol-Rez Div., Whittaker Corp.
Plexiglas	Acrylic sheets and molding powders	Rohm & Haas Co.
Plicose	Polyethylene film, sheeting, tubing, bags	Diamond Shamrock Corp.
Pliobond	Adhesive	Goodyear Tire and Rubber Co.
Pliolite	Styrene-butadiene resins	Goodyear Tire & Rubber Co., Chemical Div.
Pliothene	Polyethylene-rubber blends	Ametek/Westchester Plastics
Pliovic	PVC resins	Goodyear Tire & Rubber Co., Chemical Div.
Pluracol	Polyethers	BASF Wyandotte Corp.
Pluronic	Polyethers	BASF Wyandotte Corp.
Plyocite	Phenolic impregnated overlays	Reichhold Chemicals, Inc.
Plyophen	Phenolic resins	Reichhold Chemicals, Inc.
Polex	Oriented acrylic	Southwestern Plastics, Inc.

Tradename	Polymer	Manufacturer
Pollopas	Urea formaldehyde compounds	Dynamit Nobel of America, Inc.
Polvonite	Cellular plastic material in sheet form	Voplex Corp.
Poly-Dap	Diallyl phthalate electrical molding compounds	U. S. Polymeric, Inc.
Poly-Eth	Low-density polyethylene	Gulf Oil Corp., Plastics Div., Chemicals Dept.
Poly-Eth-Hi-D	High-density polyethylene	Gulf Oil Corp., Plastics Div., Chemicals Dept.
Polycarbafil	Fiber glass reinforced polycarbonate	Fiberfil Div., Dart Industries, Inc.
Polycure	Crosslinked polyethylene compounds	Cooke Color & Chemical Div.
Polyfoam	Polyurethane foam	General Tire & Rubber Co., Chemical/Plastics Div.
Polyimidal	Thermoplastic polyimide	Raychem Corp.
Polylite	Polyester resins	Reichhold Chemicals, Inc.
Polymet	Plastic filled sintered metal	Polymer Corp.
Polymul (Series)	Polyethylene emulsions	Nopco Chemical Div., Diamond Shamrock Chemical Co.
Polyteraglas	Polyester-coated Dacron-glass fabric	Natvar Corp.
Polywrap	Plastic film	Flex-O-Glass, Inc.
Poxy-Gard	Solventless epoxy compounds	Sterling, Div. Reichhold Chemicals, Inc.
PPO	Polyphenylene oxide	Reichhold Chemicals, Inc.
Pro-fax	Polypropylene	Hercules, Inc.
Profil	Fiber glass reinforced polypropylene	Fiberfil Div., Dart Industries, Inc.
Proglas	Fiber glass reinforced polypropylene	Fiberfil Div., Dart Industries, Inc.
Prohi	High-density polyethylene	Protective Lining Corp.
Propathene	Polypropylene polymers and compound	Imperial Chemical Industries, Ltd., Plastics Div.
Propylsar	Fiber glass reinforced polypropylene	Fiberfil Div., Dart Industries, Inc.
Propylux	Polypropylene	Westlake Plastics Co.
Protectolite	Polyethylene film	Protective Lining Corp.
Protron	Ultrahigh-strength polyethylene	Protective Lining Corp.
Quelflam	Urethanes, low surface spread flame	Baxenden Chemical Co.
Regalite	Press polished clear flexible PVC	Tenneco Advanced Materials, Inc.
REN-Shape	Epoxy material	Ren Plastics, Inc.
Ren-Thane	Urethane elastomers	Ren Plastics, Inc.
Resiglas	Polyester resins, etc.	Kristal Draft, Inc.
Resinol	Polyolefins	Allied Resinous Products, Inc.
Resorasabond	Resorcinol and phenol-resorcinol	Pacific Resins & Chemicals, Inc.
Restfoam	Urethane foam	Stauffer Chemical Co., Plastics Div.
Rexolene	Crosslinked polyolefin sheet	Brand-Rex Co.
Rexolite	Polystyrene rod & sheet stock	Brand-Rex Co.
Reynosol	Urethane, PVC	Reynolds Chemical Products Div., Hoover Ball & Bearing Co.
Rhodiod	Cellulose acetate sheet	M & B Plastics, Ltd.
Rhoplex	Acrylic emulsion	Rohm & Haas Co.
Richfoam	Urethane foam	Carpenter, E. R., Co.
Rigidite	Modified acrylic molding compound and sheet; modified polyester resin	American Cyanamid Co., Industrial Chemicals & Plastics, Div.
Rigidsol	Rigid plastisol	Watson-Standard Co.
Rolox	Two-part epoxy compounds	Hardman, Inc.
Royalex	Structural cellular thermoplastic sheet material	Uniroyal, Inc.
Royalite	Thermoplastic sheet material	Uniroyal, Inc., Uniroyal Plastic Products
Roylar	Polyurethane elastoplastic	Uniroyal, Inc.
Rucoam	Vinyl film and sheeting	Ruco Div., Hooker Chemical Corp.
Rucoblend	Vinyl compounds	Ruco Div., Hooker Chemical Corp.
Rucon	Vinyl resins	Ruco Div., Hooker Chemical Corp.
Rucothane	Polyurethanes	Ruco Div., Hooker Chemical Corp.
Santolite	Aryl sulfonamide-formaldehyde resin	Monsanto Co.
Saran	Polyvinylidene chloride resin	Dow Chemical Co.
Satin Foam	Extruded polystyrene foam	Dow Chemical Co.
Scotchpak	Heat sealable polyester film	3M Co.
Scotchpar	Polyester film	3M Co.
Selectrofoam	Urethane foam systems and polyols	PPG Industries, Inc., Resin Products Sales, C & R Div.
Selectron	Polymerizable synthetic resins: polyesters	PPG Industries, Inc., Resin Products Sales, C & R Div.
Shuvin	Vinyl molding compounds	Reichhold Chemicals, Inc.
Shuvin	Vinyl molding compounds	Blane Chemical Div., Reichhold Chemicals, Inc.
Siponate	Alkyl aryl sulfonates	Alcolac, Inc.

Tradename	Polymer	Manufacturer
Skinwich	Urethane rigid and flexible integral skinning foam	Upjohn Co.
Softlite	Ionomer foam	Gilman Brothers Co.
Solithane	Urethane prepolymers	Thiokol Chemical Corp., Chemical/Industrial Div.
Sonite	Epoxy resin compound	Smooth-On, Inc.
Spandal	Rigid urethane laminates	Baxenden Chemical Co.
Spandofoam	Rigid urethane foam board and slab	Baxenden Chemical Co.
Spandoplast	Expanded polystyrene board and slab	Baxenden Chemical Co.
Spenkel	Polyurethane resin	Spencer Kellogg Div. Textron Inc.
Starez	Polyvinyl acetate resin	Paisley, Standard Brands Chemical Industries, Inc.
Structoform	Sheet molding compounds	Fiberite Corp.
Stylafoam	Coated polystyrene sheet	Gilman Brothers Co.
Stypol	Polyesters	Freeman Chemical Corp., Div. H. H. Robertson Co.
Styrafil	Fiber glass reinforced polystyrene	Fiberfil Div., Dart Industries, Inc.
Styroflex	Biaxially oriented polystyrene film	Natvar Corp.
Styrofoam	Polystyrene foam	Dow Chemical Co.
Styrolux	Polystyrene	Westlake Plastics Co.
Styron	Polystyrene resin	Dow Chemical Co.
Styronol	Styrene	Allied Resinous Products, Inc.
Sulfasar	Fiberglass reinforced polysulfone	Fiberfil Div., Dart Industries, Inc.
Sulfil	Fiberglass reinforced polysulfone	Fiberfil Div., Dart Industries, Inc.
Sunlon	Polyamide resin	Sun Chemical Corp.
Super Aeroflex	Linear polyethylene	Anchor Plastic Co.
Super Coilife	Epoxy potting resin	Westinghouse Electric Corp.
Super Dylan	High-density polyethylene	Sinclair-Koppers Co.
Superflex	Grafted high-impact polystyrene	Gordon Chemical Co.
Superflow	Polystyrene	Gordon Chemical Co.
Sur-Flex	Ionomer film	Flex-O-Glass, Inc.
Surlyn	Ionomer resin	E. I. du Pont de Nemours & Co.
Syn-U-Tex	Urea formaldehyde and melamine formaldehyde resins	Celanese Resins, Div. Celanese Coatings Co.
Syntex	Alkyd and polyurethane ester resins	Celanese Resins, Div. Celanese Coatings Co.
Syretex	Styrenated alkyd resins	Celanese Resins, Div. Celanese Coatings Co.
TamClad	Spray or dip plastisol	Tamite Industries, Inc.
Teflon	FEP and TFE fluorocarbon resins	E. I. du Pont de Nemours & Co.
Tenite	Molding and extrusion compounds	Eastman Chemical Products, Inc., Sub. Eastman Kodak Co.
Tenn Foam	Polyurethane foam	Morristown Foam Corp.
Tere-Cast	Polyester casting compounds	Sterling, Div. Reichhold Chemicals, Inc.
Terucello	Carboxy methyl cellulose	Showa Highpolymer Company
Tetra-Phen	Phenolic type resins	Georgia-Pacific Corp., Chemical Div.,
Tetra-Ria	Amino type resins	Georgia-Pacific Corp., Chemical Div.,
Tetraloy	Filled TFE molding compounds	Whitford Chemical Corp.
Tetran	Polytetrafluoroethylene	Pennwalt Corp.
Texin	Urethane elastomer molding compound	Mobay Chemical Co.
Textolite	Industrial laminates	General Electric Co., Laminated Products Dept.
Thermalux	Polysulfone	Westlake Plastics Co.
Thermasol	Vinyl plastisols and organosols	Lakeside Plastics International
Thermco	Expanded polystyrene	Holland Plastics Co.
Thorane	Rigid polyurethane foam	Dow Chemical Co.
T-Lock	PVC sheet material	Amercoat Corp.
Tran-Stay	Flat polyester film	Transilwrap Co.
Transil GA	Precoated acetate sheets	Transilwrap Co.
Tri-Foil	TFE-coated aluminum foil	Tri-Point Industries, Inc.
Trilon	Polytetrafluoroethylene	Tri-Point Industries, Inc.
Trolen (Series)	Polyethylene and polypropylene sheets	Dynamit Nobel of America, Inc.
Trolitan (Series)	Phenol formaldehyde compounds: boron molding compounds	Dynamit Nobel of America, Inc.
Trolitax	Industrial laminates	Dynamit Nobel of America, Inc.
Trosifol	Polyvinyl butyral film	Dynamit Nobel of America, Inc.
Tuftane	Polyurethane film and sheet	B. F. Goodrich Chemical Co.
Tybrene	Acrylonitrile-butadiene-styrene resins	Dow Chemical Co.
Tynex	Nylon filaments	E. I. du Pont de Nemours & Co.

Tradename	Polymer	Manufacturer
Tyril	Styrene-acrylonitrile resin	Dow Chemical Co.
Tyrilfoam	Styrene-acrylonitrile foam	Dow Chemical Co.
Tyrin	Chlorinated polyethylene	Dow Chemical Co.
U-Thane	Rigid insulation board stock urethane	Upjohn Co., CPR Div.
Uformite	Urea and melamine resins	Rohm & Haas Co.
Ultramid	Nylon 6, 6/6, and 6/10	BASF Wyandotte Corp.
Ultrapas	Melamine formaldehyde compounds	Dynamit Nobel of America, Inc.
Ultrathene	Ethylene-vinyl acetate resins & copolymers	U.S. Industrial Chemicals Co., Div. National Distillers & Chemical Corp.
Ultron	PVC film and sheet	Monsanto Co.
Unifoam	Polyurethane foam	Burnett, William T. & Co.
Unipoxy	Epoxy resins, adhesives	Kristal Kraft, Inc.
Urafil	Fiber glass reinforced polyurethane	Fiberfil Div., Dart Industries, Inc.
Uraglas	Fiber glass reinforced polyurethane	Fiberfil Div., Dart Industries, Inc.
Uralite	Urethane compounds	Rezolin Div., Hexcel Corp.
Uramol	Urea formaldehyde molding compounds	Gordon Chemicals Co.
Urapol	Urethane elastomeric coating	Poly Resins
Uropac	Rigid urethane systems	North American Urethanes, Inc.
Uvex	Cellulose acetate butyrate sheet	Eastman Chemical Products, Inc., Sub. Eastman Kodak Co.
Valsof	Polyethylene emulsions	Valchem, Div. United Merchants & Mfrs., Inc.
Varcum	Phenolic resins	Reichhold Chemicals, Inc.
Varex	Polyester resins	McCloskey Varnish Co.
Varkyd	Alkyd and modified alkyd resins	McCloskey Varnish Co.
Varkydane	Urethane vehicles	McCloskey Varnish Co.
Varsil	Silicone coated fiberglass	New Jersey Wood Finishing Co.
Vectra	Polypropylene fibers	Enjay Chemical Co.
Velene	Styrene-foam laminate	Scott Paper Co., Foam Div.
Velon	Film and sheeting	Firestone Plastics Co., Div. Firestone Tire & Rubber Co.
Versi-Ply	Coextruded films	Pierson Industries, Inc.
Vibrathane	Polyurethane elastomer	Uniroyal, Inc.
Vibrin-Mat	Polyester-glass molding compound	Marco Chemical Div., W. R. Grace & Co.
Vibro-Flo	Epoxy and polyester coating powders	Armstrong Products Co.
Vinoflex	PVC resins	BASF Wyandotte Corp.
Vitel	Polyester resin	Goodyear Tire & Rubber Co., Chemical Div.
Vithane	Polyurethane resins	Goodyear Tire & Rubber Co., Chemical Div.
Vituf	Polyester resin	Goodyear Tire & Rubber Co., Chemical Div.
Volara	Closed-cell, low-density polyethylene foam	Voltek, Inc.
Volaron	Closed-cell, low-density polyethylene foam	Voltek, Inc.
Volasta	Closed-cell, medium-density polyethylene foam	Voltek, Inc.
Voranol	Polyurethane resins	Dow Chemical Co.
Vult-Acet	Polyvinyl acetate lattices	General Latex & Chemical Corp.
Vultafoam	Urethane foam systems	General Latex & Chemical Corp.
Vultathane	Urethane coatings	General Latex & Chemical Corp.
Vygen	PVC resin	General Tire & Rubber So., Chemical/Plastics Div.
Vynaclor	Vinyl chloride emulsion coatings/binders	National Starch & Chemical Corp.
Vynaloy	Vinyl sheet	B. F. Goodrich Chemical Co.
Vyram	Rigid PVC materials	Monsanto Co.
Weldfast	Epoxy and polyester adhesives	Fibercast Co., Div. Youngstown Sheet & Tube Co.
Wellamid (Series)	Nylon 6 and 6/6 molding resins	Wellman, Inc., Plastics Div.
Well-A-Meld	Reinforced nylon resins	Wellman, Inc.
Westcoat	Strippable coatings	Western Coating Co.
Whirlclad	Plastic coatings; coating service: coated parts	Polymer Corp.
Whitcon	Fluoroplastic lubricants	Whitford Chemical Corp.
Wicaloid	Styrene-butadiene emulsions	Wica Chemicals, Div. Ott Chemical Co.
Wicaset	Polyvinyl acetate emulsions	Wica Chemicals, Div. Ott Chemical Co.
Wilflex	Vinyl plastisols	Flexible Products Co.
Xylon	Nylon 6 and nylon 6/6	Fiberfil Div., Dart Industries, Inc.
Zelux	Polyethylene films	Union Carbide Corp., Chemicals & Plastics Div.
Zendel	Polyethylene films	Union Carbide Corp., Chemicals & Plastics Div.
Zerlon	Copolymer of acrylic and styrene	Dow Chemical Co.
Zetafin	Thylene copolymer resins	Dow Chemical Co.

Index